MW00513502

PROBLEMS IN PERTURBATION

Problems in Perturbation

ALI HASAN NAYFEH

University Distinguished Professor
Virginia Polytechnic Institute and State University
Blacksburg, Virginia

Original Edition 1985 by John Wiley & Sons, Inc.

Library of Congress Cataloging in Publication Data:

Nayfeh, Ali Hasan, 1933-
Problems in perturbation.

Includes index.
1. Differential equations-Numerical solutions.
2. Perturbation (Mathematics) I. Title.
QA371.N33 1985 515.3′5 85-3173
ISBN 0-471-82292-2

Printed in the United States of America

To my wife Samirah

Preface

Many of the problems facing physicists, engineers, and applied mathematicians involve difficulties such as nonlinear governing equations, nonlinear boundary conditions at complex known or unknown boundaries, and variable coefficients that preclude exact solutions. Consequently, solutions are approximated using numerical techniques, analytic techniques, and combinations of both. For given initial and boundary conditions and specified parameters, one can use modern computers to integrate linear and nonlinear differential equations fairly accurately. However, if one needs to obtain some insight into the character of the solutions of nonlinear problems and their dependence on certain parameters, one may need to repeat the calculations for many different values of the parameters and initial conditions. Even for simple nonlinear problems the output may be so large that it is difficult to recognize even simple general phenomena. On the other hand, analytic methods often easily delineate general phenomena, yielding useful results in closed form. In the case of nonlinear partial differential equations with variable coefficients and complicated boundaries, the combination of an analytic and a numerical method often provides an optimum procedure. The linear problem is solved using the numerical method and the Ritz–Galerkin procedure can be used to reduce the problem to solving an infinite number of coupled nonlinear ordinary differential equations, which are solved using the analytic method.

Often one is interested in a situation in which one or more of the parameters become either very large or very small. Typically these are difficult situations to treat by straightforward numerical procedures. In these situations, analytic methods can often provide an accurate approximation and even suggest a way to improve the numerical procedure.

Foremost among the analytic techniques are the systematic methods of perturbation (asymptotic expansions) in terms of a small or a large parameter or coordinate. The book *Perturbation Methods* presents in a unified way an account of most of the perturbation techniques devised by physicists, engineers, and applied mathematicians, pointing out their similarities, differences, advantages, and limitations. However, the material is concise and advanced and, therefore, is intended for researchers and advanced graduate students. In *Introduction to Perturbation Techniques* the material is presented in an elementary way, making it easily accessible to advanced undergraduates and first-year graduate students in a wide variety of scientific and engineering fields. In both books the material is presented using examples; the second volume contains more than 360 problems.

Problems in Perturbation contains detailed solutions of all the problems in *Introduction to Perturbation Techniques* and about an equal number of unsolved supplementary problems. Each chapter begins with a short introduction that gives a summary of the definitions, basic theory, and available methods. The material is self-contained for the reader who has a background in calculus and elementary ordinary-differential equations. Although the solved problems are the exercises in *Introduction to Perturbation Techniques*, the material is general and could be used to accompany any of the existing books on perturbations, as well as those on nonlinear oscillations and applied mathematics that include asymptotics and perturbations. Since perturbation techniques are best explained using examples, this book is ideal for self-study.

ALI HASAN NAYFEH

Blacksburg, Virginia
March 1985

Contents

PROBLEMS IN PERTURBATION

CHAPTER 1

Introduction

Most of the physical problems facing physicists, engineers, and applied mathematicians today exhibit certain essential features that preclude exact analytical solutions. These features include nonlinear governing equations, nonlinear boundary conditions at known or, in some cases, unknown boundaries, variable coefficients, and complex boundary shapes. Hence physicists, engineers, and applied mathematicians are forced to determine approximate solutions of the problems they are facing. The approximations may be purely numerical, purely analytical, or a combination of numerical and analytical techniques. In this book we concentrate on analytical techniques, which, when combined with a numerical method, yield very powerful and versatile techniques.

The analytical approximations can be broadly divided into rational and irrational. An irrational approximation is usually obtained by an ad hoc, mathematical-modeling process that involves keeping certain elements, neglecting some, and approximating yet others. Thus it represents a dead end, because the resulting accuracy cannot be improved by successive approximations. A rational approximation represents a systematic expansion, called asymptotic or perturbation, that can in principle be continued indefinitely.

1.1. Parameter Perturbations

The key to solving modern problems is mathematical modeling that involves deriving the governing equations and boundary and initial conditions. Then the mathematical problem should always be expressed in nondimensional or dimensionless variables before any approximations are attempted.

If the physical problem involving the dimensionless scalar or vector variable $u(x, \varepsilon)$ can be represented mathematically by the differential equation $L(y, x, \varepsilon) = 0$ and the boundary condition $B(x, \varepsilon) = 0$, where x is a scalar or vector-independent dimensionless variable and ε is a dimensionless parameter, it cannot, in general, be solved exactly. However, if there exists an $\varepsilon = \varepsilon_0$ (ε can be scaled so that $\varepsilon_0 = 0$) for which the problem given can be solved exactly, more readily, or numerically, one seeks to approximate the solution $u(x, \varepsilon)$ for small values of ε, say in an expansion in powers of ε in the form

$$u(x, \varepsilon) = u_0(x) + \varepsilon u_1(x) + \varepsilon^2 u_2(x) + \cdots$$

where the $u_n(x)$ are independent of ε and $u_0(x)$ is the solution of the problem when $\varepsilon = 0$. Such expansions are called parameter perturbations.

1.2. Coordinate Perturbations

If the physical problem is represented mathematically by the differential equation $L(u, x) = 0$ and the boundary condition $B(x) = 0$, and if $u(x)$ takes a known form u_0 as $x \to x_0$ (x_0 can be scaled to either 0 or ∞), in a coordinate perturbation, one determines the deviation of u from u_0 for x near x_0 in terms of powers of x for $x_0 = 0$, or in terms of powers of x^{-1} for $x_0 = \infty$. Examples of coordinate perturbations are

$$u = x^{\sigma} \sum_{n=1}^{\infty} a_n x^n$$

$$u = x^{\sigma} \sum_{n=1}^{\infty} a_n x^{-n}$$

$$u = e^{-2x} x^{\sigma} \sum_{n=1}^{\infty} a_n x^n$$

1.3. Gauge Functions

In both parameter and coordinate perturbations, one is interested in the behavior of functions such as $f(\varepsilon)$ as the dimensionless parameter or coordinate ε tends to a specific value ε_0 (ε can always be normalized so that $\varepsilon_0 = 0$). One way of classifying the function $f(\varepsilon)$ is based on its limit as $\varepsilon \to 0$; if this limit exists, there are three possibilities:

$$\left.\begin{array}{l} f(\varepsilon) \to 0 \\ f(\varepsilon) \to A \\ f(\varepsilon) \to \infty \end{array}\right\} \quad \text{as } \varepsilon \to 0,\ 0 < A < \infty$$

The classification given above based on the limit is not very useful because there are an infinite number of functions that tend to zero or infinity as $\varepsilon \to 0$. Therefore, to narrow down this classification, we subdivide the first and third classes according to the rate at which they tend to zero or infinity. To accomplish this, we compare the rate at which these functions tend to zero and infinity with the rate at which a set of gauge functions tends to zero and infinity. These gauge functions are so familiar that their limiting behavior is known intuitively. The simplest possible examples of gauge functions are the powers of ε.

1.4. Order Symbols

If

$$\lim_{\varepsilon \to 0} \frac{f(\varepsilon)}{g(\varepsilon)} = A$$

where $0 < A < \infty$, we write

$$f(\varepsilon) = O[g(\varepsilon)] \quad \text{as } \varepsilon \to 0$$

and say that $f(\varepsilon)$ is order $g(\varepsilon)$ as $\varepsilon \to 0$ or $f(\varepsilon)$ is big "oh" of $g(\varepsilon)$ as $\varepsilon \to 0$.

If

$$\lim_{\varepsilon \to 0} \frac{f(\varepsilon)}{g(\varepsilon)} = 0$$

we write

$$f(\varepsilon) = o[g(\varepsilon)] \quad \text{as } \varepsilon \to 0$$

and say that $f(\varepsilon)$ is little "oh" of $g(\varepsilon)$ as $\varepsilon \to 0$.

1.5. Asymptotic Series

Given a series $\sum_{n=0}^{\infty}(a_n/x^n)$, where the a_n are independent of x, we say that the series is an asymptotic series and write

$$f(x) \sim \sum_{n=0}^{\infty} \frac{a_n}{x^n} \quad \text{as } |x| \to \infty$$

if and only if

$$f(x) = \sum_{n=0}^{N} \frac{a_n}{x^n} + o\left(\frac{1}{|x|^N}\right) \quad \text{as } |x| \to \infty$$

which is equivalent to

$$f(x) = \sum_{n=0}^{N-1} \frac{a_n}{x^n} + O\left(\frac{1}{|x|^N}\right) \quad \text{as } |x| \to \infty$$

1.6. Asymptotic Sequences and Expansions

A sequence of functions $\delta_n(\varepsilon)$ is called an asymptotic sequence as $\varepsilon \to 0$ if

$$\delta_n(\varepsilon) = o[\delta_{n-1}(\varepsilon)] \quad \text{as } \varepsilon \to 0$$

Given an expansion $\sum_{n=0}^{\infty} a_n \delta_n(\varepsilon)$, where the a_n are independent of ε and $\delta_n(\varepsilon)$ is an asymptotic sequence, we say that this is an asymptotic expansion and write

$$f(\varepsilon) \sim \sum_{n=0}^{\infty} a_n \delta_n(\varepsilon) \quad \text{as } \varepsilon \to 0$$

if and only if

$$f(\varepsilon) = \sum_{n=0}^{N} a_n \delta_n(\varepsilon) + o[\delta_N(\varepsilon)] \quad \text{as } \varepsilon \to 0$$

which is equivalent to

$$f(\varepsilon) = \sum_{n=0}^{N-1} a_n \delta_n(\varepsilon) + O[\delta_N(\varepsilon)] \quad \text{as } \varepsilon \to 0$$

Clearly, an asymptotic series is a special case of an asymptotic expansion.

1.7. Convergent Versus Asymptotic Series

Let a function $f(x)$ be represented by the first N terms of a series in inverse powers of x plus a remainder $R_N(x)$ as

$$f(x) = \sum_{n=0}^{N} \frac{a_n}{x^n} + R_N(x)$$

where the a_n are independent of x. This series converges if and only if

$$\lim_{\substack{N \to \infty \\ x \text{ fixed}}} R_N(x) = 0$$

This series is an asymptotic series as $|x| \to \infty$ if and only if

$$R_N(x) = o(|x|^{-N}) \quad \text{as } |x| \to \infty$$

Clearly, a convergent series is an asymptotic series; however, an asymptotic series need not converge.

1.8. Nonuniform Expansions

An asymptotic expansion of more than one variable, such as

$$f(x,\varepsilon) \sim \sum_{n=0}^{\infty} a_n(x)\delta_n(\varepsilon) \quad \text{as } \varepsilon \to 0$$

is uniform if and only if

$$f(x,\varepsilon) = \sum_{n=0}^{N} a_n(x)\delta_n(\varepsilon) + R_N(x,\varepsilon) \quad \text{as } \varepsilon \to 0$$

where $R_N(x,\varepsilon) = o[\delta_N(\varepsilon)]$ as $\varepsilon \to 0$ for all x in the domain of interest.

Solved Exercises

1.1. For small ε determine three terms in the expansions of

(a) $(1-\frac{3}{8}a^2\varepsilon+\frac{51}{256}a^4\varepsilon^2)^{-1} = 1+(\frac{3}{8}a^2\varepsilon-\frac{51}{256}a^4\varepsilon^2)+(\frac{3}{8}a^2\varepsilon-\frac{51}{256}a^4\varepsilon^2)^2$
$+\cdots = 1+\frac{3}{8}a^2\varepsilon-\frac{51}{256}a^4\varepsilon^2+\frac{9}{64}a^4\varepsilon^2+\cdots = 1+\frac{3}{8}a^2\varepsilon-\frac{15}{256}a^4\varepsilon^2+\cdots$

(b) $\cos(\sqrt{1-\varepsilon t}) = \cos\left[\left(1-\frac{1}{2}\varepsilon t+\frac{\frac{1}{2}\times-\frac{1}{2}}{2!}\varepsilon^2 t^2+\cdots\right)\right]$
$= \cos[(1-\frac{1}{2}\varepsilon t-\frac{1}{8}\varepsilon^2 t^2)+\cdots]$
$= \cos 1-\sin 1(-\frac{1}{2}\varepsilon t-\frac{1}{8}\varepsilon^2 t^2)-\frac{1}{2!}\cos 1(-\frac{1}{2}\varepsilon t-\frac{1}{8}\varepsilon^2 t^2)^2+\cdots$
$= \cos 1+(\frac{1}{2}\varepsilon t+\frac{1}{8}\varepsilon^2 t^2)\sin 1-\frac{1}{8}\varepsilon^2 t^2\cos 1+\cdots$
$= \cos 1+\frac{1}{2}\varepsilon t\sin 1+\frac{1}{8}\varepsilon^2 t^2(\sin 1-\cos 1)+\cdots$

(c) $\sqrt{1-\frac{1}{2}\varepsilon+2\varepsilon^2} = [1-(\frac{1}{2}\varepsilon-2\varepsilon^2)]^{1/2} = 1-\frac{1}{2}(\frac{1}{2}\varepsilon-2\varepsilon^2)$
$+\frac{\frac{1}{2}\times-\frac{1}{2}}{2!}(\frac{1}{2}\varepsilon-2\varepsilon^2)^2+\cdots = 1-\frac{1}{4}\varepsilon+\varepsilon^2-\frac{1}{32}\varepsilon^2+\cdots = 1-\frac{1}{4}\varepsilon+\frac{31}{32}\varepsilon^2+\cdots$

(d) $\sin(1+\varepsilon-\varepsilon^2) = \sin 1+(\varepsilon-\varepsilon^2)\cos 1-\frac{1}{2!}(\varepsilon-\varepsilon^2)^2\sin 1+\cdots$
$= \sin 1+(\varepsilon-\varepsilon^2)\cos 1-\frac{1}{2}\varepsilon^2\sin 1+\cdots$
$= \sin 1+\varepsilon\cos 1-\varepsilon^2(\cos 1+\frac{1}{2}\sin 1)+\cdots$

1.2. Expand each of the following expressions for small ε and keep three terms:

(a) $\sqrt{1-\frac{1}{2}\varepsilon^2 t-\frac{1}{8}\varepsilon^4 t} = [1-(\frac{1}{2}\varepsilon^2 t+\frac{1}{8}\varepsilon^4 t)]^{1/2}$
$= 1-\frac{1}{2}(\frac{1}{2}\varepsilon^2 t+\frac{1}{8}\varepsilon^4 t)+\frac{\frac{1}{2}\times-\frac{1}{2}}{2!}(\frac{1}{2}\varepsilon^2 t+\frac{1}{8}\varepsilon^4 t)^2+\cdots$
$= 1-\frac{1}{4}\varepsilon^2 t-\frac{1}{16}\varepsilon^4 t-\frac{1}{32}\varepsilon^4 t^2+\cdots = 1-\frac{1}{4}\varepsilon^2 t-\frac{1}{16}\varepsilon^4 t(1+\frac{1}{2}t)+\cdots$

(b) $(1+\varepsilon\cos f)^{-1} = 1-\varepsilon\cos f+\varepsilon^2\cos^2 f+\cdots$

(c) $(1+\varepsilon\omega_1+\varepsilon^2\omega_2)^{-2} = 1-2(\varepsilon\omega_1+\varepsilon^2\omega_2)+\frac{-2\times-3}{2!}(\varepsilon\omega_1+\varepsilon^2\omega_2)^2+\cdots$
$= 1-2\varepsilon\omega_1+\varepsilon^2(-2\omega_2+3\omega_1^2)+\cdots$

(d) $\sin(s+\varepsilon\omega_1 s+\varepsilon^2\omega_2 s) = \sin s+(\varepsilon\omega_1 s+\varepsilon^2\omega_2 s)\cos s-\frac{1}{2!}(\varepsilon\omega_1 s+\varepsilon^2\omega_2 s)^2\sin s$
$+\cdots = \sin s+\varepsilon\omega_1 s\cos s+\varepsilon^2 s(\omega_2\cos s-\frac{1}{2}\omega_1^2 s\sin s)+\cdots$

(e) $\sin^{-1}\left(\frac{\varepsilon}{\sqrt{1+\varepsilon}}\right) = \sin^{-1}[\varepsilon(1+\varepsilon)^{-1/2}]$

$$= \sin^{-1}\left[\varepsilon\left(1-\tfrac{1}{2}\varepsilon+\frac{-\frac{1}{2}\times-\frac{3}{2}}{2!}\varepsilon^2\right)+\cdots\right] = \sin^{-1}(\varepsilon-\tfrac{1}{2}\varepsilon^2+\tfrac{3}{8}\varepsilon^3+\cdots)$$

$$= (\varepsilon-\tfrac{1}{2}\varepsilon^2+\tfrac{3}{8}\varepsilon^3)+\tfrac{1}{3!}(\varepsilon-\tfrac{1}{2}\varepsilon^2+\tfrac{3}{8}\varepsilon^3)^3+\cdots$$

$$= \varepsilon-\tfrac{1}{2}\varepsilon^2+\tfrac{3}{8}\varepsilon^3+\tfrac{1}{6}\varepsilon^3+\cdots = \varepsilon-\tfrac{1}{2}\varepsilon^2+\tfrac{13}{24}\varepsilon^3+\cdots$$

(f) $\ln\frac{1+2\varepsilon-\varepsilon^2}{\sqrt[3]{1+2\varepsilon}} = \ln(1+2\varepsilon-\varepsilon^2)-\tfrac{1}{3}\ln(1+2\varepsilon)$

$$= 2\varepsilon-\varepsilon^2-\tfrac{1}{2}(2\varepsilon-\varepsilon^2)^2+\tfrac{1}{3}(2\varepsilon-\varepsilon^2)^3$$

$$+\cdots-\tfrac{1}{3}[2\varepsilon-\tfrac{1}{2}(2\varepsilon)^2+\tfrac{1}{3}(2\varepsilon)^3+\cdots] = \tfrac{4}{3}\varepsilon-\tfrac{7}{3}\varepsilon^2+\tfrac{34}{9}\varepsilon^3+\cdots$$

1.3. Let $\mu = \mu_0 + e\mu_1 + e^2\mu_2$ in $h = \frac{3}{2}\left[1-\sqrt{1-3\mu(1-\mu)}\right]$, expand for small e, and keep three terms.

Using a Taylor series expansion, we have

$$h(\mu_0+e\mu_1+e^2\mu_2) = h(\mu_0)+h'(\mu_0)(e\mu_1+e^2\mu_2)+\tfrac{1}{2!}h''(\mu_0)(e\mu_1+e^2\mu_2)^2+\cdots$$

But

$$h'(\mu) = -\tfrac{3}{2}\cdot\tfrac{1}{2}[1-3\mu(1-\mu)]^{-1/2}\times(-3+6\mu)$$

$$= \tfrac{9}{4}(1-2\mu)[1-3\mu(1-\mu)]^{-1/2}$$

$$h''(\mu) = -\tfrac{9}{2}[1-3\mu(1-\mu)]^{-1/2}-\tfrac{1}{2}\times\tfrac{9}{4}(1-2\mu)[1-3\mu(1-\mu)]^{-3/2}\times(-3+6\mu)$$

$$= -\tfrac{9}{2}[1-3\mu(1-\mu)]^{-1/2}+\tfrac{27}{8}(1-2\mu)^2[1-3\mu(1-\mu)]^{-3/2}$$

Therefore,

$$h = \tfrac{3}{2}\left[1-\sqrt{1-3\mu_0(1-\mu_0)}\right]+\tfrac{9}{4}(1-2\mu_0)[1-3\mu_0(1-\mu_0)]^{-1/2}\mu_1 e$$

$$+e^2\left\{\tfrac{9}{4}(1-2\mu_0)[1-3\mu_0(1-\mu_0)]^{-1/2}\mu_2\right.$$

$$-\tfrac{9}{4}[1-3\mu_0(1-\mu_0)]^{-1/2}\mu_1^2+\tfrac{27}{16}(1-2\mu_0)^2$$

$$\left.\times[1-3\mu_0(1-\mu_0)]^{-3/2}\mu_1^2\right\}+\cdots$$

1.4. For small ε, determine the order of the functions, $\sinh(1/\varepsilon)$, $\ln(1+\sin\varepsilon)$, $\ln(2+\sin\varepsilon)$, and $e^{\ln(1-\varepsilon)}$.

Since as $\varepsilon\to 0$,

$$\sinh\frac{1}{\varepsilon} = \frac{e^{1/\varepsilon}-e^{-1/\varepsilon}}{2} \approx \tfrac{1}{2}e^{1/\varepsilon}$$

$$\sinh\frac{1}{\varepsilon} = O(e^{1/\varepsilon}) \quad \text{as } \varepsilon\to 0$$

Since as $\varepsilon \to 0$,

$$\ln(1+\sin \varepsilon) \approx \ln(1+\varepsilon) \approx \varepsilon$$

$$\ln(1+\sin \varepsilon) = O(\varepsilon) \quad \text{as } \varepsilon \to 0$$

Since as $\varepsilon \to 0$,

$$\ln(2+\sin \varepsilon) \approx \ln(2+\varepsilon) \approx \ln 2 + \ln\left(1+\tfrac{1}{2}\varepsilon\right) \approx \ln 2 + \tfrac{1}{2}\varepsilon$$

$$\ln(2+\sin \varepsilon) = O(1) \quad \text{as } \varepsilon \to 0$$

Since as $\varepsilon \to 0$,

$$e^{\ln(1-\varepsilon)} \approx e^{-\varepsilon} \approx 1-\varepsilon$$

$$e^{\ln(1-\varepsilon)} = O(1) \quad \text{as } \varepsilon \to 0$$

1.5. Determine the order of the following expressions as $\varepsilon \to 0$:

(a) Since $\sqrt{\varepsilon(1-\varepsilon)} \approx \sqrt{\varepsilon}$

$$\sqrt{\varepsilon(1-\varepsilon)} = O(\varepsilon^{1/2})$$

(b) $4\pi^2\varepsilon = O(\varepsilon)$

(c) $1000\varepsilon^{1/2} = O(\varepsilon^{1/2})$

(d) Since $\ln(1+\varepsilon) \approx \varepsilon$

$$\ln(1+\varepsilon) = O(\varepsilon)$$

(e) Since $\dfrac{1-\cos\varepsilon}{1+\cos\varepsilon} \approx \dfrac{1-1+\frac{1}{2}\varepsilon^2}{1+1-\frac{1}{2}\varepsilon^2} \approx \frac{1}{4}\varepsilon^2$

$$\frac{1-\cos\varepsilon}{1+\cos\varepsilon} = O(\varepsilon^2)$$

(f) Since $\dfrac{\varepsilon^{3/2}}{1-\cos\varepsilon} \approx \dfrac{\varepsilon^{3/2}}{1-1+\frac{1}{2}\varepsilon^2} = 2\varepsilon^{-1/2}$

$$\frac{\varepsilon^{3/2}}{1-\cos\varepsilon} = O(\varepsilon^{-1/2})$$

(g) Let $\operatorname{sech}^{-1}\varepsilon = u$ so that $\operatorname{sech} u = \varepsilon$ and $\cosh u = 1/\varepsilon$. Hence

$$e^u + e^{-u} = \frac{2}{\varepsilon}$$

As $\varepsilon \to 0$, $u \to \infty$ and hence

$$e^u \approx \frac{2}{\varepsilon} \quad \text{or} \quad u \approx \ln 2 + \ln\frac{1}{\varepsilon}$$

Therefore,

$$\operatorname{sech}^{-1}\varepsilon = O\left(\ln\frac{1}{\varepsilon}\right)$$

(h) Since $e^{\tan\varepsilon} \approx e^{\varepsilon} \approx 1$

$$e^{\tan\varepsilon} = O(1)$$

(i) Since $\ln\left[1+\dfrac{\ln(1+2\varepsilon)}{\varepsilon(1-2\varepsilon)}\right] \approx \ln\left(1+\dfrac{2\varepsilon}{\varepsilon}\right) = \ln 3$

$$\ln\left[1+\frac{\ln(1+2\varepsilon)}{\varepsilon(1-2\varepsilon)}\right] = O(1)$$

(j) Since $\ln\left[1+\dfrac{\ln[(1+2\varepsilon)/\varepsilon]}{1-2\varepsilon}\right] \approx \ln\left[1+\dfrac{\ln(1/\varepsilon)}{1}\right] \approx \ln\left(\ln\dfrac{1}{\varepsilon}\right)$

$$\ln\left[1+\frac{\ln[(1+2\varepsilon)/\varepsilon]}{1-2\varepsilon}\right] = O\left[\ln\left(\ln\frac{1}{\varepsilon}\right)\right]$$

(k) Since $e^{-\cosh(1/\varepsilon)} \approx \exp\left(-\frac{1}{2}\exp\dfrac{1}{\varepsilon}\right)$

$$e^{-\cosh(1/\varepsilon)} = O\left[\exp\left(-\tfrac{1}{2}\exp\frac{1}{\varepsilon}\right)\right]$$

(l) Since $\int_0^\varepsilon e^{-s^2}\,ds \approx \int_0^\varepsilon (1-s^2)\,ds \approx \varepsilon$

$$\int_0^\varepsilon e^{-s^2}\,ds = O(\varepsilon)$$

1.6. Determine the order of the following expressions as $\varepsilon \to 0$:

(a) Since $\ln(1+5\varepsilon) \approx 5\varepsilon$

$$\ln(1+5\varepsilon) = O(\varepsilon)$$

(b) Since $\sin^{-1}\dfrac{\varepsilon}{\sqrt{1+\varepsilon}} \approx \sin^{-1}\varepsilon \approx \varepsilon$

$$\sin^{-1}\frac{\varepsilon}{\sqrt{1+\varepsilon}} = O(\varepsilon)$$

(c) Since $\dfrac{\sqrt{\varepsilon}}{\sin\varepsilon} \approx \dfrac{\sqrt{\varepsilon}}{\varepsilon} = \varepsilon^{-1/2}$

$$\frac{\sqrt{\varepsilon}}{\sin\varepsilon} = O(\varepsilon^{-1/2})$$

(d) Since $1-\frac{1}{2}\varepsilon^2-\cos\varepsilon \approx 1-\frac{1}{2}\varepsilon^2-(1-\frac{1}{2!}\varepsilon^2+\frac{1}{4!}\varepsilon^4) = -\frac{1}{4!}\varepsilon^4$

$$1-\tfrac{1}{2}\varepsilon^2-\cos\varepsilon = O(\varepsilon^4)$$

(e) Since $\ln\left(\sinh\frac{1}{\varepsilon}\right) \approx \ln\left[\frac{e^{1/\varepsilon}-e^{-1/\varepsilon}}{2}\right] \approx \ln\left(\frac{1}{2}e^{1/\varepsilon}\right) \approx \frac{1}{\varepsilon}$

$$\ln\left(\sinh\frac{1}{\varepsilon}\right) = O\left(\frac{1}{\varepsilon}\right)$$

1.7. Determine the order of the following as $\varepsilon \to 0$:

(a) Since $\ln(\cot\varepsilon) = \ln\left(\frac{\cos\varepsilon}{\sin\varepsilon}\right) \approx \ln\left(\frac{1}{\varepsilon}\right)$

$$\ln(\cot\varepsilon) = O\left(\ln\frac{1}{\varepsilon}\right)$$

(b) Since $\sinh\frac{1}{\varepsilon} = \frac{e^{1/\varepsilon}-e^{-1/\varepsilon}}{2} \approx \frac{1}{2}e^{1/\varepsilon}$

$$\sinh\frac{1}{\varepsilon} = O(e^{1/\varepsilon})$$

(c) Since $\coth\left(\frac{1}{\varepsilon}\right) = \frac{\cosh(1/\varepsilon)}{\sinh(1/\varepsilon)} = \frac{e^{1/\varepsilon}+e^{-1/\varepsilon}}{e^{1/\varepsilon}-e^{-1/\varepsilon}} \approx 1$

$$\coth\left(\frac{1}{\varepsilon}\right) = O(1)$$

(d) Since $\frac{\varepsilon^{3/4}}{1-\cos\varepsilon} \approx \frac{\varepsilon^{3/4}}{1-\left(1-\frac{1}{2}\varepsilon^2\right)} \approx \frac{\varepsilon^{3/4}}{\frac{1}{2}\varepsilon^2} = 2\varepsilon^{-5/4}$

$$\frac{\varepsilon^{3/4}}{1-\cos\varepsilon} = O(\varepsilon^{-5/4})$$

(e) Since $\ln\left[1+\ln\frac{1+2\varepsilon}{\varepsilon}\right] \approx \ln\left[1+\ln\frac{1}{\varepsilon}\right] \approx \ln\left(\ln\frac{1}{\varepsilon}\right)$

$$\ln\left[1+\ln\frac{1+2\varepsilon}{\varepsilon}\right] = O\left[\ln\left(\ln\frac{1}{\varepsilon}\right)\right]$$

1.8. Arrange the following in descending order for small ε:

$$\varepsilon^2,\ \varepsilon^{1/2},\ \ln\left(\ln\frac{1}{\varepsilon}\right),\ 1,\ \varepsilon^{1/2}\ln\frac{1}{\varepsilon},\ \varepsilon\ln\frac{1}{\varepsilon},\ e^{-1/\varepsilon},\ \ln\frac{1}{\varepsilon},\ \varepsilon^{3/2},\ \varepsilon,\ \varepsilon^2\ln\frac{1}{\varepsilon}$$

Answer:

$$\ln\frac{1}{\varepsilon} > \ln\left(\ln\frac{1}{\varepsilon}\right) > 1 > \varepsilon^{1/2}\ln\frac{1}{\varepsilon} > \varepsilon^{1/2} > \varepsilon\ln\frac{1}{\varepsilon} > \varepsilon > \varepsilon^{3/2} > \varepsilon^2\ln\frac{1}{\varepsilon} > \varepsilon^2 > e^{-1/\varepsilon}$$

1.9. Arrange the following in descending order for small ε:

$$e^{-1/\varepsilon},\ \ln\frac{1}{\varepsilon},\ \varepsilon^{-0.01},\ \cot\varepsilon, \text{ and } \sinh\frac{1}{\varepsilon}$$

Answer:

$$\sinh\frac{1}{\varepsilon} > \cot\varepsilon > \varepsilon^{-0.01} > \ln\frac{1}{\varepsilon} > e^{-1/\varepsilon}$$

1.10. Arrange the following in descending order for small ε:

$$\ln(1+\varepsilon),\ \cot\varepsilon,\ \tanh\frac{1}{\varepsilon},\ \frac{\sin\varepsilon}{\varepsilon^{3/4}},\ \varepsilon\ln\varepsilon,\ e^{-1/\varepsilon},\ \sinh\frac{1}{\varepsilon},\ \frac{1}{\ln(1/\varepsilon)}$$

Since

$$\ln(1+\varepsilon) = O(\varepsilon)$$

$$\cot\varepsilon = O(\varepsilon^{-1})$$

$$\tanh\frac{1}{\varepsilon} = O(1)$$

$$\frac{\sin\varepsilon}{\varepsilon^{3/4}} = O(\varepsilon^{1/4})$$

$$\sinh\frac{1}{\varepsilon} = O(e^{1/\varepsilon})$$

$$\sinh\frac{1}{\varepsilon} > \cot\varepsilon > \tanh\frac{1}{\varepsilon} > \frac{1}{\ln(1/\varepsilon)} > \frac{\sin\varepsilon}{\varepsilon^{3/4}} > \varepsilon\ln\varepsilon > \ln(1+\varepsilon) > e^{-1/\varepsilon}$$

1.11. Arrange the following in descending order for small ε:

$$e^{-1/\varepsilon},\ \ln\frac{1}{\varepsilon},\ \frac{1}{\varepsilon},\ \varepsilon^{1/2},\ \left(\ln\frac{1}{\varepsilon}\right)^2,\ \varepsilon^{1/2}\ln\frac{1}{\varepsilon},\ e^{1/\varepsilon},\ \frac{1}{\varepsilon^{3/2}},\ 1,\ \varepsilon,\ \varepsilon^{3/2},\ \frac{1}{\varepsilon^{1/2}},$$

$$e^{e^{1/\varepsilon}},\ e^{-e^{1/\varepsilon}},\ \ln\left(\ln\frac{1}{\varepsilon}\right),\ \varepsilon^{0.0001},\ \varepsilon^{-0.0001},\ \varepsilon^{0.0001}\ln\frac{1}{\varepsilon},\ 5^{1/\varepsilon},\ 5^{-1/\varepsilon}$$

Answer:

$$e^{e^{1/\varepsilon}} > 5^{1/\varepsilon} > e^{1/\varepsilon} > \frac{1}{\varepsilon^{3/2}} > \frac{1}{\varepsilon} > \frac{1}{\varepsilon^{1/2}} > \varepsilon^{-0.0001} > \left(\ln\frac{1}{\varepsilon}\right)^2$$

$$> \ln\frac{1}{\varepsilon} > \ln\left(\ln\frac{1}{\varepsilon}\right) > 1 > \varepsilon^{0.0001}\ln\frac{1}{\varepsilon} > \varepsilon^{0.0001} > \varepsilon^{1/2}\ln\frac{1}{\varepsilon}$$

$$> \varepsilon^{1/2} > \varepsilon > \varepsilon^{3/2} > e^{-1/\varepsilon} > 5^{-1/\varepsilon} > e^{-e^{1/\varepsilon}}$$

1.12. Arrange the following terms in descending order for small ε:

$$\varepsilon^{\nu},\ \varepsilon^{-\nu},\ \varepsilon^{\mu},\ \varepsilon^{-\mu},\ 1,\ \varepsilon^{3/2},\ \varepsilon^{-3/2},\ e^{1/\varepsilon},\ \ln\frac{1}{\varepsilon},\ \ln\left(\ln\frac{1}{\varepsilon}\right),$$

$$\left(\ln\frac{1}{\varepsilon}\right)^2,\ \varepsilon\ln\frac{1}{\varepsilon},\ \frac{\varepsilon}{\ln 1/\varepsilon},\ e^{-1/\varepsilon},\ \varepsilon$$

where $\nu = 10^{-100}$ and $\mu = 10^{100}$.

$$e^{1/\varepsilon} > \varepsilon^{-\mu} > \varepsilon^{-3/2} > \varepsilon^{-\nu} > \left(\ln\frac{1}{\varepsilon}\right)^2 > \ln\left(\frac{1}{\varepsilon}\right) > \ln\left(\ln\frac{1}{\varepsilon}\right) > 1 > \varepsilon^{\nu}$$

$$> \varepsilon \ln\frac{1}{\varepsilon} > \varepsilon > \frac{\varepsilon}{\ln(1/\varepsilon)} > \varepsilon^{3/2} > \varepsilon^{\mu} > e^{-1/\varepsilon}$$

1.13. Arrange the following in descending order for small ε:

$$\ln(1+\varepsilon),\ \operatorname{sech}^{-1}\varepsilon,\ \frac{1-\cos\varepsilon}{1+\cos\varepsilon},\ \sqrt{\varepsilon(1-\varepsilon)}\,,\ e^{-\cosh(1/\varepsilon)},$$

$$\ln\left[1+\frac{\ln[(1+2\varepsilon)/\varepsilon]}{1-2\varepsilon}\right],\ \ln\left[1+\frac{\ln(1+2\varepsilon)}{\varepsilon(1-2\varepsilon)}\right],\ \frac{\varepsilon^{1/2}}{1-\cos\varepsilon}$$

Since

$$\ln(1+\varepsilon) = O(\varepsilon)$$

$$\operatorname{sech}^{-1}\varepsilon = O\left(\ln\frac{1}{\varepsilon}\right)$$

$$\frac{1-\cos\varepsilon}{1+\cos\varepsilon} = O(\varepsilon^2)$$

$$\sqrt{\varepsilon(1-\varepsilon)} = O(\varepsilon^{1/2})$$

$$e^{-\cosh(1/\varepsilon)} = O\left[\exp\left(-\tfrac{1}{2}\exp\frac{1}{\varepsilon}\right)\right]$$

$$\ln\left[1+\frac{\ln[(1+2\varepsilon)/\varepsilon]}{1-2\varepsilon}\right] = O\left[\ln\left(\ln\frac{1}{\varepsilon}\right)\right]$$

$$\ln\left[1+\frac{\ln(1+2\varepsilon)}{\varepsilon(1-2\varepsilon)}\right] = O(1)$$

$$\frac{\varepsilon^{1/2}}{1-\cos\varepsilon} = O(\varepsilon^{-3/2})$$

$$\frac{\varepsilon^{1/2}}{1-\cos\varepsilon} > \operatorname{sech}^{-1}\varepsilon > \ln\left[1+\frac{\ln[(1+2\varepsilon)/\varepsilon]}{1-2\varepsilon}\right] > \ln\left[1+\frac{\ln(1+2\varepsilon)}{\varepsilon(1-2\varepsilon)}\right]$$

$$> \sqrt{\varepsilon(1-\varepsilon)} > \ln(1+\varepsilon) > \frac{1-\cos\varepsilon}{1+\cos\varepsilon} > e^{-\cos(1/\varepsilon)}$$

1.14. Which of the following expansions is nonuniformly valid and what are its regions of nonuniformity?

(a) $\sqrt{2}\,t = \frac{2}{3}x^{3/2} + \varepsilon\left(\frac{2}{3}x^{3/2} + \sqrt{x} - \frac{1}{2}\ln\frac{1+\sqrt{x}}{1-\sqrt{x}}\right) + O\left(\frac{\varepsilon^2}{1-x}\right)$ is nonuniform as $x \to 1$.

Comparing the second and third terms shows that the region of nonuniformity is $x=1+O(\varepsilon)$.

(b) $\eta=\varepsilon\cos x+\frac{1}{2}\varepsilon^2\dfrac{1+\gamma^2}{1-2\gamma^2}\cos 2x+\frac{3}{16}\varepsilon^3\dfrac{2\gamma^4+7\gamma^2+2}{(1-2\gamma^2)(1-3\gamma^2)}\cos 3x+O(\varepsilon^4)$ is non-uniform as $\gamma\to 1/\sqrt{2}$ or $1/\sqrt{3}$. Comparing the first and second terms shows that one of the regions of nonuniformity is $\gamma=1/\sqrt{2}+O(\varepsilon)$, whereas comparing the first and third terms shows that a second region of nonuniformity is $\gamma=1/\sqrt{3}+O(\varepsilon^2)$.

(c) $\sigma=\sqrt{k^2-1}-\dfrac{\varepsilon^2k^2}{\sqrt{k^2-1}}+O(\varepsilon^3)$ is nonuniform as $k\to 1$. Comparing the first and second terms shows that the region of nonuniformity is $k=1+O(\varepsilon^2)$.

(d) $f=1-\varepsilon x+\varepsilon^2x^2-\varepsilon^3x^3+O(\varepsilon^4)$ is nonuniform when $x\ge O(1/\varepsilon)$.

(e) $u=a\cos(1+\frac{3}{8}\varepsilon a^2)t+\dfrac{\varepsilon a^3}{32}\cos 3(1+\frac{3}{8}\varepsilon a^2)t+O(\varepsilon^2)$ is uniformly valid.

(f) $u=a\cos t+\dfrac{\varepsilon a^3}{8}(\frac{1}{4}\cos 3t-3t\sin t)+O(\varepsilon^2)$ is nonuniform when $t\ge O(1/\varepsilon)$.

(g) $c=1+\dfrac{2\varepsilon}{\tau-1}+\dfrac{3\varepsilon^2}{(\tau-1)^2}+O(\varepsilon^3)$ is nonuniform when $\tau-1=O(\varepsilon)$.

(h) $y=\dfrac{1}{\sqrt[4]{x^2(1-x)}}\cos\left[\lambda\int\sqrt{x(1-x)}\,dx\right]+O(1)$ as $\lambda\to\infty$ breaks down as $x\to 0$ and 1. The regions of nonuniformity cannot be determined from the information given in the expansion.

(i) $f=\sin x+\varepsilon\cos x-\frac{1}{2}\varepsilon^2\sin x-\frac{1}{6}\varepsilon^3\cos x+O(\varepsilon^4)$ is uniformly valid for all x.

Supplementary Exercises

1.15. For small ε, determine three terms in the expansions of

(a) $\sin\left(\sqrt{1-\varepsilon^2}+\beta\right)$

(b) $(1-\frac{1}{2}\varepsilon^2+2\varepsilon^2)^{-1/2}$

(c) $\cos\left(\dfrac{\tau}{\omega_0+\varepsilon\omega_1+\varepsilon^2\omega_2}\right)$

(d) $\dfrac{-1\mp\sqrt{1-2\varepsilon}}{2\varepsilon}$

(e) $e^{-\varepsilon t}\cos[(1-\varepsilon^2)t+\beta]$

(f) $\dfrac{1-\varepsilon+2\varepsilon^2}{1+\varepsilon+3\varepsilon^2}$

(g) $(1-m\varepsilon)^{-3/2}$

(h) $\dfrac{\ln(1+\varepsilon)}{1-\varepsilon}$

(i) $-\dfrac{x}{\varepsilon}\mp\sqrt{x^2/\varepsilon^2+2/\varepsilon-1}$

1.16. Determine the order of the following expressions as $\varepsilon\to 0$:

$$\frac{1-\cosh\varepsilon}{1+\cosh\varepsilon},\ -500\varepsilon,\ \frac{\sin\varepsilon}{1-\cos\varepsilon},\ \ln[\ln(1+\varepsilon)],\ \tanh\frac{1}{\varepsilon},\ \ln(2+\varepsilon),\ \ln(1+2\varepsilon),$$

$$\frac{\varepsilon}{\sin\varepsilon},\ \frac{\sinh\varepsilon}{\varepsilon},\ \frac{\sin^{-1}\varepsilon}{\varepsilon+\frac{1}{3}\varepsilon^3},\ \sinh^{-1}\frac{1}{\varepsilon}$$

1.17. Arrange the following in descending order for small ε:

$$\varepsilon^{5/2},\ \ln\frac{1}{\varepsilon},\ \varepsilon^{-1},\ \left(\ln\frac{1}{\varepsilon}\right)^{10},\ \left(\ln\frac{1}{\varepsilon}\right)^{-2},\ 10,\ \frac{1}{\varepsilon}\ln\frac{1}{\varepsilon},\ \varepsilon\ln\frac{1}{\varepsilon},\ e^{-1/\varepsilon}$$

1.18. Arrange the following in descending order for small ε:

$$1,\ \varepsilon^{-1/2},\ \varepsilon^{1/2},\ \varepsilon^{1/2}\ln\frac{1}{\varepsilon},\ \varepsilon^{3/4},\ \varepsilon^{-3/4},\ \varepsilon^{-3/4}\ln\frac{1}{\varepsilon},\ \varepsilon^{-10},\ \varepsilon^{10},\ \ln\left(\ln\frac{1}{\varepsilon}\right),\ \left(\ln\frac{1}{\varepsilon}\right)^{2}$$

1.19. Arrange the following in descending order for small ε:

$$e^{1/\varepsilon},\ 3^{1/\varepsilon},\ 2^{1/\varepsilon},\ \left(\frac{1}{\varepsilon}\right)^{100},\ (\varepsilon)^{100},\ \left(\frac{1}{\varepsilon}\right)^{\varepsilon},\ \varepsilon^{0.001},\ \ln\frac{1}{\varepsilon},\ \varepsilon^{-0.001}$$

1.20. Which of the following expansions is nonuniformly valid and what are its regions of nonuniformity?

(a) $u = a\cos(t+\beta) + \varepsilon\left[\tfrac{1}{2}\left(1-\tfrac{1}{4}a^2\right)at\cos(t+\beta)+\tfrac{1}{96}a^3\cos(3t+3\beta)\right]+\cdots, \qquad \varepsilon \ll 1$

(b) $$u = a\cos(t+\beta)+\frac{F}{1-\omega^2}\cos\omega t + \varepsilon\left\{-\mu a t\cos(t+\beta)+\frac{F^3}{4(1-\omega^2)^3(1-9\omega^2)}\cos 3\omega t + \frac{3a^2F}{4(1-\omega^2)(3-\omega)(1-\omega)}\cos[(2-\omega)t+2\beta]\right\}+\cdots, \qquad \varepsilon \ll 1$$

(c) $$u = a\cos(t+\beta) + \varepsilon\left\{\frac{a\cos[(\omega+2)t+\beta]}{4(1+\omega)}+\frac{a\cos[(\omega-2)t+\beta]}{4(1-\omega)}\right\}+\cdots, \qquad \varepsilon \ll 1$$

(d) $$y = \frac{e^{-x}}{x^2}+\frac{\varepsilon e^{-x}}{x^2}\int_1^x \frac{e^{-t}}{t^3}\left(1+\frac{2}{t}\right)dt+\cdots, \qquad \varepsilon \ll 1$$

(e) $$y = x+\frac{\varepsilon}{x}+\frac{\varepsilon^2}{x^3}+\cdots+\frac{\varepsilon^n}{x^{2n-1}}+\cdots, \qquad \varepsilon \ll 1$$

CHAPTER 2

Algebraic and Transcendental Equations

We are frequently faced with the problem of determining the roots of algebraic and transcendental equations. There are many computer codes that can be used to extract the roots of such equations. Several codes exploit the presence of small parameters and use a perturbation analysis, as described here. Many also use asymptotic analysis to obtain the large roots. This is certainly true for transcendental equations.

In this chapter, we discuss some of the methods that approximate solutions of algebraic and transcendental equations. We start with polynomial equations that depend on a small parameter of the form

$$P_n(x) = \varepsilon P_m(x)$$

where x and ε are dimensionless, P_n and P_m are polynomials in x of order n and m, respectively, and $n \geq m$. When $\varepsilon = 0$, the problem reduces to

$$P_n(x) = 0$$

which is called the reduced or unperturbed equation. Let $\alpha_1, \alpha_2, \ldots, \alpha_n$ be the roots of the reduced equation and consider first the case of α_i being distinct. If $P_m(\alpha_s) = 0$, $x = \alpha_s$ is an exact solution of the perturbed equation (i.e., original equation with ε different from zero). Assume that $P_m(\alpha_s) \neq 0$ for all s; otherwise, we first divide both sides of the perturbed equation by $x - \alpha_s$. When $\varepsilon \neq 0$, we seek an approximation to the root α_i in a power series of the form

$$x = \alpha_i + \varepsilon x_1 + \cdots$$

where the ellipses stand for all terms with powers of ε^s for which $s \geq 2$. Substituting the assumed expansion into the perturbed equation yields

$$P_n(\alpha_i + \varepsilon x_1 + \cdots) = \varepsilon P_m(\alpha_i + \varepsilon x_1 + \cdots)$$

which upon using a Taylor-series expansion gives

$$P_n(\alpha_i)+\varepsilon P_n'(\alpha_i)x_1+\cdots=\varepsilon P_m(\alpha_i)+\cdots$$

Equating the coefficients of each power of ε on both sides yields

$$P_n(\alpha_i)=0$$

$$P_n'(\alpha_i)x_1=P_m(\alpha_i)$$

The first equation is satisfied because α_i is a root of $P_n(x)=0$ and the second equation yields

$$x_1=\frac{P_m(\alpha_i)}{P_n'(\alpha_i)}$$

or

$$x_1=\frac{P_m(\alpha_i)}{(\alpha_i-\alpha_1)(\alpha_i-\alpha_2)\cdots(\alpha_i-\alpha_{i-1})(\alpha_i-\alpha_{i+1})\cdots(\alpha_i-\alpha_n)}$$

Hence to the first approximation

$$x=\alpha_i+\frac{\varepsilon P_m(\alpha_i)}{(\alpha_i-\alpha_1)(\alpha_i-\alpha_2)\cdots(\alpha_i-\alpha_{i-1})(\alpha_i-\alpha_{i+1})\cdots(\alpha_i-\alpha_n)}+\cdots$$

It is clear that this expansion breaks down when $\alpha_i=\alpha_s$ for any $s\neq i$ because the correction (second term) to the solution of the reduced equation is infinite. In fact, α_i need not be exactly equal to α_s for the perturbation expansion to break down. The expansion breaks down when $\alpha_s=\alpha_i+\varepsilon\delta$, where $\delta=O(1)$, because the correction has what is called a small-divisor term that makes it $O(1)$ and hence the order of α_i. This is contrary to the assumption underlying the method; namely, the correction is small compared with α_i.

To obtain a valid expansion when $\alpha_{i+1}=\alpha_i+\varepsilon\delta$, we first rewrite the perturbed equation as

$$(x-\alpha_1)(x-\alpha_2)\cdots(x-\alpha_{i-1})(x-\alpha_i)(x-\alpha_i-\varepsilon\delta)\cdots(x-\alpha_n)=\varepsilon P_m(x)$$

or

$$(x-\alpha_i)(x-\alpha_i-\varepsilon\delta)$$

$$=\frac{\varepsilon P_m(x)}{(x-\alpha_1)(x-\alpha_2)\cdots(x-\alpha_{i-1})(x-\alpha_{i+2})\cdots(x-\alpha_n)}$$

or

$$(x-\alpha_i)^2 = \varepsilon\delta(x-\alpha_i) + \frac{\varepsilon P_m(x)}{(x-\alpha_1)(x-\alpha_2)\cdots(x-\alpha_{i-1})(x-\alpha_{i+2})\cdots(x-\alpha_n)}$$

By iteration one finds that

$$(x-\alpha_i)^2 \approx \frac{\varepsilon P_m(\alpha_i)}{(\alpha_i-\alpha_1)(\alpha_i-\alpha_2)\cdots(\alpha_i-\alpha_{i-1})(\alpha_i-\alpha_{i+2})\cdots(\alpha_i-\alpha_n)}$$

Hence to the first approximation

$$x = \alpha_i + \frac{\varepsilon^{1/2}[P_m(\alpha_i)]^{1/2}}{[(\alpha_i-\alpha_1)(\alpha_i-\alpha_2)\cdots(\alpha_i-\alpha_{i-1})(\alpha_i-\alpha_{i+2})\cdots(\alpha_i-\alpha_n)]^{1/2}} + \cdots$$

To continue the expansion further, one needs to expand x in powers of $\varepsilon^{1/2}$ rather than ε.

Next we consider the case

$$P_m(x) = \varepsilon P_n(x)$$

where $m < n$. When $\varepsilon = 0$, our equation reduces to $P_m(x) = 0$. Let us denote its roots by $\alpha_1, \alpha_2, \alpha_3, \ldots, \alpha_m$. When $\varepsilon \neq 0$, one can improve these roots by letting

$$x = \alpha_i + \varepsilon x_1 + \varepsilon^2 x_2 + \cdots$$

if α_i is away from all α_s, where $s \neq i$, and by letting

$$x = \alpha_i + \varepsilon^{\nu} x_1 + \varepsilon^{2\nu} x_2 + \cdots$$

if α_i is near one or more roots α_s of $P_m(x)$. The approach given above yields approximations to only m of the n roots of the perturbed equation. To determine approximations to the other $n-m$ roots, we let

$$x = \frac{u}{\varepsilon^{\lambda}}, \qquad \lambda > 0$$

in the perturbed equation and obtain

$$P_m\left(\frac{u}{\varepsilon^{\lambda}}\right) = \varepsilon P_n\left(\frac{u}{\varepsilon^{\lambda}}\right)$$

We keep the dominant terms on both sides of this equation and obtain

$$\frac{a_m u^m}{\varepsilon^{m\lambda}} = \varepsilon \frac{a_n u^n}{\varepsilon^{n\lambda}}$$

where a_m is the coefficient of x^m in $P_m(x)$ and a_n is the coefficient of x^n in $P_n(x)$. We choose $m\lambda = n\lambda - 1$ or $\lambda = 1/(n-m)$. Then we seek approximations to the $n - m$ roots by letting

$$x = \frac{u_0}{\varepsilon^\lambda} + u_1 + \varepsilon^\lambda u_2 + \cdots$$

Last, we consider determination of the large roots of transcendental equations such as

$$\tan x = f(x)$$

If $f(x) \to 0$ as $x \to \infty$, first approximations to the roots of this equation are given by

$$\tan x = 0 \quad \text{or} \quad x = n\pi, \qquad n = 1, 2, 3, \ldots$$

To improve these approximations, let $x = n\pi + \delta$ in the original equation and obtain

$$\tan(n\pi + \delta) = f(n\pi + \delta)$$

or

$$\tan\delta = f(n\pi + \delta)$$

$$\delta + \tfrac{1}{6}\delta^3 + \cdots = f(n\pi + \delta)$$

Then determine an approximate solution to the resulting equation for small δ.

Solved Exercises

2.1. For small ε, determine two terms in the expansion of each root of the following equations:

(a)
$$x^3 - (2+\varepsilon)x^2 - (1-\varepsilon)x + 2 + 3\varepsilon = 0 \tag{1}$$

Let

$$x = x_0 + \varepsilon x_1 + \cdots \tag{2}$$

in (1), equate coefficients of equal powers of ε, and obtain

$$x_0^3 - 2x_0^2 - x_0 + 2 = 0 \tag{3}$$

$$\left(3x_0^2 - 4x_0 - 1\right)x_1 - x_0^2 + x_0 + 3 = 0 \tag{4}$$

The solutions of (3) are

$$x_0 = 1, -1, 2 \tag{5}$$

It follows from (4) that

$$x_1 = \frac{x_0^2 - x_0 - 3}{3x_0^2 - 4x_0 - 1} \tag{6}$$

Hence

$$x_1 = \tfrac{3}{2} \quad \text{when } x_0 = 1$$

$$x_1 = -\tfrac{1}{6} \quad \text{when } x_0 = -1$$

$$x_1 = -\tfrac{1}{3} \quad \text{when } x_0 = 2$$

Therefore

$$x = 1 + \tfrac{3}{2}\varepsilon + \cdots \tag{7}$$

$$x = -1 - \tfrac{1}{6}\varepsilon + \cdots \tag{8}$$

$$x = 2 - \tfrac{1}{3}\varepsilon + \cdots \tag{9}$$

(b)
$$x^3 - (3+\varepsilon)x - 2 + \varepsilon = 0 \tag{10}$$

Substituting (2) into (10) and equating coefficients of equal powers of ε, we have

$$x_0^3 - 3x_0 - 2 = 0 \tag{11}$$

$$\left(3x_0^2 - 3\right)x_1 - x_0 + 1 = 0 \tag{12}$$

The solutions of (11) are

$$x_0 = -1, -1, 2 \tag{13}$$

It follows from (12) that

$$x_1 = \frac{x_0 - 1}{3(x_0^2 - 1)} = \frac{1}{3(x_0 + 1)} \tag{14}$$

Hence $x_1 = \infty$ when $x_0 = -1$, indicating that the form (2) is incorrect in this case. When $x_0 = 2$, $x_1 = \frac{1}{9}$. Therefore, one of the roots is

$$x = 2 + \tfrac{1}{9}\varepsilon + \cdots \tag{15}$$

Because of the multiplicity of $x_0 = -1$, we replace (2) by

$$x = -1 + \varepsilon^{\nu} x_1 + \varepsilon^{2\nu} x_2 + \cdots \tag{16}$$

Substituting (16) into (10), we have

$$3\varepsilon^{\nu}x_1+3\varepsilon^{2\nu}x_2-3\varepsilon^{2\nu}x_1^2-3\varepsilon^{\nu}x_1-3\varepsilon^{2\nu}x_2+\varepsilon-\varepsilon^{1+\nu}x_1+\varepsilon+\cdots=0 \tag{17}$$

Extracting the dominant terms in (17), we have

$$-3x_1^2\varepsilon^{2\nu}+2\varepsilon=0 \tag{18}$$

Hence

$$\nu=\tfrac{1}{2} \quad \text{and} \quad x_1=\pm\sqrt{\tfrac{2}{3}}$$

Therefore, the other two roots are

$$x=-1\pm\sqrt{\tfrac{2}{3}\varepsilon}+\cdots \tag{19}$$

(c)

$$x^3+(3-2\varepsilon)x^2+(3+\varepsilon)x+1-2\varepsilon=0 \tag{20}$$

Substituting (2) into (20) and equating coefficients of equal powers of ε, we have

$$x_0^3+3x_0^2+3x_0+1=0 \tag{21}$$

$$\left(3x_0^2+6x_0+3\right)x_1=2x_0^2-x_0+2 \tag{22}$$

The solutions of (21) are

$$x_0=-1,-1,-1 \tag{23}$$

It follows from (22) that $x_1=\infty$, indicating that none of the roots of (20) has the form (2).

To determine the roots of (20), we need to use the form (16). Instead of substituting (16) into (20) as it stands, we can simplify the algebra by rewriting it as

$$x^3+3x^2+3x+1=\varepsilon(2x^2-x+2)$$

or

$$(x+1)^3=\varepsilon(2x^2-x+2) \tag{24}$$

Substituting (16) into (24) and extracting the dominant terms, we have

$$\varepsilon^{3\nu}x_1^3=5\varepsilon$$

Hence

$$\nu=\tfrac{1}{3}, \qquad x_1=\sqrt[3]{5}\,\omega, \qquad \sqrt[3]{5}\,\omega^2, \qquad \sqrt[3]{5}\,\omega^3$$

where

$$\omega=e^{2i\pi/3}$$

Therefore, the roots of (20) are

$$x = -1 + (5\varepsilon)^{1/3}\omega + \cdots \tag{25}$$

$$x = -1 + (5\varepsilon)^{1/3}\omega^2 + \cdots \tag{26}$$

$$x = -1 + (5\varepsilon)^{1/3}\omega^3 + \cdots \tag{27}$$

(d)
$$x^4 + (2 - 3\varepsilon)x^3 - (2 - \varepsilon)x - 1 + 4\varepsilon = 0 \tag{28}$$

Substituting (2) into (28) and equating coefficients of like powers of ε, we have

$$x_0^4 + 2x_0^3 - 2x_0 - 1 = 0 \tag{29}$$

$$\left(4x_0^3 + 6x_0^2 - 2\right)x_1 = 3x_0^3 - x_0 - 4 \tag{30}$$

The solutions of (29) are

$$x_0 = 1, -1, -1, -1 \tag{31}$$

It follows from (30) that

$$x_1 = \frac{3x_0^3 - x_0 - 4}{4x_0^3 + 6x_0^2 - 2} = \frac{-2}{8} = -\tfrac{1}{4} \quad \text{when } x_0 = 1$$

Hence one of the roots is

$$x = 1 - \tfrac{1}{4}\varepsilon + \cdots \tag{32}$$

Since $x_0 = -1$ with a multiplicity of 3, the other three roots do not have the form (2) but the form (16). To determine these roots, it is convenient to express (28) as

$$x^4 + 2x^3 - 2x - 1 = \varepsilon(3x^3 - x - 4)$$

or

$$(x+1)^3(x-1) = \varepsilon(3x^3 - x - 4)$$

or

$$(x+1)^3 = \frac{\varepsilon(3x^3 - x - 4)}{x - 1} \tag{33}$$

Substituting (16) into (33) and extracting the dominant terms, we have

$$\varepsilon^{3\nu}x_1^3 = 3\varepsilon$$

Hence

$$\nu = \tfrac{1}{3} \quad \text{and} \quad x_1 = \sqrt[3]{3}\exp\left(\frac{2in\pi}{3}\right), \qquad n = 0, 1, 2$$

Therefore, the other three roots are

$$x=-1+\sqrt[3]{3\varepsilon}\exp\left(\frac{2in\pi}{3}\right)+\cdots \tag{34}$$

for $n=0$, 1, and 2.

(e)
$$x^4+(4-\varepsilon)x^3+(6+2\varepsilon)x^2+(4+\varepsilon)x+1-\varepsilon^2=0 \tag{35}$$

The reduced equation is

$$x^4+4x^3+6x^2+4x+1=0$$

or

$$(x+1)^4=0$$

Hence $x=-1$ with a multiplicity of 4 and none of the roots of (35) has the form (2). They have the form (16). To simplify the algebra, we express (35) as

$$x^4+4x^3+6x^2+4x+1=\varepsilon(x^3-2x^2-x)+\varepsilon^2$$

or

$$(x+1)^4=\varepsilon(x^3-2x^2-x)+\varepsilon^2 \tag{36}$$

Substituting (16) into (36) and extracting the dominant terms, we obtain

$$\varepsilon^{4\nu}x_1^4=-2\varepsilon$$

Hence $\nu=\frac{1}{4}$ and

$$x_1=\sqrt[4]{2}\exp\left[\tfrac{1}{4}(2n+1)i\pi\right] \quad \text{for } n=0,1,2,\text{ and } 3$$

Therefore, the roots of (35) are

$$x=-1+\sqrt[4]{2\varepsilon}\exp\left[\tfrac{1}{4}(2n+1)i\pi\right]+\cdots \quad \text{for } n=0,1,2,\text{ and } 3$$

2.2. For small ε, determine two terms in the expansion of each root of the following equations:

(a)
$$\varepsilon(u^3+u^2)+4u^2-3u-1=0 \tag{37}$$

Letting

$$u=u_0+\varepsilon u_1+\cdots \tag{38}$$

in (37) and equating coefficients of like powers of ε, we have

$$4u_0^2-3u_0-1=0 \tag{39}$$

$$(8u_0-3)u_1=-(u_0^3+u_0^2) \tag{40}$$

The solutions of (39) are

$$u_0 = -\tfrac{1}{4}, 1$$

It follows from (40) that

$$u_1 = -\frac{u_0^3 + u_0^2}{8u_0 - 3}$$

Hence

$$u_1 = \tfrac{3}{320} \quad \text{when } u_0 = -\tfrac{1}{4}$$

$$u_1 = -\tfrac{2}{5} \quad \text{when } u_0 = 1$$

Therefore, two of the roots of (37) are

$$u = -\tfrac{1}{4} + \tfrac{3}{320}\varepsilon + \cdots$$

$$u = 1 - \tfrac{2}{5}\varepsilon + \cdots$$

We seek the third root in the form

$$u = \frac{y}{\varepsilon^\nu} + u_0 + \cdots \tag{41}$$

Substituting (41) into (37), we have

$$\varepsilon\left(\frac{y^3}{\varepsilon^{3\nu}} + \frac{3y^2u_0}{\varepsilon^{2\nu}} + \frac{y^2}{\varepsilon^{2\nu}} + \cdots\right) + 4\left(\frac{y^2}{\varepsilon^{2\nu}} + \frac{2yu_0}{\varepsilon^\nu}\right) - \frac{3y}{\varepsilon^\nu} - 3u_0 - 1 + \cdots = 0 \tag{42}$$

Extracting the dominant terms in (42), we have

$$\varepsilon^{1-3\nu}y^3 + 4\varepsilon^{-2\nu}y^2 = 0$$

Hence $\nu = 1$ and $y = -4$. Extracting the dominant terms in what is left of (42), we obtain

$$-4 - 12u_0 + 8u_0 - 3 = 0$$

or

$$u_0 = -\tfrac{7}{4}$$

Hence the third root is

$$u = -\frac{4}{\varepsilon} - \tfrac{7}{4} + \cdots$$

(b)

$$\varepsilon u^3 + u - 2 = 0 \tag{43}$$

Substituting (38) into (43) and equating coefficients of like powers of ε, we have

$$u_0 - 2 = 0$$

$$u_1 = -u_0^3$$

Hence $u_0 = 2$ and $u_1 = -8$. One of the roots is

$$u = 2 - 8\varepsilon + \cdots$$

To determine the other two roots, we substitute (41) into (43) and obtain

$$\varepsilon\left(\frac{y^3}{\varepsilon^{3\nu}} + \frac{3y^2u_0}{\varepsilon^{2\nu}} + \cdots\right) + \frac{y}{\varepsilon^{\nu}} + u_0 + \cdots - 2 = 0 \tag{44}$$

Extracting the dominant terms in (44), we have

$$\varepsilon^{1-3\nu}y^3 + \varepsilon^{-\nu}y = 0$$

Hence

$$\nu = \tfrac{1}{2} \quad \text{and} \quad y = \pm i$$

Extracting the dominant terms from what is left in (44) yields

$$3y^2u_0 + u_0 - 2 = 0$$

or

$$u_0 = -1$$

Hence the other two roots are

$$u = \frac{\pm i}{\sqrt{\varepsilon}} - 1 + \cdots$$

(c)

$$\varepsilon u^3 + (u-2)^2 = 0 \tag{45}$$

In this case the reduced equation has the root 2 with a multiplicity of 2. Hence the roots close to 2 do not have the form (38) but the form

$$u = 2 + \varepsilon^{\nu}u_1 + \cdots \tag{46}$$

Putting (46) in (45) and extracting the dominant terms, we have

$$8\varepsilon + \varepsilon^{2\nu}u_1^2 = 0$$

Hence

$$\nu = \tfrac{1}{2} \quad \text{and} \quad u_1 = \pm i\sqrt{8}$$

Therefore, two of the roots of (45) are

$$u = 2 \pm i\sqrt{8\varepsilon} + \cdots$$

To determine the third root, we substitute (41) into (45) and obtain

$$\varepsilon\left(\frac{y^3}{\varepsilon^{3\nu}} + \frac{3y^2 u_0}{\varepsilon^{2\nu}} + \cdots\right) + \frac{y^2}{\varepsilon^{2\nu}} + \frac{2yu_0}{\varepsilon^{\nu}} - \frac{4y}{\varepsilon^{\nu}} - 4u_0 + 4 + \cdots = 0 \tag{47}$$

Extracting the dominant terms in (47) yields

$$\varepsilon^{1-3\nu} y^3 + \varepsilon^{-2\nu} y^2 = 0$$

Hence

$$\nu = 1 \quad \text{and} \quad y = -1$$

Extracting the dominant terms in what is left in (47), we obtain

$$3u_0 - 2u_0 + 4 = 0$$

or

$$u_0 = -4$$

Hence the third root is

$$u = -1/\varepsilon - 4 + \cdots$$

(d) $u^2 - u - 2 + \frac{1}{2}\varepsilon(u^3 + 2u + 3) = 0$

$$u = -1$$

$$u = 2 - \tfrac{5}{2}\varepsilon + \cdots$$

$$u = -2/\varepsilon - 1 + \cdots$$

(e) $\varepsilon u^4 + u^3 - 2u^2 - u + 2 = 0$

$$u = 1 + \tfrac{1}{2}\varepsilon + \cdots$$

$$u = -1 - \tfrac{1}{6}\varepsilon + \cdots$$

$$u = 2 - \tfrac{16}{3}\varepsilon + \cdots$$

$$u = -1/\varepsilon - 2 + \cdots$$

(f) $\varepsilon u^4 - u^3 + 3u - 2 = 0$

$$u = -2 + \tfrac{16}{9}\varepsilon + \cdots$$

$$u = 1 \pm \sqrt{\tfrac{1}{3}\varepsilon} + \cdots$$

$$u = 1/\varepsilon - 3\varepsilon + \cdots$$

(g) $\varepsilon u^4 - u^2 + 3u - 2 = 0$

$$u = 1 - \varepsilon + \cdots$$

$$u = 2 + 16\varepsilon + \cdots$$

$$u = \pm 1/\sqrt{\varepsilon} - \tfrac{3}{2} + \cdots$$

(h) $\varepsilon u^4 + u^2 - 3u + 2 = 0$

$$u = 1 + \varepsilon + \cdots$$

$$u = 2 - 16\varepsilon + \cdots$$

$$u = \pm i/\sqrt{\varepsilon} - \tfrac{3}{2} + \cdots$$

(i) $\varepsilon u^4 - u^2 + 2u - 1 = 0$

$$u = 1 \pm \sqrt{\varepsilon} + \cdots$$

$$u = \pm 1/\sqrt{\varepsilon} - 1 + \cdots$$

(j) $\varepsilon u^4 + u^2 - 2u + 1 = 0$

$$u = 1 \pm i\sqrt{\varepsilon} + \cdots$$

$$u = \pm i/\sqrt{\varepsilon} - 1 + \cdots$$

(k) $\varepsilon(u^4 + u^3) - u^2 + 3u - 2 = 0$

$$u = 1 - 2\varepsilon + \cdots$$

$$u = 2 + 24\varepsilon + \cdots$$

$$u = \pm 1/\sqrt{\varepsilon} - 2 + \cdots$$

(l) $\varepsilon(u^5 + u^4 - 2u^3) + 2u^2 - 3u + 1 = 0$

$$u = \tfrac{1}{2} - \tfrac{5}{32}\varepsilon + \cdots$$

$$u = 1$$

$$u = \sqrt[3]{(2/\varepsilon)}\exp\left[\tfrac{1}{3}i(1+2n)\pi\right] - \tfrac{5}{6} + \cdots, \qquad n = 0,1,2$$

(m) $\varepsilon(u^5 + u^4 - 2u^3) - 4u^2 + 4u - 1 = 0$

$$u = \tfrac{1}{2} \mp (i/8)\sqrt{\tfrac{5}{2}\varepsilon} + \cdots$$

$$u = \sqrt[3]{(4/\varepsilon)}\exp\left(\tfrac{2}{3}in\pi\right) - \tfrac{2}{3} + \cdots$$

(n) $\varepsilon^2 u^6 - \varepsilon u^4 - u^3 + 2u^2 + u - 2 = 0$

$$u = -1 - \tfrac{1}{6}\varepsilon + \cdots$$

$$u = 1 + \tfrac{1}{2}\varepsilon + \cdots$$

$$u = 2 - \tfrac{16}{3}\varepsilon + \cdots$$

$$u = \varepsilon^{-(2/3)}\exp(\tfrac{2}{3}in\pi) + \tfrac{1}{3}\varepsilon^{-1/3}\exp(\tfrac{4}{3}in\pi) + \cdots$$

for $n = 0$, 1, and 2.

2.3. For small ε, determine two-term expansions for the solutions of

(a)
$$s - \frac{\varepsilon}{3s^2} - \frac{3\varepsilon^2}{10s^4} = 0 \tag{48}$$

(b)
$$1 - \frac{\varepsilon}{3\sqrt{s}} + \frac{21\varepsilon^2}{5\sqrt{s}} = 0 \tag{49}$$

(a) Equation (48) can be written as

$$s^5 - \tfrac{1}{3}\varepsilon s^2 - \tfrac{3}{10}\varepsilon^2 = 0 \tag{50}$$

Balancing the first two terms, we find that

$$s = O(\varepsilon^{1/3})$$

Then we let

$$s = s_1\varepsilon^{1/3} + s_2\varepsilon^{2/3} + \cdots$$

in (50), equate coefficients of like powers of ε, and obtain

$$s_1^5 = \tfrac{1}{3}s_1^2 \tag{51}$$

$$5s_1^4 s_2 - \tfrac{2}{3}s_1 s_2 - \tfrac{3}{10} = 0 \tag{52}$$

It follows from (51) that for a nontrivial solution

$$s_1 = (\tfrac{1}{3})^{1/3}\exp(\tfrac{2}{3}in\pi)$$

It follows from (52) that

$$s_2 = \frac{3}{10s_1} = \frac{3(3)^{1/3}}{10}\exp(-\tfrac{2}{3}in\pi)$$

Hence three of the solutions are

$$s = \left(\frac{\varepsilon}{3}\right)^{1/3}\exp(\tfrac{2}{3}in\pi) + \frac{3(3\varepsilon^2)^{1/3}}{10}\exp(-\tfrac{2}{3}in\pi) + \cdots$$

for $n = 0$, 1, and 2.

To determine the other two roots, we need to investigate balancing other terms. The first and last terms cannot be the dominant terms, but the last two terms can dominate. Thus

$$s = s_1 \varepsilon^{1/2} + s_2 \varepsilon + \cdots \tag{53}$$

Substituting (53) into (50) and equating coefficients of like powers of ε, we have

$$-\tfrac{1}{3}s_1^2 - \tfrac{3}{10} = 0 \tag{54}$$

$$-\tfrac{2}{3}s_1 s_2 + s_1^5 = 0 \tag{55}$$

It follows from (54) that

$$s_1 = \frac{\pm 3i}{\sqrt{10}}$$

whereas it follows from (55) that

$$s_2 = \tfrac{3}{2}s_1^4 = \tfrac{243}{200}$$

Hence the other two solutions are

$$s = \frac{\pm 3i}{\sqrt{10}}\varepsilon^{1/2} + \tfrac{243}{200}\varepsilon + \cdots$$

(b) We rewrite (49) as

$$\sqrt{s} = \tfrac{1}{3}\varepsilon - \tfrac{21}{5}\varepsilon^2$$

Thus

$$s = \left(\tfrac{1}{3}\varepsilon - \tfrac{21}{5}\varepsilon^2\right)^2 = \tfrac{1}{9}\varepsilon^2 - \tfrac{14}{5}\varepsilon^3 + \cdots$$

2.4. Determine two-term expansions for the large roots of

(a) $$x \tan x = 1 \tag{56}$$

(b) $$x \cot x = 1 \tag{57}$$

(a) It follows from (56) that

$$\tan x = \frac{1}{x} \tag{58}$$

As $x \to \infty$, $\tan x \to 0$ so that $x \to n\pi$ where n is an integer. Let

$$x = n\pi + \delta$$

so that (58) becomes

$$\tan(n\pi + \delta) = \frac{1}{n\pi + \delta}$$

or

$$\frac{\tan n\pi + \tan\delta}{1-\tan n\pi \tan\delta} = \frac{1}{n\pi+\delta}$$

or

$$\tan\delta = \frac{1}{n\pi+\delta}$$

For small δ, $\tan\delta \approx \delta$ and hence

$$\delta = \frac{1}{n\pi}$$

Therefore

$$x = n\pi + \frac{1}{n\pi} + \cdots$$

where n is an integer.

(b) It follows from (57) that

$$\cot x = \frac{1}{x} \tag{59}$$

As $x \to \infty$, $\cot x \to 0$ so that $x \to (n+\frac{1}{2})\pi$ where n is an integer. Let

$$x = \left(n+\tfrac{1}{2}\right)\pi + \delta$$

in (59) and obtain

$$\cot\left[\left(n+\tfrac{1}{2}\right)\pi + \delta\right] = \frac{1}{\left(n+\frac{1}{2}\right)\pi+\delta}$$

or

$$\frac{\cot\left(n+\frac{1}{2}\right)\pi \cot\delta - 1}{\cot\left(n+\frac{1}{2}\right)\pi + \cot\delta} = \frac{1}{\left(n+\frac{1}{2}\right)\pi+\delta}$$

or

$$-\frac{1}{\cot\delta} = \frac{1}{\left(n+\frac{1}{2}\right)\pi+\delta} = -\tan\delta$$

Hence

$$\delta \approx -\frac{1}{\left(n+\frac{1}{2}\right)\pi}$$

Therefore

$$x = \left(n+\tfrac{1}{2}\right)\pi - \frac{1}{\left(n+\frac{1}{2}\right)\pi} + \cdots$$

2.5. The asymptotic expansion of Bessel's function $J_0(x)$ for large x is

$$J_0(x) = \sqrt{\frac{2}{\pi}}\left[\frac{1}{x^{1/2}}\cos(x - \tfrac{1}{4}\pi) + \frac{1}{8x^{3/2}}\sin(x - \tfrac{1}{4}\pi)\right] + \cdots \tag{60}$$

Show that the large roots of $J_0'(x)$ are approximately given by

$$\tan(x - \tfrac{1}{4}\pi) = -\frac{3}{8x} + \cdots \tag{61}$$

Then show that

$$x = (n + \tfrac{1}{4})\pi - \frac{3}{2\pi(4n+1)} + \cdots \tag{62}$$

Compare this result with those tabulated for the first seven roots.

Differentiating (60) with respect to x yields

$$J_0'(x) = \sqrt{\frac{2}{\pi}}\left[-\frac{1}{2x^{3/2}}\cos(x - \tfrac{1}{4}\pi) - \frac{1}{x^{1/2}}\sin(x - \tfrac{1}{4}\pi) + \frac{1}{8x^{3/2}}\cos(x - \tfrac{1}{4}\pi)\right] + \cdots$$

or

$$J_0'(x) = \sqrt{\frac{2}{\pi}}\left[-\frac{3}{8x^{3/2}}\cos(x - \tfrac{1}{4}\pi) - \frac{1}{x^{1/2}}\sin(x - \tfrac{1}{4}\pi)\right] + \cdots \tag{63}$$

Putting $J_0'(x) = 0$, we obtain (61).

For large x, the right-hand side of (61) can be neglected. Then

$$x - \tfrac{1}{4}\pi = n\pi \quad \text{or} \quad x = (n + \tfrac{1}{4})\pi$$

Thus we let

$$x = (n + \tfrac{1}{4})\pi + \delta \tag{64}$$

in (61) and obtain

$$\frac{\tan n\pi + \tan\delta}{1 - \tan n\pi \tan\delta} = -\frac{3}{8(n + \tfrac{1}{4})\pi + 8\delta}$$

or

$$\tan\delta = -\frac{3}{2(4n+1)\pi + 8\delta}$$

which to the first approximation gives

$$\delta = -\frac{3}{2(4n+1)\pi}$$

Substituting for δ in (64) yields (62).

Table 1. Roots of $J_0'(x)=0$

Root Number	1	2	3	4	5	6	7
Perturbation	3.83150	7.01553	10.17345	13.32368	16.47063	19.61586	22.76008
Tabulated	3.83171	7.01559	10.17347	13.32369	16.47063	19.61586	22.76008

The first seven roots obtained from (62) are compared with the corresponding tabulated values in Table 1.

2.6. The asymptotic expansion of Bessel's function of the second kind of order zero is

$$Y_0(x) \sim \sqrt{\frac{2}{\pi x}}\left[\sin\left(x-\tfrac{1}{4}\pi\right)-\frac{1}{8x}\cos\left(x-\tfrac{1}{4}\pi\right)\right] \quad \text{as } x\to\infty \tag{65}$$

Show that the roots of $Y_0(x)=0$ are given by

$$x=\left(n+\tfrac{1}{4}\right)\pi+\frac{1}{2\pi(4n+1)}+\cdots \tag{66}$$

Compare this result with those tabulated for the first seven roots.

Putting $Y_0(x)=0$ in (65) gives

$$\sin\left(x-\tfrac{1}{4}\pi\right)=\frac{1}{8x}\cos\left(x-\tfrac{1}{4}\pi\right)+\cdots$$

or

$$\tan\left(x-\tfrac{1}{4}\pi\right)=\frac{1}{8x}+\cdots \tag{67}$$

As $x\to\infty$, (67) reduces to

$$\tan\left(x-\tfrac{1}{4}\pi\right)=0$$

whose solution is

$$x-\tfrac{1}{4}\pi=n\pi \quad \text{or} \quad x=\left(n+\tfrac{1}{4}\right)\pi$$

Therefore, we seek the solutions of (67) in the form

$$x=\left(n+\tfrac{1}{4}\right)\pi+\delta \tag{68}$$

Substituting (68) into (67), we have

$$\frac{\tan n\pi+\tan\delta}{1-\tan n\pi\tan\delta}=\frac{1}{2\pi(4n+1)+8\delta}+\cdots$$

or

$$\tan\delta=\frac{1}{2\pi(4n+1)+8\delta}+\cdots$$

Table 2. Roots of $Y_0(x)=0$

Root Number	1	2	3	4	5	6	7
Perturbation	0.94455	3.95882	7.08627	10.22242	13.36113	16.50094	19.64132
Tabulated	0.89358	3.95768	7.08605	10.22235	13.36110	16.50092	19.64131

Hence

$$\delta \simeq \frac{1}{2\pi(4n+1)}$$

and x is given by (66).

Table 2 compares the first seven roots calculated from (66) with the corresponding tabulated values.

2.7. Using the asymptotic expansion of $Y_0(x)$ in the previous exercise, show that the roots of $Y_0'(x)=0$ are given by

$$x=\left(n+\tfrac{3}{4}\right)\pi-\frac{3}{2\pi(4n+3)}+\cdots \tag{69}$$

It follows from (65) that

$$Y_0'(x)\sim\sqrt{\frac{2}{\pi}}\left[\frac{-3}{8x^{3/2}}\sin\left(x-\tfrac{1}{4}\right)\pi+\frac{1}{x^{1/2}}\cos\left(x-\tfrac{1}{4}\pi\right)\right] \quad \text{as } x\to\infty \tag{70}$$

Putting $Y_0'(x)=0$ in (70) yields

$$\cot\left(x-\tfrac{1}{4}\pi\right)=\frac{3}{8x}+\cdots \tag{71}$$

Letting $x\to\infty$ in (71) gives

$$x-\tfrac{1}{4}\pi=\left(n+\tfrac{1}{2}\right)\pi \quad \text{or} \quad x=\left(n+\tfrac{3}{4}\right)\pi$$

This suggests seeking an approximate solution of (71) in the form

$$x=\left(n+\tfrac{3}{4}\right)\pi+\delta$$

Then (71) becomes

$$\frac{\cot\left(n+\frac{1}{2}\right)\pi\cot\delta-1}{\cot\left(n+\frac{1}{2}\right)\pi+\cot\delta}=\frac{3}{2\pi(4n+3)+8\delta}+\cdots$$

or

$$-\tan\delta=\frac{3}{2\pi(4n+3)+8\delta}+\cdots$$

Table 3. Roots of $Y_0'(x)=0$

Root Number	1	2	3	4	5	6	7
Perturbation	2.19704	5.42958	8.59597	11.74914	14.89744	18.04340	21.18807
Tabulated	2.19714	5.42968	8.59601	11.74915	14.89744	18.04340	21.18807

Hence

$$\delta \approx -\frac{3}{2\pi(4n+3)}$$

and x is given by (69).

Table 3 compares the first seven roots calculated from (69) with their corresponding tabulated values.

2.8. The asymptotic expansion of $J_\nu(x)$ is

$$J_\nu(x) \sim \sqrt{\frac{2}{\pi x}}\left[\cos\left(x-\tfrac{1}{2}\nu\pi-\tfrac{1}{4}\pi\right)-\frac{4\nu^2-1}{8x}\sin\left(x-\tfrac{1}{2}\nu\pi-\tfrac{1}{4}\pi\right)\right] \quad \text{as } x\to\infty \tag{72}$$

(a) Show that the roots of $J_\nu(x)=0$ are given by

$$x=\left(n+\tfrac{3}{4}+\tfrac{1}{2}\nu\right)\pi-\frac{4\nu^2-1}{2\pi(4n+3+2\nu)}+\cdots \tag{73}$$

Putting $J_\nu(x)=0$ in (72) gives

$$\cot\left(x-\tfrac{1}{2}\nu\pi-\tfrac{1}{4}\pi\right)=\frac{4\nu^2-1}{8x}+\cdots \tag{74}$$

As $x\to\infty$, (74) reduces to

$$\cot\left(x-\tfrac{1}{2}\nu\pi-\tfrac{1}{4}\pi\right)=0$$

whose solution is

$$x-\tfrac{1}{2}\nu\pi-\tfrac{1}{4}\pi=\left(n+\tfrac{1}{2}\right)\pi \quad \text{or} \quad x=\left(n+\tfrac{3}{4}+\tfrac{1}{2}\nu\right)\pi$$

This suggests that

$$x=\left(n+\tfrac{3}{4}+\tfrac{1}{2}\nu\right)\pi+\delta$$

Then (74) becomes

$$\frac{\cot\left(n+\tfrac{1}{2}\right)\pi\cot\delta-1}{\cot\left(n+\tfrac{1}{2}\right)\pi+\cot\delta}=\frac{4\nu^2-1}{2\pi(4n+3+2\nu)+8\delta}+\cdots$$

or

$$-\tan\delta = \frac{4\nu^2-1}{2\pi(4n+3+2\nu)+8\delta} + \cdots$$

Hence

$$\delta \approx -\frac{4\nu^2-1}{2\pi(4n+3+2\nu)} + \cdots$$

and x is given by (73).

(b) Show that the roots of $J_\nu'(x)=0$ are given by

$$x = \left(n+\tfrac{1}{4}+\tfrac{1}{2}\nu\right)\pi - \frac{3+4\nu^2}{2\pi(4n+1+2\nu)} + \cdots \tag{75}$$

It follows from (72) that

$$J_\nu'(x) \sim \sqrt{\frac{2}{\pi}}\left[-\frac{1}{x^{1/2}}\sin\left(x-\tfrac{1}{2}\nu\pi-\tfrac{1}{4}\pi\right) - \frac{4\nu^2+3}{8x^{3/2}}\cos\left(x-\tfrac{1}{2}\nu\pi-\tfrac{1}{4}\pi\right)\right] \quad \text{as } x\to\infty \tag{76}$$

Putting $J_\nu'(x)=0$ in (76) yields

$$\tan\left(x-\tfrac{1}{2}\nu\pi-\tfrac{1}{4}\pi\right) = -\frac{4\nu^2+3}{8x} + \cdots \tag{77}$$

As $x\to\infty$, (77) reduces to

$$\tan\left(x-\tfrac{1}{2}\nu\pi-\tfrac{1}{4}\pi\right) = 0$$

whose solution is

$$x-\tfrac{1}{2}\nu\pi-\tfrac{1}{4}\pi = n\pi \quad \text{or} \quad x = \left(n+\tfrac{1}{4}+\tfrac{1}{2}\nu\right)\pi$$

This suggests that

$$x = \left(n+\tfrac{1}{4}+\tfrac{1}{2}\nu\right)\pi + \delta$$

Then (77) becomes

$$\frac{\tan n\pi + \tan\delta}{1-\tan n\pi\tan\delta} = -\frac{4\nu^2+3}{2\pi(4n+1+2\nu)+8\delta} + \cdots$$

or

$$\tan\delta = -\frac{4\nu^2+3}{2\pi(4n+1+2\nu)+8\delta} + \cdots$$

Table 4. Roots of $J_1(x) = 0$

Root Number	1	2	3	4	5	6	7
Perturbation	3.83150	7.01553	10.17345	13.32368	16.47063	19.61586	22.76008
Tabulated	3.83171	7.01559	10.17347	13.32369	16.47063	19.61586	22.76008

Table 5. Roots of $J_1'(x) = 0$

Root Number	1	2	3	4	5	6	7
Perturbation	1.98483	5.33863	8.53810	11.70670	14.86393	18.01572	21.16449
Tabulated	1.84118	5.33144	8.53632	11.70600	14.86359	18.01553	21.16437

Hence

$$\delta = -\frac{4\nu^2+3}{2\pi(4n+1+2\nu)} + \cdots$$

and x is given by (75). When $\nu = 0$, (75) reduces to (62).

Tables 4 and 5 compare the first seven roots in (73) and (75) with their corresponding tabulated values for the case $\nu = 1$.

2.9. The asymptotic expansion of $Y_\nu(x)$ is

$$Y_\nu(x) \sim \sqrt{\frac{2}{\pi x}}\left[\sin\left(x - \tfrac{1}{2}\nu\pi - \tfrac{1}{4}\pi\right) + \frac{4\nu^2-1}{8x}\cos\left(x - \tfrac{1}{2}\nu\pi - \tfrac{1}{4}\pi\right)\right] \quad \text{as } x \to \infty \tag{78}$$

(a) Show that the roots of $Y_\nu(x) = 0$ are given by

$$x = \left(n + \tfrac{1}{4} + \tfrac{1}{2}\nu\right)\pi - \frac{4\nu^2-1}{2\pi(4n+1+2\nu)} + \cdots \tag{79}$$

Putting $Y_\nu(x) = 0$ in (78) yields

$$\tan\left(x - \tfrac{1}{2}\nu\pi - \tfrac{1}{4}\pi\right) = -\frac{4\nu^2-1}{8x} + \cdots \tag{80}$$

As $x \to \infty$, (80) reduces to

$$\tan\left(x - \tfrac{1}{2}\nu\pi - \tfrac{1}{4}\pi\right) = 0$$

whose solution is

$$x - \tfrac{1}{2}\nu\pi - \tfrac{1}{4}\pi = n\pi \quad \text{or} \quad x = \left(n + \tfrac{1}{4} + \tfrac{1}{2}\nu\right)\pi$$

This suggest that

$$x = \left(n + \tfrac{1}{4} + \tfrac{1}{2}\nu\right)\pi + \delta$$

Then (80) becomes

$$\tan\delta = -\frac{4\nu^2 - 1}{2\pi(4n + 1 + 2\nu) + 8\delta} + \cdots$$

Hence

$$\delta = -\frac{4\nu^2 - 1}{2\pi(4n + 1 + 2\nu)} + \cdots$$

and x is given by (79). When $\nu = 0$, (79) reduces to (66).

(b) Show that the roots of $Y_\nu'(x) = 0$ are given by

$$x = \left(n + \tfrac{3}{4} + \tfrac{1}{2}\nu\right)\pi - \frac{3 + 4\nu^2}{2\pi(4n + 3 + 2\nu)} + \cdots \tag{81}$$

It follows from (78) that

$$Y_\nu'(x) \sim \sqrt{\frac{2}{\pi}}\left[\frac{1}{x^{1/2}}\cos\left(x - \tfrac{1}{2}\nu\pi - \tfrac{1}{4}\pi\right) - \frac{4\nu^2 + 3}{8x^{3/2}}\sin\left(x - \tfrac{1}{2}\nu\pi - \tfrac{1}{4}\pi\right)\right] \quad \text{as } x \to \infty \tag{82}$$

Putting $Y_\nu'(x) = 0$ in (82) gives

$$\cot\left(x - \tfrac{1}{2}\nu\pi - \tfrac{1}{4}\pi\right) = \frac{4\nu^2 + 3}{8x} + \cdots \tag{83}$$

As $x \to \infty$, (83) reduces to

$$\cot\left(x - \tfrac{1}{2}\nu\pi - \tfrac{1}{4}\pi\right) = 0$$

whose solution is

$$x - \tfrac{1}{2}\nu\pi - \tfrac{1}{4}\pi = \left(n + \tfrac{1}{2}\right)\pi \quad \text{or} \quad x = \left(n + \tfrac{3}{4} + \tfrac{1}{2}\nu\right)\pi$$

This suggests that

$$x = \left(n + \tfrac{3}{4} + \tfrac{1}{2}\nu\right)\pi + \delta$$

Then (83) becomes

$$-\tan\delta = \frac{4\nu^2 + 3}{2\pi(4n + 3 + 2\nu) + 8\delta} + \cdots$$

Hence

$$\delta = -\frac{4\nu^2 + 3}{2\pi(4n + 3 + 2\nu)} + \cdots$$

and x is given by (81). When $\nu = 0$, (81) reduces to (69).

Table 6. Roots of $Y_1(x)=0$

Root Number	1	2	3	4	5	6	7
Perturbation	2.19704	5.42958	8.59597	11.74914	14.89744	18.04340	21.18807
Tabulated	2.19714	5.42968	8.59601	11.74915	14.89744	18.04340	21.18807

Table 7. Roots of $Y_1'(x)=0$

Root Number	1	2	3	4	5	6	7
Perturbation	3.70417	6.94480	10.12448	13.28623	16.44031	19.59039	22.73813
Tabulated	3.68302	6.94150	10.12340	13.28576	16.44006	19.59024	22.73863

Tables 6 and 7 compare the first seven roots in (79) and (81) with their corresponding tabulated values for the case $\nu=1$.

Supplementary Exercises

2.10. For small ε, determine two terms in the expansion of each root of the following equations:

(a) $x^3-2x^2-x+2=\varepsilon(3x^2+2x+2)$
(b) $x^3-2x^2-x+2=\varepsilon(3x^2+2x-1)$
(c) $x^3-x^2-x+1=\varepsilon(3x^2+2x+2)$
(d) $x^3-3x^2+3x-1=\varepsilon(3x^2+2x+2)$
(e) $\varepsilon x^3+2x^2+3x+1=0$
(f) $\varepsilon x^3+2x^2+4x+2=0$
(g) $\varepsilon x^4+x^3+x^2-4x-4=0$
(h) $\varepsilon x^4+x^3+2x^2-4x-8=0$
(i) $\varepsilon x^4+x^3+6x^2+12x+8=0$
(j) $\varepsilon(x^4+2x^3)+x^3+x^2-4x-4=0$
(k) $\varepsilon x^4+x^2+3x+2=0$
(l) $\varepsilon x^4+x^2+2x+1=0$
(m) $\varepsilon x^5-3x^2+4x-1=0$
(n) $\varepsilon x^5-3x^2+6x-3=0$

2.11. Consider the equation

$$x^2-\ln x=\sigma$$

where σ is a positive number. Show that

$$x \sim \sigma^{1/2}+\tfrac{1}{4}\sigma^{-1/2}\ln\sigma \quad \text{as } \sigma\to\infty$$

2.12. Given that Bessel's function of the first kind and $\frac{3}{2}$ order is given by

$$J_{3/2}(z)=\sqrt{\frac{2}{\pi z}}\left(-\cos z+\frac{\sin z}{z}\right)$$

determine two-term expansions for the large roots of (a) $J_{3/2}(z)=0$ and (b) $J'_{3/2}(z)=0$.

2.13. Given that Bessel's function of the first kind and order $\frac{5}{2}$ is given by

$$J_{5/2}(z) = \sqrt{\frac{2}{\pi z}}\left[\left(-1+\frac{3}{z^2}\right)\sin z - \frac{3\cos z}{z}\right]$$

determine two-term expansions for the large roots of (a) $J_{5/2}(z) = 0$ and (b) $J'_{5/2}(z) = 0$.

2.14. Given that Kelvin's function of the first kind is defined by

$$ber(x) = \frac{e^{x/\sqrt{2}}}{\sqrt{2\pi x}}\left[\cos\left(\frac{x}{\sqrt{2}} - \frac{\pi}{8}\right) + \frac{1}{8x}\sin\left(\frac{x}{\sqrt{2}} - \frac{\pi}{8}\right)\right] + \cdots$$

determine two-term expansions for the large roots of (a) $ber(x) = 0$ and (b) $ber'(x) = 0$.

2.15. Given that Kelvin's function of the second kind is defined by

$$bei(x) = \frac{e^{x/\sqrt{2}}}{\sqrt{2\pi x}}\left[\sin\left(\frac{x}{\sqrt{2}} - \frac{\pi}{8}\right) - \frac{1}{8x}\cos\left(\frac{x}{\sqrt{2}} - \frac{\pi}{8}\right)\right]$$

determine two-term expansions for the large roots of (a) $bei(x) = 0$ and (b) $bei'(x) = 0$.

2.16. Given that

$$I(x) = \int_x^\infty \frac{\cos t}{t}\,dt \sim \left(-\frac{1}{x} + \frac{2!}{x^3}\right)\sin x + \frac{1}{x^2}\cos x \quad \text{as } x \to \infty$$

determine two-term expansions for the large roots of (a) $I(x) = 0$ and (b) $I'(x) = 0$.

2.17. Given that

$$I_1(x) = \int_x^\infty \frac{\cos t}{\sqrt{t}}\,dt \sim \frac{1}{\sqrt{x}}(f\cos x - g\sin x) \quad \text{as } x \to \infty$$

$$I_2(x) = \int_x^\infty \frac{\sin x}{\sqrt{t}}\,dt \sim \frac{1}{\sqrt{x}}(f\sin x + g\cos x) \quad \text{as } x \to \infty$$

where

$$f(x) \sim \frac{1}{2x} - \frac{1\cdot 3\cdot 5}{(2x)^3}, \qquad g(x) \sim 1 - \frac{1\cdot 3}{(2x)^2}$$

determine two-term expansions for the large roots of (a) $I_1(x) = 0$ and (b) $I_2(x) = 0$.

2.18. In analyzing wave propagation in a planar uniform-lined duct, one encounters the following eigenvalue problem

$$\lambda \tan\lambda = -i\omega\beta$$

where ω is the frequency and β is the liner admittance. Show that

$$\lambda \sim n\pi - \frac{i\omega\beta}{n\pi} \quad \text{as } n \to \infty$$

2.19. Determine two-term expansions for the large roots of
(a) $x \sin x = 1$
(b) $x \cos x = 1$
(c) $x \sin x = \frac{1}{2}x + 1$
(d) $x \cos x = \frac{1}{2}x + 1$

2.20. Determine two-term expansions for the large roots of (a) $x \tan x = x + 1$ and (b) $x \cot x = x + 1$.

2.21. In analyzing wave propagation in a circular uniform-lined duct, one encounters the eigenvalue problem

$$\lambda J_m'(\lambda) = i\omega\beta J_m(\lambda)$$

where ω is the frequency and β is the admittance. Use the asymptotic expansions of J_m in Exercise 2.8 to determine two terms in the expansions of the large eigenvalues.

2.22. In analyzing the linear-free vibrations of a uniform hinged-clamped beam, one encounters the eigenvalue problem

$$\tan \lambda - \tanh \lambda = 0$$

Show that

$$\lambda \sim \left(n + \tfrac{1}{4}\right)\pi + \tfrac{1}{2}\left[\tanh\left(n + \tfrac{1}{4}\right) - 1\right] \quad \text{as } n \to \infty$$

2.23. In analyzing the linear-free vibrations of a uniform clamped-beam, one encounters the eigenvalue problem

$$\cos \lambda \cosh \lambda = -1$$

Show that

$$\lambda \sim \left(n - \tfrac{1}{2}\right)\pi + 2\sin\left(n - \tfrac{1}{2}\right)\pi \exp\left[-\left(n - \tfrac{1}{2}\right)\pi\right] \quad \text{as } n \to \infty$$

CHAPTER 3

Integrals

There are many differential and difference equations whose solutions cannot be expressed in terms of elementary functions but can be expressed in the form of integrals. Among the methods used to represent the solutions of differential equations as integrals, we mention the Laplace, Fourier, Hankel, and Mellin transformations. There are several techniques for obtaining asymptotic expansions of functions defined by definite integrals. They include expansion of integrands, integration by parts, Laplace's method, the method of stationary phase, and the method of steepest descent.

3.1. Expansion of Integrands

If the integrand contains a small parameter or if the limits of integration are small, one may be able to determine an asymptotic expansion of the integral by expanding the integrand and then integrating the result term by term. For example, an asymptotic expansion of

$$I(m)=\int_0^\pi \frac{d\theta}{\sqrt{1-m\sin^2\theta}}$$

for small m can be obtained by expanding the integrand for small m and then integrating the result term by term. As another example, an asymptotic expansion of

$$I(x)=\int_0^x t^{-1/4}e^{-t^2}\,dt$$

for small x can be obtained by expanding $\exp(-t^2)$ in a Taylor series and then integrating the result term by term.

3.2. Integration by Parts

Frequently asymptotic expansions of integrals may be obtained by repeated integration by parts. For example, application of integration by parts to the integral

$$I(x) = \int_0^\infty e^{-xt} f(t)\, dt$$

yields

$$I(x) = \frac{f(0)}{x} + \frac{f'(0)}{x^2} + \frac{f''(0)}{x^3} + O\left(\frac{1}{x^4}\right) \quad \text{as } x \to \infty$$

As another example, application of integration by parts to the integral

$$I(x) = \int_a^b e^{xh(t)} f(t)\, dt, \qquad b > a$$

yields

$$I(x) = \frac{e^{xh(b)} f(b)}{xh'(b)} - \frac{e^{xh(a)} f(a)}{xh'(a)} + O\left(\frac{1}{x^2}\right) \quad \text{as } x \to \infty$$

if $h'(t) \neq 0$ in the interval $[a, b]$.

3.3. Laplace's Method

If $h'(t) = 0$ in the interval $[a, b]$, integration by parts for the above integral breaks down, because the integrals in the successive integration by parts fail to exist. In this case, if $h(t)$ has a relative maximum at $t = c$, where $a \le c \le b$, an asymptotic expansion of the integral for large x can be obtained by expanding both $h(t)$ and $f(t)$ around $t = c$. This is Laplace's method, which is based on the idea that the major contribution to the value of the integral arises from the neighborhoods of the points in the interval $[a, b]$ at which the integrand has its maximum value. Thus if $h(t)$ does not have a relative maximum in $[a, b]$, the major contribution to the integral arises from the end points, and an asymptotic expansion of the integral can be obtained by a succession of integration by parts. If $h(t)$ has a finite number of maxima, the interval of integration can be broken up to a finite number of intervals so that $h(t)$ has only one maximum in each interval. An asymptotic expansion of the integral can then be obtained as the sum of asymptotic expansions of the resulting integrals.

3.4. The Method of Stationary Phase

The major contribution to the value of the generalized Fourier integral

$$I(\alpha) = \int_a^b f(t) e^{i\alpha h(t)}, \qquad b > a$$

for large positive α when $h(t)$ is real arises from the immediate neighborhoods of the end points of the interval and the points at which $h(t)$ is stationary, that is, $h'(t) = 0$. To the first approximation, the contribution from the neighborhoods of the stationary points is more important than the contribution from the neighborhoods of the end points of the interval. Thus if $h'(t) \neq 0$ in $[a, b]$, an asymptotic expansion of this integral can be obtained by integration by parts. The result is

$$I(\alpha) = \frac{i}{\alpha}\left[\frac{f(a)e^{i\alpha h(a)}}{h'(a)} - \frac{f(b)e^{i\alpha h(b)}}{h'(b)}\right] + O\left(\frac{1}{\alpha^2}\right) \quad \text{as } \alpha \to \infty$$

If $h'(c) = 0$ where $a \le c \le b$, and $h'(t)$ does not vanish at any other point, the dominant term in the asymptotic expansion of this integral can be obtained by using the method of stationary phase, according to which one keeps the first terms in the expansions of $h(t)$ and $f(t)$ around $t = c$, that is,

$$I(\alpha) \sim f(c)e^{i\alpha h(c)} \int_{c-\delta}^{c+\delta} e^{i\alpha h''(c)(t-c)^2/2}\, dt \quad \text{if } a \le c \le b$$

or

$$I(\alpha) \sim f(a)e^{i\alpha h(a)} \int_{a}^{a+\delta} e^{i\alpha h''(a)(t-a)^2/2}\, dt \quad \text{if } c = a$$

or

$$I(\alpha) \sim f(b)e^{i\alpha h(b)} \int_{b-\delta}^{b} e^{i\alpha h''(b)(t-b)^2/2}\, dt \quad \text{if } c = b$$

These integrals can be evaluated by using Cauchy's theorem and deforming the contour of integration so that the Fourier integral is transformed into a Laplace integral and the interval of integration is replaced by $[-\infty \text{ to } \infty]$ when $a < c < b$, by $[0, \infty]$ when $c = a$, and by $[-\infty, 0]$ when $c = b$.

3.5. The Method of Steepest Descent

The method of steepest descent is used to determine approximations for large positive values of α to integrals of the form

$$I(\alpha) = \int_C f(z)e^{\alpha h(z)}\, dz$$

when $f(z)$ and $h(z)$ are analytic functions of z and C is a contour of integration in the complex z plane. According to this method, one uses the analyticity of the integrand and appeals to Cauchy's theorem to deform the contour of integration C into a new contour C' on which the phase $\psi(x, y)$ is a constant, where $h(z)=\phi(x, y)+i\psi(x, y)$ and $z=x+iy$, thereby transforming the integral into a Laplace integral whose asymptotic development can be obtained using Laplace's method.

Solved Exercises

3.1. Show that as $\varepsilon \to 0$

$$\int_0^1 \frac{\sin \varepsilon t}{t}\, dt \sim \varepsilon - \tfrac{1}{18}\varepsilon^3 + \tfrac{1}{600}\varepsilon^5 \tag{1}$$

Since

$$\sin \varepsilon t = \varepsilon t - \frac{\varepsilon^3 t^3}{3!} + \frac{\varepsilon^5 t^5}{5!} + \dots$$

$$\int_0^1 \frac{\sin \varepsilon t}{t}\, dt \sim \varepsilon \int_0^1 dt - \tfrac{1}{6}\varepsilon^3 \int_0^1 t^2\, dt + \tfrac{1}{120}\varepsilon^5 \int_0^1 t^4\, dt$$

which yields (1).

3.2. Show that as $x \to 0$

$$\int_0^x t^{-3/4} e^{-t}\, dt \sim 4x^{1/4} - \tfrac{4}{5}x^{5/4} + \tfrac{2}{9}x^{9/4} \tag{2}$$

Since

$$e^{-t} = 1 - t + \tfrac{1}{2}t^2 + \dots$$

$$\int_0^x t^{-3/4} e^{-t}\, dt \sim \int_0^x t^{-3/4}\, dt - \int_0^x t^{1/4}\, dt + \tfrac{1}{2}\int_0^x t^{5/4}\, dt$$

which yields (2).

3.3. Show that as $x \to \infty$

(a)
$$\int_x^\infty \frac{e^{-t}}{t}\, dt \sim e^{-x}\left(\frac{1}{x} - \frac{1!}{x^2} + \frac{2!}{x^3} - \frac{3!}{x^4}\right) \tag{3}$$

(b)
$$\int_x^\infty \frac{e^{-t}}{t^n}\, dt \sim \frac{e^{-x}}{x^n}\left[1 - \frac{n}{x} + \frac{n(n+1)}{t^2} - \frac{n(n+1)(n+2)}{x^3}\right] \tag{4}$$

(a) Using integration by parts, we let

$$u = \frac{1}{t}, \qquad dv = e^{-t}\, dt$$

Hence

$$du = -\frac{1}{t^2}\, dt, \qquad v = -e^{-t}$$

Then

$$\int_x^\infty \frac{e^{-t}}{t}\,dt = -\frac{e^{-t}}{t}\bigg|_x^\infty - \int_x^\infty \frac{e^{-t}}{t^2}\,dt = \frac{e^{-x}}{x} + \frac{e^{-t}}{t^2}\bigg|_x^\infty + 2\int_x^\infty \frac{e^{-t}}{t^3}\,dt$$

$$= \frac{e^{-x}}{x} - \frac{e^{-x}}{x^2} + 2\left[\frac{-e^{-t}}{t^3}\bigg|_x^\infty - 3\int_x^\infty \frac{e^{-t}}{t^4}\,dt\right] = \frac{e^{-x}}{x} - \frac{e^{-x}}{x^2} + \frac{2e^{-x}}{x^3}$$

$$+3!\left[\frac{e^{-t}}{t^4}\bigg|_x^\infty + 4\int_x^\infty \frac{e^{-t}}{t^5}\,dt\right]$$

which yields (3).

(b) Using integration by parts, we let

$$u = \frac{1}{t^n}, \qquad dv = e^{-t}\,dt$$

Hence

$$du = -\frac{n}{t^{n+1}}\,dt, \qquad v = -e^{-t}$$

and

$$\int_x^\infty \frac{e^{-t}}{t^n}\,dt = -\frac{e^{-t}}{t^n}\bigg|_x^\infty - n\int_x^\infty \frac{e^{-t}}{t^{n+1}}\,dt$$

$$= \frac{e^{-x}}{x^n} - n\left[\frac{-e^{-t}}{t^{n+1}}\bigg|_x^\infty - (n+1)\int_x^\infty \frac{e^{-t}}{t^{n+2}}\,dt\right]$$

$$= \frac{e^{-x}}{x^n} - \frac{ne^{-x}}{x^{n+1}} + n(n+1)\left[\frac{-e^{-t}}{t^{n+2}}\bigg|_x^\infty - (n+2)\int_x^\infty \frac{e^{-t}}{t^{n+3}}\,dt\right]$$

$$= \frac{e^{-x}}{x^n} - \frac{ne^{-x}}{x^{n+1}} + \frac{n(n+1)e^{-x}}{x^{n+2}} - n(n+1)(n+2)$$

$$\times\left[\frac{-e^{-t}}{t^{n+3}}\bigg|_x^\infty - (n+3)\int_x^\infty \frac{e^{-t}}{t^{n+4}}\,dt\right]$$

which yields (4).

3.4. Show that as $x \to \infty$

$$\int_x^\infty e^{-t}t^{\lambda-1}\,dt \sim x^\lambda e^{-x}\left[\frac{1}{x} + \frac{\lambda-1}{x^2} + \frac{(\lambda-1)(\lambda-2)}{x^3}\right] \tag{5}$$

Using integration by parts, we let

$$u = t^{\lambda-1}, \qquad dv = e^{-t}\,dt$$

Hence

$$du = (\lambda-1)\,t^{\lambda-2}\,dt, \qquad v = -e^{-t}$$

and

$$\int_x^\infty e^{-t}t^{\lambda-1}\,dt = -e^{-t}t^{\lambda-1}\Big|_x^\infty + (\lambda-1)\int_x^\infty t^{\lambda-2}e^{-t}\,dt$$

$$= x^{\lambda-1}e^{-x} + (\lambda-1)\left[-e^{-t}t^{\lambda-2}\Big|_x^\infty + (\lambda-2)\int_x^\infty t^{\lambda-3}e^{-t}\,dt\right]$$

$$= x^{\lambda-1}e^{-x} + (\lambda-1)\,x^{\lambda-2}e^{-x} + (\lambda-1)(\lambda-2)$$

$$\times\left[-e^{-t}t^{\lambda-3}\Big|_x^\infty + (\lambda-3)\int_x^\infty t^{\lambda-4}e^{-t}\,dt\right]$$

which yields (5).

3.5. Show that as $x \to \infty$

$$\int_x^\infty e^{-t^2}\,dt \sim e^{-x^2}\left(\frac{1}{2x} - \frac{1}{2^2x^3} + \frac{1\cdot 3}{2^3x^5} - \frac{1\cdot 3\cdot 5}{2^4x^7}\right) \tag{6}$$

Using integration by parts, we let

$$dv = -2te^{-t^2}\,dt, \qquad u = \frac{-1}{2t}$$

Hence

$$v = e^{-t^2}, \qquad du = \frac{1}{2t^2}\,dt$$

and

$$\int_x^\infty e^{-t^2}\,dt = \frac{-e^{-t^2}}{2t}\Bigg|_x^\infty - \tfrac{1}{2}\int_x^\infty \frac{e^{-t^2}}{t^2}\,dt = \frac{e^{-x^2}}{2x} - \tfrac{1}{2}\left[\frac{-e^{-t^2}}{2t^3}\Bigg|_x^\infty - \tfrac{3}{2}\int_x^\infty \frac{e^{-t^2}}{t^4}\,dt\right]$$

$$= \frac{e^{-x^2}}{2x} - \frac{e^{-x^2}}{4x^3} + \tfrac{3}{4}\left[\frac{-e^{-t^2}}{2t^5}\Bigg|_x^\infty - \tfrac{5}{2}\int_x^\infty \frac{e^{-t^2}}{t^6}\,dt\right]$$

$$= \frac{e^{-x^2}}{2x} - \frac{e^{-x^2}}{4x^3} + \frac{3e^{-x^2}}{8x^5} - \tfrac{15}{8}\left[\frac{-e^{-t^2}}{2t^7}\Bigg|_x^\infty - \tfrac{7}{2}\int_x^\infty \frac{e^{-t^2}}{t^8}\,dt\right]$$

which yields (6).

3.6. Consider the complete elliptic integral of the first kind

$$I(m)=\int_0^{\pi/2}\sqrt{1-m\sin^2\theta}\,d\theta \tag{7}$$

Show that

$$I(m)=\tfrac{1}{2}\pi\left[1-\tfrac{1}{4}m-\tfrac{3}{64}m^2-\tfrac{5}{256}m^3-\tfrac{175}{16384}m^4+\cdots\right] \tag{8}$$

Using the binomial theorem, we have

$$\sqrt{1-m\sin^2\theta}=1-\tfrac{1}{2}m\sin^2\theta+\frac{(\frac{1}{2})(-\frac{1}{2})}{2!}m^2\sin^4\theta$$

$$+\frac{(\frac{1}{2})(-\frac{1}{2})(-\frac{3}{2})}{3!}(-m\sin^2\theta)^3$$

$$+\frac{(\frac{1}{2})(-\frac{1}{2})(-\frac{3}{2})(-\frac{5}{2})}{4!}(-m\sin^2\theta)^4+\cdots$$

$$=1-\tfrac{1}{2}m\sin^2\theta-\tfrac{1}{8}m^2\sin^4\theta-\tfrac{1}{16}m^3\sin^6\theta-\tfrac{5}{128}m^4\sin^8\theta+\cdots$$

Hence

$$I(m)=\int_0^{\pi/2}\sqrt{1-m\sin^2\theta}\,d\theta=\int_0^{\pi/2}d\theta-\tfrac{1}{2}m\int_0^{\pi/2}\sin^2\theta\,d\theta$$

$$-\tfrac{1}{8}m^2\int_0^{\pi/2}\sin^4\theta\,d\theta-\tfrac{1}{16}m^3\int_0^{\pi/2}\sin^6\theta\,d\theta-\tfrac{5}{128}\int_0^{\pi/2}\sin^8\theta\,d\theta+\cdots \tag{9}$$

Using $\int_0^{\pi/2}\sin^{2n}\theta\,d\theta=[(2n)!\pi/(n!)^2 2^{n+1}]$ in (9), we obtain (8).

3.7. Show that as $x\to\infty$

$$\int_x^\infty\frac{\cos t}{t}\,dt\sim\left(-\frac{1}{x}+\frac{2!}{x^3}-\frac{4!}{x^5}\right)\sin x+\left(\frac{1}{x^2}-\frac{3!}{x^4}+\frac{5!}{x^6}\right)\cos x \tag{10}$$

Using integration by parts, we let

$$u=\frac{1}{t},\qquad dv=\cos t\,dt$$

Then

$$du = -\frac{1}{t^2}\,dt, \qquad v = \sin t$$

$$\int_x^\infty \frac{\cos t}{t}\,dt = \frac{\sin t}{t}\bigg|_x^\infty + \int_x^\infty \frac{\sin t}{t^2}\,dt = -\frac{\sin x}{x} + \left[\frac{-\cos t}{t^2}\bigg|_x^\infty - 2\int_x^\infty \frac{\cos t}{t^3}\,dt\right]$$

$$= -\frac{\sin x}{x} + \frac{\cos x}{x^2} - 2\left[\frac{\sin t}{t^3}\bigg|_x^\infty + 3\int_x^\infty \frac{\sin t}{t^4}\,dt\right]$$

$$= -\frac{\sin x}{x} + \frac{\cos x}{x^2} + \frac{2\sin x}{x^3} - 6\left[\frac{-\cos t}{t^4}\bigg|_x^\infty - 4\int_x^\infty \frac{\cos t}{t^5}\,dt\right]$$

$$= -\frac{\sin x}{x} + \frac{\cos x}{x^2} + \frac{2\sin x}{x^3} - \frac{6\cos x}{x^4} + 24\left[\frac{\sin t}{t^5}\bigg|_x^\infty + 5\int_x^\infty \frac{\sin t}{t^6}\,dt\right]$$

$$= -\frac{\sin x}{x} + \frac{\cos x}{x^2} + \frac{2\sin x}{x^3} - \frac{6\cos x}{x^4} - \frac{24\sin x}{x^5}$$

$$+120\left[\frac{-\cos t}{t^6}\bigg|_x^\infty - 6\int_x^\infty \frac{\cos t}{t^7}\,dt\right]$$

which yields (10).

3.8. Show that as $x \to \infty$

(a)
$$\int_x^\infty \frac{\cos t}{\sqrt{t}}\,dt \sim \frac{1}{\sqrt{x}}(f\cos x - g\sin x) \tag{11}$$

(b)
$$\int_x^\infty \frac{\sin t}{\sqrt{t}}\,dt \sim \frac{1}{\sqrt{x}}(f\sin x + g\cos x) \tag{12}$$

where

$$f \sim \frac{1}{2x} - \frac{1\cdot3\cdot5}{(2x)^3} + \frac{1\cdot3\cdot5\cdot7\cdot9}{(2x)^5}$$

$$g \sim 1 - \frac{1\cdot3}{(2x)^2} + \frac{1\cdot3\cdot5\cdot7}{(2x)^4}$$

One can either determine the asymptotic expressions for the integrals given above by integrating them by parts, as in the preceding exercise, or by first expressing $\cos t$ as the real part of $\exp(it)$. We use the latter approach because the algebra is simpler. Thus we consider the integral

$$I = \int_x^{\infty} \frac{e^{it}}{\sqrt{t}}\, dt \tag{13}$$

Using integration by parts, we let

$$u = -\frac{i}{\sqrt{t}}, \qquad ie^{it}\, dt = dv$$

Hence

$$du = \tfrac{1}{2} i t^{-3/2}\, dt, \qquad v = e^{it}$$

and

$$I = -\, ie^{it} t^{-1/2}\Big|_x^{\infty} - \tfrac{1}{2}\int_x^{\infty} ie^{it} t^{-3/2}\, dt = ie^{ix} x^{-1/2} - \tfrac{1}{2}\left[e^{it} t^{-3/2}\Big|_x^{\infty} + \tfrac{3}{2}\int_x^{\infty} e^{it} t^{-5/2}\, dt \right]$$

$$= ie^{ix} x^{-1/2} + \tfrac{1}{2} e^{ix} x^{-3/2} - \tfrac{3}{4}\left[-\, ie^{it} t^{-5/2}\Big|_x^{\infty} - \tfrac{5}{2}\int_x^{\infty} ie^{it} t^{-7/2}\, dt \right]$$

$$= ie^{ix} x^{-1/2} + \tfrac{1}{2} e^{ix} x^{-3/2} - \tfrac{3}{4} ie^{ix} x^{-5/2} + \tfrac{15}{8}\left[e^{it} t^{-7/2}\Big|_x^{\infty} + \tfrac{7}{2}\int_x^{\infty} e^{it} t^{-9/2}\, dt \right]$$

$$= ie^{ix} x^{-1/2} + \tfrac{1}{2} e^{ix} x^{-3/2} - \tfrac{3}{4} ie^{ix} x^{-5/2} - \tfrac{15}{8} e^{ix} x^{-7/2}$$

$$+ \tfrac{105}{16}\left[-\, ie^{-it} t^{-9/2}\Big|_x^{\infty} - \tfrac{9}{2}\int_x^{\infty} ie^{it} t^{-11/2}\, dt \right]$$

$$= ie^{ix} x^{-1/2} + \tfrac{1}{2} e^{ix} x^{-3/2} - \tfrac{3}{4} ie^{ix} x^{-5/2} - \tfrac{15}{8} e^{ix} x^{-7/2} + \tfrac{105}{16} ie^{ix} x^{-9/2}$$

$$- \tfrac{945}{32}\left[e^{it} t^{-11/2}\Big|_x^{\infty} + \tfrac{11}{2}\int_x^{\infty} e^{it} t^{-13/2}\, dt \right]$$

Hence

$$I(x) \sim \frac{1}{\sqrt{x}} e^{ix} \left\{ \left[1 - \frac{1\cdot 3}{(2x)^2} + \frac{1\cdot 3\cdot 5\cdot 7}{(2x)^4} \right] i + \left[\frac{1}{2x} - \frac{1\cdot 3\cdot 5}{(2x)^3} + \frac{1\cdot 3\cdot 5\cdot 7\cdot 9}{(2x)^5} \right] \right\}$$

or

$$I(x) \sim \frac{1}{\sqrt{x}} e^{ix} (f + ig)$$

$$= \frac{1}{\sqrt{x}} (\cos x + i \sin x)(f + ig) = \frac{1}{\sqrt{x}} [\, f\cos x - g\sin x + i(f\sin x + g\cos x)] \tag{14}$$

Substituting (14) into (13) and separating real and imaginary parts, we obtain (11) and (12).

3.9. Show that as $x \to \infty$

(a)
$$\int_x^\infty \frac{\cos(t-x)}{t}\,dt \sim \frac{1}{x^2} - \frac{3!}{x^4} + \frac{5!}{x^6} \tag{15}$$

(b)
$$\int_x^\infty \frac{\sin(t-x)}{t}\,dt \sim \frac{1}{x} - \frac{2!}{x^3} + \frac{4!}{x^5} \tag{16}$$

As in the preceding example, we consider the complex integral

$$I(x) = \int_x^\infty \frac{e^{i(t-x)}}{t}\,dt \tag{17}$$

Using integration by parts, we let

$$u = -it^{-1}, \qquad dv = ie^{i(t-x)}\,dt$$

Hence

$$du = it^{-2}\,dt, \qquad v = e^{i(t-x)}$$

and

$$I(x) = -it^{-1}e^{i(t-x)}\Big|_x^\infty - \int_x^\infty it^{-2}e^{i(t-x)}\,dt$$

$$= ix^{-1} - \left[t^{-2}e^{i(t-x)}\Big|_x^\infty + 2\int_x^\infty t^{-3}e^{i(t-x)}\,dt\right]$$

$$= ix^{-1} + x^{-2} - 2\left[-it^{-3}e^{i(t-x)}\Big|_x^\infty - 3\int_x^\infty it^{-4}e^{i(t-x)}\,dt\right]$$

$$= ix^{-1} + x^{-2} - 2ix^{-3} + 3!\left[t^{-4}e^{i(t-x)}\Big|_x^\infty + 4\int_x^\infty t^{-5}e^{i(t-x)}\,dt\right]$$

$$= ix^{-1} + x^{-2} - 2ix^{-3} - 3!\,x^{-4} + 4!\left[-it^{-5}e^{i(t-x)}\Big|_x^\infty - 5\int_x^\infty it^{-6}e^{i(t-x)}\,dt\right]$$

$$= ix^{-1} + x^{-2} - 2ix^{-3} - 3!\,x^{-4} + 4!\,ix^{-5}$$

$$-5!\left[t^{-6}e^{i(t-x)}\Big|_x^\infty + 6\int_x^\infty t^{-7}e^{i(t-x)}\,dt\right]$$

Therefore

$$I(x) \sim \frac{1}{x^2} - \frac{3!}{x^4} + \frac{5!}{x^6} + i\left(\frac{1}{x} - \frac{2!}{x^3} + \frac{4!}{x^5}\right) \tag{18}$$

Substituting (18) into (17) and separating real and imaginary parts, we obtain (15) and (16).

3.10. Show that as $x \to \infty$

(a)
$$\int_x^\infty \frac{dt}{t^2 \ln t} \sim \frac{1}{x \ln x} \tag{19}$$

(b)
$$\int_2^x \frac{dt}{t \ln \ln t} \sim \frac{\ln x}{\ln \ln x} \tag{20}$$

(a) Let

$$\tau = \ln t \quad \text{so that } d\tau = \frac{dt}{t}$$

Then

$$\int_x^\infty \frac{dt}{t^2 \ln t} = \int_{\ln x}^\infty \frac{d\tau}{t \ln t} = \int_{\ln x}^\infty \frac{e^{-\tau}\, d\tau}{\tau}$$

Using integration by parts, we let

$$e^{-\tau}\, d\tau = dv, \qquad u = \frac{1}{\tau}$$

Then

$$v = -e^{-\tau}, \qquad du = -\frac{1}{\tau^2}\, d\tau$$

and

$$\int_x^\infty \frac{dt}{t^2 \ln t} = \int_{\ln x}^\infty \frac{e^{-\tau}\, d\tau}{\tau} = -\frac{e^{-\tau}}{\tau}\bigg|_{\ln x}^\infty - \int_{\ln x}^\infty \frac{e^{-\tau}}{\tau^2}\, d\tau$$

$$= \frac{e^{-\ln x}}{\ln x} - \int_{\ln x}^\infty \frac{e^{-\tau}}{\tau^2}\, d\tau \sim \frac{1}{x \ln x}$$

(b) Using integration by parts, we let

$$u = \frac{1}{\ln \ln t}, \qquad dv = \frac{dt}{t}$$

Hence

$$du = -\frac{1}{(\ln \ln t)^2} \cdot \frac{1}{\ln t} \cdot \frac{dt}{t}, \qquad v = \ln t$$

and

$$\int_2^x \frac{dt}{t\ln\ln t} = \frac{\ln t}{\ln\ln t}\bigg|_2^x + \int_2^x \frac{dt}{t(\ln\ln t)^2}$$

which yields (20).

3.11. Show that as $x \to \infty$

(a)
$$\int_x^\infty \frac{t^{1/2}}{1+t^2}\,dt \sim 2x^{-1/2} - \tfrac{2}{5}x^{-5/2} + \tfrac{2}{9}x^{-9/2} \tag{21}$$

Since $(1+t^2)^{-1} = \dfrac{1}{t^2}\left(1+\dfrac{1}{t^2}\right)^{-1} = \dfrac{1}{t^2}\left[1 - \dfrac{1}{t^2} + \dfrac{(-1)(-2)}{2!t^4} + \cdots\right]$

$$\int_x^\infty \frac{t^{1/2}}{1+t^2}\,dt \sim \int_x^\infty t^{-3/2}\,dt - \int_x^\infty t^{-7/2}\,dt + \int_x^\infty t^{-11/2}\,dt$$

which yields (21).

(b)
$$\int_0^x (t^3+t^2)^{1/2}\,dt \sim \tfrac{2}{5}x^{5/2} + \tfrac{1}{3}x^{3/2} \tag{22}$$

Since the integrand is regular at the origin, the major contribution to the asymptotic development of the integral comes from the neighborhood of $t = x$. Thus we expand the integrand for large t and obtain

$$\int_0^x (t^3+t^2)^{1/2}\,dt = \int_0^x t^{3/2}\left[1+\frac{1}{t}\right]^{1/2} dt \sim \int_0^x t^{3/2}\,dt + \tfrac{1}{2}\int_0^x t^{1/2}\,dt$$

which yields (22).

(c)
$$\int_a^x \frac{dt}{\ln t} \sim \frac{x}{\ln x} \tag{23}$$

Using integration by parts, we let

$$u = \frac{1}{\ln t}, \qquad dv = dt$$

Hence

$$\int_a^x \frac{dt}{\ln t} = \frac{t}{\ln t}\bigg|_a^x - \int_a^x t\left[-\frac{1}{(\ln t)^2}\right]\frac{dt}{t} = \frac{x}{\ln x} - \frac{a}{\ln a} + \int_a^x \frac{dt}{(\ln t)^2}$$

which yields (23).

3.12. Show that as $x \to \infty$

(a)
$$\int_0^1 e^{-xt} \ln(2+t)\, dt \sim \frac{\ln 2}{x} \tag{24}$$

For large positive x, the major contribution to the asymptotic expansion of the integral arises from the neighborhood of $t=0$. Then, using Watson's lemma, we expand $\ln(2+t)$ for small t as

$$\ln(2+t) = \ln 2 + \ln\left(1+\tfrac{1}{2}t\right) = \ln 2 + \tfrac{1}{2}t + \cdots$$

Then

$$\int_0^1 e^{-xt} \ln(2+t)\, dt \sim \int_0^\infty e^{-xt} \ln 2\, dt = \ln 2 \left.\frac{e^{-xt}}{-x}\right|_0^\infty = \frac{\ln 2}{x}$$

(b)
$$\int_0^1 e^{-xt} \ln(1+t)\, dt \sim \frac{1}{x^2} \tag{25}$$

We note that

$$\ln(1+t) = t - \frac{t^2}{2} + \frac{t^3}{3} - \cdots \quad \text{as } t \to 0$$

Hence using Watson's lemma, we have

$$\int_0^1 e^{-xt} \ln(1+t)\, dt \sim \int_0^\infty e^{-xt} t\, dt = \frac{1}{x^2} \int_0^\infty \tau e^{-\tau}\, d\tau = \frac{\Gamma(2)}{x^2} = \frac{1}{x^2}$$

(c)
$$\int_0^1 e^{-xt} \sin t\, dt \sim \frac{1}{x^2} \tag{26}$$

We note that

$$\sin t = t - \frac{t^3}{3!} + \cdots \quad \text{as } t \to 0$$

Hence using Watson's lemma, we have

$$\int_0^1 e^{-xt} \sin t\, dt \sim \int_0^\infty e^{-xt} t\, dt = \frac{1}{x^2}$$

as in the preceding case.

(d)
$$\int_0^1 e^{-(x/t)+t+xt}\, dt \sim \frac{e}{2x} \tag{27}$$

We need to determine the location where the exponent attains its maximum. In this case,

$$h(t) = xt + t - \frac{x}{t}$$

$$h'(t) = x + 1 + \frac{x}{t^2}$$

does not vanish anywhere in the interval of interest. Hence the maximum occurs at the ends of the interval. Since $h(0) = -\infty$, whereas $h(1) = 1 + x - x = 1$. The maximum occurs at $t = 1$ and the major contribution to the asymptotic expansion of the integral arises from the neighborhood of $t = 1$. Thus we let $\tau = 1 - t$ and obtain

$$\int_0^1 e^{-(x/t)+t+xt}\,dt = -\int_1^0 \exp\left[x - x\tau + 1 - \tau - \frac{x}{1-\tau}\right] d\tau$$

$$= \int_0^1 \exp[x - x\tau + 1 - \tau - x - x\tau + \cdots]\,d\tau$$

$$= e\int_0^1 e^{-2x\tau} e^{-\tau + \cdots}\,d\tau$$

Since $e^{-\tau} = 1 - \tau + \cdots$, using Watson's lemma, we have

$$\int_0^1 e^{-(x/t)+t+xt} \sim e\int_0^1 e^{-2x\tau}e^{-\tau}\,d\tau \sim e\int_0^\infty e^{-2x\tau}\,d\tau = \frac{e}{2x}$$

3.13. Show that as $\omega \to \infty$

$$\int_0^\infty \frac{e^{-x}}{\omega + x + x\sqrt{\omega}}\,dx \sim \frac{1}{\omega} - \frac{1}{\omega^{3/2}} \tag{28}$$

The major contribution to the integral arises from the neighborhood of the origin. Since

$$(\omega + x + x\sqrt{\omega})^{-1} = \omega^{-1}\left(1 + \frac{x + x\sqrt{\omega}}{\omega}\right)^{-1} = \frac{1}{\omega} - \frac{x + x\sqrt{\omega}}{\omega^2} + \frac{(x + x\sqrt{\omega})^2}{\omega^3} + \cdots$$

Using Watson's lemma, we have

$$\int_0^\infty \frac{e^{-x}}{\omega + x + x\sqrt{\omega}}\,dx \sim \frac{1}{\omega}\int_0^\infty e^{-x}\,dx - \frac{1}{\omega^{3/2}}\int_0^\infty xe^{-x}\,dx$$

$$= \frac{1}{\omega} - \frac{\Gamma(2)}{\omega^{3/2}} = \frac{1}{\omega} - \frac{1}{\omega^{3/2}}$$

3.14. Show that as $\omega \to \infty$

(a)
$$\int_0^\infty e^{-\omega(x^2+2x)}(1+x)^{5/2}\,dx \sim \frac{1}{2\omega} \tag{29}$$

The maximum of $-(x^2+2x)$ occurs at $x=0$. Hence the major contribution to the integral arises from the neighborhood of $x=0$. Then application of Watson's lemma gives

$$I(\omega)=\int_0^{\infty} e^{-\omega(x^2+2x)}(1+x)^{5/2}\,dx=\int_0^{\infty} e^{-2\omega x}e^{-\omega x^2}(1+x)^{5/2}\,dx$$

$$\sim\int_0^{\infty} e^{-2\omega x}(1-\omega x^2+\cdots)(1+\tfrac{5}{2}x+\cdots)\,dx$$

Hence

$$I(\omega)\sim\int_0^{\infty} e^{-2\omega x}\,dx=\frac{1}{2\omega}$$

(b) $$I(\omega)=\int_0^{\infty} e^{-\omega(x^2+2x)}\ln(2+x)\,dx\sim\frac{\ln 2}{2\omega} \tag{30}$$

As in the preceding case, application of Watson's lemma gives

$$I(\omega)=\int_0^{\infty} e^{-2\omega x}e^{-\omega x^2}\ln(2+x)\,dx\sim\int_0^{\infty} e^{-2\omega x}(1-\omega x^2+\cdots)(\ln 2+\tfrac{1}{2}x+\cdots)\,dx$$

Hence

$$I(\omega)\sim\int_0^{\infty}\ln 2\,e^{-2\omega x}\,dx=\frac{\ln 2}{2\omega}$$

(c) $$I(\omega)=\int_0^{\infty} e^{-\omega(x^2+2x)}\ln(1+x)\,dx\sim\frac{1}{4\omega^2} \tag{31}$$

As in the preceding two cases, application of Watson's lemma yields

$$I(\omega)=\int_0^{\infty} e^{-2\omega x}e^{-\omega x^2}\ln(1+x)\,dx\sim\int_0^{\infty} e^{-2\omega x}(1-\omega x^2+\cdots)\left(x-\frac{x^2}{2}+\cdots\right)$$

Hence

$$I(\omega)\sim\int_0^{\infty} xe^{-2\omega x}\,dx=\frac{1}{4\omega^2}\int_0^{\infty}\tau e^{-\tau}\,d\tau=\frac{\Gamma(2)}{4\omega^2}=\frac{1}{4\omega^2}$$

(d) $$I(\omega)=\int_0^{\infty}\frac{e^{-\omega(x^2+2x)}}{\sqrt{x+3x^2}}\,dx\sim\sqrt{\frac{\pi}{2\omega}} \tag{32}$$

As in the preceding three cases, application of Watson's lemma yields

$$I(\omega)=\int_0^\infty e^{-2\omega x}e^{-\omega x^2}x^{-1/2}(1+3x)^{-1/2}\,dx$$

$$\sim\int_0^\infty x^{-1/2}e^{-2\omega x}(1-\omega x^2+\cdots)(1-\tfrac{3}{2}x+\cdots)\,dx$$

Hence

$$I(\omega)\sim\int_0^\infty x^{-1/2}e^{-2\omega x}\,dx=\frac{1}{\sqrt{2\omega}}\int_0^\infty \tau^{-1/2}e^{-\tau}\,d\tau=\frac{\Gamma(\frac{1}{2})}{\sqrt{2\omega}}=\sqrt{\frac{\pi}{2\omega}}$$

3.15. Show that as $\omega\to\infty$

$$I(\omega)=\int_a^b (x-a)^\lambda e^{-\omega h(x)}\,dx\sim\frac{e^{-\omega h(a)}\Gamma(\lambda+1)}{[\omega h'(a)]^{\lambda+1}} \tag{33}$$

where $\lambda>-1$, $b>a$, and $h(x)>h(a)$. The latter condition implies that the maximum of $-h(x)$ occurs at $x=a$. If $h'(a)\neq 0$, we have

$$h(x)=h(a)+h'(a)(x-a)+\cdots \tag{34}$$

Substituting (34) into (33) and applying Watson's lemma, we obtain

$$I(\omega)\sim\int_a^\infty (x-a)^\lambda e^{-\omega h(a)}e^{-\omega h'(a)(x-a)}\,dx \tag{35}$$

We let

$$\tau=\omega h'(a)(x-a)\quad\text{so that } d\tau=\omega h'(a)\,dx$$

Then (35) becomes

$$I(\omega)\sim\frac{e^{-\omega h(a)}}{[\omega h'(a)]^{\lambda+1}}\int_0^\infty \tau^\lambda e^{-\tau}\,d\tau=\frac{e^{-\omega h(a)}\Gamma(\lambda+1)}{[\omega h'(a)]^{\lambda+1}}$$

3.16. Show that as $\omega\to\infty$

(a)
$$I(\omega)=\int_1^\infty e^{-\omega x^2}x^{5/2}\ln(1+x)\,dx\sim\frac{e^{-\omega}\ln 2}{2\omega} \tag{36}$$

The maximum value of $-x^2$ occurs at $x=1$ and hence the major contribution to the asymptotic development of I arises from the neighborhood of $x=1$. Hence we put $x-1=\tau$ and obtain

$$I(\omega)=\int_0^\infty e^{-\omega(1+2\tau+\tau^2)}(1+\tau)^{5/2}\ln(2+\tau)\,d\tau \tag{37}$$

Applying Watson's lemma to (37), we have

$$I(\omega) = e^{-\omega}\int_0^\infty e^{-2\omega\tau}e^{-\omega\tau^2}(1+\tau)^{5/2}\ln(2+\tau)\,d\tau$$

$$\sim e^{-\omega}\int_0^\infty e^{-2\omega\tau}(1-\omega\tau^2+\cdots)(1+\tfrac{5}{2}\tau+\cdots)(\ln 2+\tfrac{1}{2}\tau+\cdots)\,d\tau$$

Hence

$$I(\omega) \sim e^{-\omega}\ln 2\int_0^\infty e^{-2\omega\tau}\,d\tau = \frac{e^{-\omega}\ln 2}{2\omega}$$

(b)

$$I(\omega) = \int_1^\infty e^{-\omega x^2}x^{5/2}\ln x\,dx \sim \frac{e^{-\omega}}{4\omega^2} \tag{38}$$

As in the preceding case, we let $x-1=\tau$ and obtain

$$I(\omega) = \int_0^\infty e^{-\omega(1+2\tau+\tau^2)}(1+\tau)^{5/2}\ln(1+\tau)\,d\tau$$

Applying Watson's lemma yields

$$I(\omega) = e^{-\omega}\int_0^\infty e^{-2\omega\tau}(1-\omega\tau^2+\cdots)(1+\tfrac{5}{2}\tau+\cdots)\left(\tau-\frac{\tau^2}{2}+\cdots\right)d\tau$$

Hence

$$I(\omega) \sim e^{-\omega}\int_0^\infty e^{-2\omega\tau}\tau\,d\tau = \frac{e^{-\omega}}{4\omega^2}\int_0^\infty te^{-t}\,dt = \frac{e^{-\omega}}{4\omega^2}\Gamma(2) = \frac{e^{-\omega}}{4\omega^2}$$

3.17. Show that as $\omega \to \infty$

(a)

$$I(\omega) = \int_0^\infty \frac{e^{-\omega x^2}}{\sqrt{x+x^2}}\,dx \sim \frac{\Gamma(\frac{1}{4})}{2\omega^{1/4}} \tag{39}$$

Since the exponent has its maximum at $x=0$, the major contribution to I arises from the neighborhood of $x=0$. Hence, applying Watson's lemma gives

$$I(\omega) = \int_0^\infty \frac{e^{-\omega x^2}}{x^{1/2}}(1+x)^{-1/2}\,dx \sim \int_0^\infty x^{-1/2}e^{-\omega x^2}(1-\tfrac{1}{2}x+\cdots)\,dx \tag{40}$$

We let

$$\omega x^2 = \tau \quad \text{so that } 2\omega x\,dx = d\tau \tag{41}$$

It follows from (40) that

$$I(\omega) \sim \int_0^\infty x^{-1/2} e^{-\omega x^2}\, dx = \int_0^\infty \frac{\omega^{1/4}\tau^{-1/4} e^{-\tau}\, d\tau}{2\omega \tau^{1/2}\omega^{-1/2}}$$

$$= \frac{1}{2\omega^{1/4}} \int_0^\infty \tau^{-3/4} e^{-\tau}\, d\tau = \frac{\Gamma(\frac{1}{4})}{2\omega^{1/4}}$$

(b)
$$I(\omega) = \int_0^\infty e^{-\omega x^2} x^{5/2} \ln(1+x)\, dx \sim \frac{\Gamma(\frac{9}{4})}{2\omega^{9/4}} \tag{42}$$

As in the preceding case, the major contribution to I arises from the neighborhood of $x = 0$. Hence

$$I(\omega) = \int_0^\infty e^{-\omega x^2} x^{5/2} \left(x - \frac{x^2}{2} + \cdots \right) dx \tag{43}$$

Substituting (41) into (43), we have

$$I(\omega) \sim \int_0^\infty e^{-\omega x^2} x^{7/2}\, dx = \int_0^\infty x^{5/2} \frac{e^{-\tau}\, d\tau}{2\omega} = \int_0^\infty \frac{\tau^{5/4} e^{-\tau}}{2\omega^{9/4}}\, d\tau = \frac{\Gamma(\frac{9}{4})}{2\omega^{9/4}}$$

3.18. Show that as $\omega \to \infty$

(a)
$$I(\omega) = \int_{-\infty}^\infty e^{-\omega x^2} \ln(2+x^2)\, dx \sim \frac{\sqrt{\pi}\ln 2}{\sqrt{\omega}} \tag{44}$$

Since the maximum of the exponent occurs at $x = 0$, the major contribution to the integral arises from the neighborhood of $x = 0$. Applying Watson's lemma yields

$$I(\omega) \sim \int_{-\infty}^\infty e^{-\omega x^2} (\ln 2 + \tfrac{1}{2}x^2 + \cdots)\, dx \tag{45}$$

We substitute (41) into (45) and obtain

$$I(\omega) \sim \ln 2 \int_{-\infty}^\infty e^{-\omega x^2}\, dx = 2\ln 2 \int_0^\infty \frac{e^{-\tau}\, d\tau}{2\omega x} = \ln 2 \int_0^\infty \frac{\tau^{-1/2} e^{-\tau}}{\omega^{1/2}}\, d\tau$$

$$= \frac{\ln 2}{\omega^{1/2}} \Gamma(\tfrac{1}{2}) = \frac{\sqrt{\pi}\ln 2}{\omega^{1/2}}$$

(b)
$$I(\omega) = \int_{-\infty}^\infty e^{-\omega x^2} \ln(1+x^2)\, dx \sim \frac{\sqrt{\pi}}{2\omega^{3/2}} \tag{46}$$

As in the preceding case,

$$I(\omega) \sim \int_{-\infty}^{\infty} e^{-\omega x^2}\left(x^2 - \frac{x^4}{2} + \cdots\right) dx \tag{47}$$

Substituting (41) into (47), we have

$$I(\omega) \sim \int_{-\infty}^{\infty} x^2 e^{-\omega x^2}\, dx = 2\int_0^{\infty} x^2 e^{-\omega x^2}\, dx = 2\int_0^{\infty} \frac{\tau}{\omega} e^{-\tau} \frac{d\tau}{2\omega x}$$

$$= \frac{1}{\omega^{3/2}} \int_0^{\infty} \tau^{1/2} e^{-\tau}\, d\tau = \frac{\Gamma(\frac{3}{2})}{\omega^{3/2}} = \frac{\sqrt{\pi}}{2\omega^{3/2}}$$

3.19. Show that as $x \to \infty$

(a)

$$I(x) = \int_1^2 \exp\left[-x\left(t + \frac{1}{t}\right)\right] dt \sim \frac{\sqrt{\pi}}{2\sqrt{x}} e^{-2x} \tag{48}$$

Here

$$h(t) = -\left(t + \frac{1}{t}\right)$$

$$h'(t) = -1 + \frac{1}{t^2}$$

Thus $h'(t) = 0$ when $t = \pm 1$, and the maximum of $h(t)$ occurs at $t = 1$ in the interval of interest. We expand $h(t)$ in the neighborhood of $t = 1$ as

$$h(t) = -2 - (t-1)^2 + \cdots$$

Applying Laplace's method gives

$$I(x) \sim \int_1^{\infty} e^{-2x} e^{-x(t-1)^2}\, dt \tag{49}$$

We let

$$\tau = x(t-1)^2 \quad \text{so that } d\tau = 2x(t-1)\, dt \tag{50}$$

and obtain from (49) that

$$I(x) \sim e^{-2x} \int_0^{\infty} \frac{e^{-\tau}\, d\tau}{2x(t-1)} = \frac{e^{-2x}}{2x^{1/2}} \int_0^{\infty} \tau^{-1/2} e^{-\tau}\, d\tau = \frac{\Gamma(\frac{1}{2})\, e^{-2x}}{2x^{1/2}}$$

$$= \frac{\sqrt{\pi}\, e^{-2x}}{2x^{1/2}}$$

(b)

$$I(x) = \int_1^2 \exp\left[-x\left(t + \frac{1}{t}\right)\right] \ln(1+t)\, dt \sim \frac{\sqrt{\pi}\ln 2}{2\sqrt{x}} e^{-2x} \tag{51}$$

As in the preceding case, we use Laplace's method and obtain

$$I(x) \sim \int_1^\infty e^{-2x} e^{-x(t-1)^2} \ln[2+(t-1)]\, dt$$

or

$$I(x) \sim e^{-2x} \int_1^\infty e^{-x(t-1)^2} [\ln 2 + \tfrac{1}{2}(t-1) + \cdots]\, dt \tag{52}$$

Putting (50) in (52), we obtain

$$I(x) \sim \frac{e^{-2x} \ln 2}{2x^{1/2}} \int_0^\infty \tau^{-1/2} e^{-\tau}\, d\tau = \frac{\Gamma(\frac{1}{2}) e^{-2x} \ln 2}{2\sqrt{x}} = \frac{\sqrt{\pi}\, e^{-2x} \ln 2}{2\sqrt{x}}$$

(c)

$$I(x) = \int_1^2 \exp\left[-x\left(t+\frac{1}{t}\right)\right] \ln t\, dt \sim \frac{e^{-2x}}{2x} \tag{53}$$

As in the preceding two cases, applying Laplace's method gives

$$I(x) \sim \int_0^\infty e^{-2x} e^{-x(t-1)^2} \ln[1+(t-1)]\, dt$$

Hence

$$I(x) \sim e^{-2x} \int_0^\infty e^{-x(t-1)^2} (t-1)\, dt \tag{54}$$

Substituting (50) into (54) gives

$$I(x) \sim \frac{e^{-2x}}{2x} \int_0^\infty e^{-\tau}\, d\tau = \frac{e^{-2x}}{2x}$$

(d)

$$I(x) = \int_1^2 \frac{\exp[-x(t+1/t)]}{\sqrt{t^2-1}}\, dt \sim \frac{\Gamma(\frac{1}{4})}{2\sqrt{2}\, x^{1/4}} e^{-2x} \tag{55}$$

As in the preceding cases, applying Laplace's method gives

$$I(x) \sim \int_0^\infty e^{-2x} e^{-x(t-1)^2} (t-1)^{-1/2} [(t-1)+2]^{-1/2}\, dt$$

Hence

$$I(x) \sim \frac{e^{-2x}}{\sqrt{2}} \int_0^\infty e^{-x(t-1)^2} (t-1)^{-1/2}\, dt \tag{56}$$

Substituting (50) into (56), we have

$$I(x) \sim \frac{e^{-2x}}{2\sqrt{2}\, x^{1/4}} \int_0^\infty \tau^{-3/4} e^{-\tau}\, d\tau = \frac{e^{-2x} \Gamma(\frac{1}{4})}{2\sqrt{2}\, x^{1/4}}$$

3.20. Show that as $x \to \infty$

(a)
$$I(x) = \int_0^1 \frac{e^{-xt^n}}{1+t}\,dt \sim \frac{\Gamma(1/n)}{nx^{1/n}} \tag{57}$$

The maximum of the exponent occurs at $t = 0$. Hence applying Laplace's method gives

$$I(x) \sim \int_0^\infty e^{-xt^n}(1 - t + \cdots)\,dt \tag{58}$$

We put

$$xt^n = \tau \quad \text{so that } nxt^{n-1}\,dt = d\tau \tag{59}$$

and obtain from (58) that

$$I(x) \sim \int_0^\infty \frac{e^{-\tau}\,d\tau}{nxt^{n-1}} = \frac{1}{nx^{1/n}} \int_0^\infty e^{-\tau}\tau^{1/n-1}\,d\tau = \frac{\Gamma(1/n)}{nx^{1/n}}$$

(b)
$$I(x) = \int_0^1 e^{-xt^n} \ln(1+t)\,dt \sim \frac{\Gamma(2/n)}{nx^{1/n}} \tag{60}$$

Using Laplace's method, we have

$$I(x) \sim \int_0^\infty e^{-xt^n} t\,dt \tag{61}$$

Substituting (59) into (61) yields

$$I(x) \sim \frac{1}{nx^{2/n}} \int_0^\infty e^{-\tau}\tau^{2/n-1}\,d\tau = \frac{\Gamma(2/n)}{nx^{2/n}}$$

(c)
$$I(x) = \int_0^1 \frac{e^{-xt^n}}{\sqrt{t}}\,dt \sim \frac{\Gamma(1/2n)}{nx^{1/2n}} \tag{62}$$

Using Laplace's method, we obtain

$$I(x) \sim \int_0^\infty t^{-1/2} e^{-xt^n}\,dt \tag{63}$$

Substituting (59) into (63) gives

$$I(x) \sim \frac{1}{nx^{1/2n}} \int_0^\infty e^{-\tau}\tau^{1/2n-1}\,d\tau = \frac{\Gamma(1/2n)}{nx^{1/2n}}$$

3.21. Show that as $\alpha \to \infty$

$$Ai(-\alpha) = \frac{1}{\pi} \int_0^\infty \cos\left(\tfrac{1}{3}t^3 - \alpha t\right)\,dt \sim \frac{1}{\sqrt{\pi}\,\alpha^{1/4}} \sin\left(\tfrac{2}{3}\alpha^{3/2} + \tfrac{1}{4}\pi\right) \tag{64}$$

We note that

$$Ai(-\alpha) = \frac{1}{\pi}\,\text{Real}\int_0^{\infty} e^{i(t^3/3 - \alpha t)}\,dt$$

We let $t = \alpha^{1/2}\tau$ and obtain

$$Ai(-\alpha) = \frac{\alpha^{1/2}}{\pi}\,\text{Real}\int_0^{\infty} e^{i\alpha^{3/2}(\tau^3/3 - \tau)}\,d\tau \tag{65}$$

Here

$$h(\tau) = \tfrac{1}{3}\tau^3 - \tau$$

$$h'(\tau) = \tau^2 - 1$$

Hence $h'(\tau) = 0$ when $\tau = \pm 1$ and there is a stationary point at $\tau = 1$ in the interval of interest. We let $\tau = 1 + x$ and obtain

$$h(\tau) = h(1 + x) = -\tfrac{2}{3} + x^2 + \cdots$$

Applying the method of stationary phase to (65) gives

$$Ai(-\alpha) \sim \frac{\alpha^{1/2}}{\pi}\,\text{Real}\int_{-\infty}^{\infty} e^{-2i\alpha^{3/2}/3} e^{i\alpha^{3/2}x^2}\,dx$$

Using Cauchy's theorem, we rotate the contour of integration by $\frac{1}{4}\pi$ and obtain

$$Ai(-\alpha) \sim \frac{\alpha^{1/2}}{\pi}\,\text{Real}\, e^{-2i\alpha^{3/2}/3}\int_{-\infty e^{i\pi/4}}^{\infty e^{i\pi/4}} e^{i\alpha^{3/2}z^2}\,dz \tag{66}$$

To evaluate the integral in (66), we let

$$z = r\alpha^{-3/4}e^{i\pi/4} \quad \text{so that } \alpha^{3/2}z^2 = r^2 e^{i\pi/2} = ir^2$$

Then (66) becomes

$$Ai(-\alpha) \sim \frac{\alpha^{1/2}}{\alpha^{3/4}\pi}\,\text{Real}\,\exp\left[-\tfrac{2}{3}i\alpha^{2/3} + \tfrac{1}{4}i\pi\right]\int_{-\infty}^{\infty} e^{-r^2}\,dr$$

$$= \frac{1}{\sqrt{\pi}\,\alpha^{1/4}}\cos\left(\tfrac{2}{3}\alpha^{3/2} - \tfrac{1}{4}\pi\right) = \frac{1}{\sqrt{\pi}\,\alpha^{1/4}}\sin\left(\tfrac{2}{3}\alpha^{3/2} + \tfrac{1}{4}\pi\right)$$

3.22. Show that as $\alpha \to \infty$

$$I(\alpha) = \int_0^{\infty} e^{i\alpha(t^3/3 + t)}\,dt \sim \frac{i}{\alpha} \tag{67}$$

Since $h(t) = \frac{1}{3}t^3 + t$, $h'(t) = t^2 + 1$. Hence there are no stationary points and integration by parts is applicable. Thus we let

$$dv = i\alpha(t^2+1)\, e^{i\alpha(t^3/3+t)}\, dt, \qquad u = \frac{1}{i\alpha(t^2+1)}$$

Hence

$$v = e^{i\alpha(t^3/3+t)} \quad \text{and} \quad du = -\frac{2t\,dt}{i\alpha(t^2+1)^2}$$

Then

$$I(\alpha) = \left.\frac{e^{i\alpha(t^3/3+t)}}{i\alpha(t^2+1)}\right|_0^\infty + \int_0^\infty \frac{2te^{i\alpha(t^3/3+t)}}{i\alpha(t^2+1)^2}\, dt$$

Consequently,

$$I(\alpha) \sim -\frac{1}{i\alpha} = \frac{i}{\alpha}$$

3.23. Show that as $x \to \infty$

$$J_0(x) = \frac{1}{\pi}\int_0^\pi \cos(x\cos\theta)\, d\theta \sim \sqrt{\frac{2}{\pi x}}\cos(x - \tfrac{1}{4}\pi) \tag{68}$$

We note that

$$J_0(x) = \frac{1}{\pi}\text{Real}\int_0^\pi e^{ix\cos\theta}\, d\theta$$

Here

$$h(\theta) = \cos\theta, \qquad h'(\theta) = -\sin\theta$$

Hence there are two stationary points, one at $\theta = 0$ and the other at $\theta = \pi$. Near these points, we write

$$h(\theta) = 1 - \tfrac{1}{2}\theta^2 + \cdots$$

$$h(\theta) = -1 + \tfrac{1}{2}(\pi - \theta)^2 + \cdots$$

Applying the method of stationary phase, we have

$$J_0(x) \sim \frac{1}{\pi}\text{Real}\int_0^\delta e^{ix - ix\theta^2/2}\, d\theta + \frac{1}{\pi}\text{Real}\int_{\pi-\delta}^\pi e^{-ix+ix(\pi-\theta)^2/2}\, d\theta \tag{69}$$

where δ is a small positive number. We let $\tau = \pi - \theta$ in the second integral in (69) and obtain

$$J_0(x) \sim \frac{1}{\pi}\text{Real}\int_0^\delta e^{ix-ix\theta^2/2}\,d\theta - \frac{1}{\pi}\text{Real}\int_\delta^0 e^{-ix+ix\tau^2/2}\,d\tau$$

$$\sim \frac{1}{\pi}\text{Real}\int_0^\infty e^{ix-ix\theta^2/2}\,d\theta + \frac{1}{\pi}\text{Real}\int_0^\infty e^{-ix+ix\tau^2/2}\,d\tau$$

We rotate the contour of integration in the first integral by $-\frac{1}{4}\pi$ and in the second integral by $\frac{1}{4}\pi$ and obtain

$$J_0(x) \sim \frac{1}{\pi}\text{Real}\int_0^{\infty e^{-i\pi/4}} e^{ix-ix\theta^2/2}\,d\theta + \frac{1}{\pi}\text{Real}\int_0^{\infty e^{i\pi/4}} e^{-ix+ix\tau^2/2}\,d\tau \tag{70}$$

To evaluate the integrals in (70), we change the variable in the first integral to

$$x^{1/2}\theta = \sqrt{2}\,re^{-i\pi/4} \quad \text{so that } x\theta^2 = 2r^2e^{-i\pi/2} = -2ir^2$$

and change the variable in the second integral to

$$x^{1/2}\tau = \sqrt{2}\,ue^{i\pi/4} \quad \text{so that } x\tau^2 = 2u^2e^{i\pi/2} = 2iu^2$$

Then we rewrite (70) as

$$J_0(x) \sim \frac{\sqrt{2}}{\pi\sqrt{x}}\text{Real}[\,e^{ix-i\pi/4} + e^{-ix+1/4i\pi}]\int_0^\infty e^{-r^2}\,dr$$

$$= \sqrt{\frac{2}{\pi x}}\cos(x - \tfrac{1}{4}\pi)$$

3.24. Show that as $\alpha \to \infty$

(a)
$$I(\alpha) = \int_0^1 e^{i\alpha t^3}\,dt \sim \frac{\Gamma(\frac{1}{3})\,e^{i\pi/6}}{3\alpha^{1/3}} \tag{71}$$

In this case, there is a stationary point at $t = 0$, and the major contribution arises from its neighborhood. Using the method of stationary phase, we have

$$I(\alpha) \sim \int_0^\infty e^{i\alpha t^3}\,dt$$

We rotate the contour of integration by $\frac{1}{6}\pi$ and obtain

$$I(\alpha) \sim \int_0^{\infty e^{i\pi/6}} e^{i\alpha t^3}\,dt \tag{72}$$

We let

$$t = re^{i\pi/6} \quad \text{so that } t^3 = r^3e^{i\pi/2} = ir^3 \tag{73}$$

and rewrite (72) as

$$I(\alpha) \sim e^{i\pi/6} \int_0^\infty e^{-\alpha r^3}\, dr \tag{74}$$

To evaluate the integral in (74), we let

$$\alpha r^3 = \tau \quad \text{so that } r = \left(\frac{\tau}{\alpha}\right)^{1/3} \tag{75}$$

and obtain from (74) that

$$I(\alpha) \sim \frac{e^{i\pi/6}}{3\alpha^{1/3}} \int_0^\infty \tau^{-2/3} e^{-\tau}\, d\tau = \frac{e^{i\pi/6}\Gamma(\frac{1}{3})}{3\alpha^{1/3}}$$

(b)

$$I(\alpha) = \int_0^1 \frac{e^{i\alpha t^3}}{\sqrt{t}}\, dt \sim \frac{\Gamma(\frac{1}{6})\, e^{i\pi/12}}{3\alpha^{1/6}} \tag{76}$$

As in the preceding case, we use the method of stationary phase and write

$$I(\alpha) \sim \int_0^\infty \frac{e^{i\alpha t^3}}{\sqrt{t}}\, dt$$

We rotate the contour of integration by $\frac{1}{6}\pi$ and obtain

$$I(\alpha) \sim \int_0^{\infty e^{i\pi/6}} \frac{e^{i\alpha t^3}}{\sqrt{t}}\, dt \tag{77}$$

Substituting (73) into (77) gives

$$I(\alpha) \sim e^{i\pi/12} \int_0^\infty \frac{e^{-r^3}}{\sqrt{r}}\, dr$$

which upon using (75) becomes

$$I(\alpha) \sim \frac{e^{i\pi/12}}{3\alpha^{1/6}} \int_0^\infty \tau^{-5/6} e^{-\tau} = \frac{e^{i\pi/12}\Gamma(\frac{1}{6})}{3\alpha^{1/6}}$$

(c)

$$I(\alpha) = \int_0^1 e^{i\alpha t^3} \ln(1+t)\, dt \sim \frac{\Gamma(\frac{2}{3})\, e^{i\pi/3}}{3\alpha^{2/3}} \tag{78}$$

As in the preceding two cases, we use the method of stationary phase and obtain

$$I(\alpha) \sim \int_0^\infty e^{i\alpha t^3}\left(t - \frac{t^2}{2} + \cdots\right) dt$$

We rotate the contour of integration by $\frac{1}{6}\pi$, then use (73), and obtain

$$I(\alpha) \sim e^{i\pi/3} \int_0^\infty r e^{-\alpha r^3} \, dr \tag{79}$$

Substituting (75) into (79) gives

$$I(\alpha) \sim \frac{e^{i\pi/3}}{3\alpha^{2/3}} \int_0^\infty \tau^{-1/3} e^{-\tau} \, d\tau = \frac{\Gamma(\frac{2}{3})\, e^{i\pi/3}}{3\alpha^{2/3}}$$

(d)

$$\int_0^1 e^{i\alpha t^3} \ln(2+t) \, dt \sim \frac{\Gamma(\frac{1}{3})\, e^{i\pi/6} \ln 2}{3\alpha^{1/3}} \tag{80}$$

Using the method of stationary phase, we have

$$I(\alpha) \sim \int_0^\infty e^{i\alpha t^3} (\ln 2 + \tfrac{1}{2}t + \cdots) \, dt \sim \ln 2 \int_0^\infty e^{i\alpha t^3} \, dt$$

Evaluating the integral as in part (a), we obtain (80).

3.25. Show that as $x \to \infty$

(a)

$$K_0(x) = \int_1^\infty \frac{e^{-xt}}{\sqrt{t^2-1}} \, dt \sim \sqrt{\frac{\pi}{2x}}\, e^{-x} \tag{81}$$

The maximum of the exponent occurs at $t=1$. We let $t=1+\tau$ in (81) and obtain

$$K_0(x) = \int_0^\infty \frac{e^{-x-x\tau}}{\sqrt{2\tau+\tau^2}} \, d\tau$$

Applying Watson's lemma gives

$$K_0(x) \sim \frac{e^{-x}}{\sqrt{2}} \int_0^\infty \tau^{-1/2} e^{-x\tau} \, d\tau = \frac{e^{-x}}{\sqrt{2x}} \int_0^\infty u^{-1/2} e^{-u} \, du = \frac{\Gamma(\frac{1}{2})\, e^{-x}}{\sqrt{2x}}$$

$$= \sqrt{\frac{\pi}{2x}}\, e^{-x}$$

(b)

$$H_0^{(1)}(x) = -\frac{2}{\pi} \int_1^\infty \frac{e^{ixt}}{\sqrt{1-t^2}} \, dt \sim \sqrt{\frac{2}{\pi x}}\, e^{i(x-\pi/4)} \tag{82}$$

Since there are no stationary points, we use the method of steepest descent and deform the contour of integration into a constant-phase contour. Here $h(t) = it$ so that the phase χ at $t=1$ is 1. Thus we deform the contour into one that consists of a line segment that runs up from 1 to $1+i\infty$ and a quarter circle at ∞ joining the ends ∞ and $+i\infty$. The contribution from the circle vanishes because the integrand vanishes there. Therefore,

$$H_0^{(1)}(x) = -\frac{2}{\pi}\int_1^{1+i\infty} \frac{e^{ixt}}{\sqrt{1-t^2}}\, dt$$

Then we let

$$t = 1 + \frac{ir}{x}$$

and obtain

$$H_0^{(1)}(x) \sim -\frac{2i}{\pi\sqrt{x}}\int_0^{\infty} \frac{e^{ix}e^{-r}}{\sqrt{-2ir + r^2/x}}\, dr$$

Applying Watson's lemma yields

$$H_0^{(1)}(x) \sim \frac{\sqrt{2}}{\pi\sqrt{x}} e^{i(x-\pi/4)}\int_0^{\infty} r^{-1/2}e^{-r}\, dr = \sqrt{\frac{2}{\pi x}}\, e^{i(x-\pi/4)}$$

(c)
$$J_0(x) = \frac{2}{\pi}\int_1^{\infty} \frac{\sin xt}{\sqrt{t^2-1}}\, dt \sim \sqrt{\frac{2}{\pi x}} \cos\left(x - \tfrac{1}{4}\pi\right) \tag{83}$$

We note that

$$J_0(x) = \frac{2}{\pi} \operatorname{Im} \int_1^{\infty} \frac{e^{ixt}}{\sqrt{t^2-1}}\, dt$$

To apply the method of steepest descent, we deform the contour of integration as in the preceding case and obtain

$$J_0(x) \sim \frac{2}{\pi} \operatorname{Im} \int_1^{1+i\infty} \frac{e^{ixt}}{\sqrt{t^2-1}}\, dt$$

Then we let

$$t = 1 + \frac{i\tau}{x}$$

and obtain

$$J_0(x) \sim \frac{2}{\pi} \operatorname{Im} \int_0^{\infty} \frac{ie^{ix-\tau}}{x\sqrt{2i\tau/x - \tau^2/x^2}}\, d\tau \tag{84}$$

Applying Watson's lemma to (84) gives

$$J_0(x) \sim \frac{\sqrt{2}}{\sqrt{x\pi}} \operatorname{Im} e^{i(x+\pi/4)} \int_0^\infty \tau^{-1/2} e^{-\tau}\, d\tau = \sqrt{\frac{2}{\pi x}} \operatorname{Im} e^{i(x+\pi/4)}$$

$$= \sqrt{\frac{2}{\pi x}} \sin(x + \tfrac{1}{4}\pi) = \sqrt{\frac{2}{\pi x}} \cos(x - \tfrac{1}{4}\pi)$$

(d)

$$Y_0(x) = -\frac{2}{\pi} \int_1^\infty \frac{\cos xt}{\sqrt{t^2-1}}\, dt \sim \sqrt{\frac{2}{\pi x}} \sin(x - \tfrac{1}{4}\pi) \tag{85}$$

We note that

$$Y_0(x) = -\frac{2}{\pi} \operatorname{Real} \int_1^\infty \frac{e^{ixt}}{\sqrt{t^2-1}}\, dt$$

But

$$\frac{2}{\pi} \int_1^\infty \frac{e^{ixt}}{\sqrt{t^2-1}}\, dt \sim \sqrt{\frac{2}{\pi x}}\, e^{i(x+\pi/4)}$$

according to the preceding case. Hence

$$Y_0(x) \sim -\sqrt{\frac{2}{\pi x}} \operatorname{Real} e^{i(x+\pi/4)} = -\sqrt{\frac{2}{\pi x}} \cos(x + \tfrac{1}{4}\pi)$$

$$= \sqrt{\frac{2}{\pi x}} \sin(x - \tfrac{1}{4}\pi)$$

(e)

$$I_n(x) = \frac{1}{\pi} \int_0^\pi e^{x \cos\theta} \cos n\theta\, d\theta \sim \frac{e^x}{\sqrt{2\pi x}} \tag{86}$$

In this case, $h(\theta) = \cos\theta$ and its maximum occurs at $\theta = 0$. We expand $\cos\theta$ in a Taylor series around $\theta = 0$ and obtain

$$\cos\theta = 1 - \tfrac{1}{2}\theta^2 + \cdots$$

Applying Laplace's method, we have

$$I_n(x) \sim \frac{1}{\pi} \int_0^\infty e^{x - x\theta^2/2} (1 - \tfrac{1}{2} n^2 \theta^2 + \cdots)\, d\theta \sim \frac{e^x}{\pi} \int_0^\infty e^{-x\theta^2/2}\, d\theta \tag{87}$$

We let

$$\tfrac{1}{2} x\theta^2 = \tau^2 \quad \text{or} \quad \theta = \sqrt{\frac{2}{x}}\, \tau$$

in (87) and obtain

$$I_n(x) \sim \sqrt{\frac{2}{x}} \frac{e^x}{\pi} \int_0^\infty e^{-\tau^2} d\tau = \frac{e^x}{\sqrt{2\pi x}}$$

3.26. Show that as $z \to \infty$

$$\Gamma(z) = \int_0^\infty t^{z-1} e^{-t} \, dt \sim \sqrt{\frac{2\pi}{z}} z^z e^{-z} \tag{88}$$

Since z is large, we need to express t^z as an exponential. We put

$$t^z = e^{z \ln t}$$

Then

$$\Gamma(z) = \int_0^\infty \frac{1}{t} e^{z \ln t - t} \, dt \tag{89}$$

In this case, the maximum of the exponent is a function of z and we need a transformation to make it independent of z. We put

$$t = z\tau$$

in (89) and obtain

$$\Gamma(z) = \int_0^\infty \frac{1}{\tau} e^{z \ln z + z(\ln \tau - \tau)} \, d\tau = z^z \int_0^\infty \frac{1}{\tau} e^{z(\ln \tau - \tau)} \, d\tau$$

Here

$$h(\tau) = \ln \tau - \tau \quad \text{and} \quad h'(\tau) = \frac{1}{\tau} - 1$$

The maximum of $h(\tau)$ occurs at $\tau = 1$ and we expand it in a Taylor series around $\tau = 1$ as

$$h(\tau) = (-1) - \tfrac{1}{2}(\tau - 1)^2 + \cdots$$

Applying Laplace's method, we have

$$\Gamma(z) \sim z^z \int_{1-\delta}^{1+\delta} e^{-z} e^{-z(\tau-1)^2/2} \, d\tau$$

or

$$\Gamma(z) \sim z^z e^{-z} \int_{-\infty}^\infty e^{-1/2zu^2} \, du = \sqrt{\frac{2}{z}} z^z e^{-z} \int_{-\infty}^\infty e^{-r^2} \, dr = \sqrt{\frac{2\pi}{z}} z^z e^{-z}$$

3.27. Show that as $x \to \infty$

$$I(x) = \int_{-1}^{1} e^{ixt}(1-t^2)^{n-1/2}\,dt \sim \left(\frac{2}{x}\right)^{n+1/2} \Gamma\left(n+\tfrac{1}{2}\right)\cos\left[x - \tfrac{1}{2}\pi\left(n+\tfrac{1}{2}\right)\right] \quad (90)$$

Here $h(t) = it$ and there are no stationary points. To apply the method of steepest descent, we deform the contour of integration as in Figure 3.1.

Then

$$I(x) = \int_{-1}^{-1+i\infty} e^{ixt}(1-t^2)^{n-1/2}\,dt + \int_{1+i\infty}^{1} e^{ixt}(1-t^2)^{n-1/2}\,dt \quad (91)$$

Making the substitution $t = -1 + i\tau/x$ in the first integral and $t = 1 + i\tau/x$ in the second integral, we rewrite (91) as

$$I(x) = \frac{i}{x}\int_0^{\infty} e^{-ix}e^{-\tau}\left[\frac{2i\tau}{x} + \frac{\tau^2}{x^2}\right]^{n-1/2} d\tau + \frac{i}{x}\int_{\infty}^{0} e^{ix}e^{-\tau}\left[\frac{-2i\tau}{x} + \frac{\tau^2}{x^2}\right]^{n-1/2} d\tau$$

(92)

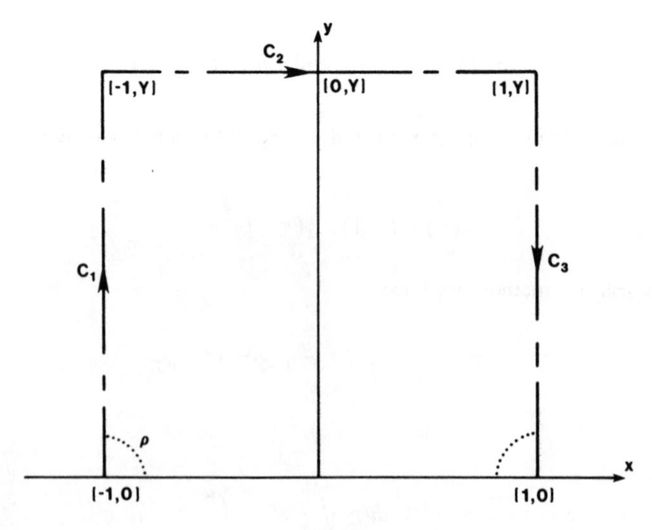

Figure 3.1

Applying Watson's lemma to (92) yields

$$I(x) \sim \tfrac{1}{2}\left(\frac{2}{x}\right)^{n+1/2}\{ie^{-i[x-(n-1/2)/2\pi]} - ie^{i[x-(n-1/2)/2\pi]}\}$$

$$\times \int_0^\infty \tau^{n-1/2}e^{-\tau}\,d\tau = \left(\frac{2}{x}\right)^{n+1/2}\Gamma(n+\tfrac{1}{2})\sin[x-\tfrac{1}{2}(n-\tfrac{1}{2})\pi]$$

$$= \left(\frac{2}{x}\right)^{n+1/2}\Gamma(n+\tfrac{1}{2})\cos[x-\tfrac{1}{2}(n+\tfrac{1}{2})\pi]$$

3.28. Show that as $\omega \to \infty$

$$I(\omega) = \int_0^1 \ln t e^{i\omega t}\,dt \sim \frac{-i\ln\omega}{\omega} - \frac{i\gamma + \frac{1}{2}\pi}{\omega} \tag{93}$$

To apply the method of steepest descent, we deform the contour of integration as in Figure 3.1, but with C_1 running up from 0 to $0+iY$. Then

$$I(\omega) = \int_0^{0+i\infty} \ln t e^{i\omega t}\,dt + \int_{1+i\infty}^1 \ln t e^{i\omega t}\,dt \tag{94}$$

Making the substitution $t = i\tau/\omega$ in the first integral and $t = 1 + i\tau/\omega$ in the second integral, we rewrite (94) as

$$I(\omega) = \frac{i}{\omega}\int_0^\infty e^{-\tau}\ln\left(\frac{i\tau}{\omega}\right)d\tau + \frac{i}{\omega}e^{i\omega}\int_\infty^0 \ln\left(1+\frac{i\tau}{\omega}\right)e^{-\tau}\,d\tau$$

or

$$I(\omega) = \frac{i}{\omega}(\tfrac{1}{2}i\pi - \ln\omega)\int_0^\infty e^{-\tau}\,d\tau + \frac{i}{\omega}\int_0^\infty e^{-\tau}\ln\tau\,d\tau$$

$$-\frac{i}{\omega}e^{i\omega}\int_0^\infty e^{-\tau}\left(\frac{i\tau}{\omega} + \frac{\tau^2}{2\omega^2} + \cdots\right)d\tau \tag{95}$$

Applying Watson's lemma to (95) gives

$$I(\omega) \sim \frac{-i\ln\omega}{\omega} - \frac{\frac{1}{2}\pi + i\gamma}{\omega}$$

where

$$\gamma = -\int_0^\infty e^{-\tau}\ln\tau\,d\tau$$

is Euler's constant.

3.29. Show that as $\alpha \to \infty$

$$I(\alpha) = \int_0^\infty e^{\alpha(z - iz^3/3)}\, dz \sim \frac{\sqrt{\pi}\exp\left[-\frac{1}{8}i\pi + \frac{2}{3}\alpha e^{-i\pi/4}\right]}{\sqrt{\alpha}} \tag{96}$$

Here

$$h(z) = z - \tfrac{1}{3}iz^3, \qquad h' = 1 - iz^2$$

Hence there are two saddle points at

$$z^2 = -i \quad \text{or} \quad z = \pm e^{-i\pi/4}$$

To apply the saddle-point method, we deform the contour of integration to pass through the saddle point $z = \exp(-\frac{1}{4}i\pi)$ and consider a short segment $\hat{C}$ along it. Thus

$$I(\alpha) \sim \int_{\hat{C}} e^{\alpha(z - iz^3/3)}\, dz \tag{97}$$

Near the saddle point,

$$z - \tfrac{1}{3}iz^3 = \tfrac{2}{3}e^{-i\pi/4} - e^{i\pi/4}\left(z - e^{-i\pi/4}\right)^2 + \cdots$$

Hence we rewrite (97) as

$$I(\alpha) \sim \exp\left(\tfrac{2}{3}\alpha e^{-i\pi/4}\right)\int_{\hat{C}} \exp\left[-\alpha e^{i\pi/4}\left(z - e^{-i\pi/4}\right)^2\right] dz \tag{98}$$

Then we let

$$e^{i\pi/4}\left(z - e^{-i\pi/4}\right)^2 = \frac{\tau^2}{\alpha} \tag{99}$$

so that

$$z - e^{-i\pi/4} = \frac{\tau}{\sqrt{\alpha}}e^{-i\pi/8} \tag{100}$$

Substituting (99) and (100) into (98) and replacing the limits of integration with $-\infty$ and $+\infty$ according to Watson's lemma, we obtain

$$I(\alpha) \sim \frac{1}{\sqrt{\alpha}}\exp\left[\tfrac{2}{3}\alpha e^{-i\pi/4} - \tfrac{1}{8}i\pi\right]\int_{-\infty}^{\infty} e^{-\tau^2}\, d\tau$$

$$= \frac{\sqrt{\pi}}{\sqrt{\alpha}}\exp\left[\tfrac{2}{3}\alpha e^{-i\pi/4} - \tfrac{1}{8}i\pi\right]$$

3.30. Consider the integral representation of the Legendre polynomial of order n

$$P_n(\mu) = \frac{1}{\pi}\int_0^{\pi}\left[\mu + \sqrt{\mu^2 - 1}\cos\theta\right]^n d\theta \tag{101}$$

Show that

$$P_n(\mu) \sim \frac{1}{\sqrt{2n\pi}}\frac{\left(\mu + \sqrt{\mu^2-1}\right)^{n+1/2}}{(\mu^2-1)^{1/4}} \quad \text{as } n \to \infty \tag{102}$$

To determine the asymptotic expansion of $P_n(\mu)$ for large n, we express the integrand in (101) as an exponential, that is,

$$P_n(\mu) = \frac{1}{\pi}\int_0^{\pi}\exp\left[n\ln\left(\mu + \sqrt{\mu^2-1}\cos\theta\right)\right] d\theta$$

In this case,

$$h(\theta) = \ln\left(\mu + \sqrt{\mu^2-1}\cos\theta\right), \qquad h'(\theta) = \frac{-\sqrt{\mu^2-1}\sin\theta}{\mu + \sqrt{\mu^2-1}\cos\theta} \tag{103}$$

It follows from (103) that $h(\theta)$ has an extremum at $\theta = 0$ and an extremum at $\theta = \pi$. The extremum at $\theta = 0$ is a maximum, whereas the other extremum is a minimum. Hence the major contribution to the integral arises from the neighborhood of $\theta = 0$. We expand $h(\theta)$ in a Taylor series around $\theta = 0$ as

$$h(\theta) = \ln\left(\mu + \sqrt{\mu^2-1}\right) - \frac{\sqrt{\mu^2-1}}{2\left(\mu + \sqrt{\mu^2-1}\right)}\theta^2 + \cdots$$

Using Laplace's method, we have

$$P_n(\mu) \sim \frac{1}{\pi}\exp\left[n\ln\left(\mu + \sqrt{\mu^2-1}\right)\right]\int_0^{\infty}\exp\left[\frac{-n\sqrt{\mu^2-1}}{2\left(\mu + \sqrt{\mu^2-1}\right)}\theta^2\right] d\theta \tag{104}$$

To evaluate the integral in (104), we let

$$\frac{n\sqrt{\mu^2-1}}{2\left(\mu + \sqrt{\mu^2-1}\right)}\theta^2 = \tau^2$$

and obtain

$$P_n(\mu) \sim \frac{\sqrt{2}\left(\mu + \sqrt{\mu^2-1}\right)^{n+1/2}}{\pi\sqrt{n}\,(\mu^2-1)^{1/4}}\int_0^{\infty} e^{-\tau^2}\, d\tau$$

Using

$$\int_0^\infty e^{-\tau^2}\,d\tau = \tfrac{1}{2}\sqrt{\pi}$$

we obtain (102).

3.31. Consider the Airy function $Bi(z)$ of the second kind

$$Bi(z) = \frac{1}{\pi}\int_0^\infty \left[\exp\left(-\tfrac{1}{3}t^3 + zt\right) + \sin\left(\tfrac{1}{3}t^3 + zt\right)\right] dt \tag{105}$$

Determine the leading term in the asymptotic expansion of $Bi(z)$ as $|z| \to \infty$.

First, we express $Bi(z)$ as the sum of the following integrals

$$I_1(z) = \frac{1}{\pi}\int_0^\infty \exp\left(-\tfrac{1}{3}t^3 + zt\right) dt \tag{106}$$

$$I_2(z) = \frac{1}{\pi}\int_0^\infty \sin\left(\tfrac{1}{3}t^3 + zt\right) dt \tag{107}$$

(a) For large positive z, I_1 is a Laplace integral whose exponent has a stationary point. To put it in the standard form for the application of Laplace's method, we let

$$t = z^{1/2}\tau$$

and obtain

$$I_1(z) = \frac{\sqrt{z}}{\pi}\int_0^\infty \exp\left[z^{3/2}\left(\tau - \tfrac{1}{3}\tau^3\right)\right] d\tau$$

In this case,

$$h(\tau) = \tau - \tfrac{1}{3}\tau^3, \qquad h'(\tau) = 1 - \tau^2$$

Hence $h(\tau)$ has its maximum at $\tau = 1$ in $[0, \infty]$. We expand $h(\tau)$ near $\tau = 1$ in a Taylor series and the result is

$$h(\tau) = \tfrac{2}{3} - (\tau - 1)^2 + \cdots$$

Using Laplace's method, we have

$$I_1(z) \sim \frac{\sqrt{z}}{\pi} e^{2/3 z^{3/2}} \int_{-\infty}^\infty e^{-z^{3/2}(\tau-1)^2}\,d\tau$$

$$= \frac{1}{\pi} z^{-1/4} e^{2/3 z^{3/2}} \int_{-\infty}^\infty e^{-u^2}\,du = \frac{e^{2/3 z^{3/2}}}{\sqrt{\pi}\,z^{1/4}} \tag{108}$$

When z is positive, the argument of the sine function does not have a stationary point. Thus we use integration by parts and put

$$(z+t^2)\sin\left(\tfrac{1}{3}t^3+zt\right)dt=dv, \qquad u=\frac{1}{z+t^2}$$

so that

$$v=-\cos\left(\tfrac{1}{3}t^3+zt\right), \qquad du=-\frac{2t}{(z+t^2)^2}\,dt$$

Hence

$$I_2(z)=\frac{1}{\pi}\left[\frac{-\cos\left(\tfrac{1}{3}t^3+zt\right)}{z+t^2}\bigg|_0^\infty-\int_0^\infty\frac{2t\cos\left(\tfrac{1}{3}t^3+zt\right)}{(z+t^2)^2}\,dt\right]$$

and

$$I_2(z)\sim\frac{1}{\pi z} \tag{109}$$

Adding (108) and (109), we find that

$$Bi(z)\sim\frac{e^{2/3z^{3/2}}}{\sqrt{\pi}\,z^{1/4}} \tag{110}$$

(b) For large negative z, the exponent in (106) has its maximum at $t=0$ and it is not a relative maximum. Applying Watson's lemma yields

$$I_1(z)\sim\frac{1}{\pi}\int_0^\infty e^{-(-z)t}\left(1-\tfrac{1}{3}t^3+\cdots\right)dt$$

or

$$I_1(z)\sim\frac{1}{\pi}\int_0^\infty e^{-(-z)t}\,dt=-\frac{1}{\pi z} \tag{111}$$

For negative z, the argument of (107) has a stationary point. Hence the asymptotic expansion of (107) can be obtained using the method of stationary phase. To accomplish this, we rewrite it as

$$I_2(z)=\frac{1}{\pi}\operatorname{Im}\int_0^\infty\exp\left[i\left(\tfrac{1}{3}t^3-\alpha t\right)\right]dt \tag{112}$$

where $z=-\alpha$. To put (112) in the standard form for application of the method of stationary phase, we put

$$t=\alpha^{1/2}\tau$$

and obtain

$$I_2(-\alpha) = \frac{\sqrt{\alpha}}{\pi} \operatorname{Im} \int_0^\infty \exp\left[i\alpha^{3/2}\left(\tfrac{1}{3}\tau^3 - \tau\right)\right] d\tau$$

There is a stationary point at $\tau = 1$ in the interval of interest. Near this point, we have

$$\tfrac{1}{3}\tau^3 - \tau = -\tfrac{2}{3} + (\tau - 1)^2 + \cdots$$

Hence

$$I_2(-\alpha) \sim \frac{\sqrt{\alpha}}{\pi} \operatorname{Im} \int_{-\infty}^{\infty} \exp\left(-\tfrac{2}{3}i\alpha^{3/2}\right) \exp\left[i\alpha^{3/2}(\tau - 1)^2\right] d\tau \tag{113}$$

Rotating the contour of integration by $\frac{1}{4}\pi$ and then introducing the change of variable

$$(\tau - 1) = \alpha^{-3/4} u \exp\left(\tfrac{1}{4}i\pi\right)$$

we rewrite (113) as

$$I_2(-\alpha) \sim \frac{1}{\pi\alpha^{1/4}} \operatorname{Im}\left\{\exp\left[i\left(-\tfrac{2}{3}\alpha^{3/2} + \tfrac{1}{4}\pi\right)\right]\right\} \int_{-\infty}^{\infty} e^{-u^2} du$$

$$= -\frac{1}{\sqrt{\pi}\,\alpha^{1/4}} \sin\left(\tfrac{2}{3}\alpha^{3/2} - \tfrac{1}{4}\pi\right) = \frac{1}{\sqrt{\pi}\,\alpha^{1/4}} \cos\left(\tfrac{2}{3}\alpha^{3/2} + \tfrac{1}{4}\pi\right) \tag{114}$$

Adding (111) and (114), we find that

$$Bi(-z) \sim \frac{1}{\sqrt{\pi}\,\alpha^{1/4}} \cos\left(\tfrac{2}{3}\alpha^{3/2} + \tfrac{1}{4}\pi\right) \tag{115}$$

Supplementary Exercises

3.32. Show that

(a) $\displaystyle\int_0^\infty e^{-ax} x^{n-1}\, dx = \frac{\Gamma(n)}{a^n} \qquad (n > 0, a > 0)$

(b) $\displaystyle\int_a^\infty e^{2ax - x^2}\, dx = \tfrac{1}{2}\sqrt{\pi}\, e^{a^2}$

(c) $\displaystyle\int_0^\infty x^m e^{-x^n}\, dx = \frac{1}{n}\Gamma\left(\frac{m+1}{n}\right), \qquad (m > -1, n > 0)$

3.33. Show that as $\varepsilon \to 0$,

(a) $\displaystyle\int_0^1 (1 - \varepsilon x^3)^{3/2}\, dx \sim 1 - \tfrac{3}{8}\varepsilon + \tfrac{3}{56}\varepsilon^2$

(b) $\displaystyle\int_0^1 (1 - \varepsilon x^3)^{-1/2}\, dx \sim 1 + \tfrac{1}{8}\varepsilon + \tfrac{3}{56}\varepsilon^2$

3.34. Show that as $m \to 0$,

(a) $\displaystyle\int_0^{\pi/2} (1 - m\sin^2\theta)^{1/4}\, d\theta \sim \tfrac{1}{2}\pi\left(1 - \tfrac{1}{8}m - \tfrac{3}{256}m^2\right)$

(b) $\displaystyle\int_0^{\pi/2} (1 - m\sin^2\theta)^{-1/4}\, d\theta \sim \tfrac{1}{2}\pi\left(1 + \tfrac{1}{8}m + \tfrac{15}{256}m^2\right)$

3.35. Show that as $\alpha \to 0$

(a) $\int_0^\infty e^{i\alpha(t^2+t)}\,dt \sim \frac{i}{\alpha}$

(b) $\int_0^\infty e^{i(t^2+\alpha t)}\,dt \sim \frac{i}{\alpha}$

3.36. Show that as $x \to \infty$

(a) $\int_x^\infty \frac{dt}{t^2(\ln t)^2} \sim \frac{1}{x(\ln x)^2}$

(b) $\int_x^\infty \frac{dt}{t^2(\ln t)^n} \sim \frac{1}{x(\ln x)^n}$

3.37. Show that as $x \to \infty$

(a) $\int_2^x \frac{dt}{(\ln t)^2} \sim \frac{x}{(\ln x)^2}$

(b) $\int_2^x \frac{dt}{(\ln t)^n} \sim \frac{x}{(\ln x)^n}$

3.38. Show that as $x \to \infty$

(a) $\int_2^x \sqrt{t^5+t^2}\,dt \sim \frac{2}{7}x^{7/2} + x^{1/2}$

(b) $\int_x^\infty \frac{t^{3/4}\,dt}{1+t^2} \sim 4x^{-1/4} - \frac{4}{9}x^{-9/4}$

3.39. Show that as $x \to \infty$

(a) $\int_0^\infty \frac{e^{-xt}}{1+t} \sim \frac{1}{x} - \frac{1!}{x^2} + \frac{2!}{x^3} - \frac{3!}{x^4}$

(b) $\operatorname{erfc}(x) = \frac{2}{\sqrt{\pi}}\int_x^\infty e^{-t^2}\,dt \sim \frac{e^{-x^2}}{x\sqrt{\pi}}\left[1 - \frac{1}{2x^2} + \frac{1.3}{(2x^2)^2}\right]$

(c) $\int_0^\infty e^{-x\sinh t}\,dt \sim \frac{1}{x} - \frac{1}{x^3} + \frac{1.3^2}{x^5}$

3.40. Show that

$$\int_0^\infty e^{-xt}\,dt = \frac{1}{x}, \qquad \int_0^\infty e^{-\alpha x}\sin\beta x\,dx = \frac{\beta}{\alpha^2+\beta^2}$$

Then show that

$$\int_0^\infty e^{-ax}\frac{\sin bx}{x}\,dx = \int_0^\infty e^{-ax}\sin bx\,dx\int_0^\infty e^{-xt}\,dt$$

$$= \int_0^\infty dt\int_0^\infty e^{-(a+t)x}\sin bx\,dx = \int_0^\infty \frac{b}{(a+t)^2+b^2}\,dt$$

$$= \tfrac{1}{2}\pi - \tan^{-1}\frac{a}{b}$$

Take the limit as $a \to 0$ and obtain

$$\int_0^\infty \frac{\sin bx}{x}\,dx = \begin{cases} \frac{1}{2}\pi & \text{when } b>0 \\ -\frac{1}{2}\pi & \text{when } b<0 \\ 0 & \text{when } b=0 \end{cases}$$

3.41. Show that as $x \to \infty$

(a) $\displaystyle\int_x^\infty \frac{\sin t}{t}\,dt \sim \left(\frac{1}{x^2} - \frac{3!}{x^4} + \frac{5!}{x^6}\right)\sin x + \left(\frac{1}{x} - \frac{2!}{x^3} + \frac{4!}{x^5}\right)\cos x$

(b) $\displaystyle\int_0^x \frac{\sin t}{t}\,dt \sim \tfrac{1}{2}\pi - \left(\frac{1}{x^2} - \frac{3!}{x^4} + \frac{5!}{x^6}\right)\sin x - \left(\frac{1}{x} - \frac{2!}{x^3} + \frac{4!}{x^5}\right)\cos x$

3.42. Show that as $x \to \infty$

(a) $\displaystyle\int_x^\infty \frac{\cos(t-x)}{\sqrt{t}}\,dt \sim \frac{1}{\sqrt{x}}\left[\frac{1}{2x} - \frac{1\cdot3\cdot5}{(2x)^3} + \frac{1\cdot3\cdot5\cdot7\cdot9}{(2x)^5}\right]$

(b) $\displaystyle\int_x^\infty \frac{\sin(t-x)}{\sqrt{t}}\,dt \sim \frac{1}{\sqrt{x}}\left[1 - \frac{1\cdot3}{(2x)^2} + \frac{1\cdot3\cdot5\cdot7}{(2x)^4}\right]$

3.43. Show that as $\to \infty$

(a) $\displaystyle\int_x^\infty \frac{\cos t}{t^{1/4}}\,dt \sim \frac{1}{x^{1/4}}(f\cos x - g\sin x)$

(b) $\displaystyle\int_x^\infty \frac{\sin t}{t^{1/4}}\,dt \sim \frac{1}{x^{1/4}}(f\sin x + g\cos x)$

(c) $\displaystyle\int_x^\infty \frac{\cos(t-x)}{t^{1/4}}\,dt \sim \frac{f}{x^{1/4}}$

(d) $\displaystyle\int_x^\infty \frac{\sin(t-x)}{t^{1/4}}\,dt \sim \frac{g}{x^{1/4}}$

where

$$f \sim \frac{1}{4x} - \frac{1\cdot5\cdot9}{(4x)^3} + \frac{1\cdot5\cdot9\cdot13\cdot17}{(4x)^5}$$

$$g \sim 1 - \frac{1\cdot5}{(4x)^2} + \frac{1\cdot5\cdot9\cdot13}{(4x)^4}$$

3.44. Show that as $x \to \infty$

(a) $\displaystyle\int_x^\infty \frac{\cos t}{t^\lambda}\,dt \sim \frac{1}{x^\lambda}(f\cos x - g\sin x)$

(b) $\displaystyle\int_x^\infty \frac{\sin t}{t^\lambda}\,dt \sim \frac{1}{x^\lambda}(f\sin x + g\cos x)$

(c) $\displaystyle\int_x^\infty \frac{\cos(t-x)}{t^\lambda}\,dt \sim \frac{f}{x^\lambda}$

(d) $\displaystyle\int_x^\infty \frac{\sin(t-x)}{t^\lambda}\,dt \sim \frac{g}{x^\lambda}$

where $\lambda > 0$ and

$$f \sim \frac{\lambda}{x} - \frac{\lambda(\lambda+1)(\lambda+2)}{x^3} + \frac{\lambda(\lambda+1)(\lambda+2)(\lambda+3)(\lambda+4)}{x^5}$$

$$g \sim 1 - \frac{\lambda(\lambda+1)}{x^2} + \frac{\lambda(\lambda+1)(\lambda+2)(\lambda+3)}{x^4}$$

3.45. Show that

$$\int_0^1 \frac{1-e^{-t}}{t}\,dt - \int_1^\infty \frac{e^{-t}}{t}\,dt = \int_0^\delta \frac{1-e^{-t}}{t}\,dt + \int_\delta^1 \frac{1-e^{-t}}{t}\,dt - \int_1^\infty \frac{e^{-t}}{t}\,dt$$

$$= \int_0^\delta \frac{1-e^{-t}}{t}\,dt + \int_\delta^1 \frac{dt}{t} - \int_\delta^\infty \frac{e^{-t}}{t}\,dt$$

$$= \delta - \ln\delta + \ln\delta - \int_\delta^\infty e^{-t}\ln t\,dt + \cdots = -\int_0^\infty e^{-t}\ln t\,dt = \gamma$$

where δ is a small positive quantity and γ is Euler's constant.

3.46. Show that as $x \to 0$

(a) $Ein(x) = \int_0^x \frac{1-e^{-t}}{t}\,dt \sim x - \frac{x^2}{2\cdot 2!} + \frac{x^3}{3\cdot 3!}$

(b) $E_1(x) = \int_x^\infty \frac{e^{-t}}{t}\,dt \sim -\ln x - \gamma + x$

Note that

$$Ein(x) = \int_0^1 \frac{1-e^{-t}}{t}\,dt + \ln x - \int_1^\infty \frac{e^{-t}}{t}\,dt + \int_x^\infty \frac{e^{-t}}{t}\,dt$$

$$= \gamma + \ln x + E_1(x)$$

(c) $e^{-x}\int_0^\infty \frac{e^{-xt}}{1+t}\,dt = E_1(x) \sim -\ln x - \gamma + x$

(d) $\int_1^\infty \frac{e^{-xt}}{t}\,dt = E_1(x) \sim -\ln x - \gamma + x$

(e) $E_2(x) = \int_x^\infty \frac{e^{-t}}{t^2}\,dt = \frac{e^{-x}}{x} - E_1(x) \sim \frac{1}{x} + \ln x + \gamma - 1 - \frac{1}{2}x$

(f) $\int_0^\infty \frac{e^{-t^2}}{t+x}\,dt = -\ln x + 2\int_0^\infty \ln(t+x)te^{-t^2}\,dt \sim -\ln x + 2\int_0^\infty t\ln te^{-t^2}\,dt$
$= -\ln x - \frac{1}{2}\gamma$

3.47. Show that as $x \to 0$

(a) $Ci(x) = -\int_x^\infty \frac{\cos t}{t}\,dt \sim \ln x + \gamma - \frac{1}{4}x^2$

(b) $Cin(x) = \int_0^x \frac{1-\cos t}{t} \sim \frac{x^2}{2\cdot 2!} - \frac{x^4}{4\cdot 4!}$

(c) $Si(x) = \int_0^x \frac{\sin t}{t}\,dt \sim x - \frac{x^3}{3\cdot 3!} + \frac{x^5}{5\cdot 5!}$

(d) $si(x) = -\int_x^\infty \frac{\sin t}{t}\,dt = -\frac{1}{2}\pi + x - \frac{x^3}{3\cdot 3!} + \frac{x^5}{5\cdot 5!}$

3.48. Show that as $x \to \infty$

$$\int_0^\infty \frac{e^{-u}}{\sqrt{u}}\left(1+\frac{iu}{2x}\right)^{-1/2} du \sim \sqrt{\pi}\left[1 - \frac{i}{8x} + \frac{1^2\cdot 3^2}{2!}\frac{i^2}{(8x)^2} - \frac{1^2\cdot 3^2\cdot 5^2}{3!}\frac{i^3}{(8x)^3}\right]$$

3.49. Show that as $\omega \to \infty$

(a) $\displaystyle\int_0^\infty \frac{e^{-2\omega x}}{\sqrt[3]{x+2x^2}}\,dx \sim \frac{\Gamma(2/3)}{(2\omega)^{2/3}}$

(b) $\displaystyle\int_0^\infty e^{-2\omega x}\sqrt{x+3x^2}\,dx \sim \frac{\sqrt{\pi}}{2(2\omega)^{3/2}}$

(c) $\displaystyle\int_0^\infty e^{-2\omega x}[\ln(1+x)]^{-1/2}\,dx \sim \sqrt{\frac{\pi}{2\omega}}$

(d) $\displaystyle\int_0^\infty e^{-2\omega x}(x+3x^2)^\lambda\,dx \sim \frac{\Gamma(\lambda+1)}{(2\omega)^{\lambda+1}}$

3.50. Show that as $\omega \to \infty$

(a) $\displaystyle\int_1^2 \frac{e^{-\omega(x+x^2)}}{\sqrt{x^2-1}}\,dx \sim \sqrt{\frac{\pi}{6\omega}}\,e^{-2\omega}$

(b) $\displaystyle\int_1^2 e^{-\omega(x+x^2)}\ln(1+x)\,dx \sim \frac{\ln 2}{3\omega}e^{-2\omega}$

(c) $\displaystyle\int_1^2 \frac{e^{-\omega(x+x^2)}}{\sqrt[n]{x^2-1}}\,dx \sim \frac{\Gamma((n-1)/n)}{2^{1/n}(3\omega)^{(n-1)/n}}e^{-2\omega}$

3.51. Show that as $\omega \to \infty$

(a) $\displaystyle\int_1^\infty e^{-\omega x^2}\sqrt{x^2-1}\,dx \sim \frac{\sqrt{\pi}}{4\omega^{3/2}}e^{-\omega}$

(b) $\displaystyle\int_1^\infty e^{-\omega x^2}\sqrt[n]{x^2-1}\,dx \sim \frac{2^{1/n}\Gamma((n+1)/n)}{(2\omega)^{(n+1)/n}}e^{-\omega}$

(c) $\displaystyle\int_1^\infty \frac{e^{-\omega x^2}}{\sqrt{x^2-1}}\ln x\,dx \sim \frac{\sqrt{\pi}}{8(\omega)^{3/2}}e^{-\omega}$

3.52. Show that as $\omega \to \infty$

(a) $\displaystyle\int_0^\infty e^{-\omega x^2}\ln(1+x)\,dx \sim \frac{1}{2\omega}$

(b) $\displaystyle\int_0^\infty \frac{e^{-\omega x^2}}{\sqrt[4]{x+x^2}}\,dx \sim \frac{\Gamma(\frac{3}{8})}{2\omega^{3/8}}$

(c) $\displaystyle\int_0^\infty e^{-\omega x^2}\ln(2+x)\,dx \sim \frac{\ln 2\sqrt{\pi}}{2\sqrt{\omega}}$

(d) $\displaystyle\int_0^\infty e^{-\omega x^2}\sqrt{\sin x}\,dx \sim \frac{\Gamma(\frac{3}{4})}{2\omega^{3/4}}$

3.53. Show that as $\omega \to \infty$

(a) $\displaystyle\int_{-\infty}^\infty e^{-\omega x^2}\cos x\,dx \sim \sqrt{\frac{\pi}{\omega}}$

(b) $\displaystyle\int_{-\infty}^\infty e^{-\omega x^2}x\sin x\,dx \sim \frac{\sqrt{\pi}}{2\omega^{3/2}}$

3.54. Show that as $x \to \infty$

(a) $\displaystyle\int_1^\infty e^{-x(t+1/t)}\sqrt{t^2-1}\,dt \sim \frac{\sqrt{2}\,\Gamma(\frac{3}{4})}{2x^{3/4}}e^{-2x}$

(b) $\displaystyle\int_1^\infty e^{-x(t+1/t)}\ln(2+t)\,dt \sim \frac{\ln 3\sqrt{\pi}}{2\sqrt{x}}e^{-2x}$

(c) $\displaystyle\int_1^\infty e^{-x(t+1/t)}(t-1)^\lambda\,dt \sim \frac{\Gamma((\lambda+1)/2)}{2x^{(\lambda+1)/2}}e^{-2x}$

3.55. Show that as $x \to \infty$

(a) $\int_1^\infty e^{-x(t^2+2/t)}\,dt \sim \frac{1}{2}\sqrt{\frac{\pi}{3x}}\,e^{-3x}$

(b) $\int_1^\infty e^{-x(t^2+2/t)}\sqrt{t^2-1}\,dt \sim \frac{\sqrt{2}\,\Gamma(\frac{3}{4})}{2(3x)^{3/4}}e^{-3x}$

(c) $\int_1^\infty e^{-x(t^2+2/t)}\ln t\,dt \sim \frac{1}{6x}e^{-3x}$

(d) $\int_1^\infty e^{-x(t^2+2/t)}(t-1)^\lambda\,dt \sim \frac{\Gamma((\lambda+1)/2)}{2(3x)^{(\lambda+1)/2}}e^{-3x}$

3.56. Show that as $x \to \infty$

(a) $\int_0^1 e^{-xt^n}\sin t\,dt \sim \frac{\Gamma(2/n)}{nx^{2/n}}$

(b) $\int_0^1 e^{-xt^n}\ln(2+t)\,dt \sim \frac{\Gamma(1/n)}{nx^{1/n}}$

(c) $\int_0^1 e^{-xt^n}\sqrt{t+t^2}\,dt \sim \frac{\Gamma(3/2n)}{nx^{3/2n}}$

(d) $\int_0^1 e^{-xt^n}(t+t^2)^m\,dt \sim \frac{\Gamma((m+1)/n)}{nx^{(m+1)/n}}$

3.57. Show that as $\alpha \to \infty$

(a) $\int_0^1 \frac{e^{i\alpha t^2}}{1+t}\,dt \sim \frac{1}{2}\sqrt{\frac{\pi}{\alpha}}\,e^{i\pi/4}$

(b) $\int_0^1 e^{i\alpha t^2}\sqrt{t+t^2}\,dt \sim \frac{\Gamma(\frac{3}{4})}{2\alpha^{3/4}}e^{3i\pi/8}$

(c) $\int_0^1 \frac{e^{i\alpha t^2}}{\sqrt{t+t^2}}\,dt \sim \frac{\Gamma(\frac{1}{4})}{2\alpha^{1/4}}e^{i\pi/8}$

(d) $\int_0^1 e^{i\alpha t^2}\ln(1+t)\,dt \sim \frac{1}{2\alpha}e^{i\pi/2}$

(e) $\int_0^1 e^{i\alpha t^2}(t+t^2)^\lambda\,dt \sim \frac{\Gamma((\lambda+1)/2)}{2\alpha^{(\lambda+1)/2}}e^{i(\lambda+1)\pi/4}$

3.58. Show that as $\alpha \to \infty$

(a) $\int_0^1 e^{i\alpha t^n}\,dt \sim \frac{\Gamma(1/n)}{n\alpha^{1/n}}e^{i\pi/2n}$

(b) $\int_0^1 e^{i\alpha t^n}\sqrt{t+t^2}\,dt \sim \frac{\Gamma(3/2n)}{n\alpha^{3/2n}}e^{3i\pi/4n}$

(c) $\int_0^1 \frac{e^{i\alpha t^n}}{\sqrt{t+t^2}}\,dt \sim \frac{\Gamma(1/2n)}{n\alpha^{1/2n}}e^{i\pi/4n}$

(d) $\int_0^1 e^{i\alpha t^n}\sin t\,dt \sim \frac{\Gamma(2/n)}{n\alpha^{2/n}}e^{i\pi/n}$

(e) $\int_0^1 e^{i\alpha t^n}(t+t^2)^\lambda\,dt \sim \frac{\Gamma((\lambda+1)/n)}{n\alpha^{(\lambda+1)/n}}e^{i(\lambda+1)\pi/2n}$

CHAPTER 4

Conservative Systems with Odd Nonlinearities

In this chapter, we discuss methods of obtaining approximate solutions to equations of the form

$$\ddot{u} + \omega_0^2 u = \varepsilon f(u, \dot{u})$$

where ε is a small dimensionless quantity, u is a dimensionless variable, ω_0 is a constant, the dot denotes the derivative with respect to the dimensionless time t, and $f(u, \dot{u})$ is a function with odd nonlinearities and does not depend on the sign of $\dot{u}$. When $\varepsilon = 0$, our equation reduces to

$$\ddot{u}_0 + \omega_0^2 u_0 = 0$$

whose general solution is $u_0 = a\cos(\omega_0 t + \beta)$, where a and β are constants. When $\varepsilon \neq 0$, one seeks a so-called straightforward or pedestrian expansion of the form

$$u = u_0(t) + \varepsilon u_1(t) + \varepsilon^2 u_2(t) + \cdots$$

Substituting this expansion into the perturbed equation, performing algebraic manipulations, and equating coefficients of equal powers of ε on both sides, one obtains equations to solve successively for the u_n. The rule rather than the exception is that the straightforward expansion is nonuniform because the u_n contain so-called secular terms of the form $t^m\cos(\omega_0 t + \beta)$ and $t^m\sin(\omega_0 t + \beta)$. For example, a first-order straightforward expansion of the Duffing equation

$$\ddot{u} + u + \varepsilon u^3 = 0, \qquad \varepsilon \ll 1$$

is

$$u = a\cos(t + \beta) + \varepsilon\left[-\tfrac{3}{8}a^3 t\sin(t + \beta) + \tfrac{1}{32}a^3\cos(3t + 3\beta)\right] + \cdots$$

Thus, to the first approximation, $u = a\cos(t+\beta)$ and its first correction is

$$-\tfrac{3}{8}a^3\varepsilon t\sin(t+\beta)+\tfrac{1}{32}a^3\varepsilon\cos(3t+3\beta)$$

Note that this correction is small, as it is supposed to be, only when εt is small compared with unity. When $\varepsilon t = O(1)$, the term that is supposed to be a small correction to the first term becomes the order of the term it is supposed to correct. In fact, when $\varepsilon t > O(1)$, the small correction dominates the term that it is supposed to correct. Therefore, the straightforward expansion is valid only for times that are less than $O(\varepsilon^{-1})$ and we say that it is nonuniform or that it breaks down for long times. The reason for the breakdown of the straightforward expansion is the presence of the so-called secular term $t\sin(t+\beta)$, which is the product of an algebraic and a circular function. To obtain uniform expansions for problems of this kind, the straightforward expansion needs to be modified. In this chapter, four such modifications are discussed.

4.1. The Lindstedt–Poincaré Technique

Inspection of the straightforward expansion shows that the frequency is forced to have the constant value ω_0. However, it is well known from experiments and available exact solutions that the frequency of the response to an initial disturbance is a nonlinear function of the amplitude of the response. Thus the breakdown of the straightforward expansion is due to its attempt to account for the nonlinear dependence of the frequency on the coefficient ε of the nonlinearity. One of the earliest techniques that yielded uniform expansions of the solutions of such problems is the Lindstedt–Poincaré technique or the method of strained parameters. According to this method, both the dependent variable u and the frequency ω of the system are expanded in powers of ε. Since ω does not appear explicitly in the differential equation, we first introduce the transformation $\tau = \omega t$ in the original perturbation equation and obtain

$$\omega^2 u'' + \omega_0^2 u = \varepsilon f(u, \omega u')$$

where the prime denotes the derivative with respect to τ. Next, we let

$$u(\tau;\varepsilon) = u_0(\tau) + \varepsilon u_1(\tau) + \cdots$$

$$\omega = \omega_0 + \varepsilon\omega_1 + \cdots$$

in the transformed equation, equate coefficients of like powers of ε, and obtain equations that can be successively solved for the u_n. The ω_n are chosen to eliminate the secular terms. Application of this technique to the Duffing equation yields the uniform first-order expansion

$$u = a\cos\left[\left(1+\tfrac{3}{8}\varepsilon a^2\right)t+\beta\right]+\tfrac{1}{32}\varepsilon a^3\cos\left[3\left(1+\tfrac{3}{8}\varepsilon a^2\right)t+3\beta\right]+\cdots$$

4.2. The Method of Renormalization

Instead of substituting the transformation $\tau = \omega t$ into the governing equation and expanding both ω and u as in the Lindstedt–Poincaré technique, one can substitute this transformation and the expansion for ω into the straightforward expansion, expand the result for small ε, and choose the ω_n to eliminate the secular terms. Thus one lets

$$\omega_0 t = \frac{\omega_0\tau}{\omega} = \frac{\omega_0\tau}{\omega_0+\varepsilon\omega_1+\varepsilon^2\omega_2+\cdots} = \tau\left(1-\frac{\varepsilon\omega_1}{\omega_0}+\frac{\varepsilon^2\omega_1^2}{\omega_0^2}-\frac{\varepsilon^2\omega_2}{\omega_0}+\cdots\right)$$

in the straightforward expansion, expands the result for small ε, keeping τ fixed, and chooses the ω_n to eliminate the secular terms. Substituting the above transformation with ω_0 being unity into the straightforward expansion of the Duffing equation and expanding the result for small ε yields

$$u = a\cos(\tau+\beta)+\varepsilon\left[\left(\omega_1-\tfrac{3}{8}a^2\right)a\tau\sin(\tau+\beta)+\tfrac{1}{32}a^3\cos(3\tau+3\beta)\right]+\cdots$$

Choosing ω_1 to eliminate the secular term gives $\omega_1 = \frac{3}{8}a^2$ and

$$u = a\cos(\tau+\beta)+\tfrac{1}{32}\varepsilon a^3\cos(3\tau+3\beta)+\cdots$$

where

$$\tau = \left(1+\tfrac{3}{8}\varepsilon a^2+\cdots\right)t$$

in agreement with the expansion obtained using the Lindstedt–Poincaré technique. It should be noted that, in general, the method of renormalization yields a uniform approximation with much less algebra than that involved in applying the Lindstedt–Poincaré technique.

4.3. The Method of Multiple Scales

We note that the functional dependence of the uniform expansion obtained using either the Lindstedt–Poincaré technique or the method of renormalization on the time t and the parameter ε is not disjoint, because u depends on the combination εt as well as on the individual t and ε. Carrying out the Lindstedt–Poincaré expansion to higher order, we find that the resulting expansion depends on the combinations $t, \varepsilon t, \varepsilon^2 t, \varepsilon^3 t, \ldots, \varepsilon^m t, \ldots$, called scales, as well as on ε. Hence we write

$$u(t;\varepsilon) = \hat{u}(t, \varepsilon t, \varepsilon^2 t, \varepsilon^3 t, \ldots; \varepsilon)$$
$$= \hat{u}(T_0, T_1, T_2, T_3, \ldots; \varepsilon)$$

where

$$T_0 = t, \qquad T_1 = \varepsilon t, \qquad T_2 = \varepsilon^3 t, \qquad \ldots, \qquad T_n = \varepsilon^n t, \ldots$$

Thus instead of determining u as a function of t directly, we embed it in the larger class of functions $\hat{u}$, determine a uniform expansion for $\hat{u}$ as a function of the scales $T_0, T_1, T_2, \ldots$, and then specialize $\hat{u}$ to u by setting $T_n = \varepsilon^n t$. Using the chain rule, we transform the derivative with respect to t into

$$\frac{d}{dt} = \frac{\partial}{\partial T_0} + \varepsilon\frac{\partial}{\partial T_1} + \varepsilon^2\frac{\partial}{\partial T_2} + \cdots$$

Using this transformation, we replace the original ordinary-differential equation by a partial-differential equation. Then we seek a uniform expansion for $\hat{u}$ in the form

$$\hat{u} = u_0(T_0, T_1, T_2, \ldots) + \varepsilon u_1(T_0, T_1, T_2, \ldots) + \varepsilon^2 u_2(T_0, T_1, T_2, \ldots) + \cdots$$

and determine its dependence on the scales $T_1, T_2, T_3, \ldots$ by requiring that the u_n for $n \geq 1$ should be free of secular terms. Application of this technique to the Duffing equation yields the same expansion obtained using the Lindstedt–Poincaré technique.

4.4. The Method of Averaging

To apply the method of averaging, one uses the method of variation of parameters and transforms the original equation into

$$u = a(t)\cos[\omega_0 t + \beta(t)]$$

$$\dot{u} = -\omega_0 a(t)\sin[\omega_0 t + \beta(t)]$$

where

$$\dot{a} = -\frac{\varepsilon}{\omega_0}\sin\phi f(a\cos\phi, -\omega_0 a\sin\phi)$$

$$\dot{\beta} = -\frac{\varepsilon}{\omega_0 a}\cos\phi f(a\cos\phi, -\omega_0 a\sin\phi)$$

where $\phi = \omega_0 t + \beta$. When ε is small, a and β are changing slowly and one assumes that they may be considered constant over the interval required for ϕ to change by 2π. Thus integrating over the interval 2π (i.e., averaging), one keeps in the first approximation only the terms that are independent of ϕ in the Fourier expansion of the right-hand sides of $\dot{a}$ and $\dot{\beta}$. Application of the method of variation of parameters to the Duffing equation yields

$$\dot{a} = \varepsilon a^3 \sin\phi \cos^3\phi = \tfrac{1}{8}\varepsilon a^3 (2\sin 2\phi + \sin 4\phi)$$

$$\dot{\beta} = \varepsilon a^2 \cos^4\phi = \tfrac{1}{8}\varepsilon a^2 (3 + 4\cos 2\phi + \cos 4\phi)$$

Keeping only the terms that are independent of ϕ in the Fourier expansions of the right-hand sides, one obtains

$$\dot{a} = 0 \quad \text{and} \quad \dot{\beta} = \frac{3\varepsilon a^2}{8}$$

Hence $a = a_0$ and $\beta = (3\varepsilon a_0^2/8)t + \beta_0$, where a_0 and β_0 are constants. Thus the resulting expansion is in agreement with those obtained by the other techniques.

Solved Exercises

4.1. Use the method of renormalization to render the following expansions uniformly valid:

(a)
$$u(t;\varepsilon) = a\cos(\omega_0 t + \theta) + \varepsilon a^3 t \sin(\omega_0 t + \theta) + O(\varepsilon^2) \tag{1}$$

We let

$$\tau = \omega t = (\omega_0 + \varepsilon\omega_1 + \cdots)t \tag{2}$$

so that

$$\omega_0 t = \frac{\tau}{1 + \varepsilon\omega_1/\omega_0 + \cdots} = \tau - \frac{\varepsilon\omega_1}{\omega_0}\tau + \cdots \tag{3}$$

Substituting (3) into (1) and expanding the result for small ε with τ fixed, we obtain

$$u = a\cos(\tau + \theta) + \varepsilon\frac{a^3\tau}{\omega_0}\sin(\tau + \theta) + \frac{a\omega_1}{\omega_0}\tau\sin(\tau + \theta) + O(\varepsilon^2) \tag{4}$$

We choose ω_1 to eliminate the secular terms in (4). Thus we have

$$\omega_1 = -a^2 \tag{5}$$

Hence to the first approximation

$$u = a\cos(\tau + \theta) + O(\varepsilon)$$

or

$$u = a\cos\left[\left(\omega_0 - \varepsilon a^2\right)t + \theta\right] + O(\varepsilon) \tag{6}$$

(b)

$$u(t;\varepsilon) = a\cos(\omega_0 t + \theta) + \varepsilon\left[a^2 t\sin(\omega_0 t + \theta)\right.$$
$$\left. + (1 - a^2)\,at\cos(\omega_0 t + \theta)\right] + O(\varepsilon^2) \tag{7}$$

Substituting (3) into (7) and expanding for small ε with τ fixed, we obtain

$$u = a\cos(\tau + \theta) + \varepsilon\left[\frac{a^2\tau}{\omega_0}\sin(\tau + \theta) + (1 - a^2)\frac{a\tau}{\omega_0}\cos(\tau + \theta)\right.$$
$$\left. + \frac{a\omega_1\tau}{\omega_0}\sin(\tau + \theta)\right] + O(\varepsilon^2) \tag{8}$$

To eliminate the secular terms in (8), we set each of the coefficients of $\sin(\tau + \theta)$ and $\cos(\tau + \theta)$ equal to zero. The result is

$$a^2 + a\omega_1 = 0 \tag{9}$$

$$(1 - a^2)\,a = 0 \tag{10}$$

For a nontrivial solution, (9) gives

$$\omega_1 = -a \tag{11}$$

whereas (10) gives

$$a = \pm 1 \tag{12}$$

Hence to the first approximation

$$u = \pm\cos[(\omega_0 \mp \varepsilon)t + \theta] + O(\varepsilon) \tag{13}$$

4.2. Consider the equation

$$\ddot{u} + \omega_0^2 u = \varepsilon\dot{u}^2 u \tag{14}$$

(a) Determine a two-term straightforward expansion and discuss its uniformity. We let

$$u = u_0(t) + \varepsilon u_1(t) + \cdots \tag{15}$$

in (14), expand the result for small ε, equate coefficients of like powers of ε, and obtain

$$\ddot{u}_0 + \omega_0^2 u_0 = 0 \tag{16}$$

$$\ddot{u}_1 + \omega_0^2 u_1 = \dot{u}_0^2 u_0 \tag{17}$$

The solution of (16) can be expressed as

$$u_0 = a \cos(\omega_0 t + \beta) \tag{18}$$

Then (17) becomes

$$\begin{aligned} \ddot{u}_1 + \omega_0^2 u_1 &= a^3 \omega_0^2 \cos(\omega_0 t + \beta) \sin^2(\omega_0 t + \beta) \\ &= \tfrac{1}{4} a^3 \omega_0^2 [\cos(\omega_0 t + \beta) - \cos(3\omega_0 t + 3\beta)] \end{aligned} \tag{19}$$

which has the particular solution

$$u_1 = \tfrac{1}{8} a^3 \omega_0 t \sin(\omega_0 t + \beta) + \tfrac{1}{32} a^3 \cos(3\omega_0 t + 3\beta) \tag{20}$$

Hence

$$u = a \cos(\omega_0 t + \beta) + \tfrac{1}{32} \varepsilon a^3 [4\omega_0 t \sin(\omega_0 t + \beta) + \cos(3\omega_0 t + 3\beta)] + \cdots \tag{21}$$

which is not valid for $t \geq O(\varepsilon^{-1})$ because of the presence of a mixed-secular term at $O(\varepsilon)$.

(b) Render this expansion uniformly valid by using the method of renormalization. Substituting (3) into (21) and expanding for small ε with τ fixed, we have

$$u = a \cos(\tau + \beta) + \varepsilon \left[\frac{\omega_1 a}{\omega_0} \tau \sin(\tau + \beta) + \frac{a^3}{8} \tau \sin(\tau + \beta) + \frac{a^3}{32} \cos(3\tau + 3\beta) \right] + \cdots \tag{22}$$

Choosing ω_1 to eliminate the secular terms yields

$$\omega_1 = -\frac{\omega_0 a^2}{8} \tag{23}$$

Hence

$$u = a \cos(\tau + \beta) + \cdots = a \cos\left[\left(1 - \frac{\varepsilon a^2}{8}\right) \omega_0 t + \beta \right] + \cdots \tag{24}$$

(c) Determine a first-order uniform expansion by using the Lindstedt–Poincaré technique.

Introducing the change of variable (2) into (14), we have

$$\left(\omega_0^2+2\varepsilon\omega_0\omega_1+\cdots\right)u''+\omega_0^2u=\varepsilon\omega_0^2u'^2u+\cdots \tag{25}$$

Letting

$$u=u_0(\tau)+\varepsilon u_1(\tau)+\cdots \tag{26}$$

in (25) and equating coefficients of like powers of ε, we obtain

$$u_0''+u_0=0 \tag{27}$$

$$u_1''+u_1=-\frac{2\omega_1}{\omega_0}u_0''+u_0'^2u_0 \tag{28}$$

The solution of (27) can be expressed as

$$u_0=a\cos(\tau+\beta) \tag{29}$$

Then (28) becomes

$$u_1''+u_1=\frac{2\omega_1}{\omega_0}a\cos(\tau+\beta)+a^3\sin^2(\tau+\beta)\cos(\tau+\beta)$$

$$=\left(\frac{2\omega_1}{\omega_0}a+\tfrac{1}{4}a^3\right)\cos(\tau+\beta)-\tfrac{1}{4}a^3\cos(3\tau+3\beta) \tag{30}$$

Eliminating the terms that lead to secular terms in (30) yields

$$\omega_1=-\frac{\omega_0a^2}{8} \tag{31}$$

Hence

$$u=a\cos\left[\omega_0\left(1-\frac{\varepsilon a^2}{8}\right)t+\beta\right]+\cdots \tag{32}$$

(d) Use the method of multiple scales to determine a first-order uniform expansion. We seek an expansion in the form

$$u=u_0(T_0,T_1)+\varepsilon u_1(T_0,T_1)+\cdots \tag{33}$$

where $T_0=t$ and $T_1=\varepsilon t$. Substituting (33) into (14) and equating coefficients of like powers of ε, we obtain

$$\frac{\partial^2 u_0}{\partial T_0^2} + \omega_0^2 u_0 = 0 \tag{34}$$

$$\frac{\partial^2 u_1}{\partial T_0^2} + \omega_0^2 u_1 = -2\frac{\partial^2 u_0}{\partial T_0 \partial T_1} + \left(\frac{\partial u_0}{\partial T_0}\right)^2 u_0 \tag{35}$$

The solution of (34) can be expressed as

$$u_0 = A(T_1)\, e^{i\omega_0 T_0} + \bar{A}(T_1)\, e^{-i\omega_0 T_0} \tag{36}$$

Then (35) becomes

$$\frac{\partial^2 u_1}{\partial T_0^2} + \omega_0^2 u_1 = -2i\omega_0 A' e^{i\omega_0 T_0} + \omega_0^2 A^2 \bar{A} e^{i\omega_0 T_0} - \omega_0^2 A^3 e^{3i\omega_0 T_0} + \text{cc} \tag{37}$$

Eliminating the terms that produce secular terms in (37) gives

$$2i\frac{\partial A}{\partial T_1} - \omega_0 A^2 \bar{A} = 0 \tag{38}$$

Substituting the polar form

$$A = \tfrac{1}{2} a \exp(i\beta) \tag{39}$$

into (38) and separating real and imaginary terms, we obtain

$$\frac{\partial a}{\partial T_1} = 0 \tag{40a}$$

$$-a\frac{\partial \beta}{\partial T_1} - \tfrac{1}{8}\omega_0 a^3 = 0 \tag{40b}$$

Hence a is a constant and

$$\beta = -\tfrac{1}{8}\omega_0 a^2 T_1 + \beta_0 \tag{41}$$

if $a \neq 0$. Therefore, to the first approximation,

$$\begin{aligned} u &= a\cos\left(\omega_0 T_0 - \tfrac{1}{8}\omega_0 a^2 T_1 + \beta_0\right) + \cdots \\ &= a\cos\left[\omega_0\left(1 - \tfrac{1}{8}\varepsilon a^2\right)t + \beta_0\right] + \cdots \end{aligned} \tag{42}$$

(e) Use the method of averaging to determine a first-order uniform expansion.

When $\varepsilon = 0$,

$$u = a\cos(\omega_0 t + \beta) \tag{43}$$

$$\dot{u} = -\omega_0 a \sin(\omega_0 t + \beta) \tag{44}$$

where a and β are constants. When $\varepsilon \neq 0$, we still express the solution as in (43) and (44) but a and β are functions of t. Differentiating (43) with respect to t and using (44), we obtain

$$\dot{a}\cos\phi - a\dot{\beta}\sin\phi = 0 \tag{45}$$

where $\phi = \omega_0 t + \beta$. Differentiating (44) with respect to t and then substituting the result together with (43) and (44) into (14), we have

$$-\dot{a}\sin\phi - a\dot{\beta}\cos\phi = \varepsilon\omega_0 a^3 \sin^2\phi\cos\phi \tag{46}$$

Solving (45) and (46) for $\dot{a}$ and $\dot{\beta}$ yields

$$\dot{a} = -\varepsilon\omega_0 a^3 \sin^3\phi\cos\phi = -\frac{\varepsilon\omega_0 a^3}{8}(2\sin 2\phi - \sin 4\phi) \tag{47}$$

$$a\dot{\beta} = -\varepsilon\omega_0 a^3 \sin^2\phi\cos^2\phi = -\frac{\varepsilon\omega_0 a^3}{8}(1-\cos 4\phi) \tag{48}$$

Keeping only the slowly varying parts on the right-hand sides of (47) and (48), we have

$$\dot{a} = 0 \tag{49}$$

$$a\dot{\beta} = -\frac{\varepsilon\omega_0 a^3}{8} \tag{50}$$

It follows from (49) that a is a constant and from (50) that

$$\beta = -\frac{\varepsilon\omega_0 a^2}{8}t + \beta_0 \tag{51}$$

when $a \neq 0$. Therefore, to the first approximation

$$u = a\cos\left[\omega_0\left(1 - \frac{\varepsilon a^2}{8}\right)t + \beta_0\right] + \cdots \tag{52}$$

4.3. Consider the equation

$$\ddot{u} + 4u + \varepsilon u^2 \ddot{u} = 0 \tag{53}$$

Equation (53) can be rewritten as

$$(1 + \varepsilon u^2)\ddot{u} + 4u = 0$$

or

$$\ddot{u} + \frac{4u}{1+\varepsilon u^2} = 0$$

which has the same form as the equation in the next section. To illustrate the techniques, we treat (53) as it stands.

(a) Determine a two-term straightforward expansion and discuss its uniformity. Let

$$u = u_0(t) + \varepsilon u_1(t) + \cdots \tag{54}$$

in (53), equate coefficients of like powers of ε, and obtain

$$\ddot{u}_0 + 4u_0 = 0 \tag{55}$$

$$\ddot{u}_1 + 4u_1 = -u_0^2 \ddot{u}_0 \tag{56}$$

The solution of (55) can be expressed as

$$u_0 = a\cos(2t+\beta) \tag{57}$$

where a and β are constants. Then (56) becomes

$$\ddot{u}_1 + 4u_1 = 4a^3\cos^3(2t+\beta) = 3a^3\cos(2t+\beta) + a^3\cos(6t+3\beta) \tag{58}$$

which has the particular solution

$$u_1 = \tfrac{3}{4}a^3 t\sin(2t+\beta) - \tfrac{1}{32}a^3\cos(6t+3\beta) \tag{59}$$

Hence

$$u = a\cos(2t+\beta) + \varepsilon\left[\tfrac{3}{4}a^3 t\sin(2t+\beta) - \tfrac{1}{32}a^3\cos(6t+3\beta)\right] + \cdots \tag{60}$$

which breaks down when $t \geq O(\varepsilon^{-1})$.

(b) Render this expansion uniformly valid by using the method of renormalization. We let

$$\tau = (2 + \varepsilon\omega_1 + \cdots)t, \qquad 2t = \left(1 - \tfrac{1}{2}\varepsilon\omega_1 + \cdots\right)\tau \tag{61}$$

where the first term in the expansion of the frequency is the same as the linear frequency (i.e., 2). Substituting (61) into (60) and expanding the result for small ε with τ fixed, we obtain

$$u = a\cos(\tau+\beta) + \varepsilon\left[\tfrac{1}{2}\varepsilon\omega_1\tau a\sin(\tau+\beta) + \tfrac{3}{8}a^3\tau\sin(\tau+\beta) - \tfrac{1}{32}a^3\cos(3\tau+3\beta)\right] + \cdots \tag{62}$$

Eliminating the secular terms in (62) demands that

$$\omega_1 = -\tfrac{3}{4}a^2 \tag{63}$$

Hence to the first approximation

$$u = a\cos(\tau+\beta)+\cdots = a\cos\left[\left(2-\tfrac{3}{4}\varepsilon a^2\right)t+\beta\right]+\cdots \tag{64}$$

(c) Determine a first-order uniform expansion by using the Lindstedt–Poincaré technique.

We seek an expansion in the form

$$u = u_0(\tau)+\varepsilon u_1(\tau)+\cdots \tag{65}$$

where τ is given by (61). Substituting (61) and (65) into (53) and equating coefficients of like powers of ε, we obtain

$$u_0'' + u_0 = 0 \tag{66}$$

$$u_1'' + u_1 = -\omega_1 u_1'' - u_0^2 u_0'' \tag{67}$$

The solution of (66) can be expressed as

$$u_0 = a\cos(\tau+\beta) \tag{68}$$

Then (67) becomes

$$\begin{aligned} u_1'' + u_1 &= a\omega_1\cos(\tau+\beta)+a^3\cos^3(\tau+\beta) \\ &= \left(a\omega_1+\tfrac{3}{4}a^3\right)\cos(\tau+\beta)+\tfrac{1}{4}a^3\cos(3\tau+3\beta) \end{aligned} \tag{69}$$

Eliminating the terms that produce secular terms in (69) yields

$$\omega_1 = -\tfrac{3}{4}a^2 \tag{70}$$

in agreement with the result obtained using the method of renormalization.

(d) Determine a first-order expansion by using the method of multiple scales.

Let

$$u = u_0(T_0,T_1)+\varepsilon u_1(T_0,T_1)+\cdots \tag{71}$$

where $T_0 = t$ and $T_1 = \varepsilon t$ in (53), equate coefficients of like powers of ε, and obtain

$$\frac{\partial^2 u_0}{\partial T_0^2} + 4u_0 = 0 \tag{72}$$

$$\frac{\partial^2 u_1}{\partial T_0^2} + 4u_1 = -2\frac{\partial^2 u_0}{\partial T_0\,\partial T_1} - u_0^2\frac{\partial^2 u_0}{\partial T_0^2} \tag{73}$$

The solution of (72) can be expressed as

$$u_0 = A(T_1)\,e^{2iT_0} + \bar{A}(T_1)\,e^{-2iT_0} \tag{74}$$

Then (73) becomes

$$\frac{\partial^2 u_1}{\partial T_0^2} + 4u_1 = -4iA'e^{2iT_0} + 4A^3 e^{6iT_0} + 12A^2\bar{A}e^{2iT_0} + \text{cc} \tag{75}$$

Eliminating the terms that lead to secular terms in (75) yields

$$iA' - 3A^2\bar{A} = 0 \tag{76}$$

Substituting (39) into (76) and separating real and imaginary parts, we have

$$\frac{\partial a}{\partial T_1} = 0 \tag{77}$$

$$a\frac{\partial \beta}{\partial T_1} = -\tfrac{3}{4}a^3 \tag{78}$$

It follows from (77) that a is a constant and from (78) that if $a \neq 0$,

$$\beta = -\tfrac{3}{4}a^2 T_1 + \beta_0 \tag{79}$$

where β_0 is a constant. Hence to the first approximation we obtain (64).

(e) Use the method of averaging to determine a first-order uniform expansion.

When $\varepsilon = 0$, the solution of (53) can be expressed as

$$u = a\cos(2t + \beta) \tag{80}$$

where a and β are constant. Then

$$\dot{u} = -2a\sin(2t + \beta) \tag{81}$$

When $\varepsilon \neq 0$, we still seek the solution as in (80) and (81) but with time-varying a and β. Differentiating (80) once with respect to t and using (81), we have

$$\dot{a}\cos\phi - a\dot{\beta}\sin\phi = 0 \tag{82}$$

where $\phi = 2t + \beta$. Then, differentiating (81) with respect to t and substituting the result together with (80) into (53), we obtain

$$\dot{a}\sin\phi + a\dot{\beta}\cos\phi = -2\varepsilon a^3\cos^3\phi \tag{83}$$

It follows from (82) and (83) that

$$\dot{a} = -2\varepsilon a^3\sin\phi\cos^3\phi = -\tfrac{1}{4}\varepsilon a^3(2\sin 2\phi + \sin 4\phi) \tag{84}$$

$$a\dot{\beta} = -2\varepsilon a^3\cos^4\phi = -\tfrac{1}{4}\varepsilon a^3(3 + 4\cos 2\phi + \cos 4\phi) \tag{85}$$

Keeping only the slowly varying terms on the right-hand sides of (84) and (85), we obtain

$$\dot{a} = 0 \tag{86}$$

$$a\dot{\beta} = -\tfrac{3}{4}\varepsilon a^3 \tag{87}$$

Hence a is a constant and

$$\beta = -\tfrac{3}{4}\varepsilon a^2 t + \beta_0 \tag{88}$$

if $a \neq 0$. Substituting (88) into (80), we obtain (64).

4.4. Consider the equation

$$\ddot{u} + \frac{\omega^2 u}{1 + u^2} = 0 \tag{89}$$

(a) Determine a two-term straightforward expansion and discuss its uniformity.

Since there is no small parameter in the problem, a solution for small but finite amplitudes is implied. Hence the term $(1 + u^2)^{-1}$ is first expanded in powers of u^2 and (89) can be rewritten as

$$\ddot{u} + \omega^2 u(1 - u^2 + \cdots) = 0$$

or

$$\ddot{u} + \omega^2 u - \alpha\omega^2 u^3 + \cdots = 0, \qquad \alpha = 1 \tag{90}$$

For bookkeeping, we use a power ε^λ, where $\lambda > 0$, of a small parameter ε to measure the amplitude. Thus we put

$$u = \varepsilon^\lambda v$$

in (90) and obtain

$$\ddot{v} + \omega^2 v - \omega^2 \alpha \varepsilon^{2\lambda} v^3 + \cdots = 0 \tag{91}$$

If we choose $\lambda = \frac{1}{2}$, the coefficient of the nonlinearity will be ε and (91) becomes

$$\ddot{v} + \omega^2 v - \varepsilon\omega^2 \alpha v^3 + \cdots = 0 \tag{92}$$

We seek a straightforward expansion for the solution of (92) in the form

$$v = v_0(t) + \varepsilon v_1(t) + \cdots \tag{93}$$

Substituting (93) into (92) and equating coefficients of like powers of ε, we obtain

$$\ddot{v}_0 + \omega^2 v_0 = 0 \tag{94}$$

$$\ddot{v}_1 + \omega^2 v_1 = \alpha\omega^2 v_0^3 \tag{95}$$

The solution of (94) can be expressed as

$$v = a\cos(\omega t + \beta) \tag{96}$$

where a and β are constants. Then (95) becomes

$$\ddot{v}_1 + \omega^2 v_1 = \alpha\omega^2 a^3 \cos^3(\omega t + \beta) = \tfrac{3}{4}\alpha\omega^2 a^3 \cos(\omega t + \beta) + \tfrac{1}{4}\alpha\omega^2 a^3 \cos(3\omega t + 3\beta) \tag{97}$$

which possesses the particular solution

$$v_1 = \tfrac{3}{8}\alpha\omega a^3 t \sin(\omega t + \beta) - \tfrac{1}{32}\alpha a^3 \cos(3\omega t + 3\beta) \tag{98}$$

Hence

$$u = \varepsilon^{1/2} a\cos(\omega t + \beta) + \varepsilon^{3/2}\left[\tfrac{3}{8}\alpha\omega a^3 t \sin(\omega t + \beta) - \tfrac{1}{32}\alpha a^3 \cos(3\omega t + 3\beta)\right] + \cdots \tag{99}$$

which breaks down for $t \geq O(\varepsilon^{-1})$.

(b) Render this expansion uniformly valid by using the method of renormalization. We let

$$\tau = \Omega t = (\omega + \varepsilon\Omega_1 + \cdots)t, \qquad \omega t = \left(1 - \frac{\varepsilon\Omega_1}{\omega} + \cdots\right)\tau \tag{100}$$

where the nonlinear frequency is denoted by Ω instead of ω so that it will not be confused with the linear frequency ω. Substituting (100) into (99) and expanding the result for small ε with τ fixed, we have

$$u = \varepsilon^{1/2} a\cos(\tau + \beta) + \varepsilon^{3/2}\left[\frac{\Omega_1 a}{\omega}\tau\sin(\tau + \beta) + \tfrac{3}{8}\alpha a^3 \tau \sin(\tau + \beta) - \tfrac{1}{32}\alpha a^3 \cos(3\tau + 3\beta)\right] + \cdots \tag{101}$$

Choosing Ω_1 to eliminate the secular terms in (101), we obtain

$$\Omega_1 = -\tfrac{3}{8}\alpha a^2 \omega \tag{102}$$

Therefore, to the first approximation

$$u = \varepsilon^{1/2} a\cos(\tau + \beta) + \cdots = \varepsilon^{1/2} a\cos\left[\left(1 - \tfrac{3}{8}\varepsilon\alpha a^2\right)\omega t + \beta\right] + \cdots \tag{103}$$

(c) Determine a first-order uniform expansion using the Lindstedt–Poincaré technique.

Let

$$v = v_0(\tau) + \varepsilon v_1(\tau) + \cdots \tag{104}$$

in (92), use (100), equate coefficients of like powers of ε, and obtain

$$v_0'' + v_0 = 0 \tag{105}$$

$$v_1'' + v_1 = -\frac{2\Omega_1}{\omega} v_0'' + \alpha v_0^3 \tag{106}$$

The solution of (105) can be expressed as

$$v_0 = a\cos(\tau + \beta) \tag{107}$$

where a and β are constants. Then (106) becomes

$$\begin{aligned} v_1'' + v_1 &= \frac{2\Omega_1}{\omega} a\cos(\tau+\beta) + \alpha a^3 \cos^3(\tau+\beta) \\ &= \left(\frac{2\Omega_1}{\omega} a + \tfrac{3}{4}\alpha a^3\right)\cos(\tau+\beta) + \tfrac{1}{4}\alpha a^3 \cos(3\tau + 3\beta) \end{aligned} \tag{108}$$

To eliminate the terms that lead to secular terms, we put

$$\Omega_1 = -\tfrac{3}{8}\alpha\omega a^2 \tag{109}$$

in agreement with that obtained by using the method of renormalization.

(d) Use the method of multiple scales to determine a first-order uniform expansion. We let

$$v = v_0(T_0, T_1) + \varepsilon v_1(T_0, T_1) + \cdots \tag{110}$$

in (92), equate coefficients of like powers of ε, and obtain

$$\frac{\partial^2 v_0}{\partial T_0^2} + \omega^2 v_0 = 0 \tag{111}$$

$$\frac{\partial^2 v_1}{\partial T_0^2} + \omega^2 v_1 = -2\frac{\partial^2 v_0}{\partial T_0 \partial T_1} + \alpha\omega^2 v_0^3 \tag{112}$$

The solution of (111) can be expressed as

$$v_0 = A(T_1)e^{i\omega T_0} + \bar{A}(T_1)e^{-i\omega T_0} \tag{113}$$

Then (112) becomes

$$\frac{\partial^2 v_1}{\partial T_0^2} + \omega^2 v_1 = -2i\omega A' e^{i\omega T_0} + 3\alpha\omega^2 A^2 \bar{A} e^{i\omega T_0} + \alpha\omega^2 A^3 e^{3i\omega T_0} + \text{cc} \tag{114}$$

Eliminating the terms that produce secular terms in (114) demands that

$$2iA' = 3\alpha\omega A^2 \bar{A} \tag{115}$$

Putting (39) in (115) and separating the result into real and imaginary parts, we have

$$\frac{\partial a}{\partial T_1} = 0 \tag{116}$$

$$a\frac{\partial \beta}{\partial T_1} = -\tfrac{3}{8}\alpha\omega a^3 \tag{117}$$

It follows from (116) that a is a constant and from (117) that if $a \neq 0$,

$$\beta = -\tfrac{3}{8}\alpha\omega a^2 T_1 + \beta_0 \tag{118}$$

where β_0 is a constant. Hence to the first approximation we obtain (103).

(e) Use the method of averaging to determine a first-order uniform expansion.

As before, we seek the solution of (92) in the form

$$v = a(t)\cos[\omega t + \beta(t)] \tag{119}$$

subject to the constraint

$$\dot{v} = -\omega a(t)\sin[\omega t + \beta(t)] \tag{120}$$

Differentiating (119) with respect to t and using (120) gives

$$\dot{a}\cos\phi - a\dot{\beta}\sin\phi = 0 \tag{121}$$

where $\phi = \omega t + \beta$. Differentiating (120) with respect to t and substituting the result together with (119) into (92), we obtain

$$\dot{a}\sin\phi + a\dot{\beta}\cos\phi = -\varepsilon\alpha\omega a^3\cos^3\phi \tag{122}$$

It follows from (121) and (122) that

$$\dot{a} = -\varepsilon\alpha\omega a^3 \sin\phi\cos^3\phi = -\tfrac{1}{8}\varepsilon\alpha\omega a^3(2\sin\phi + \sin 4\phi) \tag{123}$$

$$\dot{\beta} = -\varepsilon\alpha\omega a^2\cos^4\phi = -\tfrac{1}{8}\varepsilon\alpha a^2(3 + 4\cos 2\phi + \cos 4\phi) \quad \text{if } a \neq 0 \tag{124}$$

Keeping only the slowly varying parts on the right-hand sides of (123) and (124), we have

$$\dot{a} = 0 \tag{125}$$

$$\dot{\beta} = -\tfrac{3}{8}\varepsilon\alpha a^2 \tag{126}$$

It follows from (125) that a is a constant and from (126) that

$$\beta = -\tfrac{3}{8}\varepsilon\alpha a^2 t + \beta_0 \tag{127}$$

where β_0 is a constant. Substituting (127) into (119) and recalling that $u = \varepsilon^{1/2}v$, we obtain (103).

4.5. Consider the equation

$$\ddot{u} + \omega_0^2 u = \varepsilon u^5, \qquad \varepsilon \ll 1 \tag{128}$$

(a) Determine a two-term straightforward expansion and discuss its uniformity.
Let

$$u = u_0(t) + \varepsilon u_1(t) + \cdots \tag{129}$$

in (128), equate coefficients of like powers of ε, and obtain

$$\ddot{u}_0 + \omega_0^2 u_0 = 0 \tag{130}$$

$$\ddot{u}_1 + \omega_0^2 u_1 = u_0^5 \tag{131}$$

The solution of (130) can be expressed as

$$u_0 = a\cos(\omega_0 t + \beta) \tag{132}$$

where a and β are constants. Then (131) becomes

$$\ddot{u}_1 + \omega_0^2 u_1 = a^5\cos^5(\omega_0 t + \beta) = \tfrac{5}{8}a^5\cos(\omega_0 t + \beta) + \tfrac{5}{16}a^5\cos(3\omega_0 t + 3\beta) + \tfrac{1}{16}a^5\cos(5\omega_0 t + 5\beta) \tag{133}$$

which has the particular solution

$$u_1 = \frac{5}{16\omega_0}a^5 t\sin(\omega_0 t + \beta) - \frac{5}{128\omega_0^2}a^5\cos(3\omega_0 t + 3\beta) - \frac{1}{384\omega_0^2}a^5\cos(5\omega_0 t + 5\beta) \tag{134}$$

Hence

$$u = a\cos(\omega_0 t + \beta) + \frac{\varepsilon a^5}{16\omega_0^2}\left[5\omega_0 t \sin(\omega_0 t + \beta) - \tfrac{5}{8}\cos(3\omega_0 t + 3\beta) - \tfrac{1}{24}\cos(5\omega_0 t + 5\beta)\right] + \cdots \tag{135}$$

which breaks down when $t \geq O(\varepsilon^{-1})$.

(b) Render this expansion uniformly valid by using the method of renormalization. We introduce the transformation

$$\tau = (\omega_0 + \varepsilon\omega_1 + \cdots)t \quad \text{or} \quad \omega_0 t = \left(1 - \frac{\varepsilon\omega_1}{\omega_0} + \cdots\right)\tau \tag{136}$$

in (135), expand for small ε with τ fixed, and obtain

$$u = a\cos(\tau + \beta) + \varepsilon\left[\frac{\omega_1}{\omega_0} a\tau\sin(\tau + \beta) + \frac{5}{16\omega_0^2} a^5\tau\sin(\tau + \beta) + \text{NST}\right] + \cdots \tag{137}$$

Choosing ω_1 to eliminate the secular terms from u, we have

$$\omega_1 = -\frac{5}{16\omega_0} a^4 \tag{138}$$

Hence to the first approximation

$$u = a\cos(\tau + \beta) + \cdots = a\cos\left[\left(1 - \frac{5}{16\omega_0^2}\varepsilon a^4\right)\omega_0 t + \beta\right] + \cdots \tag{139}$$

(c) Determine a first-order uniform expansion using the Lindstedt–Poincaré technique.

Let

$$u = u_0(\tau) + \varepsilon u_1(\tau) + \cdots \tag{140}$$

in (128), use (136), equate coefficients of like powers of ε, and obtain

$$u_0'' + u_0 = 0 \tag{141}$$

$$\omega_0^2(u_1'' + u_1) = -2\omega_0\omega_1 u_0'' + u_0^5 \tag{142}$$

We express the solution of (141) as

$$u_0 = a\cos(\tau + \beta) \tag{143}$$

where a and β are constants. Then (142) becomes

$$\omega_0^2(u_1''+u_1)=2\omega_0\omega_1 a\cos(\tau+\beta)+a^5\cos^5(\tau+\beta)$$

$$=\left(2\omega_0\omega_1 a+\tfrac{5}{8}a^5\right)\cos(\tau+\beta)+\text{NST} \tag{144}$$

Eliminating the terms that lead to secular terms in u_1 gives (138). The resulting expansion would be the same as (139).

(d) Use the method of multiple scales to determine a first-order uniform expansion. Let

$$u=u_0(T_0,T_1)+\varepsilon u_1(T_0,T_1)+\cdots \tag{145}$$

in (128), equate coefficients of like powers of ε, and obtain

$$\frac{\partial^2 u_0}{\partial T_0^2}+\omega_0^2 u_0=0 \tag{146}$$

$$\frac{\partial^2 u_1}{\partial T_0^2}+\omega_0^2 u_1=-2\frac{\partial^2 u_0}{\partial T_0\,\partial T_1}+u_0^5 \tag{147}$$

The solution of (146) can be expressed as

$$u_0=A(T_1)\,e^{i\omega_0 T_0}+\bar{A}(T_1)\,e^{-i\omega_0 T_0} \tag{148}$$

Then (147) becomes

$$\frac{\partial^2 u_1}{\partial T_0^2}+\omega_0^2 u_1=-2i\omega_0 A' e^{i\omega_0 T_0}+10A^3\bar{A}^2 e^{i\omega_0 T_0}+\text{cc}+\text{NST} \tag{149}$$

Eliminating the terms that lead to secular terms in u_1, we have

$$i\omega_0 A'=5A^3\bar{A}^2 \tag{150}$$

Putting (39) in (150) and separating real and imaginary parts, we obtain

$$\frac{\partial a}{\partial T_1}=0 \tag{151}$$

$$-\omega_0\frac{\partial\beta}{\partial T_1}=\tfrac{5}{16}a^4 \quad \text{if } a\neq 0 \tag{152}$$

It follows from (151) that a is a constant and from (152) that

$$\beta=-\frac{5}{16\omega_0}T_1 a^4+\beta_0 \tag{153}$$

Therefore, to the first approximation, the resulting expansion is the same as (139).

(e) Use the method of averaging to determine a first-order uniform expansion.

As before, we use the method of variation of parameters and transform (128) into

$$u = a(t)\cos[\omega_0 t + \beta(t)] \tag{154}$$

$$\dot{u} = -\omega_0 a(t)\sin[\omega_0 t + \beta(t)] \tag{155}$$

where

$$\dot{a} = -\frac{\varepsilon}{\omega_0} a^5 \sin\phi \cos^5\phi = -\frac{\varepsilon}{32\omega_0} a^5(\sin 6\phi + 4\sin 4\phi + 5\sin 2\phi) \tag{156}$$

$$a\dot{\beta} = -\frac{\varepsilon}{\omega_0} a^5 \cos^6\phi = -\frac{\varepsilon}{32\omega_0} a^5(\cos 6\phi + 6\cos 4\phi + 15\cos 2\phi + 10) \tag{157}$$

where $\phi = \omega_0 t + \beta$. Keeping only the slowly varying terms on the right-hand sides of (156) and (157), we have

$$\dot{a} = 0 \tag{158}$$

$$\dot{\beta} = -\frac{5\varepsilon}{16\omega_0} a^4 \quad \text{if } a \neq 0 \tag{159}$$

It follows from (158) that a is a constant and from (159) that

$$\beta = -\frac{5\varepsilon}{16\omega_0} a^4 t + \beta_0 \tag{160}$$

Substituting (160) into (154) yields (139).

4.6. The motion of a simple pendulum is governed by

$$\ddot{\theta} + \frac{g}{l}\sin\theta = 0 \tag{161}$$

(a) Expand for small θ and keep up to cubic terms.

$$\ddot{\theta} + \omega_0^2\left(\theta - \frac{\theta^3}{3!} + \cdots\right) = 0$$

or

$$\ddot{\theta} + \omega_0^2\theta - \alpha\omega_0^2\theta^3 + \cdots = 0 \tag{162}$$

where

$$\omega_0^2 = \frac{g}{l}, \qquad \alpha = \tfrac{1}{6}$$

(b) Determine a first-order uniform expansion for small but finite θ.

As in Exercise 4.4, we scale θ with the factor ε^λ, where $\lambda > 0$ and ε is a measure of the amplitude. Thus we put $\theta = \varepsilon^\lambda u$ in (162) and obtain

$$\ddot{u} + \omega_0^2 u - \alpha\omega_0^2\varepsilon^{2\lambda}u^3 + \cdots = 0 \tag{163}$$

We choose $\lambda = \frac{1}{2}$ so that the nonlinearity will be proportional to ε and (163) becomes

$$\ddot{u} + \omega_0^2 u - \alpha\varepsilon\omega_0^2 u^3 + \cdots = 0 \tag{164}$$

Equation (164) is the same as equation (92) of Exercise 4.4. Hence its approximate solution can be obtained by using any one of the four techniques discussed in this chapter. The result is

$$u = a\cos\left[\left(1 - \tfrac{3}{8}\alpha\varepsilon a^2\right)\omega_0 t + \beta\right] + \cdots \tag{165}$$

But $\alpha = \frac{1}{6}$, hence

$$\theta = \varepsilon^{1/2} a \cos\left[\left(1 - \frac{\varepsilon a^2}{16}\right)\omega_0 t + \beta\right] + \cdots \tag{166}$$

Setting $\varepsilon = 1$ because it was introduced as a bookkeeping device, we have

$$\theta = a\cos\left[\left(1 - \frac{a^2}{16}\right)\omega_0 t + \beta\right] + \cdots \tag{167}$$

4.7. Consider the equation

$$\ddot{\theta} = \Omega^2 \sin\theta\cos\theta - \frac{g}{R}\sin\theta \tag{168}$$

(a) Expand for small θ and keep up to cubic terms.

$$\ddot{\theta} = \Omega^2\left(\theta - \frac{\theta^3}{3!} + \cdots\right)\left(1 - \frac{\theta^2}{2!} + \cdots\right) - \frac{g}{R}\left(\theta - \frac{\theta^3}{3!} + \cdots\right)$$

or

$$\ddot{\theta} + \left(\frac{g}{R} - \Omega^2\right)\theta = \left(\frac{g}{6R} - \frac{2\Omega^2}{3}\right)\theta^3 + \cdots \tag{169}$$

or

$$\ddot{\theta} + \omega_0^2\theta - \alpha\omega_0^2\theta^3 + \cdots = 0 \tag{170}$$

where

$$\omega_0^2 = \frac{g}{R} - \Omega^2, \qquad \alpha\omega_0^2 = \frac{g}{6R} - \frac{2\Omega^2}{3} \tag{171}$$

(b) Determine a first-order uniform expansion for small but finite θ.

As in the preceding exercise, we scale θ with $\varepsilon^{1/2}$, where ε is a measure of the amplitude. Thus we put $\theta = \varepsilon^{1/2} u$ in (170) and obtain

$$\ddot{u} + \omega_0^2 u = \varepsilon\alpha\omega_0^2 u^3 + \cdots \tag{172}$$

Equation (172) is the same as equation (92) of Exercise 4.4. Hence its approximate solution can be obtained as in Exercise 4.4 to be

$$u = a\cos\left[\omega_0\left(1 - \frac{3\varepsilon\alpha a^2}{8}\right)t + \beta\right] + \cdots \tag{173}$$

Then

$$\theta = a\cos\left[\omega_0\left(1 - \frac{3\alpha a^2}{8}\right)t + \beta\right] + \cdots \tag{174}$$

where ε is set equal to unity because it was introduced as a bookkeeping device.

4.8. The motion of a particle on a rotating parabola is governed by

$$(1 + 4p^2x^2)\ddot{x} + \Lambda x + 4p^2\dot{x}^2x = 0 \tag{175}$$

where p and Λ are constants. Determine a first-order uniform expansion for small but finite x.

To express the smallness of x, we scale it with the factor ε^λ, where $\lambda > 0$ and ε is a measure of the amplitude. Thus we put $x = \varepsilon^\lambda u$ in (175) and obtain

$$\ddot{u} + \omega_0^2 u + 4p^2\varepsilon^{2\lambda}(u^2\ddot{u} + u\dot{u}^2) = 0 \tag{176}$$

where $\Lambda = \omega_0^2$. We choose $\lambda = \frac{1}{2}$ so that the coefficient of the nonlinearity is ε and (176) becomes

$$\ddot{u} + \omega_0^2 u = -4p^2\varepsilon(u^2\ddot{u} + u\dot{u}^2) \tag{177}$$

It turns out that one can use any of the techniques described in this chapter to treat this problem. As a variation from the preceding two exercises, we use the method of multiple scales.

We seek a first-order expansion of (177) in the form

$$u = u_0(T_0, T_1) + \varepsilon u_1(T_0, T_1) + \cdots \tag{178}$$

Substituting (178) into (177) and equating coefficients of like powers of ε, we obtain

$$\frac{\partial^2 u_0}{\partial T_0^2} + \omega_0^2 u_0 = 0 \tag{179}$$

$$\frac{\partial^2 u_1}{\partial T_0^2} + \omega_0^2 u_1 = -\frac{2\,\partial^2 u_0}{\partial T_0\,\partial T_1} - 4p^2\left[u_0^2\frac{\partial^2 u_0}{\partial T_0^2} + u_0\left(\frac{\partial u_0}{\partial T_0}\right)^2\right] \tag{180}$$

The solution of (179) can be expressed as

$$u_0 = A(T_1)e^{i\omega_0 T_0} + \bar{A}(T_1)e^{-i\omega_0 T_0} \tag{181}$$

Then (180) becomes

$$\frac{\partial^2 u_1}{\partial T_0^2} + \omega_0^2 u_1 = -2i\omega_0 A' e^{i\omega_0 T_0} + 8p^2\omega_0^2 A^2\bar{A} e^{i\omega_0 T_0} + \text{cc} + \text{NST} \tag{182}$$

Eliminating the terms that lead to secular terms in (182), we have

$$2iA' = 8p^2\omega_0 A^2\bar{A} \tag{183}$$

Putting (39) in (183) and separating the result into real and imaginary parts, we obtain

$$\frac{\partial a}{\partial T_1} = 0 \tag{184}$$

$$\frac{\partial \beta}{\partial T_1} = -\,\omega_0 p^2 a^2 \quad \text{if } a \neq 0 \tag{185}$$

It follows from (184) that a is a constant and from (185) that

$$\beta = -\,\omega_0 p^2 a^2 T_1 + \beta_0 \tag{186}$$

where β_0 is a constant. Therefore,

$$u = a\cos\left(\omega_0 T_0 - \omega_0 p^2 a^2 T_1 + \beta_0\right) + \cdots$$

and

$$x = a\cos\left[\left(1 - p^2a^2\right)\omega_0 t + \beta_0\right] + \cdots \tag{187}$$

where ε is set equal to unity because it was introduced as a bookkeeping device.

4.9. Consider the equation

$$\left(1 + \frac{u^2}{1-u^2}\right)\ddot{u} + \frac{u\dot{u}^2}{(1-u^2)^2} + \omega_0^2 u + \frac{g}{l}\frac{u}{\sqrt{1-u^2}} = 0 \tag{188}$$

(a) Expand for small u and keep up to cubic terms.

$$\left[1 + u^2(1 + u^2 + \cdots)\right]\ddot{u} + u\dot{u}^2(1 + 2u^2 + \cdots) + \omega_0^2 u + \frac{g}{l}u\left(1 + \tfrac{1}{2}u^2 + \cdots\right)\right] = 0$$

or

$$(1 + u^2)\ddot{u} + u\dot{u}^2 + \left(\omega_0^2 + \frac{g}{l}\right)u + \frac{g}{2l}u^3 + \cdots = 0$$

or

$$\ddot{u} + \omega^2 u + \frac{g}{2l}u^3 + u\dot{u}^2 + u^2\ddot{u} + \cdots = 0 \tag{189}$$

where

$$\omega^2 = \omega_0^2 + \frac{g}{l}$$

(b) Determine a first-order uniform expansion for small but finite u.

As before, since u is small, we scale it with ε^λ, where $\lambda > 0$ and ε is a measure of the amplitude. Thus we put $u = \varepsilon^\lambda v$ in (189) and obtain

$$\ddot{v} + \omega^2 v + \varepsilon^{2\lambda}\left(\frac{g}{2l}v^3 + v\dot{v}^2 + v^2\ddot{v}\right) + \cdots = 0 \tag{190}$$

We choose $\lambda = \frac{1}{2}$ so that the coefficient of the nonlinearity is ε and (190) becomes

$$\ddot{v} + \omega^2 v + \varepsilon\left(\frac{g}{2l}v^3 + v\dot{v}^2 + v^2\ddot{v}\right) + \cdots = 0 \tag{191}$$

Approximations to the solution of (191) can be obtained by using any of the techniques described in this chapter because (191) represents a conservative system. As a variation from Exercise 4.8, we use the Lindstedt–Poincaré technique.

Let

$$v = v_0(\tau) + \varepsilon v_1(\tau) + \cdots \tag{192}$$

$$\tau = \Omega t = (\omega + \varepsilon\Omega_1 + \cdots)t \tag{193}$$

in (191), equate coefficients of like powers of ε, and obtain

$$v_0'' + v_0 = 0 \tag{194}$$

$$\omega^2(v_1'' + v_1) = -2\omega\Omega_1 v_0'' - \frac{g}{2l}v_0^3 - \omega^2 v_0 v_0'^2 - \omega^2 v_0^2 v_0'' \tag{195}$$

The solution of (194) can be expressed as

$$v_0 = a\cos(\tau + \beta) \tag{196}$$

where a and β are constants. Then (195) becomes

$$\begin{aligned}
\omega^2(v_1'' + v_1) &= 2\omega\Omega_1 a\cos(\tau+\beta) - \frac{g}{2l}a^3\cos^3(\tau+\beta) \\
&\quad - \omega^2 a^3\sin^2(\tau+\beta)\cos(\tau+\beta) + \omega^2 a^3\cos^3(\tau+\beta) \\
&= \left[2\omega\Omega_1 a - \frac{3g}{8l}a^3 + \tfrac{1}{2}\omega^2 a^3\right]\cos(\tau+\beta) + \text{NST}
\end{aligned} \tag{197}$$

Eliminating the terms that produce secular terms in v_1, we have

$$\Omega_1 = \left(\frac{3g}{16\omega l} - \tfrac{1}{4}\omega\right)a^2 \tag{198}$$

Therefore,

$$u = a\cos(\tau+\beta) + \cdots = a\cos\left\{\left[1+\left(\frac{3g}{16\omega^2 l} - \frac{1}{4}\right)a^2\right]\omega t + \beta\right\} + \cdots \tag{199}$$

where ε is set equal to unity because it was introduced as a bookkeeping device.

4.10. Consider the equation

$$(l^2 + r^2 - 2rl\cos\theta)\ddot{\theta} + rl\sin\theta\dot{\theta}^2 + gl\sin\theta = 0 \tag{200}$$

where g, r, and l are constants. Determine a first-order uniform expansion for small but finite θ.

We first expand (200) for small θ as

$$\left[l^2 + r^2 - 2rl\left(1 - \frac{\theta^2}{2!} + \cdots\right)\right]\ddot{\theta} + rl\left(\theta - \frac{\theta^3}{3!} + \cdots\right)\dot{\theta}^2 + gl\left(\theta - \frac{\theta^3}{3!} + \cdots\right) = 0$$

or

$$(l-r)^2\ddot{\theta} + gl\theta + rl\theta^2\ddot{\theta} + rl\theta\dot{\theta}^2 - \frac{gl}{6}\theta^3 + \cdots = 0$$

or

$$\ddot{\theta} + \omega_0^2\theta + \alpha_1(\theta^2\ddot{\theta} + \theta\dot{\theta}^2) - \alpha_2\omega_0^2\theta^3 + \cdots = 0 \tag{201}$$

where

$$\omega_0^2 = \frac{gl}{(l-r)^2}, \qquad \alpha_1 = \frac{rl}{(l-r)^2}, \qquad \alpha_2 = \frac{1}{6}$$

As in the preceding section, we scale θ by letting $\theta = \varepsilon^{1/2}u$ in (201) and obtain

$$\ddot{u} + \omega_0^2 u + \varepsilon\left[\alpha_1(u^2\ddot{u} + u\dot{u}^2) - \alpha_2\omega_0^2 u^3\right] + \cdots = 0 \tag{202}$$

We use the Lindstedt–Poincaré technique and let

$$u = u_0(\tau) + \varepsilon u_1(\tau) + \cdots \tag{203}$$

$$\tau = \omega t = (\omega_0 + \varepsilon\omega_1 + \cdots)t \tag{204}$$

Substituting (203) and (204) into (202) and equating coefficients of like powers of ε, we obtain

$$u_0'' + u_0 = 0 \tag{205}$$

$$\omega_0^2(u_1'' + u_1) = -2\omega_0\omega_1 u_0'' - \omega_0^2\alpha_1(u_0^2 u_0'' + u_0 u_0'^2) + \alpha_2\omega_0^2 u_0^3 \tag{206}$$

The solution of (205) can be expressed as

$$u_0 = a\cos(\tau + \beta) \tag{207}$$

where a and β are constants. Then (206) becomes

$$\begin{aligned}\omega_0^2(u_1'' + u_1) &= 2\omega_0\omega_1 a\cos(\tau+\beta) - \omega_0^2\alpha_1 a^3[-\cos^3(\tau+\beta) \\ &\quad + \sin^2(\tau+\beta)\cos(\tau+\beta)] + \alpha_2\omega_0^2 a^3\cos^3(\tau+\beta) \\ &= [2\omega_0\omega_1 a + \tfrac{3}{4}\omega_0^2\alpha_1 a^3 - \tfrac{1}{4}\omega_0^2\alpha_1 a^3 + \tfrac{3}{4}\alpha_2\omega_0^2 a^3]\cos(\tau+\beta) + \text{NST}\end{aligned} \tag{208}$$

Eliminating the terms that lead to secular terms in u_1, we have

$$\omega_1 = -\tfrac{1}{8}(3\alpha_2 + 2\alpha_1)\,\omega_0 a^2 \tag{209}$$

Therefore

$$\theta = a\cos(\tau+\beta) + \cdots = a\cos\{[1 - \tfrac{1}{8}(2\alpha_1 + 3\alpha_2)a^2]\,\omega_0 t + \beta\} + \cdots \tag{210}$$

where again ε is set equal to unity because it was introduced as a bookkeeping device.

4.11. Consider the equation

$$(\tfrac{1}{12}l^2 + r^2\theta^2)\ddot\theta + r^2\theta\dot\theta^2 + gr\theta\cos\theta = 0 \tag{211}$$

where r, l, and g are constants. Determine a first-order uniform expansion for small but finite θ.

First, we expand (211) for small θ as

$$(\tfrac{1}{12}l^2 + r^2\theta^2)\ddot\theta + r^2\theta\dot\theta^2 + gr\theta\left(1 - \frac{\theta^2}{2!} + \cdots\right) = 0$$

or

$$\tfrac{1}{12}l^2\ddot\theta + gr\theta + r^2(\theta^2\ddot\theta + \theta\dot\theta^2) - \tfrac{1}{2}gr\theta^3 + \cdots = 0$$

or

$$\ddot\theta + \omega_0^2\theta + \alpha_1(\theta^2\ddot\theta + \theta\dot\theta^2) - \alpha_2\omega_0^2\theta^3 + \cdots = 0 \tag{212}$$

where

$$\omega_0^2 = \frac{12gr}{l^2}, \qquad \alpha_1 = \frac{12r^2}{l^2}, \qquad \alpha_2 = \tfrac{1}{2}$$

Equation (212) is the same as equation (201) in the preceding exercise. Hence an approximate solution of (212) is given by (210).

4.12. Consider the equation

$$m\ddot x + kx(x^2 + l^2)^{-1/2}[(x^2 + l^2)^{1/2} - \tfrac{1}{2}l] = 0 \tag{213}$$

Determine a first-order uniform expansion for small but finite x.

First, we expand (213) for small x. To this end, we note that

$$(x^2+l^2)^{-1/2}\left[(x^2+l^2)^{1/2}-\tfrac{1}{2}l\right]=1-\tfrac{1}{2}l(x^2+l^2)^{-1/2}$$

$$=1-\tfrac{1}{2}\left(1+\frac{x^2}{l^2}\right)^{-1/2}=1-\tfrac{1}{2}\left(1-\frac{x^2}{2l^2}+\cdots\right)$$

$$=\tfrac{1}{2}+\frac{x^2}{4l^2}+\cdots$$

Hence (213) becomes

$$m\ddot{x}+\tfrac{1}{2}kx+\frac{k}{4l^2}x^3+\cdots=0$$

or

$$\ddot{u}+\omega_0^2 u-\alpha\omega_0^2 u^3=0 \tag{214}$$

where

$$\omega_0^2=\frac{k}{2m},\qquad u=\frac{x}{l},\qquad \alpha=-\tfrac{1}{2}$$

Equation (214) is the same as equation (162) of Exercise 4.6. Hence an approximate solution of (214) is given by (166) with ε being replaced with unity.

4.13. Expand the integrand in

$$T=\frac{2}{\sqrt{1+\varepsilon x_0^2}}\int_0^{\pi}\frac{d\theta}{\sqrt{1-m\sin^2\theta}} \tag{215a}$$

up to $O(m^2)$ and obtain

$$T_a=\frac{2\pi}{\sqrt{1+\varepsilon x_0^2}}\left(1+\tfrac{1}{4}m+\tfrac{9}{64}m^2+\cdots\right) \tag{215b}$$

It follows from (215a) that

$$T=\frac{4}{\sqrt{1+\varepsilon x_0^2}}\int_0^{\pi/2}\left[1+\tfrac{1}{2}m\sin^2\theta+\tfrac{3}{8}m^2\sin^4\theta+\cdots\right]d\theta$$

Hence

$$T=\frac{4}{\sqrt{1+\varepsilon x_0^2}}\left[\int_0^{\pi/2}d\theta+\frac{1}{2}m\int_0^{\pi/2}\sin^2\theta\,d\theta+\frac{3}{8}m^2\int_0^{\pi/2}\sin^4\theta\,d\theta\right]+\cdots$$

or

$$T_a = \frac{4}{\sqrt{1+\varepsilon x_0^2}}\left[\tfrac{1}{2}\pi + \tfrac{1}{2}m\cdot\tfrac{1}{4}\pi + \tfrac{3}{8}m^2\cdot\tfrac{3}{16}\pi + \cdots\right] \tag{216}$$

Then (215b) follows immediately from (216).

Let $m = \varepsilon x_0^2/2(1+\varepsilon x_0^2)$ and express T_a in terms of m. Solving for εx_0^2 in terms of m, we have

$$\varepsilon x_0^2 = 2m + 2\varepsilon x_0^2 m$$

Hence

$$\varepsilon x_0^2 = \frac{2m}{1-2m}$$

and

$$1+\varepsilon x_0^2 = \frac{2m}{1-2m} + 1 = \frac{1}{1-2m} \tag{217}$$

Using (217) in (215b), we have

$$T_a = 2\pi\sqrt{1-2m}\left(1 + \tfrac{1}{4}m + \tfrac{9}{64}m^2 + \cdots\right) \tag{218}$$

Express T_a in terms of x_0. Substituting for m into (215b) yields

$$T_a = \frac{2\pi}{\sqrt{1+\varepsilon x_0^2}}\left[1 + \frac{\varepsilon x_0^2}{8(1+\varepsilon x_0^2)} + \frac{9\varepsilon^2 x_0^4}{256(1+\varepsilon x_0^2)^2} + \cdots\right]$$

$$= 2\pi\left(1 - \tfrac{1}{2}\varepsilon x_0^2 + \tfrac{3}{8}\varepsilon^2 x_0^4 + \cdots\right)\left(1 + \tfrac{1}{8}\varepsilon x_0^2 - \tfrac{1}{8}\varepsilon^2 x_0^4 + \tfrac{9}{256}\varepsilon^2 x_0^4 + \cdots\right)$$

$$= 2\pi\left(1 - \tfrac{3}{8}\varepsilon x_0^2 + \tfrac{57}{256}\varepsilon^2 x_0^4 + \cdots\right). \tag{219}$$

Supplementary Exercises

4.14. Consider the equation

$$\ddot{u} + \omega_0^2 u = \varepsilon \dot{u}^2 u^3, \qquad \varepsilon \ll 1$$

(a) Determine a two-term straightforward expansion and discuss its uniformity.

(b) Render this expansion uniformly valid by using the method of renormalization.

(c) Determine a first-order uniform expansion by using the Lindstedt–Poincaré technique.

(d) Use the method of multiple scales to determine a first-order uniform expansion.

(e) Use the method of averaging to determine a first-order uniform expansion.

4.15. Consider the equation

$$\ddot{u} + \omega_0^2 u + \varepsilon \dot{u}^4 \ddot{u} = 0, \qquad \varepsilon \ll 1$$

(a) Determine a two-term straightforward expansion and discuss its uniformity.
(b) Render this expansion uniformly valid by using the method of renormalization.
(c) Determine a first-order uniform expansion by using the Lindstedt–Poincaré technique.
(d) Use the method of multiple scales to determine a first-order uniform expansion.
(e) Use the method of averaging to determine a first-order uniform expansion.

4.16. Consider the equation

$$\ddot{u} + \omega_0^2 u + \varepsilon \dot{u}^4 u = 0, \qquad \varepsilon \ll 1$$

(a) Determine a two-term straightforward expansion and discuss its uniformity.
(b) Render this expansion uniformly valid by using the method of renormalization.
(c) Determine a first-order uniform expansion by using the Lindstedt–Poincaré technique.
(d) Use the method of multiple scales to determine a first-order uniform expansion.
(e) Use the method of averaging to determine a first-order uniform expansion.

4.17. Consider the equation

$$\ddot{u} + \omega_0^2 u + \varepsilon\left(\alpha_1 u^5 + \alpha_2 \dot{u}^2 u^3 + \alpha_3 \dot{u}^4 u + \alpha_4 u^4 \ddot{u}\right) = 0, \qquad \varepsilon \ll 1$$

(a) Determine a two-term straightforward expansion and discuss its uniformity.
(b) Render this expansion uniformly valid by using the method of renormalization.
(c) Determine a first-order uniform expansion by using the Lindstedt–Poincaré technique.
(d) Use the method of multiple scales to determine a first-order uniform expansion.
(e) Use the method of averaging to determine a first-order uniform expansion.

4.18. Determine a first-order uniform expansion for small but finite amplitude for the solution of each of the following equations:

(a) $\ddot{u} + \omega_0^2 u = u^7$
(b) $\ddot{u} + \omega_0^2 u = u^6 \ddot{u}$
(c) $\ddot{u} + \omega_0^2 u = \dot{u}^4 u^3$
(d) $\ddot{u} + \omega_0^2 u = \dot{u}^2 u^5$
(e) $\ddot{u} + \omega_0^2 u = \dot{u}^2 u^4 \ddot{u}$
(f) $\ddot{u} + \dfrac{\omega^2 u}{1 + u^4} = 0$
(g) $\ddot{u} + \dfrac{\omega^2 \dot{u}^2}{1 + u^3} = 0$
(h) $\ddot{u} + \omega_0^2 u = u^n, \qquad n$ is a positive odd integer.

CHAPTER 5

Free Oscillations of Positively Damped Systems

In the preceding chapter, we considered methods of obtaining approximations to the free oscillations of conservative single-degree-of-freedom systems. However, most physical systems possess damping forces (i.e., forces that are functions of the velocity). Damping that causes the free-oscillation amplitude to decrease is called positive, whereas damping that causes the free-oscillation amplitude to increase is called negative. Positive damping mechanisms include Columb damping of the form $\mu \operatorname{sgn} \dot{u}$, linear viscous damping $2\mu\dot{u}$, nonlinear damping $\mu|\dot{u}|^{\alpha}\dot{u}$ or $\mu f(u)\dot{u}$, and hysteretic damping. Positively damped systems are treated in this chapter and negatively damped systems are treated in the next chapter.

In this chapter, we consider the problem of obtaining approximations to the solutions of the equation

$$\ddot{u} + \omega_0^2 u = \varepsilon f(u, \dot{u})$$

when $f(u, \dot{u})$ contains positive damping terms. For such systems, the free-oscillation amplitude is not constant but decreases continuously with time. Although both the Lindstedt–Poincaré technique and the method of renormalization account for the shift in frequency caused by the perturbation, they are not expected to yield approximations to the transient free oscillations because neither accounts for variations in the amplitude. In contrast, both the method of multiple scales and the method of averaging yield uniform expansions for the transient oscillations.

Solved Exercises

5.1. Consider the equation

$$\ddot{u} + 2\varepsilon\mu\dot{u} + u + \varepsilon u^3 = 0, \qquad \varepsilon \ll 1 \tag{1}$$

Use the methods of multiple scales and averaging to determine a first-order uniform expansion for u.

(a) *The Method of Multiple Scales*

We seek a first-order uniform expansion in the form

$$u = u_0(T_0, T_1) + \varepsilon u_1(T_0, T_1) + \cdots \tag{2}$$

Substituting (2) into (1) and equating coefficients of like powers of ε, we obtain

$$D_0^2 u_0 + u_0 = 0 \tag{3}$$

$$D_0^2 u_1 + u_1 = -2D_0 D_1 u_0 - 2\mu D_0 u_0 - u_0^3 \tag{4}$$

The solution of (3) can be expressed as

$$u_0 = A(T_1) e^{iT_0} + \bar{A}(T_1) e^{-iT_0} \tag{5}$$

Then (4) becomes

$$D_0^2 u_1 + u_1 = -2i(D_1 A + \mu A) e^{iT_0} - 3A^2 \bar{A} e^{iT_0} + \text{cc} + \text{NST} \tag{6}$$

Eliminating the secular terms from (6), we have

$$2i(A' + \mu A) + 3A^2 \bar{A} = 0 \tag{7}$$

Expressing A in the polar form

$$A = \tfrac{1}{2} a e^{i\beta} \tag{8}$$

and separating (7) into real and imaginary parts, we obtain

$$a' = -\mu a \tag{9}$$

$$a\beta' = \tfrac{3}{8} a^3 \tag{10}$$

The solution of (9) can be written as

$$a = a_0 e^{-\mu T_1} \tag{11}$$

where a_0 is a constant. Then (10) becomes

$$\beta' = \tfrac{3}{8} a_0^2 e^{-2\mu T_1} \tag{12}$$

whose solution is

$$\beta = -\frac{3}{16\mu} a_0^2 e^{-2\mu T_1} + \beta_0 \tag{13}$$

where β_0 is a constant. Therefore, to the first approximation,

$$u = \tfrac{1}{2}a_0 e^{-\mu T_1}\exp\left[iT_0 - \frac{3i}{16\mu}a_0^2 e^{-2\mu T_1} + i\beta_0\right] + \text{cc} + \cdots$$

$$= a_0 e^{-\epsilon\mu t}\cos\left(t - \frac{3a_0^2}{16\mu}e^{-2\epsilon\mu t} + \beta_0\right) + \cdots \tag{14}$$

(b) ***The Method of Averaging***

We introduce a transformation from $u(t)$ to $a(t)$ and $\beta(t)$ defined by

$$u = a(t)\cos[t + \beta(t)] \tag{15}$$

$$\dot{u} = -a(t)\sin[t + \beta(t)] \tag{16}$$

which coincides with the solution of (1) when $\varepsilon = 0$. Differentiating (15) once and comparing the result with (16) yields

$$\dot{a}\cos\phi - a\dot{\beta}\sin\phi = 0 \tag{17}$$

where $\phi = t + \beta$. Differentiating (16) once and substituting the result together with (15) and (16) into (1), we obtain

$$-\dot{a}\sin\phi - a\dot{\beta}\cos\phi = \varepsilon(2\mu a\sin\phi - a^3\cos^3\phi) \tag{18}$$

Solving (17) and (18) for $\dot{a}$ and $\dot{\beta}$ yields

$$\dot{a} = -\varepsilon(2\mu a\sin^2\phi - a^3\sin\phi\cos^3\phi)$$

$$= -\varepsilon(\mu a - \mu a\cos 2\phi - \tfrac{1}{4}a^3\sin 2\phi - \tfrac{1}{8}a^3\sin 4\phi) \tag{19}$$

$$a\dot{\beta} = -\varepsilon(2\mu a\sin\phi\cos\phi - a^3\cos^4\phi)$$

$$= -\varepsilon(\mu a\sin 2\phi - \tfrac{3}{8}a^3 - \tfrac{1}{2}a^3\cos 2\phi - \tfrac{1}{8}a^3\cos 4\phi) \tag{20}$$

Keeping only the slowly varying parts on the right-hand sides of (19) and (20), we obtain

$$\dot{a} = -\varepsilon\mu a \tag{21}$$

$$a\dot{\beta} = \tfrac{3}{8}\varepsilon a^3 \tag{22}$$

in agreement with (9) and (10), obtained by using the method of multiple scales since $T_1 = \varepsilon t$.

5.2. Consider the equation

$$\ddot{u} + \omega_0^2 u + \varepsilon\dot{u}^3 = 0, \qquad \varepsilon \ll 1 \tag{23}$$

Use the methods of multiple scales and averaging to determine a first-order uniform expansion for u.

(a) *The Method of Multiple Scales*

Substituting (2) into (23) and equating coefficients of like powers of ε, we obtain

$$D_0^2 u_0 + \omega_0^2 u_0 = 0 \tag{24}$$

$$D_0^2 u_1 + \omega_0^2 u_1 = -2 D_0 D_1 u_0 - (D_0 u_0)^3 \tag{25}$$

The solution of (24) can be expressed as

$$u_0 = A(T_1) e^{i\omega_0 T_0} + \bar{A}(T_1) e^{-i\omega_0 T_0} \tag{26}$$

Then (25) becomes

$$D_0^2 u_1 + \omega_0^2 u_1 = -2i\omega_0 A' e^{i\omega_0 T_0} - 3i\omega_0^3 A^2 \bar{A} e^{i\omega_0 T_0} + \text{cc} + \text{NST} \tag{27}$$

Eliminating the terms that lead to secular terms in u_1, we have

$$2iA' + 3i\omega_0^2 A^2 \bar{A} = 0 \tag{28}$$

Substituting (8) into (28) and separating the result into real and imaginary parts, we obtain

$$a' = -\tfrac{3}{8}\omega_0^2 a^3 \tag{29}$$

$$\beta' = 0 \tag{30}$$

It follows from (30) that β is a constant. Separating variables in (29) gives

$$\frac{da}{a^3} = -\tfrac{3}{8}\omega_0^2 \, dT_1$$

Hence

$$-\frac{1}{2a^2} = -\tfrac{3}{8}\omega_0^2 T_1 + c \tag{31}$$

where c is a constant. If $a = a_0$ at $t = 0$,

$$-\frac{1}{2a_0^2} = c$$

and (31) becomes

$$\frac{1}{2a_0^2} - \frac{1}{2a^2} = -\tfrac{3}{8}\omega_0^2 T_1 \tag{32}$$

Hence

$$a^2 = \frac{a_0^2}{1 + \frac{3}{4} a_0^2 \omega_0^2 T_1} \tag{33}$$

and to the first approximation,

$$u = a\cos(\omega_0 T_0 + \beta) + \cdots = \frac{a_0}{\sqrt{1 + \frac{3}{4}\varepsilon a_0^2 \omega_0^2 t}} \cos(\omega_0 t + \beta) + \cdots \tag{34}$$

(b) *The Method of Averaging*

We introduce a transformation from $u(t)$ to $a(t)$ and $\beta(t)$ according to

$$u = a(t)\cos[\omega_0 t + \beta(t)] \tag{35}$$

$$\dot{u} = -\omega_0 a(t)\sin[\omega_0 t + \beta(t)] \tag{36}$$

Differentiating (35) once and comparing the result with (36), we have

$$\dot{a}\cos\phi - a\dot{\beta}\sin\phi = 0 \tag{37}$$

where $\phi = \omega_0 t + \beta$. Differentiating (36) once and substituting the result together with (35) and (36) into (23), we obtain

$$-\dot{a}\sin\phi - a\dot{\beta}\cos\phi = \varepsilon\omega_0^2 a^3 \sin^3\phi \tag{38}$$

Solving (37) and (38) for $\dot{a}$ and $\dot{\beta}$ yields

$$\dot{a} = -\varepsilon\omega_0^2 a^3 \sin^4\phi = -\frac{\varepsilon}{8}\omega_0^2 a^3(\cos 4\phi - 4\cos 2\phi + 3) \tag{39}$$

$$a\dot{\beta} = -\varepsilon\omega_0^2 a^3 \cos\phi \sin^3\phi = -\frac{\varepsilon}{8}\omega_0^2 a^3(2\sin 2\phi - \sin 4\phi) \tag{40}$$

Keeping the slowly varying terms on the right-hand sides of (39) and (40) yields

$$\dot{a} = -\tfrac{3}{8}\varepsilon\omega_0^2 a^3 \tag{41}$$

$$\dot{\beta} = 0 \tag{42}$$

in agreement with (29) and (30), obtained by using the method of multiple scales.

5.3. Consider the following equation:

$$\ddot{u} + \omega_0^2 u + 2\varepsilon\mu u^2 \dot{u} + \varepsilon\alpha u^3 = 0 \tag{43}$$

Show that to the first approximation

$$u = a\cos(\omega_0 t + \beta) + O(\varepsilon) \tag{44}$$

and determine the equations governing a and β by using the methods of multiple scales and averaging.

(a) *The Method of Multiple Scales*

Substituting (2) into (43) and equating coefficients of like powers of ε, we obtain

$$D_0^2 u_0 + \omega_0^2 u_0 = 0 \tag{45}$$

$$D_0^2 u_1 + \omega_0^2 u_1 = -2 D_0 D_1 u_0 - 2\mu u_0^2 D_0 u_0 - \alpha u_0^3 \tag{46}$$

The solution of (45) is given by (26) and then (46) becomes

$$D_0^2 u_1 + \omega_0^2 u_1 = -2 i \omega_0 A' e^{i\omega_0 T_0} - 2 i \mu \omega_0 A^2 \bar{A} e^{i\omega_0 T_0} - 3\alpha A^2 \bar{A} e^{i\omega_0 T_0} + \text{cc} + \text{NST} \tag{47}$$

Eliminating the terms that produce secular terms in u_1, we have

$$2 i \omega_0 A' + 2 i \mu \omega_0 A^2 \bar{A} + 3\alpha A^2 \bar{A} = 0 \tag{48}$$

Substituting (8) into (48) and separating the result into real and imaginary parts, we obtain

$$a' = -\tfrac{1}{4}\mu a^3 \tag{49}$$

$$\beta' = \frac{3\alpha}{8\omega_0} a^2 \quad \text{if } a \neq 0 \tag{50}$$

Equation (49) is the same as (29) if $\frac{3}{8}\omega_0^2$ is identified with $\frac{1}{4}\mu$. Hence it follows from (33) that the solution of (49) is

$$a = \frac{a_0}{\sqrt{1 + \frac{1}{2}\mu a_0^2 T_1}} \tag{51}$$

To solve (50), we change the independent variable from T_1 to a by using (49). Thus dividing (50) by (49) gives

$$\frac{\beta'}{a'} = -\frac{3\alpha}{2\omega_0 \mu a}$$

or

$$\frac{d\beta}{da} = -\frac{3\alpha}{2\omega_0 \mu a}$$

or

$$d\beta = -\frac{3\alpha}{2\omega_0 \mu} \frac{da}{a}$$

Hence

$$\beta = -\frac{3\alpha}{2\omega_0\mu}\ln a + \hat{\beta}_0 \tag{52}$$

where $\hat{\beta}_0$ is a constant. Substituting (51) into (52) gives

$$\beta = \frac{3\alpha}{4\omega_0\mu}\ln\left(1+\tfrac{1}{2}\mu a_0^2 T_1\right)+\beta_0 \tag{53}$$

Therefore, to the first approximation,

$$u = \frac{a_0}{\sqrt{1+\frac{1}{2}\varepsilon\mu a_0^2 t}}\cos\left[\omega_0 t + \frac{3\alpha}{4\omega_0\mu}\ln\left(1+\tfrac{1}{2}\varepsilon\mu a_0^2 t\right)+\beta_0\right]+\cdots \tag{54}$$

(b) *The Method of Averaging*

We introduce the transformation (35) and (36). Differentiating (35) and comparing the result with (36) yields (37). Differentiating (36) once and substituting the result together with (35) and (36) into (43), we obtain

$$-\dot{a}\sin\phi - a\dot{\beta}\cos\phi = 2\varepsilon\mu a^3\sin\phi\cos^2\phi - \frac{\varepsilon\alpha a^3}{\omega_0}\cos^3\phi \tag{55}$$

Solving (37) and (55) for $\dot{a}$ and $\dot{\beta}$, we obtain

$$\dot{a} = -2\varepsilon\mu a^3\sin^2\phi\cos^2\phi + \frac{\varepsilon\alpha a^3}{\omega_0}\sin\phi\cos^3\phi = -\tfrac{1}{4}\varepsilon\mu a^3(1-\cos 4\phi)$$

$$+\frac{\varepsilon\alpha a^3}{8\omega_0}(2\sin 2\phi+\sin 4\phi) \tag{56}$$

$$a\dot{\beta} = -2\varepsilon\mu a^3\sin\phi\cos^3\phi + \frac{\varepsilon\alpha a^3}{\omega_0}\cos^4\phi = -\tfrac{1}{4}\varepsilon\mu a^3(2\sin 2\phi+\sin 4\phi)$$

$$+\frac{\varepsilon\alpha a^3}{8\omega_0}(3+4\cos 2\phi+\cos 4\phi) \tag{57}$$

Keeping the slowly varying terms on the right-hand sides of (56) and (57), we obtain

$$\dot{a} = -\tfrac{1}{4}\varepsilon\mu a^3 \tag{58}$$

$$\dot{\beta} = \frac{3\varepsilon\alpha}{8\omega_0}a^2 \quad \text{if } a\neq 0 \tag{59}$$

in agreement with (49) and (50), obtained by using the method of multiple scales.

5.4. Use the methods of multiple scales and averaging to determine a first-order uniform expansion for the solution of

$$\ddot{\theta} + \omega^2 \sin\theta + \frac{4\sin^2\theta}{1+4(1-\cos\theta)}\dot{\theta} = 0 \tag{60}$$

for small but finite θ.

We first expand (60) for small θ. Thus we put

$$\ddot{\theta} + \omega^2\left(\theta - \frac{\theta^3}{3!} + \cdots\right) + \frac{4(\theta - \theta^3/3! + \cdots)^2\dot{\theta}}{1+4(1-1+\theta^2/2! + \cdots)} = 0$$

or

$$\ddot{\theta} + \omega^2\theta - \tfrac{1}{6}\omega^2\theta^3 + 4\theta^2\dot{\theta} + \cdots = 0 \tag{61}$$

Since θ is small, we scale it with ε^λ, where $\lambda > 0$ and ε is a measure of the amplitude. Thus we put $\theta = \varepsilon^\lambda u$ in (61) and obtain

$$\ddot{u} + \omega^2 u + \varepsilon^{2\lambda}\left(4u^2\dot{u} - \tfrac{1}{6}\omega^2 u^3\right) = 0 \tag{62}$$

We choose $\lambda = \frac{1}{2}$ so that the coefficient of the nonlinearity is ε and (62) becomes

$$\ddot{u} + \omega^2 u + \varepsilon\left(4u^2\dot{u} - \tfrac{1}{6}\omega^2 u^3\right) + \cdots = 0 \tag{63}$$

Equation (63) is a special case of (43) in which $\mu = 2$ and $\alpha = -\frac{1}{6}\omega^2$. Thus an approximation to its solution can be obtained as in Exercise 5.3. It follows from (54) that the solution of (63) is

$$\theta = u = \frac{a_0}{\sqrt{1+a_0^2 t}}\cos\left[\omega t - \frac{\omega}{16}\ln\left(1+a_0^2 t\right) + \beta_0\right] + \cdots \tag{64}$$

where ε is set equal to unity because it was introduced as a bookkeeping device.

5.5. Consider the equation

$$\ddot{u} + \omega_0^2 u + \frac{\mu\dot{u}}{1-u^2} = 0 \tag{65}$$

Determine a first-order uniform expansion for small u.

First, we expand (65) for small u as

$$\ddot{u} + \omega_0^2 u + \mu\dot{u}(1+u^2+\cdots) = 0 \tag{66}$$

We take μ as the small parameter in an expansion using the method of multiple scales, or we take the term proportional to μ as the perturbation in an expansion using the method of averaging. To apply the method of multiple scales, we let

$$u = u_0(T_0, T_1) + \varepsilon u_1(T_0, T_1) + \cdots \tag{67}$$

where $T_0 = t$ and $T_1 = \mu t$. Substituting (67) into (66) and equating coefficients of like powers of μ, we obtain

$$D_0^2 u_0 + \omega_0^2 u_0 = 0 \tag{68}$$

$$D_0^2 u_1 + \omega_0^2 u_1 = -2 D_0 D_1 u_0 - \left(1 + u_0^2\right) D_0 u_0 \tag{69}$$

The solution of (68) can be expressed as in (26) and then (69) becomes

$$D_0^2 u_1 + \omega_0^2 u_1 = -2i\omega_0 A' e^{i\omega_0 T_0} - i\omega_0 A e^{i\omega_0 T_0} - i\omega_0 A^2 \bar{A} e^{i\omega_0 T_0} + \text{cc} + \text{NST} \tag{70}$$

Eliminating the terms that lead to secular terms in u_1, we have

$$2A' + A + A^2\bar{A} = 0 \tag{71}$$

Substituting (8) into (71) and separating real and imaginary parts, we obtain

$$a' = -\tfrac{1}{2}\left(a + \tfrac{1}{4}a^3\right) \tag{72}$$

$$\beta' = 0 \tag{73}$$

The solution of (73) is that β is a constant, whereas the solution of (72) can be obtained by separation of variables. Thus

$$-dT_1 = \frac{8\,da}{a(4 + a^2)} = \frac{2\,da}{a} - \frac{2a\,da}{(4 + a^2)}$$

Hence

$$-T_1 + c = 2\ln a - \ln(4 + a^2)$$

or

$$\frac{a^2}{4 + a^2} = e^{c - T_1} = \frac{1}{e^{T_1 - c}} \tag{74}$$

Solving (74) for a^2 gives

$$a^2 = \frac{4}{e^{T_1 - c} - 1} \tag{75}$$

If $a(0) = a_0$, it follows from (74) that

$$e^{-c} = \frac{4 + a_0^2}{a_0^2} = 1 + \frac{4}{a_0^2}$$

and (75) becomes

$$a^2 = \frac{4}{\left(1+4/a_0^2\right)e^{\mu t} - 1} \tag{76}$$

Therefore

$$u = \frac{2}{\sqrt{\left(1+4/a_0^2\right)e^{\mu t} - 1}} \cos(\omega_0 t + \beta) + \cdots \tag{77}$$

5.6. Use the methods of multiple scales and averaging to determine a first-order uniform expansion for

$$\ddot{u} + u + \varepsilon \dot{u}^5 = 0, \qquad \varepsilon \ll 1 \tag{78}$$

(a) *The Method of Multiple Scales*

Substituting (2) into (78) and equating coefficients of like powers of ε, we obtain

$$D_0^2 u_0 + u_0 = 0 \tag{79}$$

$$D_0^2 u_1 + u_1 = -2 D_0 D_1 u_0 - (D_0 u_0)^5 \tag{80}$$

The solution of (79) can be expressed as in (5) and hence (80) becomes

$$D_0^2 u_1 + u_1 = -2iA' e^{iT_0} - 10iA^3 \bar{A}^2 e^{iT_0} + \text{cc} + \text{NST} \tag{81}$$

Eliminating the terms that produce secular terms yields

$$A' + 5A^3 \bar{A}^2 = 0 \tag{82}$$

Substituting (8) into (82) and separating real and imaginary parts, we obtain

$$a' = -\tfrac{5}{16} a^5 \tag{83}$$

$$\beta' = 0 \tag{84}$$

The solution of (84) is that β is a constant, whereas the solution of (83) can be obtained by separation of variables. Thus we write

$$-\tfrac{5}{16} dT_1 = \frac{da}{a^5}$$

Hence

$$-\tfrac{5}{16} T_1 + c = -\frac{1}{4a^4} \tag{85}$$

If $a = a_0$ when $t = 0$, then

$$c = -\frac{1}{4a_0^4}$$

and

$$\frac{1}{4a^4} = \tfrac{5}{16}T_1 + \frac{1}{4a_0^4}$$

or

$$a^4 = \frac{a_0^4}{1 + \tfrac{5}{4}a_0^4 T_1} \tag{86}$$

Therefore

$$u = a_0\left(1 + \tfrac{5}{4}\varepsilon t a_0^4\right)^{-1/4}\cos(t + \beta) + \cdots \tag{87}$$

(b) *The Method of Averaging*

We introduce the transformation (15) and (16). Differentiating (15) once and comparing the result with (16) gives (17). Differentiating (16) once and substituting the result together with (15) and (16) into (78), we obtain

$$-\dot{a}\sin\phi - a\dot{\beta}\cos\phi = \varepsilon a^5 \sin^5\phi \tag{88}$$

Solving (17) and (88) for $\dot{a}$ and $\dot{\beta}$ yields

$$\dot{a} = -\varepsilon a^5 \sin^6\phi = \tfrac{1}{32}\varepsilon a^5(\cos 6\phi - 6\cos 4\phi + 15\cos 2\phi - 10) \tag{89}$$

$$a\dot{\beta} = -\varepsilon a^5 \sin^5\phi\cos\phi = -\tfrac{1}{32}\varepsilon a^5(\sin 6\phi - 4\sin 4\phi + 5\sin 2\phi) \tag{90}$$

Keeping only the slowly varying terms on the right-hand sides of (89) and (90) yields

$$\dot{a} = -\tfrac{5}{16}\varepsilon a^5 \tag{91}$$

$$\dot{\beta} = 0 \tag{92}$$

in agreement with (83) and (84), obtained by using the method of multiple scales.

Supplementary Exercises

5.7. Consider the equation

$$\ddot{u} + \omega_0^2 u + \varepsilon u^4 \dot{u} = 0, \qquad \varepsilon \ll 1$$

Use the methods of multiple scales and averaging to determine a first-order uniform expansion.

5.8. Consider the equation

$$\ddot{u} + \omega^2 u + \epsilon u^2 \dot{u}^3 = 0, \qquad \epsilon \ll 1$$

Use the methods of multiple scales and averaging to determine a first-order uniform expansion.

5.9. Consider the equation

$$\ddot{u} + \omega^2 u + \epsilon \dot{u} u \ddot{u} = 0, \qquad \epsilon \ll 1$$

Use the methods of multiple scales and averaging to determine a first-order uniform expansion.

5.10. Consider the equation

$$\ddot{u} + \omega^2 u + \epsilon \dot{u}^3 \ddot{u}^2 = 0, \qquad \epsilon \ll 1$$

Use the methods of multiple scales and averaging to determine a first-order uniform expansion.

5.11. Consider the equation

$$\ddot{u} + \omega^2 u + \epsilon \dot{u}^n = 0, \qquad \epsilon \ll 1$$

where n is a positive odd integer. Use the methods of multiple scales and averaging to determine a first-order uniform expansion.

5.12. Consider the equation

$$\ddot{u} + \omega^2 u + \epsilon u^m \dot{u}^n = 0, \qquad \epsilon \ll 1$$

where m is a positive even integer and n is a positive odd integer. Use the methods of multiple scales and averaging to determine a first-order uniform expansion.

CHAPTER 6

Self-Excited Oscillators

In this chapter, we consider methods of obtaining approximations to the solutions of equations having the form

$$\ddot{u} + \omega_0^2 u = \varepsilon f(u, \dot{u})$$

when $f(u, \dot{u})$ contains damping terms that are negative for small amplitudes but become positive for large amplitudes. Thus $f(u, \dot{u})$ contains terms of the form

$$\mu \dot{u} - h(u, \dot{u}) \dot{u}$$

where $\mu > 0$ and $h(u, \dot{u})$ is positive-definite. Systems governed by equations of this type are called self-excited or self-sustained oscillators and they are abundant in nature. They occur in mechanical systems such as violin strings, wheels of railroad cars, hearts, plasmas, liquid layers, plates and shells adjacent to supersonic flow, Q machines, oil film journal bearings, telegraph lines, and aircraft at high angles of attack.

Such systems are called self-excited oscillators because any small disturbance causes the amplitude to continuously increase until it achieves a finite value. The resulting finite-amplitude oscillation is called a limit cycle. If the initial amplitude is large, it may continuously decrease until it reaches the same limit cycle. If $h(u, \dot{u})$ changes sign, the system might possess more than one limit cycle.

Since the Lindstedt–Poincaré technique and the method of renormalization account for the shift in frequency caused by the perturbation but do not account for any time variations of the amplitude, they are not expected to yield any approximations except to the limit cycles. Both the method of multiple scales and the method of averaging yield valid approximations to the transient responses as well as the limit cycles.

Solved Exercises

6.1. Consider van der Pol's equation

$$\ddot{u} + u = \varepsilon(1 - u^2) \dot{u} \tag{1}$$

(a) Determine two terms in the straightforward expansion and discuss its nonuniformity.

Let

$$u(t;\varepsilon) = u_0(t) + \varepsilon u_1(t) + \cdots \tag{2}$$

Substituting (2) into (1) and equating coefficients of like powers of ε, we obtain

$$\ddot{u}_0 + u_0 = 0 \tag{3}$$

$$\ddot{u}_1 + u_1 = (1 - u_0^2)\,\dot{u}_0 \tag{4}$$

The solution of (3) can be expressed as

$$u_0 = a\cos(t+\beta) \tag{5}$$

where a and β are constants. Then (4) becomes

$$\ddot{u}_1 + u_1 = \left[1 - a^2\cos^2(t+\beta)\right] \cdot -a\sin(t+\beta)$$

or

$$\ddot{u}_1 + u_1 = \tfrac{1}{4}(a^3 - 4a)\sin(t+\beta) + \tfrac{1}{4}a^3\sin(3t+3\beta) \tag{6}$$

A particular solution of (6) is

$$u_1 = -\tfrac{1}{8}(a^3 - 4a)\,t\cos(t+\beta) - \tfrac{1}{32}a^3\sin(3t+3\beta) \tag{7}$$

Hence

$$u = a\cos(t+\beta) - \tfrac{1}{32}\varepsilon a\left[4(a^2-4)\,t\cos(t+\beta) + a^2\sin(3t+3\beta)\right] + \cdots \tag{8}$$

which breaks down for $t \geq O(\varepsilon^{-1})$.

(b) Use the method of renormalization to render the straightforward expansion uniformly valid.

Let

$$\tau = \omega t = (1 + \varepsilon\omega_1 + \cdots)\,t \tag{9}$$

so that

$$t = (1 - \varepsilon\omega_1 + \cdots)\,\tau \tag{10}$$

Substituting (10) into (8) and expanding the result for small ε, we obtain

$$u = a\cos(\tau+\beta) - \varepsilon a\Big[-\omega_1\tau\sin(\tau+\beta) + \tfrac{1}{8}(a^2-4)\,\tau\cos(\tau+\beta) + \tfrac{1}{32}a^2\sin(3\tau+3\beta)\Big] + \cdots \tag{11}$$

Eliminating the secular terms from (11) demands that

$$\omega_1 = 0 \quad \text{and} \quad a(a^2-4)=0$$

For a nontrivial solution, $a = \pm 2$. If the amplitude is defined to be positive, $a = 2$ and (11) becomes

$$u = 2\cos(\tau+\beta)+\cdots = 2\cos(t+\beta)+\cdots \tag{12}$$

(c) Use the method of multiple scales to determine a first-order uniform expansion. Let

$$u = u_0(T_0, T_1) + \varepsilon u_1(T_0, T_1) + \cdots \tag{13}$$

where $T_0 = t$ and $T_1 = \varepsilon t$ in (1), equate coefficients of like powers of ε, and obtain

$$D_0^2 u_0 + u_0 = 0 \tag{14}$$

$$D_0^2 u_1 + u_1 = -2D_0 D_1 u_0 + (1-u_0^2) D_0 u_0 \tag{15}$$

The solution of (14) can be expressed as

$$u_0 = A(T_1)e^{iT_0} + \bar{A}(T_1)e^{-iT_0} \tag{16}$$

Then (15) becomes

$$D_0^2 u_1 + u_1 = -2iA'e^{iT_0} + \text{cc} + \left[1-(Ae^{iT_0}+\bar{A}e^{-iT_0})^2\right] \cdot i(Ae^{iT_0}-\bar{A}e^{-iT_0})$$

or

$$D_0^2 u_1 + u_1 = -i(2A' - A + A^2\bar{A})e^{iT_0} - iA^3 e^{3iT_0} + \text{cc} \tag{17}$$

Eliminating the terms that lead to secular terms in (17) yields

$$2A' = A - A^2\bar{A} \tag{18}$$

Substituting the polar notation $\frac{1}{2}a\exp(i\beta)$ into (18) and separating real and imaginary parts, we obtain

$$a' = \tfrac{1}{2}a(1-\tfrac{1}{4}a^2) \tag{19}$$

$$\beta' = 0 \tag{20}$$

The solution of (20) is that β is a constant, whereas the solution of (19) is obtained by separation of variables. The result is

$$a = 2\left[1+\left(\frac{4}{a_0^2}-1\right)e^{-\varepsilon t}\right]^{-1/2}$$

where $u(0) = a_0$ and $u'(0) = 0$. Hence

$$u = 2\left[1 + \left(\frac{4}{a_0^2} - 1\right)e^{-\varepsilon t}\right]^{-1/2} \cos t + \cdots \tag{21}$$

Comparing (12) and (21), we conclude that the method of renormalization yields only the steady-state response (i.e., the response that can be achieved as $t \to \infty$), whereas the method of multiple scales yields both the steady-state and transient responses.

(d) Use the method of averaging to determine a first-order uniform expansion.

We introduce the transformation

$$u = a\cos(t + \beta) \tag{22}$$

$$\dot{u} = -a\sin(t + \beta) \tag{23}$$

into (1) and obtain

$$\dot{a}\cos\phi - a\dot{\beta}\sin\phi = 0 \tag{24}$$

$$\dot{a}\sin\phi + a\dot{\beta}\cos\phi = \varepsilon a\sin\phi(1 - a^2\cos^2\phi) \tag{25}$$

where $\phi = t + \beta$. Solving (24) and (25) yields

$$\dot{a} = \varepsilon a\sin^2\phi(1 - a^2\cos^2\phi) = \tfrac{1}{8}\varepsilon a(4 - a^2 - 4\cos 2\phi + a^2\cos 4\phi) \tag{26}$$

$$a\dot{\beta} = \varepsilon a\sin\phi\cos\phi(1 - a^2\cos^2\phi) = \tfrac{1}{8}\varepsilon a(4\sin 2\phi - 2a^2\sin 2\phi - a^2\sin 4\phi) \tag{27}$$

Keeping only the slowly varying parts in (26) and (27) yields

$$\dot{a} = \tfrac{1}{8}\varepsilon a(4 - a^2) \tag{28}$$

$$\dot{\beta} = 0 \tag{29}$$

in agreement with (19) and (20), obtained by using the method of multiple scales since $T_1 = \varepsilon t$.

6.2. Use the methods of multiple scales and averaging to determine a first-order uniform expansion including the transient response of the solution of

$$\ddot{u} + \omega_0^2 u = \varepsilon[\dot{u} - \dot{u}^3 + \dot{u}^2 u], \qquad \varepsilon \ll 1 \tag{30}$$

(a) ***The Method of Multiple Scales***

Substituting (13) into (30) and equating coefficients of like powers of ε, we have

$$D_0^2 u_0 + \omega_0^2 u_0 = 0 \tag{31}$$

$$D_0^2 u_1 + \omega_0^2 u_1 = -2D_0 D_1 u_0 + D_0 u_0 - (D_0 u_0)^3 + u_0(D_0 u_0)^2 \tag{32}$$

The solution of (31) can be expressed as

$$u_0 = A(T_1)\,e^{i\omega_0 T_0} + \bar{A}(T_1)\,e^{-i\omega_0 T_0} \tag{33}$$

Then (32) becomes

$$D_0^2 u_1 + \omega_0^2 u_1 = \left(-2i\omega_0 A' + i\omega_0 A\right)e^{i\omega_0 T_0} + \text{cc} + i\omega_0^3\left(Ae^{i\omega_0 T_0} - \bar{A}e^{-i\omega_0 T_0}\right)^3$$
$$- \omega_0^2\left(Ae^{i\omega_0 T_0} + \bar{A}e^{-i\omega_0 T_0}\right)\left(Ae^{i\omega_0 T_0} - \bar{A}e^{-i\omega_0 T_0}\right)^2$$

or

$$D_0^2 u_1 + \omega_0^2 u_1 = \left(-2i\omega_0 A' + i\omega_0 A - 3i\omega_0^3 A^2\bar{A} + \omega_0^2 A^2\bar{A}\right)e^{i\omega_0 T_0} + \text{cc} + \text{NST} \tag{34}$$

Eliminating the terms that lead to secular terms in (34) yields

$$2A' = A - 3\omega_0^2 A^2\bar{A} - i\omega_0 A^2\bar{A} \tag{35}$$

Substituting the polar form $\frac{1}{2}a\exp(i\beta)$ into (35) and separating real and imaginary parts, we obtain

$$a' = \tfrac{1}{2}a\left(1 - \tfrac{3}{4}\omega_0^2 a^2\right) \tag{36}$$

$$a\beta' = -\tfrac{1}{8}\omega_0 a^3 \tag{37}$$

The solution of (36) can be obtained by separation of variables. The result is

$$a^2 = a_0^2\left[\tfrac{3}{4}\omega_0^2 a_0^2 + \left(1 - \tfrac{3}{4}\omega_0^2 a_0^2\right)e^{-\varepsilon t}\right]^{-1} \tag{38}$$

where $a(0) = a_0$. Substituting (38) into (37) yields

$$\beta' = -\frac{\frac{1}{8}\omega_0 a_0^2}{\frac{3}{4}\omega_0^2 a_0^2 + \left(1 - \frac{3}{4}\omega_0^2 a_0^2\right)e^{-\varepsilon t}}$$

or

$$\beta' = -\frac{\frac{1}{8}\omega_0 a_0^2 e^{\varepsilon t}}{\frac{3}{4}\omega_0^2 a_0^2 e^{\varepsilon t} + 1 - \frac{3}{4}\omega_0^2 a_0^2}$$

Hence

$$\beta = -\int \frac{\frac{1}{8}\varepsilon\omega_0 a_0^2 e^{\varepsilon t}\,dt}{\frac{3}{4}\omega_0^2 a_0^2 e^{\varepsilon t} + 1 - \frac{3}{4}\omega_0^2 a_0^2}$$

Let

$$v = \tfrac{3}{4}\omega_0^2 a_0^2 e^{\varepsilon t} + 1 - \tfrac{3}{4}\omega_0^2 a_0^2$$

so that

$$dv = \tfrac{3}{4}\varepsilon\omega_0^2 a_0^2 e^{\varepsilon t}\, dt$$

Then

$$\beta = -\frac{1}{6\omega_0}\int \frac{dv}{v} = -\frac{1}{6\omega_0}\ln v + \text{constant} \tag{39}$$

Therefore

$$u = a_0\left[\tfrac{3}{4}\omega_0^2 a_0^2 + \left(1 - \tfrac{3}{4}\omega_0^2 a_0^2\right)e^{-\varepsilon t}\right]^{-1/2}$$
$$\times\cos\left[\omega_0 t - \frac{1}{6\omega_0}\ln\left(\tfrac{3}{4}\omega_0^2 a_0^2 e^{\varepsilon t} + 1 - \tfrac{3}{4}\omega_0^2 a_0^2\right) + \text{constant}\right] + \cdots \tag{40}$$

(b) ***The Method of Averaging***

We introduce the transformation

$$u = a(t)\cos[\omega_0 t + \beta(t)] \tag{41}$$

$$\dot{u} = -\omega_0 a(t)\sin[\omega_0 t + \beta(t)] \tag{42}$$

into (30) and obtain

$$\dot{a}\cos\phi - a\dot{\beta}\sin\phi = 0 \tag{43}$$

$$\dot{a}\sin\phi + a\dot{\beta}\cos\phi = \varepsilon\left[a\sin\phi - \omega_0^2 a^3\sin^3\phi - \omega_0 a^3\sin^2\phi\cos\phi\right] \tag{44}$$

where $\phi = \omega_0 t + \beta$. Solving (43) and (44) yields

$$\dot{a} = \varepsilon\left[a\sin^2\phi - \omega_0^2 a^3\sin^4\phi - \omega_0 a^3\sin^3\phi\cos\phi\right]$$
$$= \varepsilon\left[\tfrac{1}{2}a\left(1 - \tfrac{3}{4}\omega_0^2 a^2\right) - \tfrac{1}{2}a\left(1 - \omega_0^2 a^2\right)\cos 2\phi - \tfrac{1}{8}\omega_0^2 a^3\cos 4\phi\right.$$
$$\left. - \tfrac{1}{4}\omega_0 a^3\left(\sin 2\phi - \tfrac{1}{2}\sin 4\phi\right)\right] \tag{45}$$

$$\dot{\beta} = \varepsilon\left[\sin\phi\cos\phi - \omega_0^2 a^2\cos\phi\sin^3\phi - \omega_0 a^2\sin^2\phi\cos^2\phi\right]$$
$$= \varepsilon\left[-\tfrac{1}{8}\omega_0 a^2 + \tfrac{1}{8}\omega_0 a^2\cos 4\phi + \tfrac{1}{2}\sin 2\phi - \tfrac{1}{4}\omega_0^2 a^2\left(\sin 2\phi - \tfrac{1}{2}\sin 4\phi\right)\right] \tag{46}$$

where a is assumed to be different from zero. Keeping only the slowly varying terms in (45) and (46), we have

$$\dot{a} = \tfrac{1}{2}\varepsilon a\left(1 - \tfrac{3}{4}\omega_0^2 a^2\right) \tag{47}$$

$$\dot{\beta} = -\tfrac{1}{8}\varepsilon\omega_0 a^2 \tag{48}$$

in agreement with (36) and (37), obtained by using the method of multiple scales, since $T_1 = \varepsilon t$.

6.3. Consider the equation

$$\ddot{x} + x + \dot{x} - \tfrac{1}{2}(\dot{x} - |\dot{x}|)\delta(x - x_0) = 0 \tag{49}$$

where x_0 is a constant and δ is the Dirac delta function. Determine a first-order uniform expansion for small but finite x.

We use the method of averaging. To this end, we use the method of variation of parameters and introduce the transformation

$$x = a(t)\cos[t + \beta(t)] \tag{50}$$

$$\dot{x} = -a(t)\sin[t + \beta(t)] \tag{51}$$

Hence

$$\dot{a}\cos\phi - a\dot{\beta}\sin\phi = 0 \tag{52}$$

$$\dot{a}\sin\phi + a\dot{\beta}\cos\phi = -a\sin\phi + \tfrac{1}{2}a(\sin\phi + |\sin\phi|)\delta(a\cos\phi - x_0) \tag{53}$$

where $\phi = t + \beta$. Solving (52) and (53) gives

$$\dot{a} = -a\sin^2\phi + \tfrac{1}{2}a\sin\phi(\sin\phi + |\sin\phi|)\delta(a\cos\phi - x_0) \tag{54}$$

$$a\dot{\beta} = -a\sin\phi\cos\phi + \tfrac{1}{2}a\cos\phi(\sin\phi + |\sin\phi|)\delta(a\cos\phi - x_0) \tag{55}$$

Averaging (54) and (55) over one cycle yields

$$\dot{a} = -\frac{a}{2\pi}\int_0^{2\pi}\sin^2\phi\, d\phi + \frac{a}{4\pi}\int_0^{2\pi}\sin\phi(\sin\phi + |\sin\phi|)\delta(a\cos\phi - x_0)\, d\phi \tag{56}$$

$$a\dot{\beta} = -\frac{a}{4\pi}\int_0^{2\pi}\sin 2\phi\, d\phi + \frac{a}{4\pi}\int_0^{2\pi}\cos\phi(\sin\phi + |\sin\phi|)\delta(a\cos\phi - x_0)\, d\phi \tag{57}$$

In the interval $[0, \pi]$, $\sin\phi$ is positive so that

$$\sin\phi + |\sin\phi| = 2\sin\phi$$

whereas in the interval $(\pi, 2\pi)$, $\sin\phi$ is negative so that

$$\sin\phi + |\sin\phi| = 0 \quad \text{in } [\pi, 2\pi]$$

because $\sin\pi = \sin 2\pi = 0$. Hence, if $a \neq 0$, (56) and (57) can be rewritten as

$$\dot{a} = -\tfrac{1}{2}a + \frac{a}{2\pi}\int_0^{\pi}\sin^2\phi\,\delta(a\cos\phi - x_0)\, d\phi \tag{58}$$

$$\dot{\beta} = \frac{1}{2\pi}\int_0^{\pi}\sin\phi\cos\phi\,\delta(a\cos\phi - x_0)\, d\phi \tag{59}$$

To evaluate the integrals in (58) and (59), we use the following integral property of the Dirac delta function:

$$\int_a^b f(z)\delta(z-z_0)\,dz = \begin{cases} 0 & \text{if } z_0 < a \quad \text{or} \quad z_0 > b \\ f(z_0) & \text{if } b \ge z_0 \ge a \end{cases} \tag{60}$$

To apply (60), we let

$$z = a\cos\phi \quad \text{so that } dz = -a\sin\phi\, d\phi$$

Hence (58) and (59) become

$$\dot{a} = -\tfrac{1}{2}a + \frac{1}{2\pi}\int_{-a}^{a} \sin\phi\,\delta(z-x_0)\,dz \tag{61}$$

$$\dot{\beta} = \frac{1}{2\pi a}\int_{-a}^{a} \cos\phi\,\delta(z-x_0)\,dz \tag{62}$$

There are two cases depending on whether $x_0 > a$ or $a \ge x_0 \ge 0$. In the first case,

$$z - x_0 = a\cos\phi - x_0$$

does not vanish in the interval $\pi \ge \phi \ge 0$ and the integrals in (61) and (62) are zero. Hence

$$\dot{a} = -\tfrac{1}{2}a, \qquad \dot{\beta} = 0 \tag{63}$$

In the second case, $z = x_0$ at ϕ_a, where ϕ_a is the root of

$$a\cos\phi - x_0 = 0$$

Then (61) and (62) can be rewritten as

$$\dot{a} = -\tfrac{1}{2}a + \frac{1}{2\pi}\sin\phi_a = -\tfrac{1}{2}a + \frac{1}{2\pi}\sqrt{1-\frac{x_0^2}{a^2}} \tag{64}$$

$$\dot{\beta} = \frac{1}{2\pi a}\cos\phi_a = \frac{x_0}{2\pi a^2} \tag{65}$$

6.4. Consider the equations

$$\ddot{u} + \omega_0^2 u = 2\varepsilon[(1-v)\dot{u} - \dot{v}u] \tag{66}$$

$$\dot{v} + v = u^2 \tag{67}$$

Determine a first-order uniform expansion for u and v.

We use the method of multiple scales and let

$$u = u_0(T_0, T_1) + \varepsilon u_1(T_0, T_1) + \cdots \tag{68}$$

$$v = v_0(T_0, T_1) + \varepsilon v_1(T_0, T_1) + \cdots \tag{69}$$

where $T_0 = t$ and $T_1 = \varepsilon t$. Substituting (68) and (69) into (66) and (67) and equating coefficients of like powers of ε, we obtain

Order ε^0

$$D_0^2 u_0 + \omega_0^2 u_0 = 0 \tag{70}$$

$$D_0 v_0 + v_0 = u_0^2 \tag{71}$$

Order ε

$$D_0^2 u_1 + \omega_0^2 u_1 = -2 D_0 D_1 u_0 + 2[(1 - v_0) D_0 u_0 - u_0 D_0 v_0] \tag{72}$$

$$D_0 v_1 + v_1 = -D_1 v_0 + 2 u_0 u_1 \tag{73}$$

The solution of (70) can be expressed as

$$u_0 = A(T_1) e^{i\omega_0 T_0} + \bar{A}(T_1) e^{-i\omega_0 T_0} \tag{74}$$

Then (71) becomes

$$D_0 v_0 + v_0 = A^2 e^{2i\omega_0 T_0} + 2A\bar{A} + \bar{A}^2 e^{-2i\omega_0 T_0}$$

whose general solution is

$$v_0 = B(T_1) e^{-T_0} + \frac{A^2 e^{2i\omega_0 T_0}}{1 + 2i\omega_0} + 2A\bar{A} + \frac{\bar{A}^2 e^{-2i\omega_0 T_0}}{1 - 2i\omega_0} \tag{75}$$

Substituting (74) and (75) into (72) gives

$$\begin{aligned} D_0^2 u_1 + \omega_0^2 u_1 = & -2i\omega_0 A' e^{i\omega_0 T_0} + \text{cc} + 2i\omega_0 (A e^{i\omega_0 T_0} - \bar{A} e^{-i\omega_0 T_0}) \\ & \times \left[1 - 2A\bar{A} - B e^{-T_0} - \frac{A^2 e^{2i\omega_0 T_0}}{1 + 2i\omega_0} - \frac{\bar{A}^2 e^{-2i\omega_0 T_0}}{1 - 2i\omega_0} \right] \\ & -2(A e^{i\omega_0 T_0} + \bar{A} e^{-i\omega_0 T_0}) \\ & \times \left[-B e^{-T_0} + \frac{2i\omega_0}{1 + 2i\omega_0} A^2 e^{2i\omega_0 T_0} - \frac{2i\omega_0}{1 - 2i\omega_0} \bar{A}^2 e^{-2i\omega_0 T_0} \right] \end{aligned}$$

or

$$D_0^2 u_1 + \omega_0^2 u_1 = -2i\omega_0 A' e^{i\omega_0 T_0} + 2i\omega_0 \left[1 - \frac{3+4i\omega_0}{1+2i\omega_0} A\bar{A}\right] A e^{i\omega_0 T_0}$$

$$-2(i\omega_0 - 1) AB e^{(i\omega_0 - 1)T_0} - \frac{6i\omega_0}{1+2i\omega_0} A^3 e^{3i\omega_0 T_0} + \text{cc} \tag{76}$$

Eliminating the terms that lead to secular terms from (76) yields

$$A' = A - \frac{3+4i\omega_0}{1+2i\omega_0} A^2 \bar{A} \tag{77}$$

Substituting the polar notation $\frac{1}{2} a \exp(i\beta)$ into (77) and separating real and imaginary parts, we have

$$a' = a - \frac{3+8\omega_0^2}{4(1+4\omega_0^2)} a^3 \tag{78}$$

$$\beta' = \frac{\omega_0}{2(1+4\omega_0^2)} a^2 \tag{79}$$

The solution of (78) can be obtained by separation of variables. The result is

$$a^2 = \frac{a_0^2}{\alpha a_0^2 + (1 - \alpha a_0^2) e^{-2T_1}}, \qquad \alpha = \frac{3+8\omega_0^2}{4(1+4\omega_0^2)} \tag{80}$$

Then (79) becomes

$$\beta' = \frac{a_0^2 \omega_0}{2(1+4\omega_0^2)} \frac{1}{\alpha a_0^2 + (1-\alpha a_0^2) e^{-2T_1}}$$

$$= \frac{a_0^2 \omega_0}{2(1+4\omega_0^2)} \frac{e^{2T_1}}{\alpha a_0^2 e^{2T_1} + 1 - \alpha a_0^2}$$

whose solution can be obtained as in Exercise 6.2. The result is

$$\beta = \frac{\omega_0}{4\alpha(1+4\omega_0^2)} \ln(\alpha a_0^2 e^{2T_1} + 1 - \alpha a_0^2) + \beta_0 \tag{81}$$

where $\beta(0) = \beta_0$.

To determine B, we need to solve for v_1. With (77), the solution of (76) is

$$u_1 = \frac{-2(i\omega_0 - 1)}{1 - 2i\omega_0} ABe^{(i\omega_0 - 1)T_0} + \frac{3i}{4\omega_0(1 + 2i\omega_0)} A^3 e^{3i\omega_0 T_0} + \text{cc} \tag{82}$$

Substituting (74), (75), and (82) into (73) yields

$$D_0 v_1 + v_1 = -B' e^{-T_0} - \left[\frac{4(i\omega_0 - 1)}{1 - 2i\omega_0} + \frac{4(-i\omega_0 - 1)}{1 + 2i\omega_0}\right] A\bar{A}Be^{-T_0} + \text{NST} \tag{83}$$

Eliminating the terms that lead to secular terms from (83) yields

$$B' = \frac{8(1 + 2\omega_0^2)}{1 + 4\omega_0^2} A\bar{A}B$$

or

$$B' = \frac{2(1 + 2\omega_0^2)}{1 + 4\omega_0^2} a^2 B \tag{84}$$

Substituting (80) into (84) and separating variables, we obtain

$$\frac{dB}{B} = \frac{2(1 + 2\omega_0^2)a_0^2}{(1 + 4\omega_0^2)} \frac{e^{2T_1}\, dT_1}{\alpha a_0^2 e^{2T_1} + 1 - \alpha a_0^2}$$

Hence

$$\ln B = \frac{(1 + 2\omega_0^2)}{\alpha(1 + 4\omega_0^2)} \ln(\alpha a_0^2 e^{2T_1} + 1 - \alpha a_0^2) + \ln b$$

where $B(0) = b$. Therefore

$$B = b\left[\alpha a_0^2 e^{2T_1} + 1 - \alpha a_0^2\right]^{(1 + 2\omega_0^2)/\alpha(1 + 4\omega_0^2)} \tag{85}$$

6.5. Use the methods of averaging and multiple scales to determine a first-order uniform expansion for

$$\ddot{u} + u = \varepsilon(1 - u^4)\dot{u} \tag{86}$$

when $\varepsilon \ll 1$.

(a) ***The Method of Averaging***

To apply the method of averaging, we first use the method of variation of parameters and introduce the transformation

$$u = a(t)\cos[t + \beta(t)] \tag{87}$$

$$\dot{u} = -a(t)\sin[t + \beta(t)] \tag{88}$$

Hence

$$\dot{a}\cos\phi - a\dot{\beta}\sin\phi = 0 \tag{89}$$

$$\dot{a}\sin\phi + a\dot{\beta}\cos\phi = \varepsilon a\sin\phi(1 - a^4\cos^4\phi) \tag{90}$$

where $\phi = t + \beta$. Solving (89) and (90) yields

$$\dot{a} = \varepsilon a\sin^2\phi(1 - a^4\cos^4\phi)$$

$$= \tfrac{1}{2}\varepsilon a(1 - \tfrac{1}{8}a^4) - \tfrac{1}{2}\varepsilon a(1 + \tfrac{1}{16}a^4)\cos 2\phi + \tfrac{1}{16}\varepsilon a^5\cos 4\phi + \tfrac{1}{32}\varepsilon a^5\cos 6\phi$$

$$\dot{\beta} = \varepsilon\sin\phi\cos\phi(1 - a^4\cos^4\phi) \tag{91}$$

$$= \tfrac{1}{2}\varepsilon(1 - \tfrac{5}{16}a^4)\sin 2\phi - \tfrac{1}{8}\varepsilon a^4\sin 4\phi - \tfrac{1}{32}\varepsilon a^4\sin 6\phi \tag{92}$$

where a is assumed to be different from zero. Keeping only the slowly varying terms in (91) and (92) yields

$$\dot{a} = \tfrac{1}{2}\varepsilon a(1 - \tfrac{1}{8}a^4) \tag{93}$$

$$\dot{\beta} = 0 \tag{94}$$

The solution of (94) is that β is a constant, whereas the solution of (93) can be obtained by separation of variables. To this end, we let

$$z = \frac{1}{2\sqrt{2}}a^2 \quad \text{so that } \dot{z} = \frac{1}{\sqrt{2}}a\dot{a}$$

and (93) becomes

$$\dot{z} = \varepsilon z(1 - z^2) \tag{95}$$

Hence

$$\frac{dz}{z(1 - z^2)} = \varepsilon\, dt$$

which upon using partial fractions becomes

$$\frac{dz}{z} - \frac{dz}{2(1+z)} + \frac{dz}{2(1-z)} = \varepsilon\, dt$$

Integrating once, we have

$$2\ln z - \ln(1+z) - \ln(1-z) = 2\varepsilon t + \text{constant}$$

Hence

$$\frac{z^2}{1-z^2} = c^{-1}e^{2\varepsilon t} \quad \text{or} \quad \frac{1-z^2}{z^2} = ce^{-2\varepsilon t} \tag{96}$$

where c is a constant. Solving (96) yields

$$z^2 = \frac{1}{1+ce^{-2\varepsilon t}}$$

or

$$a^4 = \frac{8}{1+ce^{-2\varepsilon t}}$$

If $a(0) = a_0$, then

$$c = \frac{8}{a_0^4} - 1$$

and hence

$$a^4 = \frac{8a_0^4}{a_0^4 + (8-a_0^4)e^{-2\varepsilon t}} \tag{97}$$

Therefore, to the first approximation,

$$u = a_0\sqrt[4]{8}\left[a_0^4 + (8-a_0^4)e^{-2\varepsilon t}\right]^{-1/4}\cos(t+\beta_0) + \cdots \tag{98}$$

(b) ***The Method of Multiple Scales***

Let

$$u = u_0(T_0, T_1) + \varepsilon u_1(T_0, T_1) + \cdots \tag{99}$$

where $T_0 = t$ and $T_1 = \varepsilon t$ in (86), equate coefficients of like powers of ε, and obtain

$$D_0^2 u_0 + u_0 = 0 \tag{100}$$

$$D_0^2 u_1 + u_1 = -2D_0 D_1 u_0 + (1-u_0^4)D_0 u_0 \tag{101}$$

The solution of (100) can be expressed as

$$u_0 = A(T_1)e^{iT_0} + \bar{A}(T_1)e^{-iT_0} \tag{102}$$

Then (101) becomes

$$D_0^2 u_1 + u_1 = -2iA'e^{iT_0} + \text{cc} + i[1 - A^4e^{4iT_0} - 4A^3\bar{A}e^{2iT_0} - 6A^2\bar{A}^2 + \text{cc}](Ae^{iT_0} - \bar{A}e^{-iT_0})$$

or

$$D_0^2 u_1 + u_1 = -i(2A' - A + 2A^3\bar{A}^2)e^{iT_0} + \text{cc} + \text{NST} \tag{103}$$

Eliminating the terms that lead to secular terms from (103) yields

$$2A' = A - 2A^3\bar{A}^2 \tag{104}$$

Substituting the polar form $\frac{1}{2}a\exp(i\beta)$ into (104) and separating real and imaginary parts, we obtain

$$a' = \tfrac{1}{2}a\left(1 - \tfrac{1}{8}a^4\right) \tag{105}$$

$$\beta' = 0 \tag{106}$$

in agreement with (93) and (94), obtained by using the method of averaging.

6.6. Use the methods of multiple scales and averaging to determine a first-order uniform expansion for

$$\ddot{u} + u - \varepsilon(1 - u^2)\dot{u} + \varepsilon u^3 = 0 \tag{107}$$

when $\varepsilon \ll 1$.

(a) ***The Method of Multiple Scales***

Substituting (99) into (107) and equating coefficients of like powers of ε, we obtain

$$D_0^2 u_0 + u_0 = 0 \tag{108}$$

$$D_0^2 u_1 + u_1 = -2D_0 D_1 u_0 + \left(1 - u_0^2\right) D_0 u_0 - u_0^3 \tag{109}$$

The solution of (108) can be expressed as

$$u_0 = A(T_1)e^{iT_0} + \bar{A}(T_1)e^{-iT_0} \tag{110}$$

Then (109) becomes

$$D_0^2 u_1 + u_1 = -2iA'e^{iT_0} + \text{cc} + i[1 - A^2 e^{2iT_0} - 2A\bar{A} - \bar{A}^2 e^{-2iT_0}]$$
$$\times(Ae^{-T_0} - \bar{A}e^{IiT_0}) - (Ae^{iT_0} + \bar{A}e^{-iT_0})^3$$

or

$$D_0^2 u_1 + u_1 = (-2iA' + iA - iA^2\bar{A} - 3A^2\bar{A})e^{iT_0} + \text{cc} + \text{NST} \tag{111}$$

Eliminating the terms that produce secular terms from (111) yields

$$2iA' = iA - iA^2\bar{A} - 3A^2\bar{A} \tag{112}$$

Substituting the polar form $\frac{1}{2}a\exp(i\beta)$ into (112) and separating real and imaginary parts, we obtain

$$a' = \tfrac{1}{2}a\left(1 - \tfrac{1}{4}a^2\right) \tag{113}$$

$$\beta' = \tfrac{3}{8}a^2 \tag{114}$$

The solution of (113) can be obtained by separation of variables. The result is

$$a = 2\left[1 + \left(\frac{4}{a_0^2} - 1\right)e^{-T_1}\right]^{-1/2} \tag{115}$$

where $a(0) = a_0$. Then (114) becomes

$$\beta' = \frac{\frac{3}{2}}{1 + \left(4/a_0^2 - 1\right)e^{-T_1}} = \frac{\frac{3}{2}e^{T_1}}{e^{T_1} + 4/a_0^2 - 1}$$

whose solution can be obtained by separating variables. The result is

$$\beta = \tfrac{3}{2}\ln\left(e^{T_1} + \frac{4}{a_0^2} - 1\right) + c$$

where c is a constant. If $\beta(0) = \beta_0$, then

$$c = \beta_0 - \tfrac{3}{2}\ln\left(\frac{4}{a_0^2}\right)$$

Therefore, to the first approximation,

$$u = 2\left[1 + \left(\frac{4}{a_0^2} - 1\right)e^{-\epsilon t}\right]^{-1/2}\cos\left[t + \tfrac{3}{2}\ln\left(\frac{a_0^2}{4}e^{\epsilon t} + 1 - \frac{a_0^2}{4}\right) + \beta_0\right] + \cdots \tag{116}$$

(b) ***The Method of Averaging***

To apply the method of averaging, we first use the method of variation of parameters and introduce the transformation

$$u = a(t)\cos[t + \beta(t)] \tag{117}$$

$$\dot{u} = -a(t)\sin[t + \beta(t)] \tag{118}$$

It follows from (107), (117), and (118) that

$$\dot{a}\cos\phi - a\dot{\beta}\sin\phi = 0 \tag{119}$$

$$\dot{a}\sin\phi + a\dot{\beta}\cos\phi = \epsilon a\sin\phi\left(1 - a^2\cos^2\phi\right) + \epsilon a^3\sin^3\phi \tag{120}$$

where $\phi = t + \beta$. Solving (119) and (120) yields

$$\dot{a} = \varepsilon a \sin^2\phi(1 - a^2\cos^2\phi) + \varepsilon a^3 \sin\phi\cos^3\phi$$

$$= \tfrac{1}{8}\varepsilon a(4 - a^2 - 4\cos 2\phi + a^2\cos 4\phi) + \tfrac{1}{8}\varepsilon a^3(2\sin 2\phi + \sin 4\phi) \tag{121}$$

$$\dot{\beta} = \varepsilon \sin\phi\cos\phi(1 - a^2\cos^2\phi) + \varepsilon a^2\cos^4\phi$$

$$= \tfrac{1}{8}\varepsilon(4\sin 2\phi - 2a^2\sin 2\phi - a^2\sin 4\phi) + \tfrac{1}{8}\varepsilon a^2(\cos 4\phi + 4\cos 2\phi + 3) \tag{122}$$

where a is assumed to be different from zero. Keeping only the slowly varying terms in (121) and (122) yields

$$\dot{a} = \tfrac{1}{8}\varepsilon a(4 - a^2) \tag{123}$$

$$\dot{\beta} = \tfrac{3}{8}\varepsilon a^2 \tag{124}$$

in agreement with (113) and (114), obtained by using the method of multiple scales since $T_1 = \varepsilon t$.

Supplementary Exercises

6.7. Use the methods of multiple scales and averaging to determine a first-order uniform expansion of the solution of

$$\ddot{u} + \omega_0^2 u + \varepsilon(\dot{u} - \tfrac{1}{3}\dot{u}^5) = 0, \qquad \varepsilon \ll 1$$

6.8. Use the methods of multiple scales and averaging to determine a first-order uniform expansion for

$$\ddot{u} + \omega_0^2 u + \varepsilon(1 - \dot{u}^2 u^2)\dot{u} = 0, \qquad \varepsilon \ll 1$$

6.9. Use the methods of multiple scales and averaging to determine a first-order uniform expansion for

$$\ddot{u} + \omega_0^2 u + \varepsilon(\dot{u} - \dot{u}^3) + \varepsilon u^3 = 0, \qquad \varepsilon \ll 1$$

6.10. Use the methods of multiple scales and averaging to determine a first-order uniform expansion for small but finite amplitude for

$$\ddot{u} + \omega_0^2 u + \sin\dot{u} = 0$$

6.11. Determine a first-order uniform expansion for small but finite amplitude for

$$\ddot{\theta} + \omega_0^2\theta + \frac{\dot{\theta}}{1 + \theta^2} = 0$$

CHAPTER 7

Free Oscillations of Systems with Quadratic Nonlinearities

Governing equations with quadratic nonlinearities are associated with many physical systems, such as the betatron oscillations, the motion of a swinging spring, the motion of a ship, the motion of a liquid interface, the motion of a rotating shaft, the vibrations of composite plates and bowed structures, including shells, and the vibrations of a structure about a loaded static configuration.

The Lindstedt–Poincaré technique, the method of renormalization, and the method of multiple scales yield uniform approximations to the free-oscillation responses of conservative systems with quadratic nonlinearities. For example, application of these methods to

$$\ddot{u} + \omega_0^2 u + \alpha_2 u^2 + \alpha_3 u^3 = 0$$

yields

$$u = a\cos(\omega t + \beta) + \frac{a^2\alpha_2}{2\omega_0^2}\left[\tfrac{1}{3}\cos(2\omega t + 2\beta) - 1\right] + \cdots$$

where a and β are constants and

$$\omega = \omega_0\left[1 + \frac{9\alpha_3\omega_0^2 - 10\alpha_2^2}{24\omega_0^4}a^2\right] + \cdots$$

On the other hand, application of the method of averaging to this equation yields

$$\omega = \omega_0\left[1 + \frac{3\alpha_3}{8\omega_0^2}a^2\right] + \cdots$$

which does not account for all the nonlinear corrections to the frequency. To obtain a uniform expansion, one needs to use the generalized method of averaging by expanding a and $\phi = \omega_0 t + \beta_0$ as

$$a = a_0(t) + \varepsilon a_1(a_0, \phi_0) + \varepsilon^2 a_2(a_0, \phi_0) + \cdots$$

$$\phi = \phi_0(t) + \varepsilon \phi_1(a_0, \phi_0) + \varepsilon^2 \phi_2(a_0, \phi_0) + \cdots$$

$$\dot{a}_0 = \varepsilon A_1(a_0) + \varepsilon^2 A_2(a_0) + \cdots$$

$$\dot{\phi}_0 = \omega_0 + \varepsilon \Phi_1(a_0) + \varepsilon^2 \Phi_2(a_0) + \cdots$$

Substituting these expansions into the variational equations for a and ϕ and separating the slowly varying functions A_n and Φ_n from the rapidly varying functions a_n and ϕ_n, one determines the A_n and Φ_n and the equations governing the a_n and ϕ_n.

One can also use the Krylov–Bogoliubov technique by letting

$$u = \varepsilon a \cos\phi + \varepsilon^2 u_2(a, \phi) + \varepsilon^3 u_3(a, \phi) + \cdots$$

where

$$\dot{a} = \varepsilon A_1(a) + \varepsilon^2 A_2(a) + \cdots$$

$$\dot{\phi} = \omega_0 + \varepsilon \Phi_1(a) + \varepsilon^2 \Phi_2(a) + \cdots$$

Substituting these expansions into the original equations and eliminating the secular terms, one determines the A_n and Φ_n. One can view this technique as a variant of the generalized method of multiple scales in which the scales are $a(t)$ and $\phi(t)$.

For nonconservative systems, one can determine the limit cycles and trivial solutions using the Lindstedt–Poincaré technique or the method of renormalization. On the other hand, one can determine uniform expansions using the method of multiple scales or the generalized method of averaging or the Krylov–Bogoliubov technique.

Solved Exercises

7.1. Consider the equation

$$\ddot{x} - 2x - x^2 + x^3 = 0 \tag{1}$$

Show that the equilibrium positions are $x = 0$, -1, and 2. Put

$$x = 2 + u \tag{2}$$

and determine the equation governing u. Then, determine a second-order uniform expansion for small but finite amplitudes.

The equilibrium positions can be obtained by setting $\ddot{x} = 0$ in (1); that is, they are given by

$$x^3 - x^2 - 2x = 0$$

or

$$x(x^2 - x - 2) = x(x-2)(x+1) = 0 \tag{3}$$

Hence the equilibrium positions are

$$x = 0, \ -1, \text{ and } 2. \tag{4}$$

To investigate the behavior of the solution near $x = 2$, we substitute (2) into (1) and obtain

$$\ddot{u} - 2(2+u) - (2+u)^2 + (2+u)^3 = 0$$

or

$$\ddot{u} + 6u + 5u^2 + u^3 = 0 \tag{5}$$

Equation (5) can be put in the same form as

$$\ddot{u} + u + \alpha_2 u^2 + \alpha_3 u^3 = 0 \tag{6}$$

by changing the independent variable from t to

$$\tau = \omega_0 t = \sqrt{6}\, t$$

Thus (5) becomes

$$u'' + u + \tfrac{5}{6}u^2 + \tfrac{1}{6}u^3 = 0 \tag{7}$$

An approximate solution of (6) can be obtained by using the Lindstedt–Poincaré technique, the method of renormalization, the method of multiple scales, the generalized method of averaging, or the Krylov–Bogoliubov technique. The result is

$$u = a_0 \cos(\omega t + \beta_0) + \tfrac{1}{6} a_0^2 \alpha_2 [\cos(2\omega t + 2\beta_0) - 3] + \cdots \tag{8a}$$

where

$$\omega = 1 + \left(\tfrac{3}{8}\alpha_3 - \tfrac{5}{12}\alpha_2^2\right) a_0^2 + \cdots \tag{8b}$$

Hence the solution of (7) is

$$u = a_0 \cos(\omega \tau + \beta_0) + \tfrac{5}{36} a_0^2 [\cos(2\omega\tau + 2\beta_0) - 3] + \cdots \tag{9a}$$

where

$$\omega = 1 - \tfrac{49}{216} a_0^2 + \cdots \tag{9b}$$

7.2. Consider the equation

$$\ddot{u} - u + u^4 = 0 \tag{10}$$

Show that $u = 1$ is an equilibrium point. Determine a second-order uniform expansion for small but finite motions around $u = 1$.

The equilibrium points can be obtained by setting $\ddot{u} = 0$ in (10). The result is

$$u^4 - u = 0$$

or

$$u(u^3 - 1) = u(u - 1)(u^2 + u + 1) = 0$$

Hence $u = 1$ is an equilibrium point. To determine the motion in the neighborhood of $u = 1$, we put

$$u = 1 + v$$

in (10) and obtain

$$\ddot{v} - (1 + v) + (1 + v)^4 = 0$$

or

$$\ddot{v} + 3v + 6v^2 + 4v^3 + v^4 = 0 \tag{11}$$

To put (11) in the form (6), we introduce the transformation

$$\tau = \sqrt{3}\, t$$

and obtain

$$v'' + v + 2v^2 + \tfrac{4}{3}v^3 + \cdots = 0 \tag{12}$$

Comparing (6) and (12), we conclude that $\alpha_2 = 2$ and $\alpha_3 = \frac{4}{3}$. Hence it follows from (8) that a second-order uniform expansion to (12) is

$$v = a_0 \cos(\omega\tau + \beta_0) + \tfrac{1}{3}a_0^2[\cos(2\omega\tau + 2\beta_0) - 3] + \cdots \tag{13}$$

where

$$\omega = 1 - \tfrac{7}{6}a_0^2 + \cdots \tag{14}$$

7.3. Consider the equation

$$\ddot{u} - u + u^6 = 0 \tag{15}$$

Show that $u = 1$ is an equilibrium position. Determine a second-order expansion for small but finite motions around $u = 1$.

To determine the equilibrium positions, we put $\ddot{u}=0$ in (15) and obtain

$$u^6 - u = 0 \tag{16}$$

One can easily show that $u=1$ satisfies (16) and hence $u=1$ is an equilibrium position.
To determine the motion around $u=1$, we introduce the transformation

$$u = 1 + v$$

in (15) and obtain

$$\ddot{v} - (1+v) + (1+v)^6 = 0$$

or

$$\ddot{v} + 5v + 15v^2 + 20v^3 + \cdots = 0 \tag{17}$$

Equation (17) can be put in the form (6) by introducing the transformation

$$\tau = \sqrt{5}\, t$$

The result is

$$v'' + v + 3v^2 + 4v^3 + \cdots = 0 \tag{18}$$

Comparing (6) and (18), we conclude that $\alpha_2 = 3$ and $\alpha_3 = 4$. It follows from (8) that a second-order uniform expansion to the solution of (18) is

$$v = a_0 \cos(\omega\tau + \beta_0) + \tfrac{1}{2}a_0^2[\cos(2\omega\tau + 2\beta_0) - 3] + \cdots \tag{19}$$

where

$$\omega = 1 - \tfrac{9}{4}a_0^2 + \cdots \tag{20}$$

7.4. Consider the equation

$$\ddot{x} + x - \frac{3}{16(1-x)} = 0 \tag{21}$$

Show that the equilibrium points are $x=\frac{1}{4}$ and $\frac{3}{4}$. Examine the motion near the equilibrium points. Determine a second-order expansion around the stable equilibrium point (i.e., the one corresponding to sinusoidal motions).
To determine the equilibrium points, we put $\ddot{x}=0$ in (21) and obtain

$$x - \frac{3}{16(1-x)} = 0$$

or

$$x - x^2 - \tfrac{3}{16} = 0$$

or

$$x^2 - x + \tfrac{3}{16} = (x - \tfrac{1}{4})(x - \tfrac{3}{4}) = 0$$

Hence $x = \frac{1}{4}$ and $\frac{3}{4}$ are the equilibrium points.

To determine the motion around $x = \frac{1}{4}$, we introduce the transformation $x = \frac{1}{4} + u$ in (21) and obtain

$$\ddot{u} + \tfrac{1}{4} + u - \frac{3}{16(\frac{3}{4} - u)} = 0$$

or

$$\ddot{u} + \tfrac{1}{4} + u - \tfrac{1}{4}\left(1 - \frac{4u}{3}\right)^{-1} = 0 \tag{22}$$

Expanding (22) for small u yields

$$\ddot{u} + \tfrac{2}{3}u - \tfrac{4}{9}u^2 - \tfrac{16}{27}u^3 - \cdots = 0 \tag{23}$$

The linearized form of (23) shows that u is a sinusoidal function of t and hence $u = 0$ (i.e., $x = \frac{1}{4}$) is a stable equilibrium point. To determine a second-order uniform expansion for u, we put (23) in the form (6) by introducing the transformation

$$\tau = \sqrt{\tfrac{2}{3}}\, t$$

The result is

$$u'' + u - \tfrac{2}{3}u^2 - \tfrac{8}{9}u^3 + \cdots = 0 \tag{24}$$

Comparing (6) and (24), we conclude that $\alpha_2 = -\frac{2}{3}$ and $\alpha_3 = -\frac{8}{9}$. Hence it follows from (8) that a second-order uniform expansion for u is

$$u = a_0\cos(\omega\tau + \beta_0) - \tfrac{1}{9}a_0^2[\cos(2\omega\tau + 2\beta_0) - 3] + \cdots \tag{25}$$

where

$$\omega = 1 - \tfrac{14}{27}a_0^2 + \cdots \tag{26}$$

To determine the motion around $x = \frac{3}{4}$, we introduce the transformation $x = \frac{3}{4} + u$ into (21) and obtain

$$\ddot{u} + \tfrac{3}{4} + u - \frac{3}{16(\frac{1}{4} - u)} = 0$$

or

$$\ddot{u} + \tfrac{3}{4} + u - \tfrac{3}{4}(1 - 4u)^{-1} = 0 \tag{27}$$

Expanding (27) for small u yields

$$\ddot{u} - 2u - 12u^2 - 48u^3 - \cdots = 0 \tag{28}$$

The linearized form of (28) shows that u is an exponential function of t and hence $u = 0$ (i.e., $x = \frac{3}{4}$) is an unstable equilibrium point.

7.5. Determine a second-order uniform expansion for

$$\ddot{u} + u + \varepsilon^2 u^3 + \varepsilon \dot{u}^2 = 0, \qquad \varepsilon \ll 1 \tag{29}$$

We use the method of multiple scales and let

$$u = u_0(T_0, T_1, T_2) + \varepsilon u_1(T_0, T_1, T_2) + \varepsilon^2 u_2(T_0, T_1, T_2) + \cdots \tag{30}$$

where $T_0 = t$, $T_1 = \varepsilon t$, and $T_2 = \varepsilon^2 t$. Substituting (30) into (29) and equating coefficients of like powers of ε, we obtain

$$D_0^2 u_0 + u_0 = 0 \tag{31}$$

$$D_0^2 u_1 + u_1 = -2D_0 D_1 u_0 - (D_0 u_0)^2 \tag{32}$$

$$D_0^2 u_2 + u_2 = -2D_0 D_2 u_0 - D_1^2 u_0 - 2D_0 D_1 u_1 - u_0^3 - 2D_0 u_0 D_0 u_1 - 2D_0 u_0 D_1 u_0 \tag{33}$$

The solution of (31) can be expressed as

$$u_0 = A(T_1, T_2) e^{iT_0} + \bar{A}(T_1, T_2) e^{-iT_0} \tag{34}$$

Then (32) becomes

$$D_0^2 u_1 + u_1 = -2iD_1 A e^{iT_0} + A^2 e^{2iT_0} - A\bar{A} + \text{cc} \tag{35}$$

Eliminating the terms that lead to secular terms from (35) demands that $D_1 A = 0$ so that $A = A(T_2)$. Then, a particular solution of (35) is

$$u_1 = -\tfrac{1}{3}A^2 e^{2iT_0} - 2A\bar{A} - \tfrac{1}{3}\bar{A}^2 e^{-2iT_0} \tag{36}$$

Substituting (34) and (36) into (33) and recalling that $A = A(T_2)$, we obtain

$$D_0^2 u_2 + u_2 = \left(-2iA' - \tfrac{5}{3}A^2\bar{A}\right) e^{iT_0} + \text{cc} + \text{NST} \tag{37}$$

where the prime denotes the derivative with respect to T_2. Eliminating the terms that lead to secular terms from (37), we have

$$2iA' = -\tfrac{5}{3}A^2\bar{A} \tag{38}$$

Substituting the polar form $\frac{1}{2}a\exp(i\beta)$ into (38) and separating real and imaginary parts, we obtain

$$a' = 0 \tag{39}$$

$$\beta' = \tfrac{5}{24} a^2 \tag{40}$$

The solution of (39) is that a is a constant and hence the solution of (40) is

$$\beta = \tfrac{5}{24} a^2 T_2 + \beta_0 \tag{41}$$

Therefore, to second order,

$$u = a\cos(\omega t + \beta_0) - \tfrac{1}{6}\varepsilon a^2[\cos(2\omega t + 2\beta_0) + 3] + \cdots \tag{42}$$

where

$$\omega = 1 + \dot{\beta} = 1 + \tfrac{5}{24}\varepsilon^2 a^2 + \cdots \tag{43}$$

Since the amplitude is constant, one can use the Lindstedt–Poincaré technique to determine the expansion (42) and (43).

7.6. Determine a second-order uniform expansion for

$$\ddot{u} + u + \varepsilon u^2 + \varepsilon \dot{u}^2 = 0, \qquad \varepsilon \ll 1 \tag{44}$$

As in the preceding exercise, we use the method of multiple scales. Since the nonlinearity is quadratic, we need to carry out the expansion to second order. Substituting (30) into (44) and equating coefficients of like powers of ε, we obtain

$$D_0^2 u_0 + u_0 = 0 \tag{45}$$

$$D_0^2 u_1 + u_1 = -2D_0 D_1 u_0 - u_0^2 - (D_0 u_0)^2 \tag{46}$$

$$D_0^2 u_2 + u_2 = -2D_0 D_2 u_0 - D_1^2 u_0 - 2D_0 D_1 u_1 - 2u_0 u_1 - 2D_0 u_0 D_0 u_1 - 2D_0 u_0 D_1 u_0 \tag{47}$$

The general solution of (45) is

$$u_0 = A(T_1, T_2)\, e^{iT_0} + \bar{A}(T_1, T_2)\, e^{-iT_0} \tag{48}$$

Then (46) becomes

$$D_0^2 u_1 + u_1 = -2iD_1 A e^{iT_0} + \text{cc} - 4A\bar{A} \tag{49}$$

Eliminating the terms that lead to secular terms from (49) demands that $D_1 A = 0$ or $A = A(T_2)$. Then a particular solution of (49) is

$$u_1 = -4A\bar{A} \tag{50}$$

Substituting (48) and (50) into (47) and recalling that $A = A(T_2)$, we obtain

$$D_0^2 u_2 + u_2 = (-2iA' + 8A^2\bar{A})\, e^{iT_0} + \text{cc} + \text{NST} \tag{51}$$

Eliminating the terms that lead to secular terms from (51) yields

$$2iA' = 8A^2\bar{A} \tag{52}$$

Substituting the polar form $\frac{1}{2}a\exp(i\beta)$ into (52) and separating real and imaginary parts, we obtain

$$a' = 0 \tag{53}$$

$$\beta' = -a^2 \tag{54}$$

The solution of (53) is that a is a constant and hence the solution of (54) is

$$\beta = -a^2 T_2 + \beta_0$$

where β_0 is a constant. Therefore

$$u = a\cos(\omega t + \beta) - \varepsilon a^2 + \cdots \tag{55}$$

where

$$\omega = 1 + \dot{\beta} = 1 - \varepsilon^2 a^2 + \cdots \tag{56}$$

Since the amplitude is constant, we can use the Lindstedt–Poincaré technique to obtain the expansion (55) and (56).

Supplementary Exercises

7.7. Consider the equation

$$\ddot{x} + x + 3x^2 + 2x^3 = 0$$

Show that the equilibrium points are $x = 0$, $-\frac{1}{2}$ and -1. Examine the motion near these equilibrium points. Determine second-order uniform expansions around the stable equilibrium points.

7.8. Consider the equation

$$\ddot{x} + x + \alpha_2 x^2 + 2x^3 = 0$$

Show that there is only one equilibrium point when $\alpha_2 < 2\sqrt{2}$ and that there are three equilibrium points when $\alpha_2 > 2\sqrt{2}$. Determine second-order uniform expansions around the stable equilibrium points when $\alpha_2 > 2\sqrt{2}$.

7.9. Consider the equation

$$\ddot{x} + x - \frac{\alpha}{1-x} = 0$$

Show that equilibrium positions exist only when $\alpha \le \frac{1}{4}$. Determine a second-order uniform expansion around the stable equilibrium point when $\alpha < \frac{1}{4}$.

7.10. Consider the equation

$$\ddot{x} + x + x^3 = 2$$

Show that $x = 1$ is the only equilibrium point. Determine a second-order uniform expansion around this point.

7.11. Consider the equation

$$\ddot{x} - 3x + x^3 = -2$$

Determine the equilibrium points and their stability. Determine second-order uniform expansions around the stable points.

7.12. Show that, to the second approximation, the solution of

$$\ddot{u} + \omega_0^2 u + \varepsilon\alpha_2 u^2 + \varepsilon\alpha_4 \dot{u}^2 = 0$$

is

$$u = a\cos(\omega t + \beta_0) + \varepsilon a^2\left[\frac{\alpha_2 - \omega_0^2\alpha_4}{6\omega_0^2}\cos(2\omega t + 2\beta_0) - \frac{\alpha_2 + \omega_0^2\alpha_4}{2\omega_0^2}\right] + \cdots$$

where

$$\omega = \omega_0 - \frac{\varepsilon(5\alpha_2^2 + 5\omega_0^2\alpha_2\alpha_4 + 2\omega_0^4\alpha_4^2)}{12\omega_0^3}a^2 + \cdots$$

7.13. Show that, to the second approximation, the solution of

$$\ddot{u} + \omega_0^2 u + \varepsilon\alpha_4 \dot{u}^2 + \varepsilon^2\alpha_5 \dot{u}^2 u = 0$$

is

$$u = a\cos(\omega t + \beta_0) - \tfrac{1}{6}\varepsilon\alpha_4 a^2[\cos(2\omega t + 2\beta_0) + 3] + \cdots$$

where

$$\omega = \omega_0 + \frac{\varepsilon^2(3\alpha_5 - 4\alpha_4^2)\,\omega_0 a^2}{24} + \cdots$$

7.14. Show that, to the second approximation, the solution of

$$\ddot{u} + \omega_0^2 u + \varepsilon\alpha_2 u^2 + \varepsilon^2\alpha_3 u^3 + \varepsilon\alpha_4 \dot{u}^2 + \varepsilon^2\alpha_5 \dot{u}^2 u = 0$$

is

$$u = a\cos(\omega t + \beta_0) + \varepsilon a^2\left[\frac{\alpha_2 - \omega_0^2\alpha_4}{6\omega_0^2}\cos(2\omega t + 2\beta_0) - \frac{\alpha_2 + \omega_0^2\alpha_4}{2\omega_0^2}\right] + \cdots$$

where

$$\omega = \omega_0 + \frac{\varepsilon^2\left(9\omega_0^2\alpha_3 + 3\omega_0^4\alpha_5 - 10\alpha_2^2 - 10\omega_0^2\alpha_2\alpha_4 - 4\omega_0^4\alpha_4^2\right)a^2}{24\omega_0^3} + \cdots$$

CHAPTER 8

General Systems with Odd Nonlinearities

Application of the method of averaging or the method of multiple scales to the equation

$$\ddot{u} + \omega_0^2 u = \varepsilon f(u, \dot{u})$$

yields

$$u = a(t)\cos[\omega_0 t + \beta(t)] + \cdots$$

where

$$\dot{a} = -\frac{\varepsilon}{2\pi\omega_0}\int_0^{2\pi} f(a\cos\phi, -\omega_0 a\sin\phi)\sin\phi\, d\phi$$

$$a\dot{\beta} = -\frac{\varepsilon}{2\pi\omega_0}\int_0^{2\pi} f(a\cos\phi, -\omega_0 a\sin\phi)\cos\phi\, d\phi$$

On the other hand, application of the Lindstedt–Poincaré technique or the method of renormalization to this problem yields

$$u = a\cos(\omega t + \beta_0) + \cdots$$

where a and β_0 are constant and

$$\omega = \omega_0 - \frac{\varepsilon}{2\pi\omega_0 a}\int_0^{2\pi} f(a\cos\phi, -\omega_0 a\sin\phi)\cos\phi\, d\phi$$

$$\int_0^{2\pi} f(a\cos\phi, -\omega_0 a\sin\phi)\sin\phi\, d\phi = 0$$

It is clear that both the Lindstedt–Poincaré technique and the method of renormalization yield a constant value for the amplitude a and hence they yield only the trivial solution and limit cycles.

Solved Exercises

8.1. For small ε, determine first-order uniform expansions for each of the following problems:

(a)
$$\ddot{u} + u + \varepsilon u|u| = 0 \tag{1}$$

(b)
$$\ddot{u} + u + \varepsilon(\operatorname{sgn}\dot{u} + 2\mu_1\dot{u}) = 0 \tag{2}$$

(c)
$$\ddot{u} + u + \varepsilon(\operatorname{sgn}\dot{u} + \mu_2\dot{u}|\dot{u}|) = 0 \tag{3}$$

(d)
$$\ddot{u} + u + \varepsilon(2\mu_1\dot{u} + \mu_2\dot{u}|\dot{u}|) = 0 \tag{4}$$

(e)
$$\ddot{u} + u + \varepsilon(2\mu_1\dot{u} + \operatorname{sgn}\dot{u} + \mu_2\dot{u}|\dot{u}|) = 0 \tag{5}$$

Equations (1)–(5) have the form

$$\ddot{u} + u = \varepsilon f(u, \dot{u})$$

and hence first-order uniform expansions of their solutions can be obtained by using either the method of averaging or the method of multiple scales, as discussed in the present chapter. The result is

$$u = a(T_1)\cos[t + \beta(T_1)] + \cdots \tag{6}$$

where

$$a' = -\frac{1}{2\pi}\int_0^{2\pi} f(a\cos\phi, -a\sin\phi)\sin\phi\, d\phi \tag{7}$$

$$a\beta' = -\frac{1}{2\pi}\int_0^{2\pi} f(a\cos\phi, -a\sin\phi)\cos\phi\, d\phi \tag{8}$$

(a) In this case, $f(u, \dot{u}) = -u|u|$. Hence (7) becomes

$$a' = \frac{a^2}{2\pi}\int_0^{2\pi} \cos\phi|\cos\phi|\sin\phi\, d\phi = 0 \tag{9}$$

because the integrand is periodic and odd. Moreover, (8) becomes

$$a\beta' = \frac{a^2}{2\pi}\int_0^{2\pi} \cos^2\phi|\cos\phi|\, d\phi \tag{10}$$

To evaluate the integral in (10), we note that $\cos\phi$ is positive in the intervals $[0, \frac{1}{2}\pi]$ and $[\frac{3}{2}\pi, 2\pi]$. Hence $|\cos\phi| = \cos\phi$ in these intervals. However, $\cos\phi$ is negative in the interval $[\frac{1}{2}\pi, \frac{3}{2}\pi]$. Hence $|\cos\phi| = -\cos\phi$ in this interval. Therefore, we break the interval of integration into the above three intervals and rewrite (10), when $a \neq 0$, as

$$\beta' = \frac{a}{2\pi}\left[\int_0^{\pi/2}\cos^3\phi\,d\phi - \int_{\pi/2}^{3\pi/2}\cos^3\phi\,d\phi + \int_{3\pi/2}^{2\pi}\cos^3\phi\,d\phi\right] \tag{11}$$

But

$$\cos^3\phi = \tfrac{1}{4}(3\cos\phi + \cos 3\phi)$$

Hence

$$\beta' = \frac{a}{2\pi}\Big[\left(\tfrac{3}{4}\sin\phi + \tfrac{1}{12}\sin 3\phi\right)_0^{\pi/2} - \left(\tfrac{3}{4}\sin\phi + \tfrac{1}{12}\sin 3\phi\right)_{\pi/2}^{3\pi/2}$$

$$+\left(\tfrac{3}{4}\sin\phi + \tfrac{1}{12}\sin 3\phi\right)_{3\pi/2}^{2\pi}\Big] = \frac{4a}{3\pi} \tag{12}$$

The solution of (9) is that a is a constant and hence the solution of (12) is

$$\beta = \frac{4a}{3\pi}T_1 + \beta_0 \tag{13}$$

where β_0 is a constant. Therefore, to the first approximation,

$$u = a\cos\left[\left(1 + \frac{4\varepsilon a}{3\pi}\right)t + \beta_0\right] + \cdots \tag{14}$$

(b) In this case, $f(u, \dot{u}) = -(\operatorname{sgn}\dot{u} + 2\mu_1\dot{u})$. Hence (7) and (8) become

$$a' = \frac{1}{2\pi}\int_0^{2\pi}[\operatorname{sgn}(-a\sin\phi) - 2\mu_1 a\sin\phi]\sin\phi\,d\phi \tag{15}$$

$$a\beta' = \frac{1}{2\pi}\int_0^{2\pi}[\operatorname{sgn}(-a\sin\phi) - 2\mu_1 a\sin\phi]\cos\phi\,d\phi \tag{16}$$

To evaluate the integrals in (15) and (16), we note that

$$\operatorname{sgn}\dot{u} = \begin{cases} 1 & \text{if } \dot{u} > 0 \\ -1 & \text{if } \dot{u} < 0 \end{cases}$$

Hence

$$\int_0^{2\pi}\operatorname{sgn}(-a\sin\phi)\sin\phi\,d\phi$$

$$= \int_0^{\pi}\operatorname{sgn}(-a\sin\phi)\sin\phi\,d\phi + \int_{\pi}^{2\pi}\operatorname{sgn}(-a\sin\phi)\sin\phi\,d\phi$$

$$= -\int_0^{\pi}\sin\phi\,d\phi + \int_{\pi}^{2\pi}\sin\phi\,d\phi = \cos\phi|_0^{\pi} - \cos\phi|_{\pi}^{2\pi} = -4 \tag{17}$$

Hence it follows from (15) that

$$a' = -\frac{2}{\pi} - \mu_1 a \tag{18}$$

Moreover

$$\int_0^{2\pi} \operatorname{sgn}(-a\sin\phi)\cos\phi\, d\phi = -\int_0^{\pi}\cos\phi\, d\phi + \int_{\pi}^{2\pi}\cos\phi\, d\phi = -\sin\phi|_0^{\pi} + \sin\phi|_{\pi}^{2\pi} = 0 \tag{19}$$

Hence it follows from (16) that

$$\beta' = 0 \tag{20}$$

The solution of (20) is that β is a constant, whereas the solution of (18) is

$$a = \left(a_0 + \frac{2}{\pi\mu_1}\right)e^{-\mu_1 T_1} - \frac{2}{\pi\mu_1} \tag{21}$$

where $a(0) = a_0$. Therefore

$$u = \left[\left(a_0 + \frac{2}{\pi\mu_1}\right)e^{-\epsilon\mu_1 t} - \frac{2}{\pi\mu_1}\right]\cos(t + \beta_0) + \cdots \tag{22}$$

(c) In this case, $f(u, \dot{u}) = -\operatorname{sgn}\dot{u} - \mu_2\dot{u}|\dot{u}|$. Hence (7) and (8) become

$$a' = \frac{1}{2\pi}\int_0^{2\pi}\operatorname{sgn}(-a\sin\phi)\sin\phi\, d\phi - \frac{\mu_2 a^2}{2\pi}\int_0^{2\pi}\sin^2\phi|\sin\phi|\, d\phi \tag{23}$$

$$a\beta' = \frac{1}{2\pi}\int_0^{2\pi}\operatorname{sgn}(-a\sin\phi)\cos\phi\, d\phi - \frac{\mu_2 a^2}{2\pi}\int_0^{2\pi}\sin\phi\cos\phi|\sin\phi|\, d\phi \tag{24}$$

Carrying out the integrations in (23) and (24), we obtain

$$a' = -\frac{2}{\pi} - \frac{4\mu_2}{3\pi}a^2 \tag{25}$$

$$\beta' = 0 \tag{26}$$

The solution of (26) is that β is a constant, whereas the solution of (25) can be obtained by separation of variables. Thus (25) is separated into

$$\frac{da}{a^2 + 3/2\mu_2} = -\frac{4\mu_2}{3\pi}\, dT_1 \tag{27}$$

We let

$$a = \sqrt{\frac{3}{2\mu_2}}\tan\theta \quad \text{so that } da = \sqrt{\frac{3}{2\mu_2}}\sec^2\theta\, d\theta$$

Then (27) becomes

$$d\theta = -\sqrt{\frac{3}{2\mu_2}}\,\frac{4\mu_2}{3\pi}\,dT_1 = -\frac{1}{\pi}\sqrt{\frac{8\mu_2}{3}}\,dT_1$$

which, upon integration once, yields

$$\theta = c - \frac{1}{\pi}\sqrt{\frac{8\mu_2}{3}}\,T_1$$

where c is a constant. Hence

$$a = \sqrt{\frac{3}{2\mu_2}}\tan\left(c - \frac{\varepsilon}{\pi}\sqrt{\frac{8\mu_2}{3}}\,t\right)$$

and

$$u = \sqrt{\frac{3}{2\mu_2}}\tan\left(c - \frac{\varepsilon}{\pi}\sqrt{\frac{8\mu_2}{3}}\,t\right)\cos(t+\beta) + \cdots \tag{28}$$

(d) In this case, $f(u,\dot{u}) = -2\mu_1\dot{u} - \mu_2\dot{u}|\dot{u}|$ and (7) and (8) become

$$a' = -\frac{\mu_1 a}{\pi}\int_0^{2\pi}\sin^2\phi\, d\phi - \frac{\mu_2 a^2}{2\pi}\int_0^{2\pi}\sin^2\phi|\sin\phi|\, d\phi \tag{29}$$

$$a\beta' = -\frac{\mu_1 a}{\pi}\int_0^{2\pi}\sin\phi\cos\phi\, d\phi - \frac{\mu_2 a^2}{2\pi}\int_0^{2\pi}\sin\phi\cos\phi|\sin\phi|\, d\phi \tag{30}$$

Carrying out the integration in (29) and (30), we obtain

$$a' = -\mu_1 a - \frac{4\mu_2 a^2}{3\pi} \tag{31}$$

$$\beta' = 0 \tag{32}$$

The solution of (32) is that β is a constant, whereas the solution of (31) can be obtained by separation of variables. To this end, we write (31) as

$$\frac{da}{a + (4\mu_2/3\pi\mu_1)a^2} = -\mu_1\, dT_1$$

which, upon using partial fractions, can be rewritten as

$$\frac{da}{a} - \frac{4\mu_2}{3\pi\mu_1}\frac{da}{1+(4\mu_2/3\pi\mu_1)a} = -\mu_1\, dT_1$$

Integrating once gives

$$\ln a - \ln\left(1 + \frac{4\mu_2}{3\pi\mu_1}a\right) = -\mu_1 T_1 + \ln c$$

where c is a constant. Hence

$$\frac{a}{1+(4\mu_2/3\pi\mu_1)a} = \frac{c}{e^{\mu_1 T_1}}$$

or

$$a = \left[\frac{1}{c}e^{\mu_1 T_1} - \frac{4\mu_2}{3\pi\mu_1}\right]^{-1} \tag{33}$$

If $a(0) = a_0$, then

$$c = \frac{a_0}{1+(4\mu_2/3\pi\mu_1)a_0}$$

and it follows from (33) that

$$a = a_0\left[\left(1 + \frac{4\mu_2 a_0}{3\pi\mu_1}\right)e^{\varepsilon\mu_1 t} - \frac{4\mu_2 a_0}{3\pi\mu_1}\right]^{-1}$$

Therefore

$$u = a_0\left[e^{\varepsilon\mu_1 t} + \frac{4\mu_2 a_0}{3\pi\mu_1}(e^{\varepsilon\mu_1 t} - 1)\right]^{-1}\cos(t+\beta) + \cdots \tag{34}$$

(e) In this case,

$$f(u, \dot{u}) = -2\mu_1\dot{u} - \operatorname{sgn}\dot{u} - \mu_2\dot{u}|\dot{u}|$$

and (7) and (8) become

$$a' = -\frac{\mu_1 a}{\pi}\int_0^{2\pi}\sin^2\phi\, d\phi + \frac{1}{2\pi}\int_0^{2\pi}\operatorname{sgn}(-a\sin\phi)\sin\phi\, d\phi$$
$$-\frac{\mu_2 a^2}{2\pi}\int_0^{2\pi}\sin^2\phi|\sin\phi|\, d\phi \tag{35}$$

$$a\beta' = -\frac{\mu_1 a}{\pi}\int_0^{2\pi}\sin\phi\cos\phi\, d\phi + \frac{1}{2\pi}\int_0^{2\pi}\operatorname{sgn}(-a\sin\phi)\cos\phi\, d\phi$$
$$-\frac{\mu_2 a^2}{2\pi}\int_0^{2\pi}\sin\phi\cos\phi|\sin\phi|\, d\phi \tag{36}$$

Carrying out the integrations in (35) and (36), we obtain

$$a' = -\mu_1 a - \frac{2}{\pi} - \frac{4\mu_2}{3\pi} a^2 \tag{37}$$

$$\beta' = 0 \tag{38}$$

The solution of (38) is that β is a constant, whereas the solution of (37) can be obtained by separation of variables. To this end, we rewrite (37) as

$$\int \frac{da}{a^2 + (3\pi\mu_1/4\mu_2)a + 3/2\mu_2} = -\int \frac{4\mu_2}{3\pi} dT_1 \tag{39}$$

There are three possible forms for the integral on the left-hand side of (39) depending on whether the denominator of the integrand has no real roots, has two equal real roots, or has two unequal real roots. In the first case, we write

$$a^2 + \frac{3\pi\mu_1}{4\mu_2} a + \frac{3}{2\mu_2} \quad \text{as} \quad \left(a + \frac{3\pi\mu_1}{8\mu_2}\right)^2 + \frac{3}{2\mu_2} - \frac{9\pi^2\mu_1^2}{64\mu_2^2}$$

Then we let

$$a + \frac{3\pi\mu_1}{8\mu_2} = \frac{\sqrt{96\mu_2 - 9\pi^2\mu_1^2}}{8\mu_2} \tan\theta \tag{40}$$

and obtain from (39) that

$$\theta = -\frac{\sqrt{96\mu_2 - 9\pi^2\mu_1^2}}{6\pi} T_1 + c \tag{41}$$

where c is a constant. Hence

$$a = -\frac{3\pi\mu_1}{8\mu_2} + \frac{\sqrt{96\mu_2 - 9\pi^2\mu_1^2}}{8\mu_2} \tan\left[c - \frac{\sqrt{96\mu_2 - 9\pi^2\mu_1^2}}{6\pi} \varepsilon t\right] + \cdots \tag{42}$$

In the second case, $32\mu_2 = 3\pi^2\mu_1^2$ and we obtain from (39) that

$$\frac{1}{a + 3\pi\mu_1/8\mu_2} = \frac{4\mu_2}{3\pi} T_1 + c \tag{43}$$

where c is a constant. Hence

$$a = -\frac{3\pi\mu_1}{8\mu_2} + \frac{1}{c + (4\mu_2/3\pi)\varepsilon t} \tag{44}$$

In the third case, the roots of the denominator are

$$\alpha_1, \alpha_2 = -\frac{3\pi\mu_1}{8\mu_2} \mp \frac{\sqrt{9\pi^2\mu_1^2 - 96\mu_2}}{8\mu_2} \tag{45}$$

Then, using partial fractions, we rewrite (39) as

$$\int \frac{da}{a-\alpha_2} - \int \frac{da}{a-\alpha_1} = -\frac{4\mu_2(\alpha_2-\alpha_1)}{3\pi} T_1 - \ln c$$

where c is a constant. Hence

$$\frac{a-\alpha_2}{a-\alpha_1} = \frac{1}{c}\exp\left[-\frac{4\mu_2(\alpha_2-\alpha_1)}{3\pi}\varepsilon t\right]$$

or

$$a = \alpha_2 + \frac{\alpha_2 - \alpha_1}{c\exp\{[4\mu_2(\alpha_2-\alpha_1)/3\pi]\varepsilon t\} - 1} \tag{46}$$

8.2. Consider the free oscillations of a system governed by

$$\ddot{u} + F(u) = 0 \tag{47}$$

where $F(u)$ is defined in Figure 8.1 for three different cases.

An approximate solution of (47) can be conveniently obtained by using the method of harmonic balance. According to this method, one lets

$$u = a\sin(\omega t + \beta) \tag{48}$$

into (47) and sets the coefficient of the first harmonic equal to zero. Substituting (48) into (47) gives

$$-a\omega^2\sin(\omega t+\beta) + F[a\sin(\omega t+\beta)] = 0$$

Setting the coefficient of $\sin(\omega t+\beta)$ equal to zero yields

$$\omega^2 = \frac{1}{\pi a}\int_0^{2\pi} F(a\sin\phi)\sin\phi\, d\phi \tag{49}$$

(a) In this case

$$F(u) = \begin{cases} ka_c & \text{if } a_c \le u \\ ku & \text{if } -a_c \le u \le a_c \\ -ka_c & \text{if } u \le -a_c \end{cases} \tag{50}$$

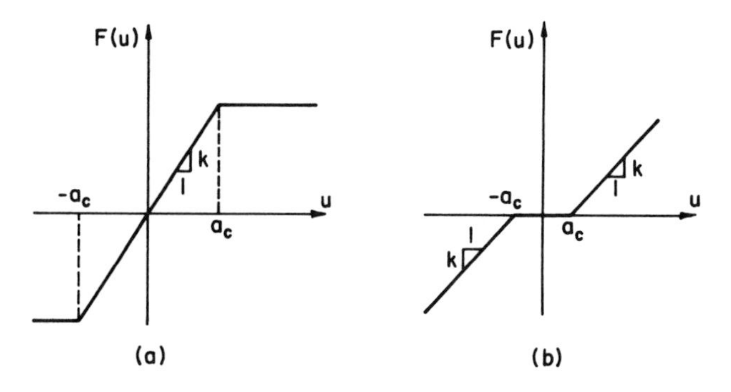

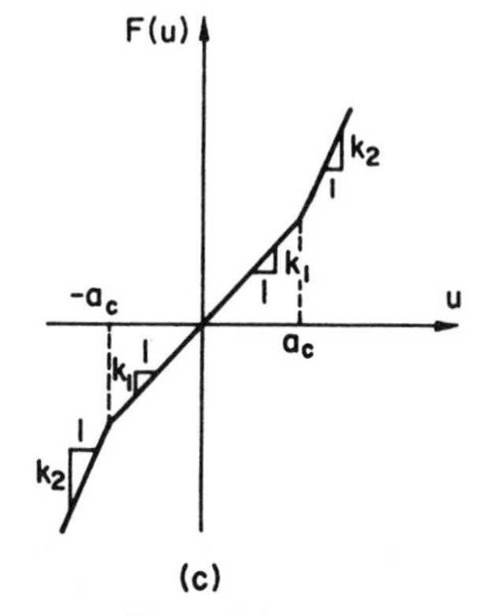

Figure 8.1

If we let ϕ_c be the root of

$$a_c = a \sin \phi \tag{51}$$

then

$$F(a \sin \phi) = \begin{cases} ka \sin \phi, & 0 \le \phi \le \phi_c \\ ka_c, & \phi_c \le \phi \le \pi - \phi_c \\ ka \sin \phi, & \pi - \phi_c \le \phi \le \pi + \phi_c \\ -ka_c, & \pi + \phi_c \le \phi \le 2\pi - \phi_c \\ ka \sin \phi, & 2\pi - \phi_c \le \phi \le 2\pi \end{cases} \tag{52}$$

Splitting the interval of integration as in (52), we rewrite (49) as

$$\omega^2 = \frac{k}{\pi} \int_0^{\phi_c} \sin^2 \phi \, d\phi + \frac{ka_c}{\pi a} \int_{\phi_c}^{\pi - \phi_c} \sin \phi \, d\phi + \frac{k}{\pi} \int_{\pi - \phi_c}^{\pi + \phi_c} \sin^2 \phi \, d\phi$$

$$- \frac{ka_c}{\pi a} \int_{\pi + \phi_c}^{2\pi - \phi_c} \sin \phi \, d\phi + \frac{k}{\pi} \int_{2\pi - \phi_c}^{2\pi} \sin^2 \phi \, d\phi$$

Hence

$$\omega^2 = \frac{k}{\pi}\Big[\left(\tfrac{1}{2}\phi - \tfrac{1}{4}\sin 2\phi\right)\Big|_0^{\phi_c} - \frac{a_c}{a}\cos\phi\Big|_{\phi_c}^{\pi-\phi_c} + \left(\tfrac{1}{2}\phi - \tfrac{1}{4}\sin 2\phi\right)\Big|_{\pi-\phi_c}^{\pi+\phi_c}$$

$$+ \frac{a_c}{a}\cos\phi\Big|_{\pi+\phi_c}^{2\pi-\phi_c} + \left(\tfrac{1}{2}\phi - \tfrac{1}{4}\sin 2\phi\right)\Big|_{2\pi-\phi_c}^{2\pi}\Big] = \frac{2k}{\pi}\left(\phi_c + \frac{a_c}{a}\cos\phi_c\right)$$

$$= \frac{2k}{\pi}\left[\sin^{-1}\left(\frac{a_c}{a}\right) + \frac{a_c}{a}\sqrt{1 - \frac{a_c^2}{a^2}}\right] \tag{53}$$

(b) In this case

$$F(u) = \begin{cases} 0 & \text{if } -a_c \le u \le a_c \\ k(u - a_c) & \text{if } a_c \le u \\ k(u + a_c) & \text{if } u \le -a_c \end{cases} \tag{54}$$

Again, if we let ϕ_c be the root of (51), then

$$F(a\sin\phi) = \begin{cases} 0 & \text{if } 0 \le \phi \le \phi_c \\ k(a\sin\phi - a_c) & \text{if } \phi_c \le \phi \le \pi - \phi_c \\ 0 & \text{if } \pi - \phi_c \le \phi \le \pi + \phi_c \\ k(a\sin\phi + a_c) & \text{if } \pi + \phi_c \le \phi \le 2\pi - \phi_c \\ 0 & \text{if } 2\pi - \phi_c \le \phi \le 2\pi \end{cases} \tag{55}$$

Splitting the interval of integration as in (55), we rewrite (49) as

$$\omega^2 = \frac{k}{\pi a}\int_{\phi_c}^{\pi-\phi_c}\left(a\sin^2\phi - a_c\sin\phi\right)d\phi + \frac{k}{\pi a}\int_{\pi+\phi_c}^{2\pi-\phi_c}\left(a\sin^2\phi + a_c\sin\phi\right)d\phi$$

$$= \frac{k}{\pi a}\Big[\tfrac{1}{2}a\left(\phi - \tfrac{1}{2}\sin 2\phi\right)\Big|_{\phi_c}^{\pi-\phi_c} + a_c\cos\phi\Big|_{\phi_c}^{\pi-\phi_c}$$

$$+ \tfrac{1}{2}a\left(\phi - \tfrac{1}{2}\sin 2\phi\right)\Big|_{\pi+\phi_c}^{2\pi-\phi_c} - a_c\cos\phi\Big|_{\pi+\phi_c}^{2\pi-\phi_c}\Big]$$

$$= k - \frac{2k}{\pi}\left(\phi_c + \frac{a_c}{a}\cos\phi_c\right) = k - \frac{2k}{\pi}\left[\sin^{-1}\left(\frac{a_c}{a}\right) + \frac{a_c}{a}\sqrt{1 - \frac{a_c^2}{a^2}}\right] \tag{56}$$

(c) In this case

$$F(u) = \begin{cases} k_1 u & \text{if } -a_c \le u \le a_c \\ k_2 u - (k_2 - k_1)a_c & \text{if } a_c \le u \\ k_2 u + (k_2 - k_1)a_c & \text{if } u \le -a_c \end{cases} \tag{57}$$

Again, if we let ϕ_c be the root of (51), then

$$F(a\sin\phi)=\begin{cases}k_1 a\sin\phi, & 0\le\phi\le\phi_c\\ k_2 a\sin\phi-(k_2-k_1)a_c, & \phi_c\le\phi\le\pi-\phi_c\\ k_1 a\sin\phi, & \pi-\phi_c\le\phi\le\pi+\phi_c\\ k_2 a\sin\phi+(k_2-k_1)a_c, & \pi+\phi_c<\phi<2\pi-\phi_c\\ k_1 a\sin\phi & 2\pi-\phi_c<\phi<2\pi\end{cases} \tag{58}$$

Splitting the interval of integration as in (58), we rewrite (49) as

$$\omega^2=\frac{k_1}{\pi}\int_0^{\phi_c}\sin^2\phi\,d\phi+\frac{k_2}{\pi}\int_{\phi_c}^{\pi-\phi_c}\sin^2\phi\,d\phi-\frac{(k_2-k_1)a_c}{\pi a}\int_{\phi_c}^{\pi-\phi_c}\sin\phi\,d\phi$$

$$+\frac{k_1}{\pi}\int_{\pi-\phi_c}^{\pi+\phi_c}\sin^2\phi\,d\phi+\frac{k_2}{\pi}\int_{\pi+\phi_c}^{2\pi-\phi_c}\sin^2\phi\,d\phi$$

$$+\frac{(k_2-k_1)a_c}{\pi a}\int_{\pi+\phi_c}^{2\pi-\phi_c}\sin\phi\,d\phi+\frac{k_1}{\pi}\int_{2\pi-\phi_c}^{2\pi}\sin^2\phi\,d\phi$$

$$=\frac{k_1}{2\pi}\left[\left(\phi-\tfrac{1}{2}\sin2\phi\right)\Big|_0^{\phi_c}+\left(\phi-\tfrac{1}{2}\sin2\phi\right)\Big|_{\pi-\phi_c}^{\pi+\phi_c}+\left(\phi-\tfrac{1}{2}\sin2\phi\right)\Big|_{2\pi-\phi_c}^{2\pi}\right]$$

$$+\frac{k_2}{2\pi}\left[\left(\phi-\tfrac{1}{2}\sin2\phi\right)\Big|_{\phi_c}^{\pi-\phi_c}+\left(\phi-\tfrac{1}{2}\sin2\phi\right)\Big|_{\pi+\phi_c}^{2\pi-\phi_c}\right]$$

$$+\frac{(k_2-k_1)a_c}{\pi a}\left[\cos\phi\Big|_{\phi_c}^{\pi-\phi_c}-\cos\phi\Big|_{\pi+\phi_c}^{2\pi-\phi_c}\right]$$

$$=k_2-\frac{2(k_2-k_1)}{\pi}\left(\phi_c+\frac{a_c}{a}\cos\phi_c\right)$$

$$=k_2-\frac{2(k_2-k_1)}{\pi}\left[\sin^{-1}\left(\frac{a_c}{a}\right)+\frac{a_c}{a}\sqrt{1-\frac{a_c^2}{a^2}}\right] \tag{59}$$

Supplementary Exercises

8.3. For small ε, determine first-order uniform expansions for each of the following problems:

(a) $\ddot{u}+u+\varepsilon u^2|u|=0$
(b) $\ddot{u}+u+\varepsilon\dot{u}^4|\dot{u}|=0$
(c) $\ddot{u}+u+\varepsilon\dot{u}^2u^2|\dot{u}|=0$
(d) $\ddot{u}+u+\varepsilon\dot{u}^2\operatorname{sgn}\dot{u}=0$
(e) $\ddot{u}+u+\varepsilon(2\mu\dot{u}+u|u|)=0$
(f) $\ddot{u}+u+\varepsilon(u^3+\dot{u}|\dot{u}|)=0$
(g) $\ddot{u}+u+\varepsilon u^4|\dot{u}|=0$

CHAPTER 9

Nonlinear Systems Subject to Harmonic Excitations

In the preceding five chapters, we considered the free oscillations of nonlinear systems; that is, the systems were initially disturbed and then allowed to respond with no further excitation. However, in many physical situations, systems are continuously disturbed or excited. Mathematically, the excitations appear either as inhomogeneous terms or as time-dependent coefficients in the governing equations. The first type of excitation is called external or direct excitation, whereas the second type of excitation is called parametric excitation. In this chapter we consider external excitations involving a single harmonic. In the following chapter we consider external excitations involving multiple independent frequencies and arbitrary amplitudes and phases, and in Chapter 11 we consider parametric excitations.

In the absence of excitations, the response decays with time in the presence of any small damping. On the other hand, the response does not decay in the presence of external excitations and it depends on the resonances that are present. The possible resonances depend on the type of nonlinearity and the frequencies involved in the excitations. They can be easily identified by carrying out a straightforward expansion. For example, the straightforward expansion of

$$\ddot{u} + \omega_0^2 u + 2\varepsilon\mu\dot{u} + \varepsilon u^3 = F\cos\Omega t$$

is

$$u = a\cos(\omega_0 t + \beta) + \frac{F}{\omega_0^2 - \Omega^2}\cos\Omega t + \varepsilon\Bigg\{ - a\mu t\cos(\omega_0 t + \beta)$$

$$+ \frac{2\mu\Omega F}{\left(\omega_0^2 - \Omega^2\right)^2}\sin\Omega t + \frac{a^3}{32\omega_0^2}\cos(3\omega_0 t + 3\beta)$$

$$-\frac{3a}{8\omega_0}\left[a^2+\frac{2F^2}{(\omega_0^2-\Omega^2)^2}\right]t\sin(\omega_0 t+\beta)$$

$$-\frac{F^3}{4(\omega_0^2-\Omega^2)^3(\omega_0^2-9\Omega^2)}\cos 3\Omega t$$

$$-\frac{3F}{2(\omega_0^2-\Omega^2)}\left[a^2+\frac{F^2}{2(\omega_0^2-\Omega^2)^2}\right]\cos\Omega t$$

$$+\frac{3a^2F}{4(\omega_0^2-\Omega^2)(\Omega+\omega_0)(\Omega+3\omega_0)}\cos[(2\omega_0+\Omega)t+2\beta]$$

$$+\frac{3a^2F}{4(\omega_0^2-\Omega^2)(\omega_0-\Omega)(3\omega_0-\Omega)}\cos[(2\omega_0-\Omega)t+2\beta]$$

$$+\frac{3aF^2}{16\Omega(\omega_0^2-\Omega^2)^2(\omega_0+\Omega)}\cos[(\omega_0+2\Omega)t+\beta]$$

$$\left.+\frac{3aF^2}{16\Omega(\omega_0^2-\Omega^2)^2(\Omega-\omega_0)}\cos[(\omega_0-2\Omega)t+\beta]\right\}+\cdots$$

In addition to the secular term, we note that the straightforward expansion contains terms whose denominators may be very small. Such terms are called small-divisor terms. If the frequencies are defined to be positive, small divisors occur when $\Omega \approx \omega_0$, $\Omega \approx 0$, $\Omega \approx \frac{1}{3}\omega_0$, and $\Omega \approx 3\omega_0$. These special frequencies are called resonant frequencies. Carrying out the expansion to higher order, one finds that small divisors also occur when $\Omega \approx n\omega_0$ and $\Omega \approx (1/n)\omega_0$, where n is an odd integer greater than 3.

When $\Omega \approx \omega_0$, small divisors first appear in the first term and hence it is called primary or main resonance. When $\Omega \approx m\omega_0$ or $\Omega \approx (1/m)\omega_0$, where m is an odd integer greater than or equal to 3, small divisors first appear in the second- and higher-order terms and hence they are called secondary resonances. We note that the resonances that occur depend on the order of nonlinearity. For a quadratic nonlinearity, resonances occur when $\Omega \approx \omega_0$, $\Omega \approx n\omega_0$, and $\Omega \approx (1/n)\omega_0$, where n is an integer greater than or equal to two.

The straightforward expansion breaks down owing to the presence of small-divisor terms as well as secular terms. To determine a uniform expansion, one needs to modify the straightforward expansion. Using the Lindstedt–Poincaré technique, one can determine periodic solutions whose stability cannot be determined directly. Instead, their stability can be investigated by analyzing the solutions of the variational equations resulting from perturbing these periodic motions. However, one can use the method of multiple scales or the method of averaging or the Krylov–Bogoliubov technique to determine the evolution of the amplitude and phase with damping, nonlinearity, and possible resonances. These equations can be used to determine steady-state motions, including periodic motions, and their stability. Moreover, integration of these equations yields transient as well as steady-state behavior.

Solved Exercises

9.1. The response of a system with quadratic nonlinearities to a sinusoidal excitation is governed by

$$\ddot{u} + \omega_0^2 u = -2\varepsilon\mu\dot{u} - \varepsilon\alpha u^2 + K\cos\Omega t \tag{1}$$

(a) Use the method of multiple scales to determine first-order uniform expansions when $\Omega \approx 2\omega_0$ and $2\Omega \approx \omega_0$.

We let

$$u = u_0(T_0, T_1) + \varepsilon u_1(T_0, T_1) + \cdots \tag{2}$$

Moreover, since Ω is away from zero, we express $\cos\Omega t$ as $\cos\Omega T_0$, where $T_0 = t$ and $T_1 = \varepsilon t$. Substituting (2) into (1) and equating coefficients of like powers of ε, we obtain

$$D_0^2 u_0 + \omega_0^2 u_0 = K\cos\Omega T_0 \tag{3}$$

$$D_0^2 u_1 + \omega_0^2 u_1 = -2D_0 D_1 u_0 - 2\mu D_0 u_0 - \alpha u_0^2 \tag{4}$$

The solution of (3) can be expressed as

$$u_0 = A(T_1)e^{i\omega_0 T_0} + \Lambda e^{i\Omega T_0} + cc \tag{5}$$

where

$$\Lambda = \frac{K}{2(\omega_0^2 - \Omega^2)}$$

Substituting (5) into (4) yields

$$D_0^2 u_1 + \omega_0^2 u_1 = -2i\omega_0(A' + \mu A)e^{i\omega_0 T_0} - 2i\mu\Omega\Lambda e^{i\Omega T_0} - \alpha[A^2 e^{2i\omega_0 T_0} + \Lambda^2 e^{2i\Omega T_0}$$

$$+ 2A\Lambda e^{i(\omega_0+\Omega)T_0} + 2\bar{A}\Lambda e^{i(\Omega-\omega_0)T_0} + A\bar{A} + \Lambda^2] + cc \tag{6}$$

Any particular solution of (6) contains secular terms due to the inhomogeneous term proportional to $\exp(i\omega_0 T_0)$ and small-divisor terms if 2Ω or $\Omega - \omega_0$ is near ω_0.

(b) *When* $2\Omega \approx \omega_0$, we introduce the detuning parameter σ defined by

$$2\Omega = \omega_0 + \varepsilon\sigma$$

and write

$$2\Omega T_0 = \omega_0 T_0 + \varepsilon\sigma T_0 = \omega_0 T_0 + \sigma T_1$$

Eliminating the terms that lead to secular terms from (6) yields

$$2i\omega_0\left(A' + \mu A\right) + \alpha\Lambda^2 e^{i\sigma T_1} = 0 \tag{7}$$

whose solution is

$$A = a_0 e^{-\mu T_1} - \frac{\alpha\Lambda^2}{2i\omega_0(\mu + i\sigma)} e^{i\sigma T_1} \tag{8}$$

where a_0 is a constant. As $T_1 \to \infty$, A tends to the steady-state solution

$$A = \frac{i\alpha\Lambda^2}{2\omega_0(\mu + i\sigma)} e^{i\varepsilon\sigma t} \tag{9}$$

(c) *When* $\Omega - \omega_0 \approx \omega_0$, we introduce the detuning parameter σ defined by

$$\Omega = 2\omega_0 + \varepsilon\sigma \tag{10}$$

and write

$$(\Omega - \omega_0) T_0 = \omega_0 T_0 + \varepsilon\sigma T_0 = \omega_0 T_0 + \sigma T_1$$

Eliminating the terms that lead to secular terms from (6) yields

$$i\omega_0\left(A' + \mu A\right) + \alpha\bar{A}\Lambda e^{i\sigma T_1} = 0 \tag{11}$$

To solve this equation, we transform it into an autonomous equation by using the change of variable

$$A = B e^{i\sigma T_1/2} \tag{12}$$

Then (11) becomes

$$\left(B' + \tfrac{1}{2}i\sigma B + \mu B\right) - \frac{i\alpha\Lambda}{\omega_0}\bar{B} = 0 \tag{13}$$

Putting $B = B_r + ib_i$, where B_r and B_i are real, in (13) and separating real and imaginary parts, we obtain

$$B_r' + \mu B_r - \left(\tfrac{1}{2}\sigma + \frac{\alpha\Lambda}{\omega_0}\right) B_i = 0 \tag{14}$$

$$B_i' + \mu B_i + \left(\tfrac{1}{2}\sigma - \frac{\alpha\Lambda}{\omega_0}\right) B_r = 0 \tag{15}$$

Since (14) and (15) are coupled equations with constant coefficients, they possess a solution of the form

$$B_r = b_r e^{\lambda T_1}, \qquad B_i = b_i e^{\lambda T_1} \tag{16}$$

Substituting (16) into (14) and (15) yields

$$(\lambda + \mu) b_r - \left(\tfrac{1}{2}\sigma + \frac{\alpha\Lambda}{\omega_0}\right) b_i = 0 \tag{17}$$

$$\left(\tfrac{1}{2}\sigma - \frac{\alpha\Lambda}{\omega_0}\right) b_r + (\lambda + \mu) b_i = 0 \tag{18}$$

For a nontrivial solution, the determinant of the coefficient matrix of (17) and (18) must vanish; that is,

$$(\lambda + \mu)^2 = \frac{\alpha^2\Lambda^2}{\omega_0^2} - \tfrac{1}{4}\sigma^2$$

Hence

$$\lambda = -\mu \mp \sqrt{\frac{\alpha^2\Lambda^2}{\omega_0^2} - \tfrac{1}{4}\sigma^2} \tag{19}$$

It follows from (12), (16), and (19) that A is unbounded if

$$\sqrt{\frac{\alpha^2\Lambda^2}{\omega_0^2} - \tfrac{1}{4}\sigma^2} > \mu$$

In this case, u is unbounded and the motion is termed unstable.

9.2. Consider

$$\ddot{u} + \omega_0^2 u = \varepsilon\left(\dot{u} - \tfrac{1}{3}\dot{u}^3\right) + \varepsilon k \cos \Omega t \tag{20}$$

When $\Omega = \omega_0 + \varepsilon\sigma$, use the methods of multiple scales and averaging to show that

$$u = a\cos(\omega_0 t + \beta) + \cdots \tag{21}$$

where

$$\dot{a} = \tfrac{1}{2}\varepsilon\left(1 - \tfrac{1}{4}\omega_0^2 a^2\right)a + \frac{\varepsilon k}{2\omega_0}\sin(\varepsilon\sigma t - \beta) \tag{22}$$

$$a\dot{\beta} = -\frac{\varepsilon k}{2\omega_0}\cos(\varepsilon\sigma t - \beta) \tag{23}$$

(a) ***The Method of Averaging***

When $\varepsilon = 0$, the solution of (20) can be expressed as

$$u = a\cos(\omega_0 t + \beta) \tag{24}$$

where a and β are constants. Hence

$$\dot{u} = -a\omega_0\sin(\omega_0 t + \beta) \tag{25}$$

When $\varepsilon \neq 0$, we continue to express u as in (24) subject to the constraint (25) but with time varying a and β. Differentiating (24) with respect to t and using (25), we conclude that

$$\dot{a}\cos\phi - a\dot{\beta}\sin\phi = 0 \tag{26}$$

where $\phi = \omega_0 t + \beta$. Substituting (24) and (25) into (20) yields

$$\dot{a}\sin\phi + a\dot{\beta}\cos\phi = \varepsilon\left(a\sin\phi - \tfrac{1}{3}a^3\omega_0^2\sin^3\phi\right) - \frac{\varepsilon k}{\omega_0}\cos\Omega t \tag{27}$$

Solving (26) and (27) yields

$$\begin{aligned}\dot{a} &= \varepsilon\sin\phi\left(a\sin\phi - \tfrac{1}{3}a^3\omega_0^2\sin^3\phi\right) - \frac{\varepsilon k}{\omega_0}\sin\phi\cos\Omega \\ &= \varepsilon\left[\tfrac{1}{2}a - \tfrac{1}{8}\omega_0^2 a^3 - \left(\tfrac{1}{2}a - \tfrac{1}{6}\omega_0^2 a^3\right)\cos 2\phi - \tfrac{1}{24}\omega_0^2 a^3\cos 4\phi\right] - \frac{\varepsilon k}{2\omega_0}\sin[(\omega_0 + \Omega)t + \beta] \\ &\quad - \frac{\varepsilon k}{2\omega_0}\sin[(\omega_0 - \Omega)t + \beta]\end{aligned} \tag{28}$$

$$\begin{aligned}a\dot{\beta} &= \varepsilon\cos\phi\left(a\sin\phi - \tfrac{1}{3}a^3\omega_0^2\sin^3\phi\right) - \frac{\varepsilon k}{\omega_0}\cos\phi\cos\Omega t \\ &= \varepsilon\left[\left(\tfrac{1}{2}a - \tfrac{1}{12}\omega_0^2 a^3\right)\sin 2\phi + \tfrac{1}{24}\omega_0^2 a^3\sin 4\phi\right] - \frac{\varepsilon k}{2\omega_0}\cos[(\omega_0 + \Omega)t + \beta] \\ &\quad - \frac{\varepsilon k}{2\omega_0}\cos[(\omega_0 - \Omega)t + \beta]\end{aligned} \tag{29}$$

where ϕ was replaced with $\omega_0 t + \beta$ in the last two terms in (28) and (29) in order to exhibit the slowly varying terms. Since $\Omega = \omega_0 + \varepsilon\sigma$, the terms proportional to

$$\sin[(\omega_0 - \Omega)t + \beta] \quad \text{and} \quad \cos[(\omega_0 - \Omega)t + \beta]$$

are slowly varying. Thus keeping only the slowly varying terms in (28) and (29), we obtain (22) and (23). Therefore, to the first approximation, u is given by (21), where a and β are defined by (22) and (23).

(b) *The Method of Multiple Scales*

Substituting (2) into (20), expressing $\cos \Omega t$ as $\cos \Omega T_0$ since $\Omega \approx \omega_0$ and ω_0 is assumed to be away from zero, and equating coefficients of like powers of ε, we obtain

$$D_0^2 u_0 + \omega_0^2 u_0 = 0 \tag{30}$$

$$D_0^2 u_1 + \omega_0^2 u_1 = -2D_0 D_1 u_0 + D_0 u_0 - \tfrac{1}{3}(D_0 u_0)^3 + k \cos \Omega T_0 \tag{31}$$

The solution of (30) can be expressed as

$$u_0 = A(T_1) e^{i\omega_0 T_0} + \bar{A}(T_1) e^{-i\omega_0 T_0} \tag{32}$$

Then (31) becomes

$$D_0^2 u_1 + \omega_0^2 u_1 = -i\omega_0\left(2A' - A + \omega_0^2 A^2 \bar{A}\right) e^{i\omega_0 T_0} + \tfrac{1}{3} i\omega_0^3 A^3 e^{3i\omega_0 T_0}$$
$$+ \tfrac{1}{2} k e^{i\Omega T_0} + \text{cc} \tag{33}$$

Since $\Omega = \omega_0 + \varepsilon\sigma$,

$$\Omega T_0 = \omega_0 T_0 + \varepsilon\sigma T_0 = \omega_0 T_0 + \sigma T_1$$

Hence eliminating the terms that lead to secular terms from (33) yields

$$i\omega_0\left(2A' - A + \omega_0^2 A^2 \bar{A}\right) - \tfrac{1}{2} k e^{i\sigma T_1} = 0 \tag{34}$$

Expressing A in the polar form

$$A = \tfrac{1}{2} a e^{i\beta} \tag{35}$$

and separating real and imaginary parts in (34), we obtain

$$a' = \tfrac{1}{2} a\left(1 - \tfrac{1}{4}\omega_0^2 a^2\right) + \frac{k}{2\omega_0} \sin(\sigma T_1 - \beta)$$

$$a\beta' = -\frac{k}{2\omega_0} \cos(\sigma T_1 - \beta) \tag{36}$$

in agreement with (22) and (23) because $T_1 = \varepsilon t$.

9.3. Consider

$$\ddot{u} + \omega_0^2 u = \varepsilon\left(\dot{u} - \tfrac{1}{3}\dot{u}^3\right) + K\cos\Omega t \tag{37}$$

Use the methods of multiple scales and averaging to determine a first-order uniform expansion when Ω is away from $3\omega_0$, $\frac{1}{3}\omega_0$, and 0.

(2) ***The Method of Multiple Scales***

Substituting (2) into (37), expressing $\cos\Omega t$ as $\cos\Omega T_0$ since Ω is away from zero, and equating coefficients of like powers of ε, we obtain

$$D_0^2 u_0 + \omega_0^2 u_0 = K\cos\Omega T_0 \tag{38}$$

$$D_0^2 u_1 + \omega_0^2 u_1 = -2D_0 D_1 u_0 + Du_0 - \tfrac{1}{3}(Du_0)^3 \tag{39}$$

The solution of (38) can be expressed as

$$u_0 = A(T_1)e^{i\omega_0 T_0} + \Lambda e^{i\Omega T_0} + \text{cc} \tag{40}$$

where

$$\Lambda = \frac{K}{2(\omega_0^2 - \Omega^2)} \tag{41}$$

Then (39) becomes

$$\begin{aligned} D_0^2 u_1 + \omega_0^2 u_1 = {} & \left[-2i\omega_0 A' + i\omega_0(1 - 2\Omega^2\Lambda^2)A - i\omega_0^3 A^2\bar{A}\right]e^{i\omega_0 T_0} \\ & + \tfrac{1}{3}i\left[\omega_0^3 A^3 e^{3i\omega_0 T_0} + \Omega^3\Lambda^3 e^{3i\Omega T_0} + 3\Omega\Lambda\left(1 - \Omega^2\Lambda^2 - 2\omega_0^2 A\bar{A}\right)e^{i\Omega T_0}\right. \\ & + 3\omega_0^2\Omega\Lambda A^2 e^{i(\Omega + 2\omega_0)T_0} + 3\omega_0^2\Omega\Lambda\bar{A}^2 e^{i(\Omega - 2\omega_0)T_0} + 3\omega_0\Omega^2\Lambda^2 A e^{i(2\Omega + \omega_0)T_0} \\ & \left. - 3\omega_0\Omega^2\Lambda^2\bar{A}e^{i(2\Omega - \omega_0)T_0}\right] + \text{cc} \end{aligned} \tag{42}$$

Since Ω is away from $3\omega_0$, $(1/3)\omega_0$, and 0, there are no terms that lead to small-divisor terms in the particular solutions of (42). Hence eliminating the terms that lead to secular terms from (42) yields

$$2A' = (1 - 2\Omega^2\Lambda^2)A - \omega_0^2 A^2\bar{A} \tag{43}$$

Expressing A in the polar form (35) and separating (43) into real and imaginary parts, we obtain

$$a' = \tfrac{1}{2}a\left(\eta - \tfrac{1}{4}\omega_0^2 a^2\right), \qquad \beta' = 0 \tag{44}$$

where

$$\eta = 1 - 2\Omega^2\Lambda^2 = 1 - \frac{K^2\Omega^2}{2(\omega_0^2 - \Omega^2)^2} \tag{45}$$

Substituting the polar form (35) of A into (40) and substituting the result into (2), we obtain

$$u = a(T_1)\cos[\omega_0 t + \beta(T_1)] - \frac{K}{\Omega^2 - \omega_0^2}\cos\Omega t + \cdots \tag{46}$$

where a and β are defined in (44).

(b) *The Method of Averaging*

When $\varepsilon = 0$, (37) reduces to

$$\ddot{u} + \omega_0^2 u = K\cos\Omega t \tag{47}$$

whose solution is

$$u = a\cos(\omega_0 t + \beta) + 2\Lambda\cos\Omega t \tag{48}$$

where a and β are constant and Λ is defined in (41). Differentiating (48) with respect to t yields

$$\dot{u} = -\omega_0 a\sin(\omega_0 t + \beta) - 2\Omega\Lambda\sin\Omega t \tag{49}$$

When $\varepsilon \neq 0$, we still express the solution of (37) as in (48), subject to the constraint (49) but with time varying a and β. Thus differentiating (48) with respect to t and using (49), we have

$$\dot{a}\cos(\omega_0 t + \beta) - a\dot{\beta}\sin(\omega_0 t + \beta) = 0 \tag{50}$$

Substituting (48) and (49) into (37) yields

$$\dot{a}\sin(\omega_0 t + \beta) + a\dot{\beta}\cos(\omega_0 t + \beta) = \varepsilon a\sin(\omega_0 t + \beta) + \frac{2\varepsilon\Omega\Lambda}{\omega_0}\sin\Omega t$$

$$-\tfrac{1}{3}\varepsilon\omega_0^2\left[a\sin(\omega_0 t + \beta) + \frac{2\Omega\Lambda}{\omega_0}\sin\Omega t\right]^3 \tag{51}$$

Solving (50) and (51), we have

$$\dot{a} = \varepsilon \sin(\omega_0 t + \beta)\left\{ a\sin(\omega_0 t + \beta) + \frac{2\Omega\Lambda}{\omega_0}\sin\Omega t - \tfrac{1}{3}\omega_0^2\left[a\sin(\omega_0 t + \beta) + \frac{2\Omega\Lambda}{\omega_0}\sin\Omega t\right]^3\right\}$$

$$= \tfrac{1}{2}\varepsilon a\left(1 - 2\Omega^2\Lambda^2 - \tfrac{1}{4}\omega_0^2 a^2\right) - \tfrac{1}{2}\varepsilon a\left(1 - 2\Omega^2\Lambda^2 - \tfrac{1}{3}\omega_0^2 a^2\right)\cos(2\omega_0 t + 2\beta)$$

$$+ \varepsilon\Omega^2\Lambda^2 a\cos 2\Omega t + \varepsilon\left(\frac{\Omega\Lambda}{\omega_0} - \frac{\Omega^3\Lambda^3}{\omega_0} - \tfrac{3}{4}\omega_0\Omega\Lambda a^2\right)$$

$$\times\left\{\cos[(\omega_0 - \Omega)t + \beta] - \cos[(\omega_0 + \Omega)t + \beta]\right\}$$

$$- \tfrac{1}{24}\varepsilon\omega_0^2 a^3\cos(4\omega_0 t + 4\beta)$$

$$+ \tfrac{1}{4}\varepsilon\omega_0\Omega\Lambda a^2\left\{\cos[(3\omega_0 - \Omega)t + 3\beta] - \cos[(3\omega_0 + \Omega)t + \beta]\right\}$$

$$- \tfrac{1}{2}\varepsilon\Omega^2\Lambda^2 a\left\{\cos[2(\omega_0 + \Omega)t + 2\beta] + \cos[2(\omega_0 - \Omega)t + 2\beta]\right\}$$

$$+ \frac{\varepsilon\Omega^3\Lambda^3}{3\omega_0}\left\{\cos[(\omega_0 - 3\Omega)t + \beta] - \cos[(\omega_0 + 3\Omega)t + \beta]\right\} \tag{52}$$

$$a\dot{\beta} = \varepsilon\cos(\omega_0 t + \beta)\left\{ a\sin(\omega_0 t + \beta) + \frac{2\Omega\Lambda}{\omega_0}\sin\Omega t - \tfrac{1}{3}\omega_0^2\left[a\sin(\omega_0 t + \beta) + \frac{2\Omega\Lambda}{\omega_0}\sin\Omega t\right]^3\right\}$$

$$= \tfrac{1}{2}\varepsilon a\left(1 - \tfrac{1}{6}\omega_0^2 a^2 - 2\Omega^2\Lambda^2\right)\sin(2\omega_0 t + 2\beta) + \tfrac{1}{24}\varepsilon\omega_0^2 a^3\sin(4\omega_0 t + 4\beta)$$

$$+ \varepsilon\left(\frac{\Omega\Lambda}{\omega_0} - \tfrac{1}{4}\omega_0\Omega\Lambda a^2 - \frac{\Omega^3\Lambda^3}{\omega_0}\right)\sin[(\omega_0 + \Omega)t + \beta]$$

$$- \varepsilon\left(\frac{\Omega\Lambda}{\omega_0} - \tfrac{1}{4}\Omega\Lambda\omega_0 a^2 - \frac{\Omega^3\Lambda^3}{\omega_0}\right)\sin[(\omega_0 - \Omega)t + \beta]$$

$$+ \tfrac{1}{4}\varepsilon\Omega\Lambda\omega_0 a^2\left\{\sin[(3\omega_0 + \Omega)t + 3\beta] - \sin[(3\omega_0 - \Omega)t + 3\beta]\right\}$$

$$+ \tfrac{1}{2}\varepsilon\Omega^2\Lambda^2 a\left\{\sin[(2\omega_0 + 2\Omega)t + 2\beta] + \sin[(2\omega_0 - 2\Omega)t + 2\beta]\right\}$$

$$+ \frac{\varepsilon\Omega^3\Lambda^3}{3\omega_0}\left\{\sin[(\omega_0 + 3\Omega)t + \beta] - \sin[(\omega_0 - 3\Omega)t + \beta]\right\} \tag{53}$$

When Ω is away from $3\omega_0$, $\frac{1}{3}\omega_0$, and 0, there are no slowly varying terms in (53) and only the first term on the right-hand side of (52) is slowly varying. Hence, keeping the slowly varying terms in (52) and (53), we obtain

$$\dot{a} = \tfrac{1}{2}\varepsilon a\left(1 - 2\Omega^2\Lambda^2 - \tfrac{1}{4}\omega_0^2 a^2\right) \tag{54}$$

$$\dot{\beta} = 0 \tag{55}$$

in agreement with (44) and (45), obtained by using the method of multiple scales, since $T_1 = \varepsilon t$.

9.4. Use the methods of multiple scales and averaging to determine a first-order uniform expansion for the solution of (37) when $3\Omega = \omega_0 + \varepsilon\sigma$.

(a) ***The Method of Multiple Scales***

In this case, Equations (38)–(42) and (46) of the preceding exercise still hold. To determine the solvability condition for (42), we note that, in addition to the term that produces secular terms, there is a term, namely, $\frac{1}{3}i\Omega^3\Lambda^3\exp(3i\Omega T_0)$, that leads to a small-divisor term. We first transform it into a term that produces a secular term by rewriting as

$$\tfrac{1}{3}i\Omega^3\Lambda^3 e^{3i\Omega T_0} = \tfrac{1}{3}i\Omega^3\Lambda^3 e^{i\omega_0 T_0 + i\sigma T_1} \tag{56}$$

Eliminating the terms that lead to secular terms from (42) yields

$$2A' = \left(1 - 2\Omega^2\Lambda^2\right)A - \omega_0^2 A^2\bar{A} + \frac{\Omega^3\Lambda^3}{3\omega_0}e^{i\sigma T_1} \tag{57}$$

Expressing A in the polar form (35) and separating real and imaginary parts in (57), we obtain

$$a' = \tfrac{1}{2}a\left(\eta - \tfrac{1}{4}\omega_0^2 a^2\right) + \Gamma\cos(\sigma T_1 - \beta) \tag{58}$$

$$a\beta' = \Gamma\sin(\sigma T_1 - \beta) \tag{59}$$

where

$$\eta = 1 - 2\Omega^2\Lambda^2 \quad \text{and} \quad \Gamma = \frac{\Omega^3\Lambda^3}{3\omega_0} \tag{60}$$

Therefore, u is given by (46) where a and β are defined by (58) and (59).

(b) ***The Method of Averaging***

In this case, Equations (47)–(53) of the preceding exercise still hold. In addition to the slowly varying terms in the preceding exercise, the term proportional to $\cos[(\omega_0 - 3\Omega)t + \beta]$ in (52) and the term proportional to $\sin[(\omega_0 - 3\Omega)t + \beta]$ in (53) are slowly varying. Hence, keeping the slowly varying terms in (52) and (53), we have

$$\dot{a} = \tfrac{1}{2}\varepsilon a\left(1 - 2\Omega^2\Lambda^2 - \tfrac{1}{4}\omega_0^2 a^2\right) + \varepsilon\Gamma\cos[(\omega_0 - 3\Omega)t + \beta] \tag{61}$$

$$a\dot{\beta} = -\varepsilon\Gamma\sin[(\omega_0 - 3\Omega)t + \beta] \tag{62}$$

in agreement with (58) and (59), obtained by using the method of multiple scales, since $T_1 = \varepsilon t$ and $3\Omega = \omega_0 + \varepsilon\sigma$.

9.5. Use the methods of multiple scales and averaging to determine a first-order uniform expansion for the solution of (37) when $\Omega = 3\omega_0 + \varepsilon\sigma$.

(a) ***The Method of Multiple Scales***

In this case, Equations (38)–(42) and (46) of Exercise 9.3 still hold. In addition to the term that leads to secular terms in (42), the term $i\omega_0^2\Omega\Lambda\bar{A}^2\exp[i(\Omega - 2\omega_0)T_0]$ leads to a small-divisor term. To determine the solvability condition for (42), we first transform this term into a term that leads to a secular term. To this end, we put

$$(\Omega - 2\omega_0)T_0 = \omega_0 T_0 + \varepsilon\sigma T_0 = \omega_0 T_0 + \sigma T_1$$

Eliminating the terms that lead to secular terms from (42) we obtain

$$2A' = (1 - 2\Omega^2\Lambda^2)A - \omega_0^2 A^2\bar{A} + \omega_0\Omega\Lambda\bar{A}^2 e^{i\sigma T_1} \tag{63}$$

Expressing A in the polar form (35) and separating (63) into real and imaginary parts, we obtain

$$a' = \tfrac{1}{2}a\left(\eta - \tfrac{1}{4}\omega_0^2 a^2\right) + \tfrac{1}{4}\omega_0\Omega\Lambda a^2\cos(\sigma T_1 - 3\beta) \tag{64}$$

$$a\beta' = \tfrac{1}{4}\omega_0\Omega\Lambda a^2\sin(\sigma T_1 - 3\beta) \tag{65}$$

Thus, to the first approximation, u is given by (46) with a and β defined by (64) and (65).

(b) ***The Method of Averaging***

In this case, Equations (47)–(53) of Exercise 9.3 still hold. In addition to the first term in (52), the term proportional to $\cos[(3\omega_0 - \Omega)t + 3\beta]$ in (52) and the term proportional to $\sin[(3\omega_0 - \Omega)t + 3\beta]$ in (53) are slowly varying because $\Omega \approx 3\omega_0$. Keeping the slowly varying terms in (52) and (53), we obtain

$$\dot{a} = \tfrac{1}{2}\varepsilon a\left(1 - 2\Omega^2\Lambda^2 - \tfrac{1}{4}\omega_0^2 a^2\right) + \tfrac{1}{4}\varepsilon\omega_0\Omega\Lambda a^2\cos[(3\omega_0 - \Omega)t + 3\beta] \tag{66}$$

$$a\dot{\beta} = -\tfrac{1}{4}\varepsilon\omega_0\Omega\Lambda a^2\sin[(3\omega_0 - \Omega) + 3\beta] \tag{67}$$

in agreement with (64) and (65), since $\Omega = 3\omega_0 + \varepsilon\sigma$.

9.6. Consider

$$\ddot{u} + u = -\varepsilon\dot{u}|\dot{u}| + 2\varepsilon k\cos\Omega t \tag{68}$$

when $\Omega = 1 + \varepsilon\sigma$. Use the methods of multiple scales and averaging to show that

$$u = a\cos(t + \beta) + \cdots \tag{69}$$

where

$$\dot{a} = -\frac{4\varepsilon a^2}{3\pi} + \varepsilon k \sin(\varepsilon\sigma t - \beta) \tag{70}$$

$$a\dot{\beta} = -\varepsilon k \cos(\varepsilon\sigma t - \beta) \tag{71}$$

(a) ***The Method of Multiple Scales***

Since $\Omega = 1 + \varepsilon\sigma$, we express $\cos\Omega t$ in terms of the fast and slow scales T_0 and T_1 and write

$$\cos\Omega t = \cos(t + \varepsilon\sigma t) = \cos(T_0 + \sigma T_1)$$

Substituting (2) into (68) and equating coefficients of like powers of ε, we obtain

$$D_0^2 u_0 + u_0 = 0 \tag{72}$$

$$D_0^2 u_1 + u_1 = -2D_0 D_1 u_0 - Du_0|Du_0| + 2k\cos(T_0 + \sigma T_1) \tag{73}$$

The solution of (72) can be written as

$$u_0 = A(T_1)e^{iT_0} + \bar{A}(T_1)e^{-iT_0} \tag{74}$$

Then (73) becomes

$$D_0^2 u_1 + u_1 = -2iA'e^{iT_0} + ke^{i(T_0+\sigma T_1)} + \text{cc} + f(iAe^{iT_0} - i\bar{A}e^{-iT_0}) \tag{75}$$

where

$$f = -i(Ae^{iT_0} - \bar{A}e^{-iT_0})|i(Ae^{iT_0} - Ae^{-iT_0})| \tag{76}$$

To eliminate the terms that lead to secular terms, we multiply the right-hand side of (75) with $\exp(-iT_0)$, integrate the result from 0 to 2π, and obtain

$$2iA' = ke^{i\sigma T_1} + \frac{1}{2\pi}\int_0^{2\pi} e^{-iT_0} f(iAe^{iT_0} - i\bar{A}e^{-iT_0})\, dT_0 \tag{77}$$

Expressing A in the polar form (35), we rewrite (77) as

$$i(a' + ia\beta') = k\cos(\sigma T_1 - \beta) + ik\sin(\sigma T_1 - \beta)$$
$$+ \frac{1}{2\pi}\int_0^{2\pi} e^{-i(T_0+\beta)} f dT_0$$

or

$$i(a' + ia\beta') = k\cos(\sigma T_1 - \beta) + ik\sin(\sigma T_1 - \beta)$$
$$+ \frac{1}{2\pi}\int_0^{2\pi} (\cos\phi - i\sin\phi) f(-a\sin\phi)\, d\phi \tag{78}$$

Separating (78) into real and imaginary parts, we obtain

$$a' = k\sin(\sigma T_i - \beta) - \frac{1}{2\pi}\int_0^{2\pi} f(-a\sin\phi)\sin\phi\,d\phi \tag{79}$$

$$a\beta' = -k\cos(\sigma T_1 - \beta) - \frac{1}{2\pi}\int_0^{2\pi} f(-a\sin\phi)\cos\phi\,d\phi \tag{80}$$

Using (76), we rewrite the integral in (78) as

$$\frac{1}{2\pi}\int_0^{2\pi} f(-a\sin\phi)\sin\phi\,d\phi = \frac{a^2}{2\pi}\int_0^{2\pi}\sin^2\phi|\sin\phi|\,d\phi$$

Since $|\sin\phi| = \sin\phi$ in $[0,\pi]$ and $|\sin\phi| = -\sin\phi$ in $[\pi,2\pi]$, we split the interval of integration and write

$$\begin{aligned}
\frac{a^2}{2\pi}\int_0^{2\pi}\sin^2\phi|\sin\phi|\,d\phi &= \frac{a^2}{2\pi}\left[\int_0^{\pi}\sin^3\phi\,d\phi - \int_{\pi}^{2\pi}\sin^3\phi\,d\phi\right] \\
&= \frac{a^2}{2\pi}\left[\tfrac{1}{4}\int_0^{\pi}(3\sin\phi - \sin 3\phi)\,d\phi - \tfrac{1}{4}\int_{\pi}^{2\pi}(3\sin\phi - \sin 3\phi)\,d\phi\right] \\
&= \frac{a^2}{8\pi}\left[(-3\cos\phi + \tfrac{1}{3}\cos 3\phi)\big|_0^{\pi} - (-3\cos\phi + \tfrac{1}{3}\cos 3\phi)\big|_{\pi}^{2\pi}\right] \\
&= \frac{4a^2}{3\pi}
\end{aligned} \tag{81}$$

Substituting (81) into (79) yields (70), because $T_1 = \varepsilon t$. Similarly, the integral in (80) can be split into two as follows:

$$\begin{aligned}
\frac{1}{2\pi}\int_0^{2\pi} f(-a\sin\phi)\cos\phi &= \frac{a^2}{2\pi}\left[\int_0^{\pi}\sin^2\phi\cos\phi\,d\phi - \int_{\pi}^{2\pi}\sin^2\phi\cos\phi\,d\phi\right] \\
&= \frac{a^2}{6\pi}\left[\sin^3\phi\big|_0^{\pi} - \sin^3\phi\big|_{\pi}^{2\pi}\right] = 0
\end{aligned} \tag{82}$$

Substituting (82) into (80) yields (71), because $T_1 = \varepsilon t$.

(b) ***The Method of Averaging***

When $\varepsilon = 0$, the solution of (68) can be expressed as

$$u = a\cos(t + \beta) \tag{83}$$

where a and β are constants. Differentiating (83) with respect to t yields

$$\dot{u} = -a\sin(t + \beta) \tag{84}$$

When $\varepsilon \neq 0$, the solution of (68) is still given by (83), subject to the constraint (84), but with time-varying a and β. Then, differentiating (83) once with respect to t and using

(84), we obtain

$$\dot{a}\cos\phi - a\dot{\beta}\sin\phi = 0 \tag{85}$$

where $\phi = t + \beta$. Substituting (83) and (84) into (68) yields

$$\dot{a}\sin\phi + a\dot{\beta}\cos\phi = -\,\varepsilon a^2 \sin\phi|\sin\phi| - 2\varepsilon k\cos\Omega t \tag{86}$$

Solving (85) and (86) yields

$$\dot{a} = -\,\varepsilon a^2 \sin^2\!\phi|\sin\phi| - 2\varepsilon k\sin\phi\cos\Omega t \tag{87}$$

$$a\dot{\beta} = -\,\varepsilon a^2 \sin\phi\cos\phi|\sin\phi| - 2\varepsilon k\cos\phi\cos\Omega t \tag{88}$$

Using (81) and recalling that $\Omega = 1 + \varepsilon\sigma$, we conclude that the slowly varying part of (87) is given by (70). Moreover, using (82) we conclude that the slowly varying part of (88) is given by (71).

9.7. Consider

$$\ddot{u} + u + 2\varepsilon u^2\dot{u} = 2\varepsilon k\cos\Omega t \tag{89}$$

when $\Omega = 1 + \varepsilon\sigma$. Use the methods of multiple scales and averaging to show that

$$u = a\cos(t+\beta) + \cdots \tag{90}$$

where

$$\dot{a} = -\tfrac{1}{4}\varepsilon a^3 + \varepsilon k\sin(\varepsilon\sigma t - \beta) \tag{91}$$

$$a\dot{\beta} = -\,\varepsilon k\cos(\varepsilon\sigma - \beta) \tag{92}$$

(a) ***The Method of Multiple Scales***

Substituting (2) into (89), using $\Omega t = T_0 + \sigma T_1$, and equating coefficients of like powers of ε, we obtain

$$D_0^2 u_0 + u_0 = 0 \tag{93}$$

$$D_0^2 u_1 + u_1 = -2D_0D_1u_0 - 2u_0^2 D_0 u_0 + 2k\cos(T_0 + \sigma T_1) \tag{94}$$

The solution of (93) can be expressed as in (74). Then (94) becomes

$$D_0^2 u_1 + u_1 = -2iA'e^{iT_0} + 2i\bar{A}'e^{-iT_0}$$

$$-2i(Ae^{iT_0} + \bar{A}e^{-iT_0})^2(Ae^{iT_0} - \bar{A}e^{-iT_0}) + ke^{i\sigma T_1}e^{iT_0} + ke^{-i\sigma T_1}e^{-iT_0}$$

or

$$D_0^2 u_1 + u_1 = (-2iA' - 2iA^2\bar{A} + ke^{i\sigma T_1})\,e^{iT_0} + \text{cc} + \text{NST} \tag{95}$$

Eliminating the terms that lead to secular terms from (95) yields

$$2iA' + 2iA^2\bar{A} - ke^{i\sigma T_1} = 0 \tag{96}$$

Substituting the polar form (35) into (96) and separating real and imaginary parts, we obtain

$$a' = -\tfrac{1}{4}a^3 + k\sin(\sigma T_1 - \beta) \tag{97}$$

$$a\beta' = -k\cos(\sigma T_1 - \beta) \tag{98}$$

in agreement with (91) and (92), since $T_1 = \varepsilon t$.

(b) ***The Method of Averaging***

When $\varepsilon = 0$, the solution of (89) can be expressed as in (83). When $\varepsilon \neq 0$, the solution of (89) can still be expressed as in (83), subject to the constraint (84), but with time-varying a and β. Differentiating (83) once with respect to t and using (84), we obtain (85). Substituting (83) and (84) into (89) yields

$$\dot{a}\sin\phi + a\dot{\beta}\cos\phi = -2\varepsilon a^3 \sin\phi\cos^2\phi - 2\varepsilon k\cos\Omega t \tag{99}$$

Solving (85) and (99), we have

$$\begin{aligned}\dot{a} &= -2\varepsilon a^3 \sin^2\phi\cos^2\phi - 2\varepsilon k\sin\phi\cos\Omega t\\ &= -\tfrac{1}{4}\varepsilon a^3(1-\cos 4\phi) - \varepsilon k\sin[(1+\Omega)t+\beta] - \varepsilon k\sin[(1-\Omega)t+\beta]\end{aligned} \tag{100}$$

$$\begin{aligned}a\dot{\beta} &= -2\varepsilon a^3 \sin\phi\cos^3\phi - 2\varepsilon k\cos\phi\cos\Omega t\\ &= -\tfrac{1}{4}\varepsilon a^3(2\sin 2\phi + \sin 4\phi) - \varepsilon k\cos[(1+\Omega)t+\beta] - \varepsilon k\cos[(1-\Omega)t+\beta]\end{aligned} \tag{101}$$

Keeping the slowly varying terms in (100) and (101), we obtain

$$\dot{a} = -\tfrac{1}{4}\varepsilon a^3 - \varepsilon k\sin[(1-\Omega)t+\beta] \tag{102}$$

$$a\dot{\beta} = -\varepsilon k\cos[(1-\Omega)t+\beta] \tag{103}$$

in agreement with (91) and (92), since $\Omega = 1 + \varepsilon\sigma$.

9.8. Consider

$$\ddot{u} + \omega_0^2 u + \varepsilon u^4 = 2K\cos\Omega t \tag{104}$$

Show that to first-order, resonances exist when $\Omega \approx 0$, $4\omega_0$, $2\omega_0$, $\frac{1}{4}\omega_0$, $\frac{1}{2}\omega_0$, $\frac{2}{3}\omega_0$, and $\frac{3}{2}\omega_0$. Use the methods of multiple scales and averaging to determine the equations describing the amplitude and phase for each case.

(a) *The Straightforward Expansion*

Putting

$$u = u_0(t) + \varepsilon u_1(t) + \cdots \tag{105}$$

in (104) and equating coefficients of like powers of ε, we obtain

$$\ddot{u}_0 + \omega_0^2 u_0 = 2K\cos\Omega t \tag{106}$$

$$\ddot{u}_1 + \omega_0^2 u_1 = -u_0^4 \tag{107}$$

The solution of (106) can be written as

$$u_0 = a\cos(\omega_0 t + \beta) + 2\Lambda\cos\Omega t \tag{108}$$

where a and β are constants and

$$\Lambda = \frac{K}{\omega_0^2 - \Omega^2} \tag{109}$$

Then (107) becomes

$$\ddot{u}_1 + \omega_0^2 u_1 = -a^4\cos^4(\omega_0 t + \beta) - 8\Lambda a^3\cos^3(\omega_0 t + \beta)\cos\Omega t$$
$$-24\Lambda^2 a^2\cos^2(\omega_0 t + \beta)\cos^2\Omega t - 32\Lambda^3 a\cos(\omega_0 t + \beta)\cos^3\Omega t$$
$$-16\Lambda^4\cos^4\Omega t$$

or

$$\ddot{u}_1 + \omega_0^2 u_1 = F \tag{110a}$$

where

$$F = -\left(\tfrac{3}{8}a^4 + 6\Lambda^2 a^2 + 6\Lambda^4\right) - a^2\left(\tfrac{1}{2}a^2 + 6\Lambda^2\right)\cos(2\omega_0 t + 2\beta)$$
$$-2\Lambda^2(3a^2 + 4\Lambda^2)\cos 2\Omega t - \tfrac{1}{8}a^4\cos(4\omega_0 t + 4\beta) - 2\Lambda^4\cos 4\Omega t$$
$$-3\Lambda a(4\Lambda^2 + a^2)\{\cos[(\omega_0 + \Omega)t + \beta] + \cos[(\omega_0 - \Omega)t + \beta]\}$$
$$-3\Lambda^2 a^2\{\cos[(2\omega_0 + 2\Omega)t + 2\beta] + \cos[(2\omega_0 - 2\Omega)t + 2\beta]\}$$
$$-\Lambda a^3\{\cos[(3\omega_0 + \Omega)t + 3\beta] + \cos[(3\omega_0 - \Omega)t + 3\beta]\}$$
$$-4\Lambda^3 a\{\cos[(\omega_0 + 3\Omega)t + \beta] + \cos[(\omega_0 - 3\Omega)t + \beta]\} \tag{110b}$$

A particular solution of (110) is

$$u_1 = -\frac{1}{\omega_0^2}\left(\tfrac{3}{8}a^4 + 6\Lambda^2 a^2 + 6\Lambda^4\right) + \frac{a^2(a^2+12\Lambda^2)}{6\omega_0^2}\cos(2\omega_0 t + 2\beta)$$

$$-\frac{2\Lambda^2(3a^2+4\Lambda^2)}{\omega_0^2 - 4\Omega^2}\cos 2\Omega t + \frac{a^4}{120\omega_0^2}\cos(4\omega_0 t + 4\beta)$$

$$-\frac{2\Lambda^4}{\omega_0^2 - 16\Omega^2}\cos 4\Omega t + \frac{3\Lambda a(4\Lambda^2 + a^2)}{\Omega(2\omega_0 + \Omega)}\cos[(\omega_0 + \Omega)t + \beta]$$

$$+\frac{3\Lambda a(4\Lambda^2 + a^2)}{\Omega(\Omega - 2\omega_0)}\cos[(\omega_0 - \Omega)t + \beta]$$

$$+\frac{3\Lambda^2 a^2}{(3\omega_0 + 2\Omega)(\omega_0 + 2\Omega)}\cos[(2\omega_0 + 2\Omega)t + 2\beta]$$

$$+\frac{3\Lambda^2 a^2}{(3\omega_0 - 2\Omega)(\omega_0 - 2\Omega)}\cos[(2\omega_0 - 2\Omega)t + 2\beta]$$

$$+\frac{\Lambda a^3}{(4\omega_0 + \Omega)(2\omega_0 + \Omega)}\cos[(3\omega_0 + \Omega)t + 3\beta]$$

$$+\frac{\Lambda a^3}{(4\omega_0 - \Omega)(2\omega_0 - \Omega)}\cos[(3\omega_0 - \Omega)t + 3\beta]$$

$$+\frac{4\Lambda^3 a}{3\Omega(2\omega_0 + 3\Omega)}\cos[(\omega_0 + 3\Omega)t + \beta]$$

$$-\frac{4\Lambda^3 a}{3\Omega(2\omega_0 - 3\Omega)}\cos[(\omega_0 - 3\Omega)t + \beta] \tag{111}$$

Inspection of (111) shows that small-divisor terms and hence resonances occur to first order when $\Omega \approx \frac{1}{2}\omega_0$, $\frac{1}{4}\omega_0$, 0, $2\omega_0$, $\frac{3}{2}\omega_0$, $4\omega_0$, and $\frac{2}{3}\omega_0$.

(b) *The Method of Multiple Scales*

For Ω away from 0, we express $\cos\Omega t$ in terms of the fast-scale T_0 and write $\cos\Omega t = \cos\Omega T_0$. Then, substituting (2) into (104) and equating coefficients of like powers of ε, we obtain

$$D_0^2 u_0 + \omega_0^2 u_0 = 2K\cos\Omega T_0 \tag{112}$$

$$D_0^2 u_1 + \omega_0^2 u_1 = -2D_0 D_1 u_0 - u_0^4 \tag{113}$$

The solution of (112) can be expressed as in (5), where Λ is defined in (109). Then (113) becomes

$$D_0^2 u_1 + \omega_0^2 u_1 = -2iA' e^{i\omega_0 T_0} + 2i\bar{A}' e^{-i\omega_0 T_0} - \left(A e^{i\omega_0 T_0} + \bar{A} e^{-i\omega_0 T_0} + \Lambda e^{i\Omega T_0} + \Lambda e^{-i\Omega T_0} \right)^4$$

or

$$\begin{aligned} D_0^2 u_1 + \omega_0^2 u_1 = &-2i\omega_0 A' e^{i\omega_0 T_0} - (6A^2\bar{A}^2 + 24A\bar{A}\Lambda^2 + 6\Lambda^4) \\ &-4A^2(A\bar{A} + 3\Lambda^2) e^{2i\omega_0 T_0} - 4\Lambda^2(3A\bar{A} + \Lambda^2) e^{2i\Omega T_0} - A^4 e^{4i\omega_0 T_0} \\ &- \Lambda^4 e^{4i\Omega T_0} - 12\Lambda A(\Lambda^2 + A\bar{A})\left[e^{i(\omega_0+\Omega)T_0} + e^{i(\omega_0-\Omega)T_0} \right] \\ &-6\Lambda^2 A^2\left[e^{i(2\omega_0+2\Omega)T_0} + e^{i(2\omega_0-2\Omega)T_0} \right] - 4\Lambda A^3\left[e^{i(3\omega_0+\Omega)T_0} \right. \\ &\left. + e^{i(3\omega_0-\Omega)T_0} \right] - 4\Lambda^3 A\left[e^{i(\omega_0+3\Omega)T_0} + e^{i(\omega_0-3\Omega)T_0} \right] + \text{cc} \end{aligned} \tag{114}$$

When $\Omega = 4\omega_0 + \varepsilon\sigma$, we write

$$(\Omega - 3\omega_0) T_0 = \omega_0 T_0 + \sigma T_1$$

Eliminating the terms that lead to secular terms from (114), we obtain

$$2i\omega_0 A' = -4\Lambda \bar{A}^3 e^{i\sigma T_1} \tag{115}$$

Substituting (35) into (115) and separating real and imaginary parts, we obtain

$$a' = -\frac{\Lambda a^3}{2\omega_0} \sin(\sigma T_1 - 4\beta) \tag{116}$$

$$a\beta' = \frac{\Lambda a^3}{2\omega_0} \cos(\sigma T_1 - 4\beta) \tag{117}$$

When $\Omega = 2\omega_0 + \varepsilon\sigma$, we write

$$(\Omega - \omega_0) T_0 = \omega_0 T_0 + \sigma T_1$$

Eliminating the terms that lead to secular terms from (114) yields

$$2i\omega_0 A' = -12\Lambda \bar{A}(\Lambda^2 + A\bar{A}) e^{i\sigma T_1} - 4\Lambda A^3 e^{-i\sigma T_1} \tag{118}$$

Substituting (35) into (118) and separating real and imaginary parts, we obtain

$$a' = -\frac{\Lambda a}{\omega_0}(6\Lambda^2 + a^2)\sin(\sigma T_1 - 2\beta) \tag{119}$$

$$a\beta' = \frac{2\Lambda a}{\omega_0}(3\Lambda^2 + a^2)\cos(\sigma T_1 - 2\beta) \tag{120}$$

When $4\Omega = \omega_0 + \varepsilon\sigma$, we write

$$4\Omega T_0 = \omega_0 T_0 + \sigma T_1$$

Then the solvability condition for (114) is

$$2i\omega_0 A' = -\Lambda^4 e^{i\sigma T_1} \tag{121}$$

Substituting (35) into (121) and separating real and imaginary parts, we obtain

$$a' = -\frac{\Lambda^4}{\omega_0}\sin(\sigma T_1 - \beta) \tag{122}$$

$$a\beta' = \frac{\Lambda^4}{\omega_0}\cos(\sigma T_1 - \beta) \tag{123}$$

When $2\Omega = \omega_0 + \varepsilon\sigma$, we put

$$2\Omega T_0 = \omega_0 T_0 + \sigma T_1$$

Then the solvability condition of (114) is

$$2i\omega_0 A' = -6\Lambda^2 A^2 e^{-i\sigma T_1} - 4\Lambda^2(3A\bar{A} + \Lambda^2)\, e^{i\sigma T_1} \tag{124}$$

Substituting (35) into (124) and separating real and imaginary parts, we obtain

$$a' = -\frac{\Lambda^2}{\omega_0}\left(\tfrac{3}{2}a^2 + 4\Lambda^2\right)\sin(\sigma T_1 - \beta) \tag{125}$$

$$a\beta' = \frac{\Lambda^2}{\omega_0}\left(\tfrac{9}{2}a^2 + 4\Lambda^2\right)\cos(\sigma T_1 - \beta) \tag{126}$$

When $3\Omega = 2\omega_0 + \varepsilon\sigma$, we put

$$3\Omega T_0 = 2\omega_0 T_0 + \sigma T_1$$

Then the solvability condition for (114) is

$$2i\omega_0 A' = -4\Lambda^3 \bar{A} e^{i\sigma T_1} \tag{127}$$

Substituting (35) into (127) and separating real and imaginary parts, we obtain

$$a' = -\frac{2\Lambda^3}{\omega_0} a \sin(\sigma T_1 - 2\beta) \tag{128}$$

$$a\beta' = \frac{2\Lambda^3}{\omega_0} a \cos(\sigma T_1 - 2\beta) \tag{129}$$

When $2\Omega = 3\omega_0 + \varepsilon\sigma$, we put

$$2\Omega T_0 = 3\omega_0 T_0 + \sigma T_1$$

Then the solvability condition of (114) is

$$2i\omega_0 A' = -6\Lambda^2 \bar{A}^2 e^{i\sigma T_1} \tag{130}$$

Substituting (35) into (130) and separating real and imaginary parts, we obtain

$$a' = -\frac{3\Lambda^2}{2\omega_0} a^2 \sin(\sigma T_1 - 3\beta) \tag{131}$$

$$a\beta' = \frac{3\Lambda^2}{2\omega_0} a^2 \cos(\sigma T_1 - 3\beta) \tag{132}$$

(c) *The Method of Averaging*

When $\varepsilon = 0$, the solution of (104) is

$$u = a\cos(\omega_0 t + \beta) + 2\Lambda\cos\Omega t \tag{133}$$

where a and β are constants. Differentiating (133) once with respect to t gives

$$\dot{u} = -\omega_0 a \sin(\omega_0 t + \beta) - 2\Omega\Lambda\sin\Omega t \tag{134}$$

When $\varepsilon \neq 0$, we still represent the solution of (104) by (133), subject to the constraint (134), but with time-varying a and β. Differentiating (133) once with respect to t and using (134) gives

$$\dot{a}\cos\phi - a\dot{\beta}\sin\phi = 0 \tag{135}$$

where $\phi = \omega_0 t + \beta$. Substituting (133) and (134) into (104), we have

$$\dot{a}\sin\phi + a\dot{\beta}\cos\phi = \frac{\varepsilon}{\omega_0}(a\cos\phi + 2\Lambda\cos\Omega t)^4 \tag{136}$$

Solving (135) and (136) yields

$$\dot{a} = \frac{\varepsilon}{\omega_0}\sin\phi(a\cos\phi + 2\Lambda\cos\Omega t)^4 = -\frac{\varepsilon}{\omega_0}F\sin\phi \tag{137}$$

$$a\dot{\beta} = \frac{\varepsilon}{\omega_0}\cos\phi(a\cos\phi + 2\Lambda\cos\Omega t)^4 = -\frac{\varepsilon}{\omega_0}F\cos\phi \tag{138}$$

where F is defined in (110b).

When $\Omega = 4\omega_0 + \varepsilon\sigma$, keeping the slowly varying terms in (137) and (138), we obtain

$$\dot{a} = -\frac{\varepsilon\Lambda}{2\omega_0} a^3 \sin(\varepsilon\sigma t - 4\beta) \tag{139}$$

$$a\dot{\beta} = \frac{\varepsilon\Lambda}{2\omega_0} a^3 \cos(\varepsilon\sigma t - 4\beta) \tag{140}$$

in agreement with (116) and (117), since $T_1 = \varepsilon t$.

When $\Omega = 2\omega_0 + \varepsilon\sigma$, keeping the slowly varying terms in (137) and (138), we have

$$\dot{a} = -\frac{\varepsilon\Lambda a}{\omega_0}(6\Lambda^2 + a^2)\sin(\varepsilon\sigma t - 2\beta) \tag{141}$$

$$a\dot{\beta} = \frac{2\varepsilon\Lambda a}{\omega_0}(3\Lambda^2 + a^2)\cos(\varepsilon\sigma t - 2\beta) \tag{142}$$

in agreement with (119) and (120), since $T_1 = \varepsilon t$.

When $4\Omega = \omega_0 + \varepsilon\sigma$, keeping the slowly varying terms in (137) and (138), we have

$$\dot{a} = -\frac{\varepsilon\Lambda^4}{\omega_0}\sin(\varepsilon\sigma t - \beta) \tag{143}$$

$$a\dot{\beta} = \frac{\varepsilon\Lambda^4}{\omega_0}\cos(\varepsilon\sigma t - \beta) \tag{144}$$

in agreement with (122) and (123), since $T_1 = \varepsilon t$.

When $2\Omega = \omega_0 + \varepsilon\sigma$, keeping the slowly varying parts in (137) and (138), we have

$$\dot{a} = -\frac{\varepsilon\Lambda^2}{\omega_0}\left(\tfrac{3}{2}a^2 + 4\Lambda^2\right)\sin(\varepsilon\sigma t - \beta) \tag{145}$$

$$a\dot{\beta} = \frac{\varepsilon\Lambda^2}{\omega_0}\left(\tfrac{9}{2}a^2 + 4\Lambda^2\right)\cos(\varepsilon\sigma t - \beta) \tag{146}$$

in agreement with (125) and (126), because $T_1 = \varepsilon t$.

When $3\Omega = 2\omega_0 + \varepsilon\sigma$, keeping the slowly varying terms in (137) and (138), we have

$$\dot{a} = -\frac{2\varepsilon\Lambda^3 a}{\omega_0}\sin(\varepsilon\sigma t - 2\beta) \tag{147}$$

$$a\dot{\beta} = \frac{2\varepsilon\Lambda^3}{\omega_0}\cos(\varepsilon\sigma t - 2\beta) \tag{148}$$

in agreement with (128) and (129), since $T_1 = \varepsilon t$.

When $2\Omega = 3\omega_0 + \varepsilon\sigma$, keeping the slowly varying terms in (137) and (138), we have

$$\dot{a} = -\frac{3\varepsilon\Lambda^2 a^2}{2\omega_0}\sin(\varepsilon\sigma t - 3\beta) \tag{149}$$

$$a\dot{\beta} = \frac{3\varepsilon\Lambda^2 a^2}{2\omega_0}\cos(\varepsilon\sigma t - 3\beta) \tag{150}$$

in agreement with (131) and (132), since $T_1 = \varepsilon t$.

Supplementary Exercises

9.9. Use the methods of multiple scales and averaging to show that, to the first approximation, the solution of

$$\ddot{u} + u + 2\varepsilon\mu\dot{u} + \varepsilon u|u| = 2\varepsilon f \cos\Omega t, \qquad \Omega = 1 + \varepsilon\sigma, \qquad \varepsilon \ll 1$$

is

$$u = a\cos(t + \beta) + \cdots$$

$$\dot{a} = -\varepsilon\mu a + \varepsilon f \sin(\varepsilon\sigma t - \beta)$$

$$a\dot{\beta} = \frac{4}{3\pi}\varepsilon a^2 - \varepsilon f \cos(\varepsilon\sigma t - \beta)$$

9.10. Use the methods of multiple scales and averaging to show that, to the first approximation, the solution of

$$\ddot{u} + u + 2\varepsilon u^2\dot{u} = K\cos\Omega t, \qquad \varepsilon \ll 1$$

is

$$u = a\cos(t + \beta) + 2\Lambda\cos\Omega t + \cdots$$

where

$$\Lambda = \frac{K}{2(1 - \Omega^2)}$$

and a and β are solutions of

(a)
$$\dot{a} = -2\varepsilon\left(\Lambda^2 + \tfrac{1}{8}a^2\right)a - \tfrac{2}{3}\varepsilon\Lambda^3\cos(\varepsilon\sigma t - \beta)$$

$$a\dot{\beta} = -\tfrac{2}{3}\varepsilon\Lambda^3\sin(\varepsilon\sigma t - \beta)$$

when $3\Omega = 1 + \varepsilon\sigma t$

(b)
$$\dot{a} = -2\varepsilon\left(\Lambda^2 + \tfrac{1}{8}a^2\right)a - \tfrac{1}{2}\varepsilon\Lambda a^2\cos(\varepsilon\sigma t - 3\beta)$$

$$a\dot{\beta} = -\tfrac{1}{2}\varepsilon\Lambda a^2\sin(\varepsilon\sigma t - 3\beta)$$

when $\Omega = 3 + \varepsilon\sigma$.

9.11. Show that, to the second approximation, the solution of

$$\ddot{u} + \omega_0^2 u + 2\varepsilon^2\mu\dot{u} + \varepsilon\alpha_2 u^2 = 2\varepsilon^2 f\cos\Omega t, \qquad \Omega = \omega_0 + \varepsilon^2\sigma, \qquad \varepsilon \ll 1$$

is

$$u = a\cos(\omega_0 t + \beta) + \frac{\varepsilon\alpha_2}{6\omega_0^2}a^2[\cos(2\omega_0 t + 2\beta) - 3] + \cdots$$

where a and β are given by

$$\dot{a} = -\varepsilon^2\mu a + \frac{\varepsilon^2 f}{\omega_0}\sin(\varepsilon^2\sigma t - \beta)$$

$$a\dot{\beta} = -\frac{5\varepsilon^2\alpha_2^2 a^3}{12\omega_0^3} - \frac{\varepsilon^2 f}{\omega_0}\cos(\varepsilon^2\sigma t - \beta)$$

9.12. Show that, to the second approximation, the solution of

$$\ddot{u} + \omega_0^2 u + 2\varepsilon^2\mu\dot{u} + \varepsilon\alpha_4\dot{u}^2 = 2\varepsilon^2 f\cos\Omega t, \qquad \Omega = \omega_0 + \varepsilon^2\sigma, \qquad \varepsilon \ll 1$$

is

$$u = a\cos(\omega_0 t + \beta) - \tfrac{1}{6}\varepsilon\alpha_4 a^2[\cos(2\omega_0 t + 2\beta) + 3] + \cdots$$

where a and β are given by

$$\dot{a} = -\varepsilon^2\mu a + \frac{\varepsilon^2 f}{\omega_0}\sin(\varepsilon^2\sigma t - \beta)$$

$$a\dot{\beta} = -\tfrac{1}{6}\varepsilon^2\omega_0\alpha_4^2 a^3 - \frac{\varepsilon^2 f}{\omega_0}\cos(\varepsilon^2\sigma t - \beta)$$

9.13. Show that, to the second approximation, the solution of

$$\ddot{u} + \omega_0^2 u + 2\varepsilon^2\mu\dot{u} + \varepsilon\alpha_2 u^2 + \varepsilon^2\alpha_3 u^3 = 2\varepsilon f\cos\Omega t, \qquad \Omega = 2\omega_0 + \varepsilon^2\sigma, \qquad \varepsilon \ll 1$$

is

$$u = a\cos(\omega_0 t + \beta) + \varepsilon\left\{\frac{\alpha_2 a^2}{6\omega_0^2}[\cos(2\omega_0 t + 2\beta) - 3] + \frac{2f}{\omega_0^2 - \Omega^2}\cos\Omega t\right\} + \cdots$$

where a and β are given by

$$\dot{a} = -\varepsilon^2\mu a + \frac{\varepsilon^2\alpha_2}{3\omega_0^3} fa\sin(\varepsilon^2\sigma t - 2\beta)$$

$$a\dot{\beta} = \frac{9\omega_0^2\alpha_3 - 10\alpha_2^2}{24\omega_0^3}a^3 - \frac{\varepsilon^2\alpha_2}{3\omega_0^3} fa\cos(\varepsilon^2\sigma t - 2\beta)$$

9.14. Consider the equation

$$\ddot{u} + \omega_0^2 u + 2\varepsilon\mu\dot{u} + \varepsilon\alpha_2 u^2 = f\cos\Omega t, \qquad 2\Omega = \omega_0 + \varepsilon\sigma, \qquad \varepsilon \ll 1$$

Seek a uniform second order expansion in the form

$$u = u_0(T_0, T_1, T_2) + \varepsilon u_1(T_0, T_1, T_2) + \varepsilon^2 u_2(T_0, T_1, T_2) + \cdots$$

Show that

$$u_0 = A(T_1, T_2)\, e^{i\omega_0 T_0} + \Lambda e^{i\Omega T_0} + \text{cc}$$

where

$$\Lambda = \frac{f}{2(\omega_0^2 - \Omega^2)}$$

Show that eliminating the secular terms from u_1 yields

$$2i\omega_0(D_1 A + \mu A) + \alpha_2 \Lambda^2 e^{i\sigma T_1} = 0$$

Then show that

$$u_1 = \frac{2i\mu\Omega\Lambda}{\Omega^2 - \omega_0^2} e^{i\Omega T_0} + \frac{\alpha_2 A^2}{3\omega_0^2} e^{2i\omega_0 T_0} + \frac{2\alpha_2 \Lambda \bar{A}}{\Omega(\Omega + 2\omega_0)} e^{i(\Omega + \omega_0)T_0}$$

$$+ \frac{2\alpha_2 \Lambda \bar{A}}{\Omega(\Omega - 2\omega_0)} e^{i(\Omega - \omega_0)T_0} - \frac{\alpha_2(A\bar{A} + \Lambda^2)}{\omega_0^2} + \text{cc}$$

Show that eliminating the secular terms from u_2 demands that

$$2i\omega_0 D_2 A + D_1^2 A + 2\mu D_1 A - 4\alpha_2^2\left(\frac{1}{\omega_0^2} - \frac{2}{\Omega^2 - 4\omega_0^2}\right)\Lambda^2 A$$

$$- \frac{10\alpha_2^2}{3\omega_0^2} A^2\bar{A} + \frac{4i\mu\alpha_2\Omega\Lambda^2}{\Omega^2 - \omega_0^2} e^{i\sigma T_1} = 0$$

9.15. Consider the problem in the preceding exercise but with different scaling as follows:

$$\ddot{u} + \omega_0^2 u + 2\varepsilon^2\mu\dot{u} + \varepsilon\alpha_2 u^2 = \varepsilon^{1/2} f\cos\Omega t, \qquad 2\Omega = \omega_0 + \varepsilon^2\sigma, \qquad \varepsilon \ll 1$$

Seek a uniform expansion in the form

$$u(t;\varepsilon) = \sum_{n=0}^{4} \varepsilon^{n/2} u_n(T_0, T_2) + \cdots, \quad T_2 = \varepsilon^2 t$$

Show that

$$u = a\cos(\omega_0 t + \beta) + \frac{\varepsilon^{1/2} f}{\omega_0^2 - \Omega^2}\cos\Omega t + \frac{\varepsilon\alpha_2 a^2}{6\omega_0^2}[\cos(2\omega_0 t + 2\beta) - 3] + \cdots$$

where

$$\dot{a} = -\varepsilon^2\mu a - \frac{\alpha_2\varepsilon^2 f^2}{4\omega_0(\omega_0^2 - \Omega^2)^2}\sin(\varepsilon^2\sigma t - \beta)$$

$$a\dot{\beta} = -\frac{5\alpha_2^2\varepsilon^2}{12\omega_0^3}a^3 + \frac{\alpha_2\varepsilon^2 f^2}{4\omega_0(\omega_0^2 - \Omega^2)^2}\cos(\varepsilon^2\sigma t - \beta)$$

CHAPTER 10

Multifrequency Excitations

In many practical situations, the load on a structure can be modeled by a finite sum of temporally harmonic terms. For example, such a situation occurs when a single structure supports several rotating machines. In other situations, the load acting on a structure, although periodic, is not harmonic, but can be expressed as the sum of a finite number of harmonic terms. Some curious phenomena can develop as a result of the interactions of the different harmonics. The devastating effects of one harmonic load having a frequency near the natural frequency (i.e., primary resonance) might be lowered to a tolerable level by shifting the natural frequency by simply adding one or more nonresonant harmonic loads. The large response produced by a primary resonant excitation can also be significantly reduced by simply adding other superharmonic-resonant excitations having the proper amplitudes and phases. When the sum of three frequencies in the load equals or nearly equals the natural frequency, the system can experience a combination-resonant response in which the peak amplitudes are several times larger than those predicted by linear theory.

The type of harmonic interactions (i.e., resonances) that occur in a nonlinear system depends on the number of harmonics in the excitation, the number of degrees of freedom, and the order of the nonlinearity. A convenient way of determining the possible resonances is to carry out a straightforward expansion and then recognize the combinations that produce secular or small-divisor terms. For a two-harmonic load having the frequencies Ω_1 and Ω_2 acting on a single-degree-of-freedom system with cubic nonlinearities, to the first approximation, resonances occur when $\Omega_n \approx \omega_0$, $\Omega_n \approx 3\omega_0$, $\Omega_n \approx \frac{1}{3}\omega_0$, $|\Omega_2 \pm 2\Omega_1| \approx \omega_0$, $|\pm 2\Omega_2 + \Omega_1| \approx \omega_0$, and $|\Omega_2 \pm \Omega_1| \approx 2\omega_0$, where ω_0 is the linear natural frequency of the system. It should be noted that more than one resonance can occur simultaneously, such as $\Omega_1 \approx \omega_0$ and $\Omega_2 \approx 3\omega_0$; $\Omega_1 \approx \omega_0$ and $\Omega_2 \approx \frac{1}{3}\omega_0$; $\Omega_1 \approx \frac{1}{3}\omega_0$ and $\Omega_2 \approx 3\omega_0$; $\Omega_1 \approx 3\omega_0$ and $\Omega_2 \approx 7\omega_0$; and $\Omega_1 \approx 3\omega_0$ and $\Omega_2 \approx 5\omega_0$. For a three-harmonic load, the combination resonance $|\Omega_3 \pm \Omega_2 \pm \Omega_1| \approx \omega_0$ may occur besides the preceding resonances. For a single-degree-of-freedom system with quadratic nonlinearities, to the second approximation, resonances occur when $\Omega_n \approx 2\omega_0$, $\Omega_n \approx \frac{1}{2}\omega_0$, and $|\Omega_n \pm \Omega_m| \approx \omega_0$, besides the resonances that occur in the case of cubic nonlinearities.

One can use the method of multiple scales, the method of averaging (including the generalized version), or the Krylov–Bogoliubov technique to determine first-order ordinary-differential equations that describe the evolution of the amplitudes and the phases with damping, nonlinearity and all possible resonances. These equations can be used to determine the long-time behavior as well as steady-state motions and their stability.

Solved Exercises

10.1. Use the methods of multiple scales and averaging to determine the equations governing the amplitudes and the phases to first order for a system governed by

$$\ddot{u} + \omega_0^2 u = \varepsilon \dot{u}^2 + K_1 \cos \Omega_1 t + K_2 \cos \Omega_2 t, \qquad \varepsilon \ll 1 \tag{1}$$

when $\Omega_2 \pm \Omega_1 \approx \omega_0$. Consider only the case when Ω_1 is away from zero.

(a) ***The Method of Multiple Scales***

We let

$$u(t;\varepsilon) = u_0(T_0, T_1) + \varepsilon u_1(T_0, T_1) + \dots \tag{2}$$

where $T_0 = t$ and $T_1 = \varepsilon t$. Since the Ω_n are away from zero, we express the excitation in terms of T_0 and put

$$\cos \Omega_n t = \cos \Omega_n T_0$$

Substituting (2) into (1) and equating coefficients of like powers of ε, we obtain

$$D_0^2 u_0 + \omega_0^2 u_0 = K_1 \cos \Omega_1 T_0 + K_2 \cos \Omega_2 T_0 \tag{3}$$

$$D_0^2 u_1 + \omega_0^2 u_1 = (D_0 u_0)^2 - 2 D_0 D_1 u_0 \tag{4}$$

The solution of (3) can be expressed as

$$u_0 = A(T_1) e^{i\omega_0 T_0} + \Lambda_1 e^{i\Omega_1 T_0} + \Lambda_2 e^{i\Omega_2 T_0} + cc \tag{5}$$

where

$$\Lambda_n = \frac{K_n}{2(\omega_0^2 - \Omega_n^2)}$$

Then (4) becomes

$$D_0^2 u_1 + \omega_0^2 u_1 = -2i\omega_0 A' e^{i\omega_0 T_0} - \omega_0^2 A^2 e^{2i\omega_0 T_0} - \Omega_1^2 \Lambda_1^2 e^{2i\Omega_1 T_0} - \Omega_2^2 \Lambda_2^2 e^{2i\Omega_2 T_0}$$

$$-2\omega_0 \Omega_1 A \Lambda_1 e^{i(\omega_0+\Omega_1)T_0} - 2\omega_0 \Omega_2 A \Lambda_2 e^{i(\omega_0+\Omega_2)T_0} + 2\omega_0 \Omega_1 A \Lambda_1 e^{i(\omega_0-\Omega_1)T_0}$$

$$+2\omega_0 \Omega_2 A \Lambda_2 e^{i(\omega_0-\Omega_2)T_0} - 2\Omega_1 \Omega_2 \Lambda_1 \Lambda_2 e^{i(\Omega_2+\Omega_1)T_0}$$

$$+2\Omega_1 \Omega_2 \Lambda_1 \Lambda_2 e^{i(\Omega_2-\Omega_1)T_0} + \omega_0^2 A\bar{A} + \Omega_1^2 \Lambda_1^2 + \Omega_2^2 \Lambda_2^2 + \text{cc} \tag{6}$$

When $\Omega_2 + \Omega_1 \approx \omega_0$, we let

$$\Omega_2 + \Omega_1 = \omega_0 + \varepsilon\sigma \tag{7}$$

so that

$$(\Omega_2 + \Omega_1) T_0 = \omega_0 T_0 + \varepsilon\sigma T_0 = \omega_0 T_0 + \sigma T_1 \tag{8}$$

Thus we write

$$e^{i(\Omega_2+\Omega_1)T_0} = e^{i\omega_0 T_0} e^{i\sigma T_1}$$

Eliminating the terms that lead to secular terms from (6) yields

$$i\omega_0 A' = -\Omega_1 \Omega_2 \Lambda_1 \Lambda_2 e^{i\sigma T_1} \tag{9}$$

Substituting the polar form

$$A = \tfrac{1}{2} a e^{i\beta} \tag{10}$$

into (9), and separating real and imaginary parts, we obtain

$$a' = -2\omega_0^{-1} \Omega_1 \Omega_2 \Lambda_1 \Lambda_2 \sin(\sigma T_1 - \beta) \tag{11}$$

$$a\beta' = 2\omega_0^{-1} \Omega_1 \Omega_2 \Lambda_1 \Lambda_2 \cos(\sigma T_1 - \beta) \tag{12}$$

When

$$\Omega_2 - \Omega_1 = \omega_0 + \varepsilon\sigma$$

we write

$$e^{i(\Omega_2-\Omega_1)T_0} = e^{i\omega_0 T_0} e^{i\sigma T_1}$$

Eliminating the terms that lead to secular terms from (6) yields

$$i\omega_0 A' = \Omega_1 \Omega_2 \Lambda_1 \Lambda_2 e^{i\sigma T_1} \tag{13}$$

Substituting (10) into (13) and separating real and imaginary parts, we obtain

$$a' = 2\omega_0^{-1}\Omega_1\Omega_2\Lambda_1\Lambda_2\sin(\sigma T_1 - \beta) \tag{14}$$

$$a\beta' = -2\omega_0^{-1}\Omega_1\Omega_2\Lambda_1\Lambda_2\cos(\sigma T_1 - \beta) \tag{15}$$

(b) *The Method of Averaging*

When $\varepsilon = 0$, the solution of (1) can be expressed as

$$u = a\cos(\omega_0 t + \beta) + 2\Lambda_1\cos\Omega_1 t + 2\Lambda_2\cos\Omega_2 t \tag{16}$$

which, upon differentiation, gives

$$\dot{u} = -\omega_0 a\sin(\omega_0 t + \beta) - 2\Omega_1\Lambda_1\sin\Omega_1 t - 2\Omega_2\Lambda_2\sin\Omega_2 t \tag{17}$$

When $\varepsilon \neq 0$, we still represent the solution of (1) as in (16), subject to the constraint (17), but with time-varying a and β. Then, differentiating (16) once with respect to t and using (17), we have

$$\dot{a}\cos\phi - a\dot{\beta}\sin\phi = 0 \tag{18}$$

where $\phi = \omega_0 t + \beta$. Substituting (16) and (17) into (1) yields

$$\dot{a}\sin\phi + a\dot{\beta}\cos\phi = -\varepsilon\omega_0^{-1}f \tag{19}$$

$$f = (\omega_0 a\sin\phi + 2\Omega_1\Lambda_1\sin\Omega_1 t + 2\Omega_2\Lambda_2\sin\Omega_2 t)^2$$

or

$$\begin{aligned} f = {} & \tfrac{1}{2}\omega_0^2 a^2 + 2\Omega_1^2\Lambda_1^2 + 2\Omega_2^2\Lambda_2^2 - \tfrac{1}{2}\omega_0^2 a^2\cos 2(\omega_0 t + \beta) - 2\Omega_1^2\Lambda_1^2\cos 2\Omega_1 t \\ & - 2\Omega_2^2\Lambda_2^2\cos 2\Omega_2 t + 2\omega_0\Omega_1 a\Lambda_1\{\cos[(\omega_0 - \Omega_1)t + \beta] - \cos[(\omega_0 + \Omega_1)t + \beta]\} \\ & + 2\omega_0\Omega_2 a\Lambda_2\{\cos[(\omega_0 - \Omega_2)t + \beta] - \cos[(\omega_0 + \Omega_2)t + \beta]\} \\ & + 4\Omega_1\Omega_2\Lambda_1\Lambda_2[\cos(\Omega_2 - \Omega_1)t - \cos(\Omega_2 + \Omega_1)t] \end{aligned} \tag{20}$$

Solving (18) and (19) yields

$$\dot{a} = -\varepsilon\omega_0^{-1}f\sin(\omega_0 t + \beta) \tag{21}$$

$$a\dot{\beta} = -\varepsilon\omega_0^{-1}f\cos(\omega_0 t + \beta) \tag{22}$$

Thus inspection of (20)–(22) shows that slowly varying terms arise only from the terms in f whose frequencies are close to ω_0. When $\Omega_2 + \Omega_1 \approx \omega_0$, only the terms proportional to $\sin(\omega_0 t + \beta)\cos(\Omega_2 + \Omega_1)t$ and $\cos(\omega_0 t + \beta)\cos(\Omega_2 + \Omega_1)t$ on the right-hand sides of (21) and (22) contain slowly varying parts; namely,

$$\tfrac{1}{2}\sin[(\omega_0 - \Omega_2 - \Omega_1)t + \beta] \quad \text{and} \quad \tfrac{1}{2}\cos[(\omega_0 - \Omega_2 - \Omega_1)t + \beta]$$

Keeping only the slowly varying terms in (21) and (22) we have

$$\dot{a} = 2\varepsilon\omega_0^{-1}\Omega_1\Omega_2\Lambda_1\Lambda_2\sin[(\omega_0 - \Omega_2 - \Omega_1)t + \beta] \tag{23}$$

$$a\dot{\beta} = 2\varepsilon\omega_0^{-1}\Omega_1\Omega_2\Lambda_1\Lambda_2\cos[(\omega_0 - \Omega_2 - \Omega_1)t + \beta] \tag{24}$$

in agreement with (11) and (12) on account of (7).

When $\Omega_2 - \Omega_1 \approx \omega_0$, only the terms proportional to

$$\sin(\omega_0 t + \beta)\cos(\Omega_2 - \Omega_1)t \quad \text{and} \quad \cos(\omega_0 t + \beta)\cos(\Omega_2 - \Omega_1)t$$

on the right-hand sides of (21) and (22) contain slowly varying parts; namely,

$$\tfrac{1}{2}\sin[(\omega_0 - \Omega_2 + \Omega_1)t + \beta] \quad \text{and} \quad \tfrac{1}{2}\cos[(\omega_0 - \Omega_2 + \Omega_1)t + \beta]$$

Keeping only the slowly varying terms in (21) and (22), we obtain

$$\dot{a} = -2\varepsilon\omega_0^{-1}\Omega_1\Omega_2\Lambda_1\Lambda_2\sin[(\omega_0 - \Omega_2 + \Omega_1)t + \beta] \tag{25}$$

$$a\dot{\beta} = -2\varepsilon\omega_0^{-1}\Omega_1\Omega_2\Lambda_1\Lambda_2\cos[(\omega_0 - \Omega_2 + \Omega_1)t + \beta] \tag{26}$$

in agreement with (14) and (15) because of $\Omega_2 - \Omega_1 = \omega_0 + \varepsilon\sigma$.

10.2. Consider the equation

$$\ddot{u} + \omega_0^2 u + \varepsilon u^3 = K_1\cos(\Omega_1 t + \theta_1) + K_2\cos(\Omega_2 t + \theta_2) \tag{27}$$

where ω_0, K_n, and θ_n are constants. When Ω_n is away from ω_0, use the method of multiple scales to determine a first-order uniform expansion.

Substituting (2) into (27) and equating coefficients of like powers of ε, we obtain

$$D_0^2 u_0 + \omega_0^2 u_0 = K_1\cos(\Omega_1 T_0 + \theta_1) + K_2\cos(\Omega_2 T_0 + \theta_2) \tag{28}$$

$$D_0^2 u_1 + \omega_0^2 u_1 = -2D_0 D_1 u_0 - u_0^3 \tag{29}$$

where Ω_n is assumed to be away from zero so that $\cos(\Omega_n t + \theta_n)$ can be expressed as $\cos(\Omega_n T_0 + \theta_n)$. The solution of (28) can be expressed as

$$u_0 = A(T_1)e^{i\omega_0 T_0} + \Lambda_1 e^{i\Omega_1 T_0} + \Lambda_2 e^{i\Omega_2 T_0} + cc \tag{30}$$

where

$$\Lambda_n = \frac{K_n}{2(\omega_0^2 - \Omega_n^2)} e^{i\theta_n} \tag{31}$$

Substituting (30) into (29) yields

$$\begin{aligned}
D_0^2 u_1 + \omega_0^2 u_1 = & -\left[2i\omega_0 A' + 3(A\bar{A} + 2\Lambda_1\bar{\Lambda}_1 + 2\Lambda_2\bar{\Lambda}_2)A\right] e^{i\omega_0 T_0} \\
& - 3(2A\bar{A} + \Lambda_1\bar{\Lambda}_1 + 2\Lambda_2\bar{\Lambda}_2)\Lambda_1 e^{i\Omega_1 T_0} - 3(2A\bar{A} + 2\Lambda_1\bar{\Lambda}_1 + \Lambda_2\bar{\Lambda}_2)\Lambda_2 e^{i\Omega_2 T_0} \\
& - A^3 e^{3i\omega_0 T_0} - \Lambda_1^3 e^{3i\Omega_1 T_0} - \Lambda_2^3 e^{3i\Omega_2 T_0} - 3A^2\Lambda_1 e^{i(2\omega_0 + \Omega_1)T_0} \\
& - 3A^2\Lambda_2 e^{i(2\omega_0 + \Omega_2)T_0} - 3A^2\bar{\Lambda}_1 e^{i(2\omega_0 - \Omega_1)T_0} - 3A^2\bar{\Lambda}_2 e^{i(2\omega_0 - \Omega_2)T_0} \\
& - 3A\Lambda_1^2 e^{i(\omega_0 + 2\Omega_1)T_0} - 3A\Lambda_2^2 e^{i(\omega_0 + 2\Omega_2)T_0} - 3A\bar{\Lambda}_1^2 e^{i(\omega_0 - 2\Omega_1)T_0} \\
& - 3A\bar{\Lambda}_2^2 e^{i(\omega_0 - 2\Omega_2)T_0} - 6A\Lambda_1\Lambda_2 e^{i(\omega_0 + \Omega_1 + \Omega_2)T_0} \\
& - 6A\bar{\Lambda}_1\bar{\Lambda}_2 e^{i(\omega_0 - \Omega_1 - \Omega_2)T_0} - 6A\bar{\Lambda}_1\Lambda_2 e^{i(\omega_0 - \Omega_1 + \Omega_2)T_0} \\
& - 6A\Lambda_1\bar{\Lambda}_2 e^{i(\omega_0 + \Omega_1 - \Omega_2)T_0} - 3\Lambda_1^2\Lambda_2 e^{i(2\Omega_1 + \Omega_2)T_0} - 3\Lambda_1^2\bar{\Lambda}_2 e^{i(2\Omega_1 - \Omega_2)T_0} \\
& - 3\Lambda_1\Lambda_2^2 e^{i(\Omega_1 + 2\Omega_2)T_0} - 3\bar{\Lambda}_1\Lambda_2^2 e^{i(2\Omega_2 - \Omega_1)T_0} + cc
\end{aligned} \tag{32}$$

(a) Show that if $\Omega_2 > \Omega_1$, resonances occur whenever

$$\omega_0 \approx 3\Omega_1 \quad \text{or} \quad 3\Omega_2 \tag{33}$$

$$\omega_0 \approx \tfrac{1}{3}\Omega_1 \quad \text{or} \quad \tfrac{1}{3}\Omega_2 \tag{34}$$

$$\omega_0 \approx \Omega_2 \pm 2\Omega_1 \quad \text{or} \quad 2\Omega_1 - \Omega_2 \tag{35}$$

$$\omega_0 \approx 2\Omega_2 \pm \Omega_1 \tag{36}$$

$$\omega_0 \approx \tfrac{1}{2}(\Omega_2 \pm \Omega_1) \tag{37}$$

If we determine particular solutions for (32), we find that they contain secular and small-divisor terms. Besides the primary resonance case $\Omega_n \approx \omega_0$ which was excluded in the exercise, a resonance (secondary resonance) occurs at first order whenever a small-divisor term occurs. To determine these resonances, we do not need to determine particular solutions for (32). It is sufficient to inspect the frequencies on the right-hand side and determine the conditions under which one or more of them are close to the natural frequency ω_0 of the system. Checking all the right-hand side terms, we put

$$\Omega_1 \approx \pm \omega_0 \tag{38}$$

$$\Omega_2 \approx \pm \omega_0 \tag{39}$$

$$3\Omega_1 \approx \pm \omega_0 \quad \text{or} \quad \Omega_1 \approx \pm \tfrac{1}{3}\omega_0 \tag{40}$$

$$3\Omega_2 \approx \pm \omega_0 \quad \text{or} \quad \Omega_2 \approx \pm \tfrac{1}{3}\omega_0 \tag{41}$$

$$2\omega_0 + \Omega_1 \approx \pm \omega_0 \quad \text{or} \quad \Omega_1 = -\omega_0, -3\omega_0 \tag{42}$$

$$2\omega_0 + \Omega_2 \approx \pm \omega_0 \quad \text{or} \quad \Omega_2 = -\omega_0, -3\omega_0 \tag{43}$$

$$2\omega_0 - \Omega_1 \approx \pm \omega_0 \quad \text{or} \quad \Omega_1 \approx \omega_0, 3\omega_0 \tag{44}$$

$$2\omega_0 - \Omega_2 \approx \pm \omega_0 \quad \text{or} \quad \Omega_2 \approx \omega_0, 3\omega_0 \tag{45}$$

$$\omega_0 + 2\Omega_1 \approx \pm \omega_0 \quad \text{or} \quad \Omega_1 \approx 0, -\omega_0 \tag{46}$$

$$\omega_0 + 2\Omega_2 \approx \pm \omega_0 \quad \text{or} \quad \Omega_2 \approx 0, -\omega_0 \tag{47}$$

$$\omega_0 - 2\Omega_1 \approx \pm \omega_0 \quad \text{or} \quad \Omega_1 \approx 0, \omega_0 \tag{48}$$

$$\omega_0 - 2\Omega_2 \approx \pm \omega_0 \quad \text{or} \quad \Omega_2 \approx 0, \omega_0 \tag{49}$$

$$\omega_0 + \Omega_1 + \Omega_2 \approx \pm \omega_0 \quad \text{or} \quad \Omega_1 + \Omega_2 \approx 0, -2\omega_0 \tag{50}$$

$$\omega_0 - \Omega_1 - \Omega_2 \approx \pm \omega_0 \quad \text{or} \quad \Omega_1 + \Omega_2 \approx 0, 2\omega_0 \tag{51}$$

$$\omega_0 - \Omega_1 + \Omega_2 \approx \pm \omega_0 \quad \text{or} \quad \Omega_2 - \Omega_1 \approx 0, -2\omega_0 \tag{52}$$

$$\omega_0 + \Omega_1 - \Omega_2 \approx \pm \omega_0 \quad \text{or} \quad \Omega_2 - \Omega_1 \approx 0, 2\omega_0 \tag{53}$$

$$2\Omega_1 + \Omega_2 \approx \pm \omega_0 \tag{54}$$

$$2\Omega_1 - \Omega_2 \approx \pm \omega_0 \tag{55}$$

$$\Omega_1 + 2\Omega_2 \approx \pm \omega_0 \tag{56}$$

$$2\Omega_2 - \Omega_1 \approx \pm \omega_0 \tag{57}$$

Besides the assumption that the Ω_n are away from ω_0, if we assume that the Ω_n are positive and away from zero, (38) and (39) should be discarded, (40) and (41) yield (33), (42) and (43) should be discarded, (44) and (45) yield (34), (46)–(50) should be discarded, (51) yields part of (37), (52) should be discarded, (53) yields the other part of (37), (54) and (55) yield (35), and (56) and (57) yield (36).

(b) Show that simultaneous resonances occur whenever

$$\Omega_2 \approx 9\Omega_1 \approx 3\omega_0 \tag{58}$$

$$\Omega_2 \approx \Omega_1 \approx 3\omega_0 \tag{59}$$

$$\Omega_2 \approx \Omega_1 \approx \tfrac{1}{3}\omega_0 \tag{60}$$

$$\Omega_2 \approx 5\Omega_1 \approx \tfrac{5}{3}\omega_0 \tag{61}$$

$$\Omega_2 \approx 7\Omega_1 \approx \tfrac{7}{3}\omega_0 \tag{62}$$

$$\Omega_2 \approx 2\Omega_1 \approx \tfrac{2}{3}\omega_0 \tag{63}$$

$$\Omega_2 \approx \tfrac{7}{3}\Omega_1 \approx 7\omega_0 \tag{64}$$

$$\Omega_2 \approx \tfrac{5}{3}\Omega_1 \approx 5\omega_0 \tag{65}$$

$$\Omega_2 \approx \tfrac{3}{2}\Omega_1 \approx 3\omega_0 \tag{66}$$

$$\Omega_2 \approx 3\Omega_1 \approx \tfrac{3}{5}\omega_0 \tag{67}$$

By definition, whenever two or more resonance conditions are satisfied simultaneously, we speak of simultaneous resonance. When both conditions in (33) are satisfied simultaneously, we obtain the simultaneous resonances in (60). Similarly, satisfaction of both conditions in (34) leads to the simultaneous resonances in (59). If $\Omega_2 > \Omega_1$, satisfaction of the first condition in (33) and the second condition in (34) yield (58). Satisfaction of (33) and (35) leads to either

$$\Omega_1 \approx \tfrac{1}{3}\omega_0 \quad \text{and} \quad \Omega_2 \approx \tfrac{1}{3}\omega_0,\ \tfrac{5}{3}\omega_0$$

or

$$\Omega_2 \approx \tfrac{1}{3}\omega_0 \quad \text{and} \quad \Omega_1 \approx \tfrac{1}{3}\omega_0,\ \tfrac{2}{3}\omega_0$$

and hence it leads to either (60) or (61). Satisfaction of (33) and (36) leads to either

$$\Omega_1 \approx \tfrac{1}{3}\omega_0, \qquad \Omega_2 \approx \tfrac{1}{3}\omega_0, \qquad \tfrac{2}{3}\omega_0$$

or

$$\Omega_2 \approx \tfrac{1}{3}\omega_0, \qquad \Omega_1 \approx \tfrac{1}{3}\omega_0$$

and hence it leads to either (60) or (63). Satisfaction of (33) and (37) leads to either

$$\Omega_1 \approx \tfrac{1}{3}\omega_0, \qquad \Omega_2 \approx \tfrac{5}{3}\omega_0, \qquad \tfrac{7}{3}\omega_0$$

or

$$\Omega_2 \approx \tfrac{1}{3}\omega_0, \qquad \Omega_1 \approx \tfrac{5}{3}\omega_0$$

and hence it leads to either (61) or (62). Satisfaction of (34) and (35) leads to either

$$\Omega_1 \approx 3\omega_0 \quad \text{and} \quad \Omega_2 \approx 7\omega_0, 5\omega_0$$

or

$$\Omega_2 \approx 3\omega_0 \quad \text{and} \quad \Omega_1 \approx \omega_0, 2\omega_0$$

and hence it leads to either (64), (65), or (66). Satisfaction of (34) and (36) leads to either

$$\Omega_1 \approx 3\omega_0 \quad \text{and} \quad \Omega_2 \approx 2\omega_0$$

or

$$\Omega_2 \approx 3\omega_0 \quad \text{and} \quad \Omega_1 \approx 5\omega_0$$

and hence neither can be satisfied because $\Omega_2 > \Omega_1$. Satisfaction of (34) and (37) leads to either

$$\Omega_1 \approx 3\omega_0 \quad \text{and} \quad \Omega_2 \approx 5\omega_0$$

or

$$\Omega_2 \approx 3\omega_0 \quad \text{and} \quad \Omega_1 \approx \omega_0$$

and hence it leads to (65). Satisfaction of (35) and (36) leads to either

$$\Omega_2 \pm 2\Omega_1 \approx 2\Omega_2 \pm \Omega_1 \quad \text{or} \quad 2\Omega_1 - \Omega_2 \approx 2\Omega_2 \pm \Omega_1$$

and hence to

$$\Omega_2 \approx 3\Omega_1 \approx \tfrac{3}{5}\omega_0$$

which is (67). Satisfaction of (35) and (37) leads to either

$$\Omega_2 \pm 2\Omega_1 \approx \tfrac{1}{2}(\Omega_2 \pm \Omega_1) \quad \text{or} \quad 2\Omega_1 - \Omega_2 \approx \tfrac{1}{2}(\Omega_2 \pm \Omega_1)$$

and hence to

$$\Omega_2 \approx 5\Omega_1 \approx \tfrac{5}{3}\omega_0 \quad \text{or} \quad \Omega_2 \approx \tfrac{5}{3}\Omega_1 \approx 5\omega_0$$

which yields (61) and (65). Satisfaction of (36) and (37) leads to

$$2\Omega_2 \pm \Omega_1 \approx \tfrac{1}{2}(\Omega_2 \pm \Omega_1)$$

or

$$\Omega_2 \approx \Omega_1 \approx \omega_0$$

which was excluded in the exercise.

(c) Determine the solvability conditions for the following cases:

(i)
$$\omega_0 = 2\Omega_1 + \Omega_2 + \varepsilon\sigma \tag{68}$$

Thus we write

$$\omega_0 T_0 = (2\Omega_1 + \Omega_2)T_0 + \varepsilon\sigma T_0 = (2\Omega_1 + \Omega_2)T_0 + \sigma T_1$$

Eliminating the terms that lead to secular terms from (32), we have

$$2i\omega_0 A' + 3(A\bar{A} + 2\Lambda_1\bar{\Lambda}_1 + 2\Lambda_2\bar{\Lambda}_2)A + 3\Lambda_1^2\Lambda_2 e^{-i\sigma T_1} = 0 \tag{69}$$

(ii)
$$3\Omega_1 = \omega_0 + \varepsilon\sigma_1 \quad \text{and} \quad \Omega_2 = 3\omega_0 + \varepsilon\sigma_2 \tag{70}$$

Thus we write

$$3\Omega_1 T_0 = \omega_0 T_0 + \sigma_1 T_1 \quad \text{and} \quad \Omega_2 T_2 = 3\omega_0 T_0 + \sigma_2 T_1$$

Eliminating the terms that lead to secular terms from (32), we have

$$2i\omega_0 A' + 3(A\bar{A} + 2\Lambda_1\bar{\Lambda}_1 + 2\Lambda_2\bar{\Lambda}_2)A + \Lambda_1^3 e^{i\sigma_1 T_1} + 3\bar{A}^2\Lambda_2 e^{i\sigma_2 T_1} = 0 \tag{71}$$

(iii)
$$3\Omega_1 = \omega_0 + \varepsilon\sigma_1 \quad \text{and} \quad \Omega_2 + \Omega_1 = 2\omega_0 + \varepsilon\sigma_2 \tag{72}$$

Thus we write

$$3\Omega_1 T_0 = \omega_0 T_0 + \sigma_1 T_1 \quad \text{and} \quad (\Omega_2 + \Omega_1)T_0 = 2\omega_0 T_0 + \sigma_2 T_1$$

Eliminating the terms that lead to secular terms from (32), we have

$$2i\omega_0 A' + 3(A\bar{A} + 2\Lambda_1\bar{\Lambda}_1 + 2\Lambda_2\bar{\Lambda}_2)A + \Lambda_1^3 e^{i\sigma_1 T_1} + 6\bar{A}\Lambda_1\Lambda_2 e^{i\sigma_2 T_1}$$

$$+ 3\Lambda_2\bar{\Lambda}_1^2 e^{i(\sigma_2 - \sigma_1)T_1} = 0 \tag{73}$$

(iv)
$$3\Omega_1 = \omega_0 + \varepsilon\sigma_1 \quad \text{and} \quad \Omega_2 - \Omega_1 = 2\omega_0 + \varepsilon\sigma_2 \tag{74}$$

Thus we write

$$3\Omega_1 T_0 = \omega_0 T_0 + \sigma_1 T_1 \quad \text{and} \quad (\Omega_2 - \Omega_1)T_0 = 2\omega_0 T_0 + \sigma_2 T_1$$

Eliminating the terms that lead to secular terms from (32), we have

$$2i\omega_0 A' + 3(A\bar{A} + 2\Lambda_1\bar{\Lambda}_1 + 2\Lambda_2\bar{\Lambda}_2)A + \Lambda_1^3 e^{i\sigma_1 T_1} + 6\bar{A}\bar{\Lambda}_1\Lambda_2 e^{i\sigma_2 T_1} = 0 \tag{75}$$

(v)
$$3\Omega_1 = \omega_0 + \varepsilon\sigma_1 \quad \text{and} \quad 2\Omega_2 - \Omega_1 = \omega_0 + \varepsilon\sigma_2 \tag{76}$$

Thus we write

$$3\Omega_1 T_0 = \omega_0 T_0 + \sigma_1 T_1 \quad \text{and} \quad (2\Omega_2 - \Omega_1)T_0 = \omega_0 T_0 + \sigma_2 T_1$$

Eliminating the terms that lead to secular terms from (32), we have

$$2i\omega_0 A' + 3(A\bar{A} + 2\Lambda_1\bar{\Lambda}_1 + 2\Lambda_2\bar{\Lambda}_2)A + \Lambda_1^3 e^{i\sigma_1 T_1} + 3\bar{\Lambda}_1\Lambda_2^2 e^{i\sigma_2 T_1} = 0 \tag{77}$$

(vi)

$$\Omega_1 = 3\omega_0 + \varepsilon\sigma_1 \quad \text{and} \quad \Omega_2 - 2\Omega_1 = \omega_0 + \varepsilon\sigma_2 \tag{78}$$

Thus we write

$$\Omega_1 T_0 = 3\omega_0 T_0 + \sigma_1 T_1 \quad \text{and} \quad (\Omega_2 - 2\Omega_1)T_0 = \omega_0 T_0 + \sigma_2 T_1$$

Eliminating the terms that lead to secular terms from (32), we have

$$2i\omega_0 A' + 3(A\bar{A} + 2\Lambda_1\bar{\Lambda}_1 + 2\Lambda_2\bar{\Lambda}_2)A + 3\bar{A}^2\Lambda_1 e^{i\sigma_1 T_1} + 3\bar{\Lambda}_1^2\Lambda_2 e^{i\sigma_2 T_1} = 0 \tag{79}$$

(vii)

$$\Omega_1 = 3\omega_0 + \varepsilon\sigma_1 \quad \text{and} \quad \Omega_2 - \Omega_1 = 2\omega_0 + \varepsilon\sigma_2 \tag{80}$$

Thus we write

$$\Omega_1 T_0 = 3\omega_0 T_0 + \sigma_1 T_1 \quad \text{and} \quad (\Omega_2 - \Omega_1)T_0 = 2\omega_0 T_0 + \sigma_2 T_1$$

Eliminating the terms that lead to secular terms from (32), we have

$$2i\omega_0 A' + 3(A\bar{A} + 2\Lambda_1\bar{\Lambda}_1 + 2\Lambda_2\bar{\Lambda}_2)A + 3\bar{A}^2\Lambda_1 e^{i\sigma_1 T_1} + 6\bar{A}\bar{\Lambda}_1\Lambda_2 e^{i\sigma_2 T_1}$$

$$+ 3\Lambda_1^2\bar{\Lambda}_2 e^{i(\sigma_1 - \sigma_2)T_1} = 0 \tag{81}$$

10.3. Consider Equation (27). Let

$$u = A(t)e^{i\omega_0 t} + \Lambda_1 e^{i\Omega_1 t} + \Lambda_2 e^{i\Omega_2 t} + \text{cc} \tag{82}$$

where Λ_n is defined in (31). Use the method of variation of parameters to determine the equation governing A.

When $\varepsilon = 0$, the solution of (27) can be expressed as in (82) if A is constant. Then, differentiating (82) once with respect to t and assuming that A is constant, we obtain

$$\dot{u} = i\omega_0 A e^{i\omega_0 t} + i\Omega_1\Lambda_1 e^{i\Omega_1 t} + i\Omega_2\Lambda_2 e^{i\Omega_2 t} + \text{cc} \tag{83}$$

When $\varepsilon \neq 0$, we still express the solution of (27) as in (82), subject to the constraint (83), but with time-varying A. Then, differentiating (82) once with respect to t and using (83), we obtain

$$\dot{A}e^{i\omega_0 t} + \dot{\bar{A}}e^{-i\omega_0 t} = 0 \tag{84}$$

Differentiating (83) once with respect to t, we have

$$\ddot{u} = -\omega_0^2 A e^{i\omega_0 t} - \Omega_1^2\Lambda_1 e^{i\Omega_1 t} - \Omega_2^2\Lambda_2 e^{i\Omega_2 t} + i\omega_0\dot{A}e^{i\omega_0 t} + \text{cc} \tag{85}$$

Substituting (82) and (85) into (27) yields

$$i\omega_0\left(\dot{A}e^{i\omega_0 t}-\dot{\bar{A}}e^{-i\omega_0 t}\right)=-\varepsilon\left[Ae^{i\omega_0 t}+\Lambda_1 e^{i\Omega_1 t}+\Lambda_2 e^{i\Omega_2 t}+\mathrm{cc}\right]^3 \tag{86}$$

Solving (84) and (86) for $\dot{A}$, we have

$$\begin{aligned}
-2i\omega_0\dot{A}=\varepsilon\Big[&3(A\bar{A}+2\Lambda_1\bar{\Lambda}_1+2\Lambda_2\bar{\Lambda}_2)Ae^{i\omega_0 t}+3(2A\bar{A}+\Lambda_1\bar{\Lambda}_1+2\Lambda_2\bar{\Lambda}_2)\Lambda_1 e^{i\Omega_1 t}\\
&+3(2A\bar{A}+2\Lambda_1\bar{\Lambda}_1+\Lambda_2\bar{\Lambda}_2)\Lambda_2 e^{i\Omega_2 t}+A^3e^{3i\omega_0 t}+\Lambda_1^3 e^{3i\Omega_1 t}+\Lambda_2^3 e^{3i\Omega_2 t}\\
&+3A^2\Lambda_1 e^{i(2\omega_0+\Omega_1)t}+3A^2\Lambda_2 e^{i(2\omega_0+\Omega_2)t}+3A^2\bar{\Lambda}_1 e^{i(2\omega_0-\Omega_1)t}\\
&+3A^2\bar{\Lambda}_2 e^{i(2\omega_0-\Omega_2)t}+3A\Lambda_1^2 e^{i(\omega_0+2\Omega_1)t}+3A\Lambda_2^2 e^{i(\omega_0+2\Omega_2)t}\\
&+3A\bar{\Lambda}_1^2 e^{i(\omega_0-2\Omega_1)t}+3A\bar{\Lambda}_2^2 e^{i(\omega_0-2\Omega_2)t}+6A\Lambda_1\Lambda_2 e^{i(\Omega_1+\Omega_2+\omega_0)t}\\
&+6A\bar{\Lambda}_1\bar{\Lambda}_2 e^{i(\omega_0-\Omega_1-\Omega_2)t}+6A\bar{\Lambda}_1\Lambda_2 e^{i(\omega_0-\Omega_1+\Omega_2)t}\\
&+6A\Lambda_1\bar{\Lambda}_2 e^{i(\omega_0+\Omega_1-\Omega_2)t}+3\Lambda_1^2\Lambda_2 e^{i(2\Omega_1+\Omega_2)t}+3\Lambda_1^2\bar{\Lambda}_2 e^{i(2\Omega_1-\Omega_2)t}\\
&+3\Lambda_1\Lambda_2^2 e^{i(\Omega_1+2\Omega_2)t}+3\bar{\Lambda}_1\Lambda_2^2 e^{i(2\Omega_2-\Omega_1)t}+\mathrm{cc}\Big]e^{-i\omega_0 t}
\end{aligned} \tag{87}$$

Average (87) for the cases in the preceding exercise.

(a) The case $\omega_0=2\Omega_1+\Omega_2+\varepsilon\sigma$. In this case, averaging (87) yields

$$2i\omega_0\dot{A}+\varepsilon\left[3(A\bar{A}+2\Lambda_1\bar{\Lambda}_1+2\Lambda_2\bar{\Lambda}_2)A+3\Lambda_1^2\Lambda_2 e^{-i\varepsilon\sigma t}\right]=0 \tag{88}$$

in agreement with (69).

(b) The case $3\Omega_1=\omega_0+\varepsilon\sigma_1$ and $\Omega_2=3\omega_0+\varepsilon\sigma_2$. In this case, averaging (87) yields

$$2i\omega_0\dot{A}+\varepsilon\left[3(A\bar{A}+2\Lambda_1\bar{\Lambda}_1+2\Lambda_2\bar{\Lambda}_2)A+\Lambda_1^3 e^{i\varepsilon\sigma_1 t}+3\bar{A}^2\Lambda_2 e^{i\varepsilon\sigma_2 t}\right]=0 \tag{89}$$

in agreement with (71).

(c) The case $3\Omega_1=\omega_0+\varepsilon\sigma_1$ and $\Omega_2+\Omega_1=2\omega_0+\varepsilon\sigma_2$. In this case, averaging (87) yields

$$\begin{aligned}
2i\omega_0\dot{A}+\varepsilon\Big[&3(A\bar{A}+2\Lambda_1\bar{\Lambda}_1+2\Lambda_2\bar{\Lambda}_2)A+\Lambda_1^3 e^{i\varepsilon\sigma_1 t}+6\bar{A}\Lambda_1\Lambda_2 e^{i\varepsilon\sigma_2 t}\\
&+3\Lambda_2\bar{\Lambda}_1^2 e^{i\varepsilon(\sigma_2-\sigma_1)t}\Big]=0
\end{aligned} \tag{90}$$

in agreement with (73).

(d) The case $3\Omega_1 = \omega_0 + \varepsilon\sigma_1$ and $\Omega_2 - \Omega_1 = 2\omega_0 + \varepsilon\sigma_2$. In this case, averaging (87) yields

$$2i\omega_0 \dot{A} + \varepsilon\left[3(A\bar{A} + 2\Lambda_1\bar{\Lambda}_1 + 2\Lambda_2\bar{\Lambda}_2)A + \Lambda_1^3 e^{i\varepsilon\sigma_1 t} + 6\bar{A}\bar{\Lambda}_1\Lambda_2 e^{i\varepsilon\sigma_2 t}\right] = 0 \quad (91)$$

in agreement with (75).

(e) The case $3\Omega_1 = \omega_0 + \varepsilon\sigma_1$ and $2\Omega_2 - \Omega_1 = \omega_0 + \varepsilon\sigma_2$. In this case, averaging (87) yields

$$2i\omega_0 \dot{A} + \varepsilon\left[3(A\bar{A} + 2\Lambda_1\bar{\Lambda}_1 + 2\Lambda_2\bar{\Lambda}_2)A + \Lambda_1^3 e^{i\varepsilon\sigma_1 t} + 3\Lambda_2^2\bar{\Lambda}_1 e^{i\varepsilon\sigma_2 t}\right] = 0 \quad (92)$$

in agreement with (77).

(f) The case $\Omega_1 = 3\omega_0 + \varepsilon\sigma_1$ and $\Omega_2 - 2\Omega_1 = \omega_0 + \varepsilon\sigma_2$. In this case, averaging (87) yields

$$2i\omega_0 \dot{A} + \varepsilon\left[3(A\bar{A} + 2\Lambda_1\bar{\Lambda}_1 + 2\Lambda_2\bar{\Lambda}_2)A + 3\Lambda_1\bar{A}^2 e^{i\varepsilon\sigma_1 t} + 3\Lambda_2\bar{\Lambda}_1^2 e^{i\varepsilon\sigma_2 t}\right] = 0 \quad (93)$$

in agreement with (79).

(g) The case $\Omega_1 = 3\omega_0 + \varepsilon\sigma_1$ and $\Omega_2 - \Omega_1 = 2\omega_0 + \varepsilon\sigma_2$. In this case, averaging (87) yields

$$2i\omega_0 \dot{A} + \varepsilon\left[3(A\bar{A} + 2\Lambda_1\bar{\Lambda}_1 + 2\Lambda_2\bar{\Lambda}_2)A + 3\Lambda_1\bar{A}^2 e^{i\varepsilon\sigma_1 t} + 6\Lambda_2\bar{\Lambda}_1\bar{A} e^{i\varepsilon\sigma_2 t}\right.$$

$$\left. + 3\bar{\Lambda}_2\Lambda_1^2 e^{i\varepsilon(\sigma_1 - \sigma_2)t}\right] = 0 \quad (94)$$

in agreement with (81).

10.4. Use the methods of multiple scales and averaging to determine a first-order uniform expansion for

$$\ddot{u}_1 + \omega_1^2 u_1 = \alpha_1 u_1 u_2 \quad (95)$$

$$\ddot{u}_2 + \omega_2^2 u_2 = \alpha_2 u_1^2 \quad (96)$$

for small but finite amplitudes when $\omega_2 \approx 2\omega_1$.

(a) *The Method of Multiple Scales*

We introduce the small parameter ε, the order of the amplitudes, which we will use as a bookkeeping device. Moreover, we seek a uniform expansion in the form

$$u_1 = \varepsilon u_{11}(T_0, T_1) + \varepsilon^2 u_{12}(T_0, T_1) + \cdots \quad (97)$$

$$u_2 = \varepsilon u_{21}(T_0, T_1) + \varepsilon^2 u_{22}(T_0, T_1) + \cdots \quad (98)$$

Substituting (97) and (98) into (95) and (96) and equating coefficients of like powers of

ε, we obtain

Order ε

$$D_0^2 u_{11} + \omega_1^2 u_{11} = 0 \tag{99}$$

$$D_0^2 u_{21} + \omega_2^2 u_{21} = 0 \tag{100}$$

Order ε^2

$$D_0^2 u_{12} + \omega_1^2 u_{12} = -2 D_0 D_1 u_{11} + \alpha_1 u_{11} u_{21} \tag{101}$$

$$D_0^2 u_{22} + \omega_2^2 u_{22} = -2 D_0 D_1 u_{21} + \alpha_2 u_{11}^2 \tag{102}$$

The solutions of (99) and (100) can be expressed as

$$u_{11} = A_1(T_1) e^{i\omega_1 T_0} + \bar{A}_1(T_1) e^{-i\omega_1 T_0} \tag{103}$$

$$u_{21} = A_2(T_1) e^{i\omega_2 T_0} + \bar{A}_2(T_1) e^{-i\omega_2 T_0} \tag{104}$$

Then (101) and (102) become

$$D_0^2 u_{12} + \omega_1^2 u_{12} = -2i\omega_1 A_1' e^{i\omega_1 T_0} + \alpha_1 A_1 A_2 e^{i(\omega_1+\omega_2)T_0} + \alpha_1 A_2 \bar{A}_1 e^{i(\omega_2-\omega_1)T_0} + \text{cc} \tag{105}$$

$$D_0^2 u_{22} + \omega_2^2 u_{22} = -2i\omega_2 A_2' e^{i\omega_2 T_0} + \alpha_2 A_1^2 e^{2i\omega_1 T_0} + \alpha_2 A_1 \bar{A}_1 + \text{cc} \tag{106}$$

To express quantitatively the nearness of ω_2 to $2\omega_1$, we introduce the detuning parameter σ defined by

$$\omega_2 - 2\omega_1 = \varepsilon\sigma$$

and write

$$(\omega_2 - \omega_1) T_0 = \omega_1 T_0 + \varepsilon\sigma T_0 = \omega_1 T_0 + \sigma T_1$$

$$2\omega_1 T_0 = \omega_2 T_0 - \varepsilon\sigma T_0 = \omega_2 T_0 - \sigma T_1$$

Eliminating the terms that lead to secular terms from (105) and (106), we obtain

$$2i\omega_1 A_1' = \alpha_1 A_2 \bar{A}_1 e^{i\sigma T_1} \tag{107}$$

$$2i\omega_2 A_2' = \alpha_2 A_1^2 e^{-i\sigma T_1} \tag{108}$$

Equations (107) and (108) are usually analyzed by expressing them in real form using the polar representation

$$A_n = \tfrac{1}{2} a_n e^{i\beta_n} \tag{109}$$

Substituting (109) into (107) and (108) and separating real and imaginary parts, we obtain

$$a_1' = \frac{\alpha_1}{4\omega_1} a_2 a_1 \sin\gamma \tag{110}$$

$$a_1\beta_1' = -\frac{\alpha_1}{4\omega_1} a_2 a_1 \cos\gamma \tag{111}$$

$$a_2' = -\frac{\alpha_2}{4\omega_2} a_1^2 \sin\gamma \tag{112}$$

$$a_2\beta_2' = -\frac{\alpha_2}{4\omega_2} a_1^2 \cos\gamma \tag{113}$$

where

$$\gamma = \beta_2 - 2\beta_1 + \sigma T_1 \tag{114}$$

An analysis of (110)–(114), including a closed-form solution, can be found in either *Perturbation Methods* or *Nonlinear Oscillations*.*

(b) *The Method of Averaging*

To apply the method of averaging, we need first to use the method of variation of parameters to determine the general equations describing the amplitudes and phases. To this end, we note that in the absence of the nonlinearity, the solutions of (95) and (96) can be expressed as

$$u_1 = a_1 \cos(\omega_1 t + \beta_1) \tag{115}$$

$$u_2 = a_2 \cos(\omega_2 t + \beta_2) \tag{116}$$

where the a_n and β_n are constants. Differentiating (115) and (116) once with respect to t yields

$$\dot{u}_1 = -\omega_1 a_1 \sin(\omega_1 t + \beta_1) \tag{117}$$

$$\dot{u}_2 = -\omega_2 a_2 \sin(\omega_2 t + \beta_2) \tag{118}$$

In the presence of the nonlinearity, we continue to express the solutions of (95) and (96) as in (115) and (116), subject to the constraints (117) and (118), but with time-varying a_n and β_n. Then, differentiating (115) and (116) once with respect to t and using (117) and (118), we find that

$$\dot{a}_1 \cos(\omega_1 t + \beta_1) - a_1 \dot{\beta}_1 \sin(\omega_1 t + \beta_1) = 0 \tag{119}$$

$$\dot{a}_2 \cos(\omega_2 t + \beta_2) - a_2 \dot{\beta}_2 \sin(\omega_2 t + \beta_2) = 0 \tag{120}$$

*A. H. Nayfeh, *Perturbation Methods*, Wiley, New York, 1973; A. H. Nayfeh and D. T. Mook, *Nonlinear Oscillations*, Wiley, New York, 1979.

Differentiating (117) and (118) once with respect to t, we have

$$\ddot{u}_1 = -\omega_1^2 a_1 \cos(\omega_1 t + \beta_1) - \omega_1 \dot{a}_1 \sin(\omega_1 t + \beta_1) - \omega_1 a_1 \dot{\beta}_1 \cos(\omega_1 t + \beta_1) \tag{121}$$

$$\ddot{u}_2 = -\omega_2^2 a_2 \cos(\omega_2 t + \beta_2) - \omega_2 \dot{a}_2 \sin(\omega_2 t + \beta_2) - \omega_2 a_2 \dot{\beta}_2 \cos(\omega_2 t + \beta_2) \tag{122}$$

Substituting (115), (116), (121), and (122) into (95) and (96), we have

$$\dot{a}_1 \sin(\omega_1 t + \beta_1) + a_1 \dot{\beta}_1 \cos(\omega_1 t + \beta_1) = -\frac{\alpha_1}{\omega_1} a_1 a_2 \cos(\omega_1 t + \beta_1) \times \cos(\omega_2 t + \beta_2) \tag{123}$$

$$\dot{a}_2 \sin(\omega_2 t + \beta_2) + a_2 \dot{\beta}_2 \cos(\omega_2 t + \beta_2) = -\frac{\alpha_2}{\omega_2} a_1^2 \cos^2(\omega_1 t + \beta_1) \tag{124}$$

Solving (119) and (123) yields

$$\begin{aligned}\dot{a}_1 &= -\frac{\alpha_1}{\omega_1} a_1 a_2 \sin(\omega_1 t + \beta_1) \cos(\omega_1 t + \beta_1) \cos(\omega_2 t + \beta_2) \\ &= -\frac{\alpha_1}{4\omega_1} a_1 a_2 \{\sin[(2\omega_1 + \omega_2) t + 2\beta_1 + \beta_2] + \sin[(2\omega_1 - \omega_2) t \\ &\quad + (2\beta_1 - \beta_2)]\}\end{aligned} \tag{125}$$

$$\begin{aligned}a_1 \dot{\beta}_1 &= -\frac{\alpha_1}{\omega_1} a_1 a_2 \cos^2(\omega_1 t + \beta_1) \cos(\omega_2 t + \beta_2) \\ &= -\frac{\alpha_1}{4\omega_1} a_1 a_2 \{2\cos(\omega_2 t + \beta_2) + \cos[(2\omega_1 + \omega_2) t + 2\beta_1 + \beta_2] \\ &\quad + \cos[(2\omega_1 - \omega_2) t + 2\beta_1 - \beta_2]\}\end{aligned} \tag{126}$$

Solving (120) and (124) yields

$$\begin{aligned}\dot{a}_2 &= -\frac{\alpha_2}{\omega_2} a_1^2 \sin(\omega_2 t + \beta_2) \cos^2(\omega_1 t + \beta_1) \\ &= -\frac{\alpha_2}{4\omega_2} a_1^2 \{2\sin(\omega_2 t + \beta_2) + \sin[(\omega_2 + 2\omega_1) t + \beta_2 + 2\beta_1] \\ &\quad + \sin[(\omega_2 - 2\omega_1) t + \beta_2 - 2\beta_1]\}\end{aligned} \tag{127}$$

$$\begin{aligned}a_2 \dot{\beta}_2 &= -\frac{\alpha_2}{\omega_2} a_1^2 \cos(\omega_2 t + \beta_2) \cos^2(\omega_1 t + \beta_1) \\ &= -\frac{\alpha_2}{4\omega_2} a_1^2 \{2\cos(\omega_2 t + \beta_2) + \cos[(\omega_2 + 2\omega_1) t + \beta_2 + 2\beta_1] \\ &\quad + \cos[(\omega_2 - 2\omega_1) t + \beta_2 - 2\beta_1]\}\end{aligned} \tag{128}$$

Keeping only the slowly varying parts in (125)–(128) and recalling that $\omega_2 \approx 2\omega_1$, we obtain

$$\dot{a}_1 = \frac{\alpha_1}{4\omega_1} a_1 a_2 \sin[(\omega_2 - 2\omega_1)t + \beta_2 - 2\beta_1] \tag{129}$$

$$a_1\dot{\beta}_1 = -\frac{\alpha_1}{4\omega_1} a_1 a_2 \cos[(\omega_2 - 2\omega_1)t + \beta_2 - 2\beta_1] \tag{130}$$

$$\dot{a}_2 = -\frac{\alpha_2}{4\omega_2} a_1^2 \sin[(\omega_2 - 2\omega_1)t + \beta_2 - 2\beta_1] \tag{131}$$

$$a_2\dot{\beta}_2 = -\frac{\alpha_2}{4\omega_2} a_1^2 \cos[(\omega_2 - 2\omega_1)t + \beta_2 - 2\beta_1] \tag{132}$$

in agreement with (110)–(114).

10.5. Use the methods of multiple scales and averaging to determine the equations describing the amplitudes and phases of the system

$$\ddot{u}_1 + \omega_1^2 u_1 = \alpha_1 u_2 u_3 \tag{133}$$

$$\ddot{u}_2 + \omega_2^2 u_2 = \alpha_2 u_1 u_3 \tag{134}$$

$$\ddot{u}_3 + \omega_3^2 u_3 = \alpha_3 u_1 u_2 \tag{135}$$

for small but finite amplitudes when $\omega_3 \approx \omega_1 + \omega_2$.

(a) ***The Method of Multiple Scales***

To determine an expansion valid for small but finite amplitudes, we introduce the small parameter ε that can be used as a bookkeeping device. Then we let

$$u_n = \varepsilon u_{n_1}(T_0, T_1) + \varepsilon^2 u_{n_2}(T_0, T_1) + \cdots \tag{136}$$

where $T_0 = t$ and $T_1 = \varepsilon t$. Substituting (136) into (133)–(135) and equating coefficients of like powers of ε, we obtain

Order ε

$$D_0^2 u_{n_1} + \omega_n^2 u_{n_1} = 0 \quad \text{for } n = 1,2,3 \tag{137}$$

Order ε^2

$$D_0^2 u_{12} + \omega_1^2 u_{12} = -2D_0 D_1 u_{11} + \alpha_1 u_{21} u_{31} \tag{138}$$

$$D_0^2 u_{22} + \omega_2^2 u_{22} = -2D_0 D_1 u_{21} + \alpha_2 u_{11} u_{31} \tag{139}$$

$$D_0^2 u_{32} + \omega_3^2 u_{32} = -2D_0 D_1 u_{31} + \alpha_3 u_{11} u_{21} \tag{140}$$

The solution of (137) can be expressed as

$$u_{n_1} = A_n(T_1)\, e^{i\omega_n T_0} + \bar{A}_n(T_1)\, e^{-i\omega_n T_0} \tag{141}$$

Then (138)–(140) become

$$D_0^2 u_{12} + \omega_1^2 u_{12} = -2i\omega_1 A_1' e^{i\omega_1 T_0} + \alpha_1 A_2 A_3 e^{i(\omega_3+\omega_2)T_0} + \alpha_1 A_3 \bar{A}_2 e^{i(\omega_3-\omega_2)T_0} + \text{cc} \tag{142}$$

$$D_0^2 u_{22} + \omega_2^2 u_{22} = -2i\omega_2 A_2' e^{i\omega_2 T_0} + \alpha_2 A_3 A_1 e^{i(\omega_3+\omega_1)T_0} + \alpha_2 A_3 \bar{A}_1 e^{i(\omega_3-\omega_1)T_0} + \text{cc} \tag{143}$$

$$D_0^2 u_{32} + \omega_3^2 u_{32} = -2i\omega_3 A_3' e^{i\omega_3 T_0} + \alpha_3 A_2 A_1 e^{i(\omega_2+\omega_1)T_0} + \alpha_3 A_2 \bar{A}_1 e^{i(\omega_2-\omega_1)T_0} + \text{cc} \tag{144}$$

To express quantitatively the nearness of ω_3 to $\omega_1+\omega_2$, we introduce the detuning parameter σ according to

$$\omega_3 = \omega_1 + \omega_2 + \varepsilon\sigma \tag{145}$$

Then we write

$$(\omega_3 - \omega_2)T_0 = (\omega_1 + \varepsilon\sigma)T_0 = \omega_1 T_0 + \sigma T_1$$

$$(\omega_3 - \omega_1)T_0 = (\omega_2 + \varepsilon\sigma)T_0 = \omega_2 T_0 + \sigma T_1$$

$$(\omega_2 + \omega_1)T_0 = (\omega_3 - \varepsilon\sigma)T_0 = \omega_3 T_0 - \sigma T_1$$

Eliminating the terms that produce secular terms from (142)–(144) yields

$$2i\omega_1 A_1' = \alpha_1 A_3 \bar{A}_2 e^{i\sigma T_1} \tag{146}$$

$$2i\omega_2 A_2' = \alpha_2 A_3 \bar{A}_1 e^{i\sigma T_1} \tag{147}$$

$$2i\omega_3 A_3' = \alpha_3 A_2 A_1 e^{-i\sigma T_1} \tag{148}$$

Introducing the polar notation

$$A_n = \tfrac{1}{2} a_n e^{i\beta_n} \tag{149}$$

into (146)–(148) and separating real and imaginary parts, we obtain

$$a_1' = \frac{\alpha_1}{4\omega_1} a_3 a_2 \sin\gamma \tag{150}$$

$$a_1 \beta_1' = -\frac{\alpha_1}{4\omega_1} a_3 a_2 \cos\gamma \tag{151}$$

$$a_2' = \frac{\alpha_2}{4\omega_2} a_3 a_1 \sin\gamma \tag{152}$$

$$a_2 \beta_2' = -\frac{\alpha_2}{4\omega_2} a_3 a_1 \cos\gamma \tag{153}$$

$$a_3' = -\frac{\alpha_3}{4\omega_3} a_2 a_1 \sin\gamma \tag{154}$$

$$a_3 \beta_3' = -\frac{\alpha_3}{4\omega_3} a_2 a_1 \cos\gamma \tag{155}$$

where

$$\gamma = \beta_3 - \beta_2 - \beta_1 + \sigma T_1 \tag{156}$$

(b) ***The Method of Averaging***

In the absence of the nonlinear terms, the solutions of (133)–(135) can be written as

$$u_n = a_n \cos(\omega_n t + \beta_n) \tag{157}$$

where the a_n and β_n are constants. Differentiating (157) once with respect to t yields

$$\dot{u}_n = -\omega_n a_n \sin(\omega_n t + \beta_n) \tag{158}$$

In the presence of the nonlinearities, we still express the solutions of (133)–(135) as in (157), subject to the constraints (158), but with time-varying a_n and β_n. Then differentiating (157) once with respect to t and using (158), we obtain

$$\dot{a}_n \cos\phi_n - a_n \dot{\beta}_n \sin\phi_n = 0 \tag{159}$$

where $\phi_n = \omega_n t + \beta_n$. Differentiating (158) once with respect to t, we have

$$\ddot{u}_n = -\omega_n^2 a_n \cos\phi_n - \omega_n \dot{a}_n \sin\phi_n - \omega_n a_n \dot{\beta}_n \cos\phi_n$$

which, upon substitution into (133)–(135), yields

$$\omega_1 \dot{a}_1 \sin\phi_1 + \omega_1 a_1 \dot{\beta}_1 \cos\phi_1 = -\alpha_1 a_2 a_3 \cos\phi_2 \cos\phi_3 \tag{160}$$

$$\omega_2 \dot{a}_2 \sin\phi_2 + \omega_2 a_2 \dot{\beta}_2 \cos\phi_2 = -\alpha_2 a_1 a_3 \cos\phi_1 \cos\phi_3 \tag{161}$$

$$\omega_3 \dot{a}_3 \sin\phi_3 + \omega_3 a_3 \dot{\beta}_3 \cos\phi_3 = -\alpha_3 a_1 a_2 \cos\phi_1 \cos\phi_2 \tag{162}$$

Solving (159)–(162) for the $\dot{a}_n$ and $\dot{\beta}_n$ yields

$$\dot{a}_1 = -\frac{\alpha_1}{\omega_1} a_2 a_3 \sin\phi_1 \cos\phi_2 \cos\phi_3$$

$$= -\frac{\alpha_1}{4\omega_1} a_2 a_3 \{ \sin[(\omega_1 + \omega_2 + \omega_3)t + \beta_1 + \beta_2 + \beta_3]$$

$$+ \sin[(\omega_1 - \omega_2 - \omega_3)t + \beta_1 - \beta_2 - \beta_3]$$

$$+ \sin[(\omega_1 + \omega_2 - \omega_3)t + \beta_1 + \beta_2 - \beta_3]$$

$$+ \sin[(\omega_1 - \omega_2 + \omega_3)t + \beta_1 - \beta_2 + \beta_3]\} \tag{163}$$

$$a_1\dot{\beta}_1 = -\frac{\alpha_1}{\omega_1} a_2 a_3 \cos\phi_1 \cos\phi_2 \cos\phi_3$$

$$= -\frac{\alpha_1}{4\omega_1} a_2 a_3 \{ \cos[(\omega_1 + \omega_2 + \omega_3)t + \beta_1 + \beta_2 + \beta_3]$$

$$+ \cos[(\omega_1 - \omega_2 - \omega_3)t + \beta_1 - \beta_2 - \beta_3]$$

$$+ \cos[(\omega_1 + \omega_2 - \omega_3)t + \beta_1 + \beta_2 - \beta_3]$$

$$+ \cos[(\omega_1 - \omega_2 + \omega_3)t + \beta_1 - \beta_2 + \beta_3]\} \tag{164}$$

$$\dot{a}_2 = -\frac{\alpha_2}{\omega_2} a_1 a_3 \sin\phi_2 \cos\phi_1 \cos\phi_3$$

$$= -\frac{\alpha_2}{4\omega_2} a_1 a_3 \{ \sin[(\omega_1 + \omega_2 + \omega_3)t + \beta_1 + \beta_2 + \beta_3]$$

$$+ \sin[(\omega_2 - \omega_1 - \omega_3)t + \beta_2 - \beta_1 - \beta_3]$$

$$+ \sin[(\omega_1 + \omega_2 - \omega_3)t + \beta_1 + \beta_2 - \beta_3]$$

$$+ \sin[(\omega_2 - \omega_1 + \omega_3)t + \beta_2 - \beta_1 + \beta_3]\} \tag{165}$$

$$a_2\dot{\beta}_2 = -\frac{\alpha_2}{\omega_2} a_1 a_3 \cos\phi_2 \cos\phi_1 \cos\phi_3$$

$$= -\frac{\alpha_2}{4\omega_2} a_1 a_3 \{ \cos[(\omega_1 + \omega_2 + \omega_3)t + \beta_1 + \beta_2 + \beta_3]$$

$$+ \cos[(\omega_1 - \omega_2 - \omega_3)t + \beta_1 - \beta_2 - \beta_3]$$

$$+ \cos[(\omega_1 + \omega_2 - \omega_3)t + \beta_1 + \beta_2 - \beta_3]$$

$$+ \cos[(\omega_1 - \omega_2 + \omega_3)t + \beta_1 - \beta_2 + \beta_3]\} \tag{166}$$

$$\dot{a}_3 = -\frac{\alpha_3}{\omega_3} a_1 a_2 \sin\phi_3 \cos\phi_1 \cos\phi_2$$

$$= -\frac{\alpha_3}{4\omega_3} a_1 a_2 \{\sin[(\omega_3 + \omega_2 + \omega_1)t + \beta_3 + \beta_2 + \beta_1]$$

$$+ \sin[(\omega_3 + \omega_2 - \omega_1)t + \beta_3 + \beta_2 - \beta_1]$$

$$+ \sin[(\omega_3 - \omega_2 - \omega_1)t + \beta_3 - \beta_2 - \beta_1]$$

$$+ \sin[(\omega_3 - \omega_2 + \omega_1)t + \beta_3 - \beta_2 + \beta_1]\} \tag{167}$$

$$a_3\dot{\beta}_3 = -\frac{\alpha_3}{\omega_3} a_1 a_2 \cos\phi_3 \cos\phi_1 \cos\phi_2$$

$$= -\frac{\alpha_3}{4\omega_3} a_1 a_2 \{\cos[(\omega_1 + \omega_2 + \omega_3)t + \beta_1 + \beta_2 + \beta_3]$$

$$+ \cos[(\omega_1 - \omega_2 - \omega_3)t + \beta_1 - \beta_2 - \beta_3]$$

$$+ \cos[(\omega_1 + \omega_2 - \omega_3)t + \beta_1 + \beta_2 - \beta_3]$$

$$+ \cos[(\omega_1 - \omega_2 + \omega_3)t + \beta_1 - \beta_2 + \beta_3]\} \tag{168}$$

When $\omega_3 \approx \omega_1 + \omega_2$, only the third terms on the right-hand sides of (163)–(168) are slowly varying. Keeping only these slowly-varying terms, we reduce (163)–(168) to the following equations:

$$\dot{a}_1 = \frac{\alpha_1}{4\omega_1} a_2 a_3 \sin[(\omega_3 - \omega_2 - \omega_1)t + \beta_3 - \beta_2 - \beta_1] \tag{169}$$

$$a_1\dot{\beta}_1 = -\frac{\alpha_1}{4\omega_1} a_2 a_3 \cos[(\omega_3 - \omega_2 - \omega_1)t + \beta_3 - \beta_2 - \beta_1] \tag{170}$$

$$\dot{a}_2 = \frac{\alpha_2}{4\omega_2} a_1 a_3 \sin[(\omega_3 - \omega_2 - \omega_1)t + \beta_3 - \beta_2 - \beta_1] \tag{171}$$

$$a_2\dot{\beta}_2 = -\frac{\alpha_2}{4\omega_2} a_1 a_3 \cos[(\omega_3 - \omega_2 - \omega_1)t + \beta_3 - \beta_2 - \beta_1] \tag{172}$$

$$\dot{a}_3 = -\frac{\alpha_3}{4\omega_3} a_1 a_2 \sin[(\omega_3 - \omega_2 - \omega_1)t + \beta_3 - \beta_2 - \beta_1] \tag{173}$$

$$a_3\dot{\beta}_3 = -\frac{\alpha_3}{4\omega_3} a_1 a_2 \cos[(\omega_3 - \omega_2 - \omega_1)t + \beta_3 - \beta_2 - \beta_1] \tag{174}$$

in agreement with (150)–(156), since $T_1 = \varepsilon t$ and $\omega_3 = \omega_2 + \omega_1 + \varepsilon\sigma$.

10.6. Use the methods of multiple scales and averaging to determine first-order uniform expansions for

$$\ddot{u}_1 + \omega_1^2 u_1 = \varepsilon \alpha_1 u_1 u_2 + \varepsilon k_1 \cos \Omega_1 t \tag{175}$$

$$\ddot{u}_2 + \omega_2^2 u_2 = \varepsilon \alpha_2 u_1^2 + \varepsilon k_2 \cos \Omega_2 t \tag{176}$$

when

(i) $$\omega_2 \approx 2\omega_1 \quad \text{and} \quad \Omega_1 \approx \omega_1$$

(ii) $$\omega_2 \approx 2\omega_1 \quad \text{and} \quad \Omega_2 \approx \omega_2$$

(a) *The Method of Multiple Scales*

When Ω_n is away from zero, we express $\cos \Omega_n t$ as $\cos \Omega_n T_0$. Substituting (136) into (175) and (176) and equating coefficients of like powers of ε, we obtain

Order ε^0

$$D_0^2 u_{11} + \omega_1^2 u_{11} = 0 \tag{177}$$

$$D_0^2 u_{21} + \omega_2^2 u_{21} = 0 \tag{178}$$

Order ε

$$D_0^2 u_{12} + \omega_1^2 u_{12} = -2D_0 D_1 u_{11} + \alpha_1 u_{11} u_{21} + k_1 \cos \Omega_1 T_0 \tag{179}$$

$$D_0^2 u_{22} + \omega_2^2 u_{22} = -2D_0 D_1 u_{21} + \alpha_2 u_{11}^2 + k_2 \cos \Omega_2 T_0 \tag{180}$$

The solutions of (177) and (178) can be expressed as

$$u_{11} = A_1(T_1)\, e^{i\omega_1 T_0} + \text{cc} \tag{181}$$

$$u_{21} = A_2(T_1)\, e^{i\omega_2 T_0} + \text{cc} \tag{182}$$

Then (179) and (180) become

$$D_0^2 u_{12} + \omega_1^2 u_{12} = -2i\omega_1 A_1' e^{i\omega_1 T_0} + \alpha_1 A_1 A_2 e^{i(\omega_1+\omega_2)T_0} + \alpha_1 A_2 \bar{A}_1 e^{i(\omega_2-\omega_1)T_0} + \tfrac{1}{2} k_1 e^{i\Omega_1 T_0} + \text{cc} \tag{183}$$

$$D_0^2 u_{22} + \omega_2^2 u_{22} = -2i\omega_2 A_2' e^{i\omega_2 T_0} + \alpha_2 A_1^2 e^{2i\omega_1 T_0} + \alpha_2 A_1 \bar{A}_1 + \tfrac{1}{2} k_2 e^{i\Omega_2 T_0} + \text{cc} \tag{184}$$

When $\omega_2 \approx 2\omega_1$ *and* $\Omega_1 \approx \omega_1$, we introduce the detuning parameters σ_1 and σ_2 defined as follows:

$$\omega_2 = 2\omega_1 + \varepsilon\sigma_1, \qquad \Omega_1 = \omega_1 + \varepsilon\sigma_2$$

Then we write

$$(\omega_2 - \omega_1) T_0 = \omega_1 T_0 + \sigma_1 T_1, \qquad \Omega_1 T_0 = \omega_1 T_0 + \sigma_2 T_1$$

$$2\omega_1 T_0 = \omega_2 T_0 - \sigma_1 T_1$$

Eliminating the terms that lead to secular terms from (183) and (184),we obtain

$$2i\omega_1 A_1' = \alpha_1 A_2 \bar{A}_1 e^{i\sigma_1 T_1} + \tfrac{1}{2} k_1 e^{i\sigma_2 T_1} \tag{185}$$

$$2i\omega_2 A_2' = \alpha_2 A_1^2 e^{-i\sigma_1 T_1} \tag{186}$$

Substituting the polar form (149) into (185) and (186) and separating real and imaginary parts, we have

$$a_1' = \frac{\alpha_1}{4\omega_1} a_1 a_2 \sin\gamma_1 + \frac{k_1}{2\omega_1} \sin\gamma_2 \tag{187}$$

$$a_1 \beta_1' = -\frac{\alpha_1}{4\omega_1} a_1 a_2 \cos\gamma_1 - \frac{k_1}{2\omega_1} \cos\gamma_2 \tag{188}$$

$$a_2' = -\frac{\alpha_2}{4\omega_2} a_1^2 \sin\gamma_1 \tag{189}$$

$$a_2 \beta_2' = -\frac{\alpha_2}{4\omega_2} a_1^2 \cos\gamma_1 \tag{190}$$

where

$$\gamma_1 = \sigma_1 T_1 + \beta_2 - 2\beta_1, \qquad \gamma_2 = \sigma_2 T_1 - \beta_1 \tag{191}$$

When $\omega_2 \approx 2\omega_1$ *and* $\Omega_2 \approx \omega_2$, we introduce the detuning parameters σ_1 and σ_2, where σ_1 is defined above and σ_2 is defined by

$$\Omega_2 = \omega_2 + \varepsilon\sigma_2$$

so that $\Omega_2 T_0$ can be written as

$$\Omega_2 T_0 = \omega_2 T_0 + \sigma_2 T_1$$

Eliminating the terms that lead to secular terms from (183) and (184), we obtain

$$2i\omega_1 A_1' = \alpha_1 A_2 \bar{A}_1 e^{i\sigma_1 T_1} \tag{192}$$

$$2i\omega_2 A_2' = \alpha_2 A_1^2 e^{-i\sigma_1 T_1} + \tfrac{1}{2} k_2 e^{i\sigma_2 T_1} \tag{193}$$

Substituting the polar form (149) into (192) and (193) and separating real and imaginary parts, we have

$$a_1' = \frac{\alpha_1}{4\omega_1} a_1 a_2 \sin\gamma_1 \tag{194}$$

$$a_1\beta_1' = -\frac{\alpha_1}{4\omega_1} a_1 a_2 \cos\gamma_1 \tag{195}$$

$$a_2' = -\frac{\alpha_2}{4\omega_2} a_1^2 \sin\gamma_1 + \frac{k_2}{2\omega_2}\sin\gamma_2 \tag{196}$$

$$a_2\beta_2' = -\frac{\alpha_2}{4\omega_2} a_1^2 \cos\gamma_1 - \frac{k_2}{2\omega_2}\cos\gamma_2 \tag{197}$$

where

$$\gamma_1 = \sigma_1 T_1 + \beta_2 - 2\beta_1 \quad \text{and} \quad \gamma_2 = \sigma_2 T_1 - \beta_2 \tag{198}$$

(b) *The Method of Averaging*

As in Exercise 10.4, we seek a solution for Equations (175) and (176) as in (115) and (116), subject to te constraints (117) and (118). Following steps similar to those in Exercise 10.4, we obtain, in place of (125)–(128), the following equations:

$$\dot{a}_1 = -\frac{\varepsilon\alpha_1}{4\omega_1} a_1 a_2 \{\sin[(2\omega_1 + \omega_2)t + 2\beta_1 + \beta_2] + \sin[(2\omega_1 - \omega_2)t + 2\beta_1 - \beta_2]\}$$

$$-\frac{\varepsilon k_1}{2\omega_1}\{\sin[(\omega_1 + \Omega_1)t + \beta_1] + \sin[(\omega_1 - \Omega_1)t + \beta_1]\} \tag{199}$$

$$a_1\dot{\beta}_1 = -\frac{\varepsilon\alpha_1}{4\omega_1} a_1 a_2 \{2\cos(\omega_2 t + \beta_2) + \cos[(2\omega_1 + \omega_2)t + 2\beta_1 + \beta_2]$$

$$+\cos[(2\omega_1 - \omega_2)t + 2\beta_1 - \beta_2]\}$$

$$-\frac{\varepsilon k_1}{2\omega_1}\{\cos[(\omega_1 + \Omega_1)t + \beta_1] + \cos[(\omega_1 - \Omega_1)t + \beta_1]\} \tag{200}$$

$$\dot{a}_2 = -\frac{\varepsilon\alpha_2}{4\omega_2} a_1^2 \{2\sin(\omega_2 t + \beta_2) + \sin[(\omega_2 + 2\omega_1)t + \beta_2 + 2\beta_1]$$

$$+\sin[(\omega_2 - 2\omega_1)t + \beta_2 - 2\beta_1]\}$$

$$-\frac{\varepsilon k_2}{2\omega_2}\{\sin[(\omega_2 + \Omega_2)t + \beta_2] + \sin[(\omega_2 - \Omega_2)t + \beta_2]\} \tag{201}$$

$$a_2\dot{\beta}_2 = -\frac{\varepsilon\alpha_2}{4\omega_2} a_1^2 \{2\cos(\omega_2 t + \beta_2) + \cos[(\omega_2 + 2\omega_1)t + \beta_2 + 2\beta_1]$$

$$+\cos[(\omega_2 - 2\omega_1)t + \beta_2 - 2\beta_1]\}$$

$$-\frac{\varepsilon k_2}{2\omega_2}\{\cos[(\omega_2 + \Omega_2)t + \beta_2] + \cos[(\omega_2 - \Omega_2)t + \beta_2]\} \tag{202}$$

When $\omega_2 = 2\omega_1 + \varepsilon\sigma_1$ *and* $\Omega_1 = \omega_1 + \varepsilon\sigma_2$, keeping only the slowly varying parts in (199)–(202) yields

$$\dot{a}_1 = \frac{\varepsilon\alpha_1}{4\omega_1} a_1 a_2 \sin\gamma_1 + \frac{\varepsilon k_1}{2\omega_1}\sin\gamma_2 \tag{203}$$

$$a_1\dot{\beta}_1 = -\frac{\varepsilon\alpha_1}{4\omega_1} a_1 a_2 \cos\gamma_1 - \frac{\varepsilon k_1}{2\omega_1}\cos\gamma_2 \tag{204}$$

$$\dot{a}_2 = -\frac{\varepsilon\alpha_2}{4\omega_2} a_1^2 \sin\gamma_1 \tag{205}$$

$$a_2\dot{\beta}_2 = -\frac{\varepsilon\alpha_2}{4\omega_2} a_1^2 \cos\gamma_1 \tag{206}$$

where

$$\gamma_1 = \varepsilon\sigma_1 t + \beta_2 - 2\beta_1 \quad \text{and} \quad \gamma_2 = \varepsilon\sigma_2 t - \beta_1 \tag{207}$$

in agreement with (187)–(191), obtained using the method of multiple scales, because $T_1 = \varepsilon t$.

When $\omega_2 = 2\omega_1 + \varepsilon\sigma_1$ *and* $\Omega_2 = \omega_2 + \varepsilon\sigma_2$, keeping only the slowly varying parts in (199)–(202) yields

$$\dot{a}_1 = \frac{\varepsilon\alpha_1}{4\omega_1} a_1 a_2 \sin\gamma_1 \tag{208}$$

$$a_1\dot{\beta}_1 = -\frac{\varepsilon\alpha_1}{4\omega_1} a_1 a_2 \cos\gamma_1 \tag{209}$$

$$\dot{a}_2 = -\frac{\varepsilon\alpha_2}{4\omega_2} a_1^2 \sin\gamma_1 + \frac{\varepsilon k_2}{2\omega_2}\sin\gamma_2 \tag{210}$$

$$a_2\dot{\beta}_2 = -\frac{\varepsilon\alpha_2}{4\omega_2} a_1^2 \cos\gamma_1 - \frac{\varepsilon k_2}{2\omega_2}\cos\gamma_2 \tag{211}$$

where

$$\gamma_1 = \varepsilon\sigma_1 t + \beta_2 - 2\beta_1 \quad \text{and} \quad \gamma_2 = \varepsilon\sigma_2 t - \beta_2 \tag{212}$$

in agreement with (194)–(198), obtained using the method of multiple scales, because $T_1 = \varepsilon t$.

Supplementary Exercises

10.7. Consider the equation

$$\ddot{u} + \omega_0^2 u + 2\varepsilon^2\mu\dot{u} + \varepsilon\alpha_2 u^2 = F_1\cos\Omega_1 t + \varepsilon F_2\cos\Omega_2 t, \qquad \Omega_2 + \Omega_1 = \omega_0 + \varepsilon\sigma, \qquad \varepsilon \ll 1$$

Show that, to the second approximation,

$$u = a\cos(\omega_0 t + \beta) + 2\Lambda_1 \cos\Omega_1 t + \varepsilon\left\{2\Lambda_2\cos\Omega_2 t + \frac{2\alpha_2\Lambda_1^2}{4\Omega_1^2 - \omega_0^2}\cos 2\Omega_1 t\right.$$

$$+ \frac{\alpha_2 a^2}{6\omega_0^2}\cos(2\omega_0 t + 2\beta) + \frac{2\alpha_2\Lambda_1 a}{\Omega_1(\Omega_1 + 2\omega_0)}\cos[(\omega_0 + \Omega_1)t + \beta]$$

$$\left. + \frac{2\alpha_2\Lambda_1 a}{\Omega_1(\Omega_1 - 2\omega_0)}\cos[(\omega_0 - \Omega_1)t + \beta] - \frac{\alpha_2 a^2}{2\omega_0^2} - \frac{2\alpha_2\Lambda_1^2}{\omega_0^2}\right\} + \cdots$$

where a and β are given by

$$\omega_0 \dot{a} = -\varepsilon^2\omega_0\mu a - 2\varepsilon^2\alpha_2\Lambda_1\Lambda_2\sin\gamma$$

$$\omega_0 a\dot{\beta} = 2\varepsilon^2\alpha_2^2\Lambda_1^2\left[\frac{2}{\Omega_1^2 - 4\omega_0^2} - \frac{1}{\omega_0^2}\right]a - \frac{5\alpha_2^2\varepsilon^2}{12\omega_0^2}a^3 + 2\varepsilon^2\alpha_2\Lambda_1\Lambda_2\cos\gamma$$

$$\gamma = \varepsilon^2\sigma t - \beta, \qquad \Lambda_n = \frac{F_n}{2(\omega_0^2 - \Omega_n^2)}$$

10.8. Consider the equation

$$\ddot{u} + \omega_0^2 u + 2\varepsilon^2\mu\dot{u} + \varepsilon\alpha_2 u^2 = \varepsilon^{1/2}F_1\cos\Omega_1 t + \varepsilon^{1/2}F_2\cos\Omega_2 t,$$

$$\Omega_2 + \Omega_1 = \omega_0 + \varepsilon^2\sigma, \qquad \varepsilon \ll 1$$

Show that

$$u = a\cos(\omega_0 t + \beta) + 2\varepsilon^{1/2}\Lambda_1\cos\Omega_1 t + 2\varepsilon^{1/2}\Lambda_2\cos\Omega_2 t$$

$$+ \frac{\varepsilon\alpha_2}{6\omega_0^2}[\cos(2\omega_0 t + 2\beta) - 3] + \cdots$$

where a and β are given by

$$\omega_0 \dot{a} = -\varepsilon^2\omega_0\mu a - 2\varepsilon^2\alpha_2\Lambda_1\Lambda_2\sin\gamma$$

$$\omega_0 a\dot{\beta} = -\frac{5\alpha_2^2\varepsilon^2}{12\omega_0^2}a^3 + 2\varepsilon^2\alpha_2\Lambda_1\Lambda_2\cos\gamma$$

$$\gamma = \varepsilon\sigma t - \beta, \qquad \Lambda_n = \frac{F_n}{2(\omega_0^2 - \Omega_n^2)}$$

10.9. Use the methods of multiple scales and averaging to show that, to the first approximation, the solution of

$$\ddot{u}_1 + \omega_1^2 u_1 = \alpha_1 \dot{u}_1 \dot{u}_2$$

$$\ddot{u}_2 + \omega_2^2 u_2 = \alpha_2 \dot{u}_1^2$$

$$\omega_2 \approx 2\omega_1$$

is

$$u_n = a_n \cos(\omega_n t + \beta_n) + \cdots$$

where the a_n and β_n are given by

$$\dot{a}_1 = \tfrac{1}{4}\alpha_1 \omega_2 a_1 a_2 \sin\gamma$$

$$a_1 \dot{\beta}_1 = -\tfrac{1}{4}\alpha_1 \omega_2 a_1 a_2 \cos\gamma$$

$$\dot{a}_2 = \frac{\alpha_2 \omega_1^2}{4\omega_2} a_1^2 \sin\gamma$$

$$a_2 \dot{\beta}_2 = \frac{\alpha_2 \omega_1^2}{4\omega_2} a_1^2 \cos\gamma$$

$$\gamma = \beta_2 - 2\beta_1 + (\omega_2 - 2\omega_1)t$$

10.10. Use the methods of multiple scales and averaging to show that, to the first approximation, the solution of

$$\ddot{u}_1 + \omega_1^2 u_1 = \varepsilon(\alpha_1 u_1^2 + \alpha_2 u_1 u_2 + \alpha_3 u_2^2)$$

$$\ddot{u}_2 + \omega_2^2 u_2 = \varepsilon(\alpha_4 u_1^2 + \alpha_5 u_1 u_2 + \alpha_6 u_2^3)$$

$$\omega_2 = 2\omega_1 + \varepsilon\sigma$$

is

$$u_n = a_n \cos(\omega_n t + \beta_n) + \cdots$$

where the a_n and β_n are given by

$$\dot{a}_1 = \frac{\varepsilon\alpha_2}{4\omega_1} a_1 a_2 \sin\gamma$$

$$a_1\dot{\beta}_1 = -\frac{\varepsilon\alpha_2}{4\omega_1} a_1 a_2 \cos\gamma$$

$$\dot{a}_2 = -\frac{\varepsilon\alpha_4}{4\omega_2} a_1^2 \sin\gamma$$

$$a_2\dot{\beta}_2 = -\frac{\varepsilon\alpha_4}{4\omega_2} a_1^2 \cos\gamma$$

$$\gamma = \beta_2 - 2\beta_1 + \varepsilon\sigma t$$

10.11. Consider the equations

$$\ddot{u}_1 + \omega_1^2 u_1 + \varepsilon\left(\delta_1 u_1^2 + 2\delta_2 u_1 u_2 + \delta_3 u_2^2\right) = F_1 \cos\Omega t$$

$$\ddot{u}_2 + \omega_2^2 u_2 + \varepsilon\left(\delta_2 u_1^2 + 2\delta_3 u_1 u_2 + \delta_4 u_2^2\right) = 0$$

where

$$\omega_2 = 2\omega_1 + \varepsilon\sigma_1 \quad \text{and} \quad \Omega = 2\omega_1 + \varepsilon\sigma_2$$

Show that, to the first approximation,

$$u_1 = a_1 \cos(\omega_1 t + \beta_1) + 2\Lambda_1 \cos\Omega t + \cdots$$

$$u_2 = a_2 \cos(\omega_2 t + \beta_2) + \cdots$$

where the a_n and β_n are given by

$$\dot{a}_1 = -\frac{\varepsilon\delta_2}{2\omega_1} a_1 a_2 \sin\gamma_1 - \frac{\varepsilon\delta_1\Lambda_1}{\omega_1} a_1 \sin\gamma_2$$

$$a_1\dot{\beta}_1 = \frac{\varepsilon\delta_2}{2\omega_1} a_1 a_2 \cos\gamma_1 + \frac{\varepsilon\delta_1\Lambda_1}{\omega_1} a_1 \cos\gamma_2$$

$$\dot{a}_2 = \frac{\varepsilon\delta_2}{4\omega_2} a_1^2 \sin\gamma_1$$

$$a_2\dot{\beta}_2 = \frac{\varepsilon\delta_2}{4\omega_2} a_1^2 \cos\gamma_1$$

$$\gamma_1 = \beta_2 - 2\beta_1 + \varepsilon\sigma_1 t, \qquad \gamma_2 = \varepsilon\sigma_2 t - 2\beta, \qquad \Lambda_1 = \frac{F_1}{2\left(\omega_1^2 - \Omega^2\right)}$$

10.12. Consider the equations

$$\ddot{u}_1 + \omega_1^2 u_1 + \varepsilon\left(\delta_1 u_1^2 + 2\delta_2 u_1 u_2 + \delta_3 u_2^2\right) = 2F_1 \cos \Omega t$$

$$\ddot{u}_2 + \omega_2^2 u_2 + \varepsilon\left(\delta_2 u_1^2 + 2\delta_3 u_1 u_2 + \delta_4 u_2^2\right) = 0$$

where

$$\omega_2 = 2\omega_1 + \varepsilon\sigma_1 \quad \text{and} \quad \Omega = \omega_2 + \omega_1 + \varepsilon\sigma_2$$

Show that, to the first approximation,

$$u_1 = a_1 \cos(\omega_1 t + \beta_1) + 2\Lambda_1 \cos \Omega t + \cdots$$

$$u_2 = a_2 \cos(\omega_2 t + \beta_2) + \cdots$$

where the a_n and β_n are given by

$$\dot{a}_1 = -\frac{\varepsilon\delta_2}{2\omega_1} a_1 a_2 \sin\gamma_1 - \frac{\varepsilon\delta_2\Lambda_1}{\omega_1} a_2 \sin\gamma_2$$

$$a_1\dot{\beta}_1 = \frac{\varepsilon\delta_2}{2\omega_1} a_1 a_2 \cos\gamma_1 + \frac{\varepsilon\delta_2\Lambda_1}{\omega_1} a_2 \cos\gamma_2$$

$$\dot{a}_2 = \frac{\varepsilon\delta_2}{4\omega_2} a_1^2 \sin\gamma_1 - \frac{\varepsilon\delta_2\Lambda_1}{\omega_2} a_1 \sin\gamma_2$$

$$a_2\dot{\beta}_2 = \frac{\varepsilon\delta_2}{4\omega_2} a_1^2 \cos\gamma_1 + \frac{\varepsilon\delta_2\Lambda_1}{\omega_2} a_1 \cos\gamma_2$$

$$\gamma_1 = \beta_2 - 2\beta_1 + \varepsilon\sigma_1 t, \qquad \gamma_2 = \varepsilon\sigma_2 t - \beta_1 - \beta_2, \qquad \Lambda_1 = \frac{F_1}{\omega_1^2 - \Omega^2}$$

CHAPTER 11

Parametric Excitations

As mentioned in Chapter 9, physical systems are often subject to excitations that appear as time-dependent coefficients in the governing equations and/or the boundary conditions. Such excitations are called parametric excitations.

Linear equations with periodic coefficients have been studied extensively as evidenced by the many available books devoted to the subject or that devote one or more chapters to the subject. The simplest equation of this type is the Mathieu equation

$$\ddot{u} + (\delta + 2\varepsilon\cos 2t)u = 0 \tag{1}$$

Using Floquet theory, one can show that these equations possess solutions of the form

$$u = e^{\gamma t}\phi(t) \tag{2}$$

where u is the dependent variable, γ is in general a complex constant whose value depends on δ and ε and is called the characteristic exponent, and $\phi(t)$ is a periodic function having the same period as the time-dependent coefficients in the equations. When the real part of γ is negative, the response decays; when the real part of γ is zero, the response is finite and bounded; and when the real part of γ is positive, the response grows. Thus the values of the parameters for which the real part of γ is zero divide the parameter space into regions of stability and instability (stability chart). In the case of the Mathieu equation, the stability chart is called the Strutt diagram and it divides the $\varepsilon\delta$-plane into regions of stability and instability. Including the damping in the analysis usually diminishes the region of instability, but sometimes it may change a stable state into an unstable one. The growth of the unstable solutions is exponential and hence unrealistic. Consequently, to determine the long-time behavior of the unstable solutions, one must include nonlinear terms.

In the linear case, one can use the method of strained parameters to determine the transition curves separating the stable and unstable solutions by expanding both the dependent variables as well as the parameters in terms of the excitation amplitude. For the Mathieu equation, one expands u and δ as

$$u(t;\varepsilon)=u_0(t)+\varepsilon u_1(t)+\varepsilon^2 u_2(t)+\cdots \tag{3a}$$

$$\delta=\delta_0+\varepsilon\delta_1+\varepsilon^2\delta_2+\cdots \tag{3b}$$

and determines the δ_n such that the u_n are free of secular terms. The resulting expansions are valid on the transition curves. To obtain expansions that are valid on or near the transition curves, one can use Whittaker's method and expand u, δ, and γ as

$$u(t;\varepsilon)=u_0(t)+\varepsilon u_1(t)+\varepsilon^2 u_2(t)+\cdots \tag{4a}$$

$$\delta=\delta_0+\varepsilon\delta_1+\varepsilon^2\delta_2+\cdots \tag{4b}$$

$$\gamma=\varepsilon\gamma_1+\varepsilon^2\gamma_2+\cdots \tag{4c}$$

and determine the γ_n given the δ_n by imposing the condition that the u_n are free of secular terms.

Although the method of strained parameters and Whittaker's technique are effective for linear equations, they cannot be applied to nonlinear equations, because Floquet theory does not apply to such equations. However, one can use the method of multiple scales, the method of averaging, or the Krylov–Bogoliubov technique to determine uniform expansions for linear and nonlinear equations with periodic as well as nonperiodic coefficients.

Solved Exercises

11.1. Consider the Mathieu equation (1). Determine a second-order uniform expansion by using the method of multiple scales when $\delta \approx 0$ and $\delta \approx 4$.

We seek a second-order expansion in the form

$$u=u_0(T_0,T_1,T_2)+\varepsilon u_1(T_0,T_1,T_2)+\varepsilon^2 u_2(T_0,T_1,T_2)+\cdots \tag{5}$$

At this point, one has a choice: either to expand δ in a power series or not. Since the latter approach was used with the method of multiple scales, we utilize the first approach and expand δ as in (4b). We should note also that one needs to treat the case $\delta \approx 0$ separately, if δ is not expanded. Substituting (5) and (4b) into (1) and equating coefficients of like powers ε, we obtain

$$D_0^2 u_0+\delta_0 u_0=0 \tag{6}$$

$$D_0^2 u_1+\delta_0 u_1=-\delta_1 u_0-2D_0D_1u_0-2u_0\cos 2T_0 \tag{7}$$

$$D_0^2 u_2+\delta_0 u_2=-\delta_2 u_0-\delta_1 u_1-2D_0D_2u_0-D_1^2u_0-2D_0D_1u_1-2u_1\cos 2T_0 \tag{8}$$

The Case $\delta \approx 0$. When $\delta \approx 0$, $\delta_0 = 0$ and the bounded solution of (6) can be expressed as

$$u_0 = A(T_1, T_2) \tag{9}$$

Then (7) becomes

$$D_0^2 u_1 = -\delta_1 A - 2A\cos 2T_0 \tag{10}$$

Eliminating secular terms from u_1 demands that $\delta_1 = 0$. Then, disregarding the homogeneous solution, we express the solution of (10) as

$$u_1 = \tfrac{1}{2}A\cos 2T_0 \tag{11}$$

Then (8) becomes

$$D_0^2 u_2 = -\delta_2 A - D_1^2 A + 2D_1 A\sin 2T_0 - A\cos^2 2T_0$$

or

$$D_0^2 u_2 = -\left(\delta_2 + \tfrac{1}{2}\right)A - D_1^2 A + 2D_1 A\sin 2T_0 - \tfrac{1}{2}A\cos 4T_0$$

Eliminating the secular terms from u_2 demands that

$$D_1^2 A + \left(\delta_2 + \tfrac{1}{2}\right)A = 0 \tag{12}$$

The solution of (12) is

$$A = a_1 e^{i\sqrt{\delta_2 + 1/2}\,T_1} + a_2 e^{-i\sqrt{\delta_2 + 1/2}\,T_1}$$

where a_1 and a_2 are functions of T_2. Therefore

$$u = \left[a_1 e^{i\varepsilon\sqrt{\delta_2 + 1/2}\,t} + a_2 e^{-i\varepsilon\sqrt{\delta_2 + 1/2}\,t}\right]\left(1 + \tfrac{1}{2}\varepsilon\cos 2t\right) + \cdots \tag{13}$$

Hence, u is bounded if $\delta_2 \geq -\frac{1}{2}$ and unbounded if $\delta_2 < -\frac{1}{2}$. Consequently, $\delta_2 = -\frac{1}{2}$ separates stable from unstable solutions and the transition curve emanating from $\delta \approx 0$ is

$$\delta = -\tfrac{1}{2}\varepsilon^2 + \cdots \tag{14}$$

The Case $\delta \approx 4$. In this case, $\delta_0 = 4$ and the solution of (6) can be expressed as

$$u_0 = A(T_1, T_2)e^{2iT_0} + \bar{A}(T_1, T_2)e^{-2iT_0} \tag{15}$$

Then (7) becomes

$$D_0^2 u_1 + 4u_1 = -\delta_1 A e^{2iT_0} - 4iD_1 A e^{2iT_0} + \text{cc}$$
$$-\left(e^{2iT_0} + e^{-2iT_0}\right)\left(Ae^{2iT_0} + \bar{A}e^{-2iT_0}\right)$$

or

$$D_0^2 u_1 + 4u_1 = -(\delta_1 A + 4iD_1 A)e^{2iT_0} - Ae^{4iT_0} - A + \text{cc} \tag{16}$$

Eliminating the secular terms from (16) demands that

$$4iD_1 A + \delta_1 A = 0 \quad \text{or} \quad D_1 A = \tfrac{1}{4} i\delta_1 A \tag{17}$$

Then the solution of (16) becomes

$$u_1 = -\tfrac{1}{4}A + \tfrac{1}{12}Ae^{4iT_0} + \text{cc} \tag{18}$$

The solution of (17) is

$$A = A_0(T_2)e^{i\delta_1 T_1/4} \tag{19}$$

Since δ_1 is real, the variation of A with T_1 is always oscillatory, and hence its value does not affect the stability. Consequently, without less of generality, we put $\delta_1 = 0$ so that $A = A(T_2)$. In other words, secular terms are eliminated from u_2 if

$$\delta_1 = 0 \quad \text{and} \quad D_1 A = 0 \tag{20}$$

Substituting (15) and (18) into (8) and using (20), we obtain

$$D_0^2 u_2 + 4u_2 = -\delta_2 Ae^{2iT_0} - 4iD_2 Ae^{2iT_0} + \text{cc} - (e^{2iT_0} + e^{-2iT_0})$$
$$\cdot\left(-\tfrac{1}{4}A + \tfrac{1}{12}Ae^{4iT_0} - \tfrac{1}{4}\bar{A} + \tfrac{1}{12}\bar{A}e^{-4iT_0}\right)$$

or

$$D_0^2 u_2 + 4u_2 = -(4iD_2 A + \delta_2 A)e^{2iT_0} + \tfrac{1}{4}(A + \bar{A})e^{2iT_0} - \tfrac{1}{12}Ae^{2iT_0} + \text{cc} + \text{NST} \tag{21}$$

Eliminating the secular terms from (21) demands that

$$4iD_2 A + (\delta_2 - \tfrac{1}{6})A - \tfrac{1}{4}\bar{A} = 0 \tag{22}$$

To solve (22), we put $A = B_r + iB_i$, separate real and imaginary parts, and obtain

$$\begin{aligned} &4B_r' + (\delta_2 + \tfrac{1}{12})B_i = 0 \\ &4B_i' - (\delta_2 - \tfrac{5}{12})B_r = 0 \end{aligned} \tag{23}$$

Eliminating B_i from (23) gives

$$B_r'' + \tfrac{1}{16}(\delta_2 + \tfrac{1}{12})(\delta_2 - \tfrac{5}{12})B_r = 0$$

Hence

$$B_r = a_1 e^{\sqrt{(1/12 + \delta_2)(5/12 - \delta_2)}\,T_2/4} + a_2 e^{-\sqrt{(1/12 + \delta_2)(5/12 - \delta_2)}\,T_2/4} \tag{24}$$

where a_1 and a_2 are constants. It follows from (23) that

$$B_i = \frac{-4B_r'}{\frac{1}{12}+\delta_2}$$

or

$$B_i = -\sqrt{\frac{\frac{5}{12}-\delta_2}{\frac{1}{12}+\delta_2}}\, a_1 e^{\sqrt{(1/12+\delta_2)(5/12-\delta_2)}\,T_2/4} + \sqrt{\frac{\frac{5}{12}-\delta_2}{\frac{1}{12}+\delta_2}}\, a_2 e^{-\sqrt{(1/12+\delta_2)(5/12-\delta_2)}\,T_2/4} \tag{25}$$

Therefore, to the second approximation,

$$u = (Ae^{2it}+\bar{A}e^{-2it}) + \varepsilon\left[-\tfrac{1}{4}(A+\bar{A}) + \tfrac{1}{12}(Ae^{4it}+\bar{A}e^{-4it})\right] + \cdots$$

$$= 2B_r\cos 2t - 2B_i\sin 2t + \varepsilon\left[-\tfrac{1}{2}B_r + \tfrac{1}{6}B_r\cos 4t - \tfrac{1}{6}B_i\sin 4t\right] + \cdots \tag{26}$$

It follows from (24)–(26) that u is bounded if

$$\left(\tfrac{1}{12}+\delta_2\right)\left(\tfrac{5}{12}-\delta_2\right) \le 0$$

otherwise, it is unbounded. Thus $\delta_2 = -\frac{1}{12}$ and $\delta_2 = \frac{5}{12}$ separate stable from unstable solutions. Consequently, the transition curves emanating from $\delta = 4$ are

$$\begin{aligned} \delta &= 4 - \tfrac{1}{12}\varepsilon^2 + \cdots \\ \delta &= 4 + \tfrac{5}{12}\varepsilon^2 + \cdots \end{aligned} \tag{27a}$$

If δ_1 were not zero, then the transition curves would be given by

$$\begin{aligned} \delta &= 4 + \varepsilon\delta_1 + \varepsilon^2\left(\frac{\delta_1^2}{16} - \tfrac{1}{12}\right) + \cdots \\ \delta &= 4 + \varepsilon\delta_1 + \varepsilon^2\left(\frac{\delta_1^2}{16} + \tfrac{5}{12}\right) + \cdots \end{aligned} \tag{27b}$$

11.2. Consider the Mathieu equation (1). Use Whittaker's method to determine second-order uniform expansions when $\delta \approx 0$ and $\delta \approx 4$.

According to Floquet theory, (1) possesses normal solutions of the form

$$u = e^{\gamma t}\phi(t) \tag{28}$$

where $\phi(t) = \phi(t+\pi)$. Putting (28) in (1) yields

$$\ddot{\phi} + 2\gamma\dot{\phi} + (\delta + \gamma^2 + 2\varepsilon\cos 2t)\phi = 0 \tag{29}$$

We seek an expansion of (29) in the form

$$\phi = \phi_0 + \varepsilon\phi_1 + \varepsilon^2\phi_2 + \cdots \tag{30}$$

$$\delta = \delta_0 + \varepsilon\delta_1 + \varepsilon^2\delta_2 + \cdots \tag{31}$$

$$\gamma = \varepsilon\gamma_1 + \varepsilon^2\gamma_2 + \cdots \tag{32}$$

Substituting (30)–(32) into (29) and equating coefficients of like powers of ε, we obtain

$$\ddot{\phi}_0 + \delta_0\phi_0 = 0 \tag{33}$$

$$\ddot{\phi}_1 + \delta_0\phi_1 = -2\gamma_1\dot{\phi}_0 - \delta_1\phi_0 - 2\phi_0\cos 2t \tag{34}$$

$$\ddot{\phi}_2 + \delta_0\phi_2 = -2\gamma_2\dot{\phi}_0 - 2\gamma_1\dot{\phi}_1 - \delta_1\phi_1 - \delta_2\phi_0 - \gamma_1^2\phi_0 - 2\phi_1\cos 2t \tag{35}$$

The Case $\delta \approx 0$. In this case, $\delta_0 = 0$ and the bounded solution of (33) can be expressed as

$$\phi_0 = a \tag{36}$$

where a is a constant. Then (34) becomes

$$\ddot{\phi}_1 = -\delta_1 a - 2a\cos 2t \tag{37}$$

Eliminating the secular terms from (37) demands that $\delta_1 = 0$ for a nontrivial solution. Then the solution of (37) is

$$\phi_1 = \tfrac{1}{2}a\cos 2t \tag{38}$$

Substituting (36) and (38) into (35) and recalling that $\delta_0 = \delta_1 = 0$, we have

$$\ddot{\phi}_2 = 2\gamma_1 a\sin 2t - (\delta_2 + \gamma_1^2)a - a\cos^2 2t$$

or

$$\ddot{\phi}_2 = 2\gamma_1 a\sin 2t - (\delta_2 + \gamma_1^2 + \tfrac{1}{2})a - \tfrac{1}{2}a\cos 4t \tag{39}$$

Eliminating the secular terms from ϕ_2 demands that

$$\delta_2 + \gamma_1^2 + \tfrac{1}{2} = 0 \quad \text{or} \quad \gamma_1 = \pm i\sqrt{\delta_2 + \tfrac{1}{2}} \tag{40}$$

Therefore

$$\phi = a + \tfrac{1}{2}\varepsilon a\cos 2t + \cdots$$

and

$$u = \left(a_1 e^{i\varepsilon t\sqrt{\delta_2 + 1/2}} + a_2 e^{-i\varepsilon t\sqrt{\delta_2 + 1/2}} \right)\left(1 + \tfrac{1}{2}\varepsilon \cos 2t\right) + \cdots \tag{41}$$

in agreement with (13), obtained using the method of multiple scales.

The Case $\delta \approx 4$. In this case, $\delta_0 = 4$ and the solution of (33) can be expressed as

$$\phi_0 = a\cos 2t + b\sin 2t \tag{42}$$

where a and b are constants. Then (34) becomes

$$\ddot{\phi}_1 + 4\phi_1 = (4\gamma_1 a - \delta_1 b)\sin 2t - (4\gamma_1 b + \delta_1 a)\cos 2t - a - a\cos 4t - b\sin 4t \tag{43}$$

Eliminating the secular terms from ϕ_1 demands that

$$\begin{aligned} 4\gamma_1 a - \delta_1 b &= 0 \\ \delta_1 a + 4\gamma_1 b &= 0 \end{aligned} \tag{44}$$

For a nontrivial solution, the determinant of the coefficient matrix in (44) must vanish. Thus

$$16\gamma_1^2 + \delta_1^2 = 0 \quad \text{or} \quad \gamma_1 = \pm\tfrac{1}{4}i\delta_1 \tag{45}$$

Since δ_1 is real, the variation of the solution with γ_1 is always bounded. Hence, without loss of generality, we put $\delta_1 = 0$ and $\gamma_1 = 0$. Then, the solution of (43) can be expressed as

$$\phi_1 = -\tfrac{1}{4}a + \tfrac{1}{12}a\cos 4t + \tfrac{1}{12}b\sin 4t \tag{46}$$

Substituting (42) and (46) into (35) and recalling that $\delta_1 = 0$ and $\gamma_1 = 0$, we have

$$\begin{aligned} \ddot{\phi}_2 + 4\phi_2 = {} & (4\gamma_2 a - \delta_2 b)\sin 2t - (4\gamma_2 b + \delta_2 a)\cos 2t \\ & - 2\cos 2t\left(-\tfrac{1}{4}a + \tfrac{1}{12}a\cos 4t + \tfrac{1}{12}b\sin 4t\right) \end{aligned}$$

or

$$\ddot{\phi}_2 + 4\phi_2 = \left(4\gamma_2 a - \delta_2 b - \tfrac{1}{12}b\right)\sin 2t - \left(4\gamma_2 b + \delta_2 a - \tfrac{5}{12}a\right)\cos 2t + \text{NST} \tag{47}$$

Eliminating the secular terms from ϕ_2 demands that

$$\begin{aligned} 4\gamma_2 a - \left(\delta_2 + \tfrac{1}{12}\right) b &= 0 \\ \left(\delta_2 - \tfrac{5}{12}\right) a + 4\gamma_2 b &= 0 \end{aligned} \tag{48}$$

For a nontrivial solution, the determinant of the coefficient matrix in (48) vanishes.

Thus

$$16\gamma_2^2 = \left(\tfrac{1}{12} + \delta_2\right)\left(\tfrac{5}{12} - \delta_2\right)$$

or

$$\gamma_2 = \pm\tfrac{1}{4}\sqrt{\left(\tfrac{1}{12} + \delta_2\right)\left(\tfrac{5}{12} - \delta_2\right)} \tag{49}$$

It follows from (48) that

$$b = \pm\sqrt{\frac{\frac{5}{12} - \delta_2}{\frac{1}{12} + \delta_2}}\; a \tag{50}$$

Therefore, the second approximation, u is given by (26), obtained using the method of multiple scales.

11.3. Consider the equation

$$\ddot{u} + \frac{\delta u}{1 + \varepsilon \cos 2t} = 0 \tag{51}$$

(a) Determine second-order expansions for the transition curves near $\delta = 0, 1$, and 4.

(b) Use Whittaker's technique to determine second-order expansions near these curves.

The Method of Strained Parameters. First, we expand (51) for small ε and obtain

$$\ddot{u} + \delta(1 - \varepsilon\cos 2t + \varepsilon^2\cos^2 2t + \cdots)u = 0 \tag{52}$$

Then we seek expansions in the form

$$u = u_0(t) + \varepsilon u_1(t) + \varepsilon^2 u_2(t) + \cdots \tag{53}$$

$$\delta = \delta_0 + \varepsilon\delta_1 + \varepsilon^2\delta_2 + \cdots \tag{54}$$

Substituting (53) and (54) into (52) and equating coefficients of like powers of ε, we obtain

$$\ddot{u}_0 + \delta_0 u_0 = 0 \tag{55}$$

$$\ddot{u}_1 + \delta_0 u_1 = -(\delta_1 - \delta_0\cos 2t)u_0 \tag{56}$$

$$\ddot{u}_2 + \delta_0 u_2 = -\left(\delta_2 + \delta_0\cos^2 2t - \delta_1\cos 2t\right)u_0 - (\delta_1 - \delta_0\cos 2t)u_1 \tag{57}$$

When $\delta \approx 0$, $\delta_0 = 0$ and the bounded solution of (55) can be written as

$$u_0 = a \tag{58}$$

where a is a constant. Then (56) becomes

$$\ddot{u}_1 = -\delta_1 a \tag{59}$$

Eliminating secular terms from (59) demands that $\delta_1 = 0$. Then the solution of (59) is $u_1 = 0$. Substituting u_0 and u_1 into (57), we have

$$\ddot{u}_2 = -\delta_2 a \tag{60}$$

Eliminating the secular terms from u_2 demands that $\delta_2 = 0$. Therefore, bounded solutions to (52) when $\delta \approx 0$ exist only for $\delta \equiv 0$.

When $\delta \approx 1$, $\delta_0 = 1$ and the solution of (55) can be expressed as

$$u_0 = a\cos t + b\sin t \tag{61}$$

where a and b are constants. Then (56) becomes

$$\ddot{u}_1 + u_1 = -\left(\delta_1 - \tfrac{1}{2}\right) a\cos t - \left(\delta_1 + \tfrac{1}{2}\right) b\sin t + \tfrac{1}{2} a\cos 3t + \tfrac{1}{2} b\sin 3t \tag{62}$$

Eliminating the secular terms from (62) demands that

$$\left(\delta_1 - \tfrac{1}{2}\right) a = 0 \quad \text{and} \quad \left(\delta_1 + \tfrac{1}{2}\right) b = 0 \tag{63}$$

Then the solution of (62) is

$$u_1 = -\tfrac{1}{16} a\cos 3t - \tfrac{1}{16} b\sin 3t \tag{64}$$

It follows from (63) that either

$$\delta_1 = \tfrac{1}{2} \quad \text{and} \quad b = 0 \tag{65}$$

or

$$\delta_1 = -\tfrac{1}{2} \quad \text{and } a = 0 \tag{66}$$

Substituting u_0 and u_1 into (57) and using (65), we have

$$\ddot{u}_2 + u_2 = -\delta_2 a\cos t - a\cos^2 2t\cos t + \tfrac{1}{2} a\cos t\cos 2t + \tfrac{1}{32} a\cos 3t - \tfrac{1}{16} a\cos 2t\cos 3t$$

or

$$\ddot{u}_2 + u_2 = -a\left(\delta_2 + \tfrac{9}{32}\right)\cos t + \text{NST} \tag{67}$$

Eliminating the terms that lead to secular terms leads to $\delta_2 = -\frac{9}{32}$. Hence one of the transition curves emanating from $\delta = 1$ is

$$\delta = 1 + \tfrac{1}{2}\varepsilon - \tfrac{9}{32}\varepsilon^2 + \cdots \tag{68}$$

Substituting u_0 and u_1 into (57) and using (66) we have

$$\ddot{u}_2 + u_2 = -\delta_2 b \sin t - b \sin t \cos^2 2t - \tfrac{1}{2} b \sin t \cos 2t - \tfrac{1}{32} b \sin 3t - \tfrac{1}{16} b \sin 3t \cos 2t$$

or

$$\ddot{u}_2 + u_2 = -b\left(\delta_2 + \tfrac{9}{32}\right) \sin t + \text{NST} \tag{69}$$

Eliminating the terms that lead to secular terms from (69) yields $\delta_2 = -\frac{9}{32}$. Hence the other transition curve emanating from $\delta = 1$ is

$$\delta = 1 - \tfrac{1}{2}\varepsilon - \tfrac{9}{32}\varepsilon^2 + \cdots \tag{70}$$

When $\delta \approx 4$, $\delta_0 = 4$ and the solution of (55) can be written as

$$u_0 = a \cos 2t + b \sin 2t \tag{71}$$

Then (56) becomes

$$\ddot{u}_1 + 4u_1 = -\delta_1 a \cos 2t - \delta_1 b \sin 2t + 2a + 2a \cos 4t + 2b \sin 4t \tag{72}$$

Eliminating the terms that lead to secular terms leads to $\delta_1 = 0$. Then the solution of (72) is

$$u_1 = \tfrac{1}{2}a - \tfrac{1}{6}a \cos 4t - \tfrac{1}{6} b \sin 4t \tag{73}$$

Substituting u_0 and u_1 into (57), we have

$$\ddot{u}_2 + 4u_2 = -\left(\delta_2 + \tfrac{4}{3}\right) a \cos 2t - \left(\delta_2 + \tfrac{4}{3}\right) b \sin 2t + \text{NST}$$

Eliminating the terms that lead to secular terms, we have $\delta_2 = -\frac{4}{3}$. Hence there is only one transition curve to this order. It is given by

$$\delta = 4 - \tfrac{4}{3}\varepsilon^2 + \cdots \tag{74}$$

Whittaker's Technique. According to Floquet theory, (52) possesses normal solutions of the form

$$u(t) = e^{\gamma t}\phi(t) \tag{75}$$

where $\phi(t) = \phi(t + \pi)$. Substituting (75) into (52) yields

$$\ddot{\phi} + 2\gamma\dot{\phi} + \left[\gamma^2 + \delta(1 - \varepsilon \cos 2t + \varepsilon^2 \cos^2 2t + \cdots)\right]\phi = 0 \tag{76}$$

We seek an approxmiate solution to (76) by expanding ϕ, δ, and γ as

$$\phi = \phi_0 + \varepsilon\phi_1 + \varepsilon^2\phi_2 + \cdots \tag{77}$$

$$\delta = \delta_0 + \varepsilon\delta_1 + \varepsilon^2\delta_2 + \cdots \tag{78}$$

$$\gamma = \varepsilon\gamma_1 + \varepsilon^2\gamma_2 + \cdots \tag{79}$$

Substituting (77)–(79) into (76) and equating coefficients of like powers of ε, we obtain

$$\ddot{\phi}_0 + \delta_0\phi_0 = 0 \tag{80}$$

$$\ddot{\phi}_1 + \delta_0\phi_1 = -2\gamma_1\dot{\phi}_0 - (\delta_1 - \delta_0\cos 2t)\phi_0 \tag{81}$$

$$\ddot{\phi}_2 + \delta_0\phi_2 = -2\gamma_2\dot{\phi}_0 - 2\gamma_1\dot{\phi}_1 - \gamma_1^2\phi_0 - \left(\delta_2 + \delta_0\cos^2 2t - \delta_1\cos 2t\right)\phi_0$$
$$- (\delta_1 - \delta_0\cos 2t)\phi_1 \tag{82}$$

When $\delta \approx 0$, $\delta_0 = 0$ and the bounded solution of (80) is

$$\phi_0 = a \tag{83}$$

where a is a constant. Then (81) becomes

$$\ddot{\phi}_1 = -\delta_1 a \tag{84}$$

In order that ϕ_1 be periodic, $\delta_1 = 0$ and hence $\phi_1 = 0$. Substituting ϕ_0, ϕ_1, and δ_1 into (82) yields

$$\ddot{\phi}_2 = -\gamma_1^2 a - \delta_2 a \tag{85}$$

In order that ϕ_2 be periodic, $\gamma_1^2 = -\delta_2$ and hence $\phi_2 = 0$. Therefore, to the first approximation

$$u = a_1 e^{\varepsilon\sqrt{-\delta_2}\,t} + a_2 e^{-\varepsilon\sqrt{-\delta_2}\,t} + O(\varepsilon^2) \tag{86}$$

Therefore, u is bounded or unbounded depending on whether δ_2 is positive or negative, hence the transition curve separating stability from instability and emanating from $\delta = 0$ is

$$\delta = 0 + O(\varepsilon^3) \tag{87}$$

When $\delta \approx 1$, $\delta_0 = 1$ and the solution of (80) can be written as

$$\phi_0 = a\cos t + b\sin t \tag{88}$$

Then (81) becomes

$$\ddot{\phi}_1 + \phi_1 = \left(2\gamma_1 a - \delta_1 b - \tfrac{1}{2}b\right)\sin t - \left(2\gamma_1 b + \delta_1 a - \tfrac{1}{2}a\right)\cos t$$
$$+ \tfrac{1}{2}a\cos 3t + \tfrac{1}{2}b\sin 3t \tag{89}$$

In order that ϕ_1 be periodic,

$$2\gamma_1 a - \left(\delta_1 + \tfrac{1}{2}\right)b = 0 \tag{90}$$

$$\left(\delta_1 - \tfrac{1}{2}\right)a + 2\gamma_1 b = 0 \tag{91}$$

Then the solution of (89) can be written as

$$\phi_1 = A\cos t + B\sin t - \tfrac{1}{16}a\cos 3t - \tfrac{1}{16}b\sin 3t \tag{92}$$

It turns out that the solution of the homogeneous equation needs to be included in this case to avoid inconsistent results at second order as shown below.

For a nontrivial solution, the determinant of the coefficient matrix in (90) and (91) must vanish; that is,

$$4\gamma_1^2 = \tfrac{1}{4} - \delta_1^2$$

or

$$\gamma_1 = \pm\tfrac{1}{2}\sqrt{\tfrac{1}{4} - \delta_1^2} \tag{93}$$

Then it follows from (91) that

$$b = \frac{\tfrac{1}{2} - \delta_1}{2\gamma_1}a \tag{94}$$

Substituting ϕ_0 and ϕ_1 into (82) yields

$$\ddot{\phi}_2 + \phi_2 = \left[2\gamma_1 A - \left(\delta_1 + \tfrac{1}{2}\right)B + 2\gamma_2 a - \left(\gamma_1^2 + \delta_2 + \tfrac{1}{2}\delta_1 + \tfrac{17}{32}\right)b\right]\sin t$$
$$- \left[2\gamma_1 B + \left(\delta_1 - \tfrac{1}{2}\right)A + 2\gamma_2 b + \left(\gamma_1^2 + \delta_2 - \tfrac{1}{2}\delta_1 + \tfrac{17}{32}\right)a\right]\cos t + \text{NST} \tag{95}$$

In order that ϕ_2 be periodic,

$$2\gamma_1 A - \left(\delta_1 + \tfrac{1}{2}\right)B = \left(\gamma_1^2 + \delta_2 + \tfrac{1}{2}\delta_1 + \tfrac{17}{32}\right)b - 2\gamma_2 a \tag{96}$$

$$\left(\delta_1 - \tfrac{1}{2}\right)A + 2\gamma_1 B = -2\gamma_2 b - \left(\gamma_1^2 + \delta_2 - \tfrac{1}{2}\delta_1 + \tfrac{17}{32}\right)a \tag{97}$$

Since the homogeneous system of equations (96) and (97) has a nontrivial solution, the inhomogeneous equations (96) and (97) have a solution if and only if their right-hand sides are orthogonal to every solution of the adjoint homogeneous problem; that is,

$$2\gamma_1\left[\left(\gamma_1^2 + \delta_2 + \tfrac{1}{2}\delta_1 + \tfrac{17}{32}\right)b - 2\gamma_2 a\right] = \left(\delta_1 + \tfrac{1}{2}\right)\left[2\gamma_2 b + \left(\gamma_1^2 + \delta_2 - \tfrac{1}{2}\delta_1 + \tfrac{17}{32}\right)a\right] \tag{98}$$

Using (93) and (94), we obtain from (98) that

$$\gamma_2 = \pm\tfrac{1}{4}\sqrt{\frac{\tfrac{1}{2} - \delta_1}{\tfrac{1}{2} + \delta_1}}\left[\delta_2 - \tfrac{1}{4}\delta_1^2 + \tfrac{1}{2}\delta_1 + \tfrac{19}{32}\right] \mp \tfrac{1}{4}\sqrt{\frac{\tfrac{1}{2} + \delta_1}{\tfrac{1}{2} - \delta_1}}\left[\delta_2 - \tfrac{1}{4}\delta_1^2 - \tfrac{1}{2}\delta_1 + \tfrac{19}{32}\right] \tag{99}$$

Since the transition curves correspond to $\gamma = 0$ or $\gamma_1 = \gamma_2 = 0$, it follows from (93) that the transition curves correspond to $\delta_1 = \tfrac{1}{2}$ or $-\tfrac{1}{2}$. When $\delta_1 = \tfrac{1}{2}$, it follows from

(99) that

$$\delta_2 = \tfrac{1}{4}\delta_1^2 + \tfrac{1}{2}\delta_1 - \tfrac{19}{32} = -\tfrac{9}{32}$$

otherwise γ_2 will be unbounded. When $\delta_1 = -\tfrac{1}{2}$, it follows from (99) that

$$\delta_2 = \tfrac{1}{4}\delta_1^2 - \tfrac{1}{2}\delta_1 - \tfrac{19}{32} = -\tfrac{9}{32}$$

otherwise γ_2 will be unbounded. Hence the transition curves emanating from $\delta = 1$ are given by

$$\delta = 1 \pm \tfrac{1}{2}\varepsilon - \tfrac{9}{32}\varepsilon^2 + \cdots \tag{100}$$

in agreement with (68) and (70), obtained using the method of strained parameters.

Had we not included the solution of the homogeneous equation (89) (i.e., $A = B = 0$) in (95), the condition that ϕ_2 is periodic would demand (96) and (97) with $A = B = 0$. These conditions yield two values for γ_2 since b/a is fixed in (94). These two values are not equal in general and hence we need to include the solution of the homogeneous equation (89).

When $\delta \approx 4$, $\delta_0 = 4$ and the solution of (80) can be written as

$$\phi_0 = a\cos 2t + b\sin 2t \tag{101}$$

Then (81) becomes

$$\ddot{\phi}_1 + 4\phi_1 = (4\gamma_1 a - \delta_1 b)\sin 2t - (4\gamma_1 b + \delta_1 a)\cos 2t + 2a + 2a\cos 4t + 2b\sin 4t \tag{102}$$

In order that ϕ_1 be periodic,

$$\begin{aligned} 4\gamma_1 a - \delta_1 b &= 0 \\ \delta_1 a + 4\gamma_1 b &= 0 \end{aligned} \tag{103}$$

For a nontrivial solution, the determinant of the coefficient matrix in (103) must vanish; that is,

$$\gamma_1 = \pm\tfrac{1}{4}i\delta_1 \tag{104}$$

Since δ_1 is real, the variation of the solution with γ_1 is always bounded. Hence, without loss of generality, we put $\delta_1 = 0$ and $\gamma_1 = 0$. Consequently, b/a is arbitrary at this level of approximation. Thus we do not need the solution of the homogeneous equation (102). Therefore

$$\phi_1 = \tfrac{1}{2}a - \tfrac{1}{6}a\cos 4t - \tfrac{1}{6}b\sin 4t \tag{105}$$

Substituting ϕ_0 and ϕ_1 into (82) and recalling the fact that $\delta_1 = \gamma_1 = 0$, we have

$$\ddot{\phi}_2 + 4\phi_2 = -\left[4\gamma_2 b + \left(\delta_2 + \tfrac{4}{3}\right)a\right]\cos 2t + \left[4\gamma_2 a - \left(\delta_2 + \tfrac{4}{3}\right)b\right]\sin 2t + \text{NST} \tag{106}$$

In order that ϕ_2 be periodic,

$$4\gamma_2 b + \left(\delta_2 + \tfrac{4}{3}\right) a = 0$$

$$\left(\delta_2 + \tfrac{4}{3}\right) b - 4\gamma_2 a = 0 \tag{107}$$

For a nontrivial solution, the determinant of the coefficient matrix in (107) must vanish; that is

$$16\gamma_2^2 + \left(\delta_2 + \tfrac{4}{3}\right)^2 = 0$$

or

$$\gamma_2 = \pm \tfrac{1}{4} i \left(\delta_2 + \tfrac{4}{3}\right) \tag{108}$$

Thus the solution is always bounded. Hence the result (74) of the method of strained parameters, that there is a transition curve separating stability from instability, is misleading.

11.4. Consider the equation

$$\ddot{u} + \frac{\delta - \varepsilon \cos^2 t}{1 - \varepsilon \cos^2 t} u = 0 \tag{109}$$

(a) Determine second-order expansions of the first three transition curves (i.e., near $\delta = 0$, 1, and 4).

(b) Use Whittaker's technique to determine u near these curves.

The Method of Strained Parameters. First, we expand (109) for small ε and obtain

$$\ddot{u} + \left[\delta + \varepsilon(\delta - 1)\cos^2 t + \varepsilon^2(\delta - 1)\cos^4 t + \cdots\right] u = 0 \tag{110}$$

Then we seek expansions in the form

$$u = u_0(t) + \varepsilon u_1(t) + \varepsilon^2 u_2(t) + \cdots \tag{111}$$

$$\delta = \delta_0 + \varepsilon\delta_1 + \varepsilon^2\delta_2 + \cdots \tag{112}$$

Substituting (111) and (112) into (110) and equating coefficients of like powers of ε, we obtain

$$\ddot{u}_0 + \delta_0 u_0 = 0 \tag{113}$$

$$\ddot{u}_1 + \delta_0 u_1 = -\delta_1 u_0 - (\delta_0 - 1) u_0 \cos^2 t \tag{114}$$

$$\ddot{u}_2 + \delta_0 u_2 = -\delta_2 u_0 - \delta_1 u_1 - (\delta_0 - 1) u_1 \cos^2 t - \delta_1 u_0 \cos^2 t - (\delta_0 - 1) u_0 \cos^4 t \tag{115}$$

When $\delta \approx 0$, $\delta_0 = 0$ and the bounded solution of (113) is

$$u_0 = a \tag{116}$$

where a is a constant. Then (114) becomes

$$\ddot{u}_1 = -\delta_1 a + a\cos^2 t = \left(\tfrac{1}{2} - \delta_1\right)a + \tfrac{1}{2}a\cos 2t \tag{117}$$

Eliminating the term that leads to a secular term from (117) demands that $\delta_1 = \frac{1}{2}$. Then the solution of (117) can be written as

$$u_1 = -\tfrac{1}{8}a\cos 2t \tag{118}$$

Substituting u_0 and u_1 into (115) yields

$$\ddot{u}_2 = -\left(\delta_2 - \tfrac{3}{32}\right)a + \text{NST} \tag{119}$$

Eliminating the term that leads to a secular term from (119) yields $\delta_2 = \frac{3}{32}$. Therefore, the transition curve that emanates from $\delta = 0$ is

$$\delta = \tfrac{1}{2}\varepsilon + \tfrac{3}{32}\varepsilon^2 + \cdots \tag{120}$$

When $\delta \approx 1$, $\delta_0 = 1$ and the solution of (113) can be written as

$$u_0 = a\cos t + b\sin t \tag{121}$$

Then (114) becomes

$$\ddot{u}_1 + u_1 = -\delta_1(a\cos t + b\sin t) \tag{122}$$

Eliminating the terms that lead to secular terms from (122) yields $\delta_1 = 0$. Then the solution of (122) is $u_1 = 0$.

Substituting u_0 and u_1 into (115) yields

$$\ddot{u}_2 + u_2 = -\delta_2(a\cos t + b\sin t) \tag{123}$$

Eliminating the terms that lead to secular terms from (123) yields $\delta_2 = 0$. Hence, to second order, the transition curve that separates stability from instability and emanating from $\delta = 1$ is

$$\delta = 1 + O(\varepsilon^3) \tag{124}$$

When $\delta \approx 4$, $\delta_0 = 4$ and the solution of (113) can be written as

$$u_0 = a\cos 2t + b\sin 2t \tag{125}$$

Then (114) becomes

$$\ddot{u}_1 + 4u_1 = -\left(\delta_1 + \tfrac{3}{2}\right)(a\cos 2t + b\sin 2t) - \tfrac{3}{4}(a + a\cos 4t + b\sin 4t) \tag{126}$$

Eliminating the terms that lead to secular terms from (126) yields $\delta_1 = -\frac{3}{2}$. Then the solution of (126) can be written as

$$u_1 = -\tfrac{3}{16}a + \tfrac{1}{16}a\cos 4t + \tfrac{1}{16}b\sin 4t \tag{127}$$

Substituting u_0 and u_1 into (115) yields

$$\ddot{u}_2 + 4u_2 = -\left(\delta_2 + \tfrac{21}{64}\right) a \cos 2t - \left(\delta_2 + \tfrac{15}{64}\right) b \sin 2t + \text{NST} \tag{128}$$

Eliminating the terms that lead to secular terms from (128), we have

$$\begin{aligned} \left(\delta_2 + \tfrac{21}{64}\right) a &= 0 \\ \left(\delta_2 + \tfrac{15}{64}\right) b &= 0 \end{aligned} \tag{129}$$

Hence either

$$\delta_2 = -\tfrac{21}{64} \quad \text{and} \quad b = 0 \tag{130}$$

or

$$\delta_2 = -\tfrac{15}{64} \quad \text{and} \quad a = 0 \tag{131}$$

Therefore, one of the transition curves emanating from $\delta = 4$ is

$$\delta = 4 - \tfrac{3}{2}\varepsilon - \tfrac{21}{64}\varepsilon^2 + \cdots \tag{132}$$

along which

$$u = a \cos 2t - \tfrac{1}{16}\varepsilon a(3 - \cos 4t) + \cdots \tag{133}$$

The other transition curve emanating from $\delta = 4$ is

$$\delta = 4 - \tfrac{3}{2}\varepsilon - \tfrac{15}{64}\varepsilon^2 + \cdots \tag{134}$$

along which

$$u = b \sin 2t + \tfrac{1}{16}\varepsilon b \sin 4t + \cdots \tag{135}$$

Whittaker's Method. According to Floquet theory, (110) possesses normal solutions of the form (75) where $\phi(t) = \phi(t + \pi)$. Substituting (75) into (110) yields

$$\ddot{\phi} + 2\gamma\dot{\phi} + \left[\gamma^2 + \delta + \varepsilon(\delta - 1)\cos^2 t + \varepsilon^2(\delta - 1)\cos^4 t + \cdots\right]\phi = 0 \tag{136}$$

Substituting (77)–(79) into (136) and equating coefficients of like powers of ε, we obtain

$$\ddot{\phi}_0 + \delta_0\phi_0 = 0 \tag{137}$$

$$\ddot{\phi}_1 + \delta_0\phi_1 = -2\gamma_1\dot{\phi}_0 - \delta_1\phi_0 - (\delta_0 - 1)\phi_0\cos^2 t \tag{138}$$

$$\begin{aligned} \ddot{\phi} + \delta_0\phi_2 = {} & -2\gamma_2\dot{\phi}_0 - 2\gamma_1\dot{\phi}_1 - \gamma_1^2\phi_0 - \delta_2\phi_0 - \delta_1\phi_1 - (\delta_0 - 1)\phi_1\cos^2 t \\ & - \delta_1\phi_0\cos^2 t - (\delta_0 - 1)\phi_0\cos^4 t \end{aligned} \tag{139}$$

When $\delta \approx 0$, $\delta_0 = 0$ and the bounded solution of (137) can be written as

$$\phi_0 = a \tag{140}$$

where a is a constant. Then (138) becomes

$$\ddot{\phi}_1 = -\delta_1 a + \tfrac{1}{2}a + \tfrac{1}{2}a\cos 2t \tag{141}$$

In order that ϕ_1 be periodic, $\delta_1 = \frac{1}{2}$. Then the solution of (141) can be written as

$$\phi_1 = -\tfrac{1}{8}a\cos 2t \tag{142}$$

Substituting ϕ_0 and ϕ_1 into (139) yields

$$\ddot{\phi}_2 = -\left(\gamma_1^2 + \delta_2 - \tfrac{3}{32}\right)a + \text{NST} \tag{143}$$

In order that ϕ_2 be periodic,

$$\gamma_1^2 = \tfrac{3}{32} - \delta_2$$

or

$$\gamma_1 = \pm\sqrt{\tfrac{3}{32} - \delta_2} \tag{144}$$

Hence the solution is bounded or unbounded depending on whether δ_2 is greater than or less than $\frac{3}{32}$. Thus $\delta_2 = \frac{3}{32}$ corresponds to the transition from stability to instability. Consequently, the transition curve emanating from $\delta = 0$ is given by (120), obtained using the method of strained parameters.

When $\delta \approx 1$, $\delta_0 = 1$ and the solution of (137) can be written as

$$\phi_0 = a\cos t + b\sin t \tag{145}$$

Then (138) becomes

$$\ddot{\phi}_1 + \phi_1 = (2\gamma_1 a - \delta_1 b)\sin t - (2\gamma_1 b + \delta_1 a)\cos t \tag{146}$$

In order that ϕ_1 be periodic,

$$\begin{aligned} 2\gamma_1 a - \delta_1 b &= 0 \\ \delta_1 a + 2\gamma_1 b &= 0 \end{aligned} \tag{147}$$

Then the solution of (146) is $\phi_1 = 0$.

For a nontrivial solution, the determinant of the coefficient matrix in (147) must vanish; that is,

$$4\gamma_1^2 + \delta_1^2 = 0$$

or

$$\gamma_1 = \pm\tfrac{1}{2}i\delta_1 \tag{148}$$

Since δ_1 is real, the variation of the solution with γ_1 is always bounded. Hence, without loss of generality, we put $\delta_1 = 0$ and $\gamma_1 = 0$.

Substituting ϕ_0 and ϕ_1 into (139), we have

$$\ddot{\phi}_2 + \phi_2 = (2\gamma_2 a - \delta_2 b)\sin t - (2\gamma_2 b + \delta_2 a)\cos t + \text{NST} \tag{149}$$

In order that ϕ_2 be periodic,

$$\begin{aligned} 2\gamma_2 a - \delta_2 b &= 0 \\ \delta_2 a + 2\gamma_2 b &= 0 \end{aligned} \tag{150}$$

For a nontrivial solution, the determinant of the coefficient matrix must vanish; that is,

$$4\gamma_2^2 + \delta_2^2 = 0$$

or

$$\gamma_2 = \pm\tfrac{1}{2}i\delta_2 \tag{151}$$

Since δ_2 is real, the variation of the solution with γ_2 is always bounded. Thus the transition curve separating stability from instability and emanating from $\delta = 1$ is

$$\delta = 1 + O(\varepsilon^3) \tag{152}$$

in agreement with (124), obtained using the method of strained parameters.

When $\delta \approx 4$, $\delta_0 = 4$ and the solution of (137) can be written as

$$\phi_0 = a\cos 2t + b\sin 2t \tag{153}$$

Then (138) becomes

$$\begin{aligned} \ddot{\phi}_1 + 4\phi_1 = {} & (4\gamma_1 a - \delta_1 b - \tfrac{3}{2}b)\sin 2t - (4\gamma_1 b + \delta_1 a + \tfrac{3}{2}a)\cos 2t - \tfrac{3}{4}a \\ & - \tfrac{3}{4}a\cos 4t - \tfrac{3}{4}b\sin 4t \end{aligned} \tag{154}$$

In order that ϕ_1 be periodic,

$$\begin{aligned} 4\gamma_1 a - (\delta_1 + \tfrac{3}{2})b &= 0 \\ (\delta_1 + \tfrac{3}{2})a + 4\gamma_1 b &= 0 \end{aligned} \tag{155}$$

Then the solution of (154) can be written as

$$\phi_1 = -\tfrac{3}{16}a + \tfrac{1}{16}a\cos 4t + \tfrac{1}{16}b\sin 4t \tag{156}$$

For a nontrivial solution, the determinant of the coefficient matrix in (155) must vanish; that is,

$$16\gamma_1^2 + \left(\delta_1 + \tfrac{3}{2}\right)^2 = 0$$

or

$$\gamma_1 = \pm\tfrac{1}{4}i\left(\delta_1 + \tfrac{3}{2}\right) \tag{157}$$

Since δ_1 is real, the variation of the solution with γ_1 is always bounded. Hence, without loss of generality, we put $\delta_1 = -\tfrac{3}{2}$ and $\gamma_1 = 0$.

Substituting ϕ_1 and ϕ_2 into (139) and recalling that $\delta_1 = -\tfrac{3}{2}$ and $\gamma_1 = 0$, we obtain

$$\ddot{\phi}_2 + 4\phi_2 = -\left[\left(\delta_2 + \tfrac{21}{64}\right)a + 4\gamma_2 b\right]\cos 2t + \left[4\gamma_2 a - \left(\delta_2 + \tfrac{15}{64}\right)b\right]\sin 2t + \text{NST} \tag{158}$$

In order that ϕ_2 be periodic,

$$\begin{aligned} \left(\delta_2 + \tfrac{21}{64}\right)a + 4\gamma_2 b &= 0 \\ 4\gamma_2 a - \left(\delta_2 + \tfrac{15}{64}\right)b &= 0 \end{aligned} \tag{159}$$

For a nontrivial solution, the determinant of the coefficient matrix in (159) must vanish; that is,

$$16\gamma_2^2 = -\left(\delta_2 + \tfrac{21}{64}\right)\left(\delta_2 + \tfrac{15}{64}\right)$$

or

$$\gamma_2 = \pm\tfrac{1}{4}\left[-\left(\delta_2 + \tfrac{21}{64}\right)\left(\delta_2 + \tfrac{15}{64}\right)\right]^{1/2} \tag{160}$$

Therefore, γ_2 is real and the solution is unbounded if $-\tfrac{21}{64} < \delta_2 < -\tfrac{15}{64}$, and γ_2 is imaginary and the solution is bounded if $\delta_2 < -\tfrac{21}{64}$ or $\delta_2 > -\tfrac{15}{64}$. The transition curves separating stable from unstable solutions correspond to $\gamma_2 = 0$ or $\delta_2 = -\tfrac{21}{64}$ or $-\tfrac{15}{64}$. These values of δ_2 lead to the transition curves given by (132) and (134), obtained using the method of strained parameters.

11.5. Consider the equation

$$\ddot{u} + \left(\delta + \varepsilon\cos^3 t\right) = 0 \tag{161}$$

Determine second-order expansions for the first three transition curves using both the method of strained parameters and Whittaker's technique.

The Method of Strained Parameters. According to Floquet theory, (161) possesses normal solutions of the form (75) where $\phi(t) = \phi(t + 2\pi)$. Hence the transition curves separating stable from unstable solutions correspond to $\gamma = 0$ or $\gamma = i\pi$ along which u is periodic with the period π or 2π. Therefore, the transition curves emanate from

$\delta = \frac{1}{4}n^2$, where $n = 0, 1, 2, \ldots$. Alternatively, one can determine the values of δ from which the transition curves emanate by determining a straightforward expansion and finding the locations of the small-divisor terms.

To determine second-order expansions for the transition curves using the method of strained coordinates, we utilize the results of the Floquet theory and seek a second-order uniform expansion in the form

$$u(t;\varepsilon) = u_0(t) + \varepsilon u_1(t) + \varepsilon^2 u_2(t) + \cdots$$
$$\delta = \tfrac{1}{4}n^2 + \varepsilon\delta_1 + \varepsilon^2\delta_2 + \cdots \tag{162}$$

Substituting (162) into (161) and equating coefficients of like powers of ε, we obtain

$$\ddot{u} + \tfrac{1}{4}n^2 u_0 = 0 \tag{163}$$

$$\ddot{u}_1 + \tfrac{1}{4}n^2 u_1 = -\delta_1 u_0 - \tfrac{1}{4}(3\cos t + \cos 3t)\, u_0 \tag{164}$$

$$\ddot{u}_2 + \tfrac{1}{4}n^2 u_2 = -\delta_2 u_0 - \delta_1 u_1 - \tfrac{1}{4}(3\cos t + \cos 3t)\, u_1 \tag{165}$$

When $n = 0$, the bounded solution of (163) can be written as

$$u_0 = a$$

where a is a constant. Then (164) becomes

$$\ddot{u}_1 = -\tfrac{1}{4}(3\cos t + \cos 3t)\, a - \delta_1 a \tag{166}$$

Eliminating the terms that lead to secular terms in u_1 demands that $\delta_1 = 0$. Then the solution of (166) becomes

$$u_1 = \tfrac{3}{4}a\cos t + \tfrac{1}{36}a\cos 3t \tag{167}$$

Substituting u_0 and u_1 into (165) yields

$$\ddot{u}_2 = -\delta_2 a - \tfrac{1}{4}a(3\cos t + \cos 3t)\left(\tfrac{3}{4}\cos t + \tfrac{1}{36}\cos 3t\right) = -\left(\delta_2 + \tfrac{41}{144}\right)a + \text{NST} \tag{168}$$

Eliminating the terms that lead to secular terms from (168) demands that $\delta_2 = -\frac{41}{144}$. Hence the transition curve emanating from $\delta = 0$ is

$$\delta = -\tfrac{41}{144}\varepsilon^2 + \cdots \tag{169}$$

When $n = 2$, the solution of (163) can be written as

$$u_0 = a\cos t + b\sin t \tag{170}$$

where a and b are constants. Then (164) becomes

$$\ddot{u}_1 + u_1 = -\delta_1(a\cos t + b\sin t) - \tfrac{1}{4}(3\cos t + \cos 3t)(a\cos t + b\sin t)$$

or

$$\ddot{u}_1 + u_1 = -\delta_1(a\cos t + b\sin t) - \tfrac{3}{8}a - \tfrac{1}{2}a\cos 2t - \tfrac{1}{8}a\cos 4t$$
$$-\tfrac{1}{4}b\sin 2t - \tfrac{1}{8}b\sin 4t \tag{171}$$

Eliminating the terms that lead to secular terms from (171) demands that $\delta_1 = 0$. Then the solution of (171) can be written as

$$u_1 = -\tfrac{3}{8}a + \tfrac{1}{6}a\cos 2t + \tfrac{1}{120}a\cos 4t + \tfrac{1}{12}b\sin 2t + \tfrac{1}{120}b\sin 4t \tag{172}$$

Substituting u_0 and u_1 into (165) and recalling that $\delta_1 = 0$, we obtain

$$\ddot{u}_2 + u_2 = -\delta_2(a\cos t + b\sin t) - \tfrac{1}{4}(3\cos t + \cos 3t)$$
$$\times\left[-\tfrac{3}{8}a + \tfrac{1}{6}a\cos 2t + \tfrac{1}{120}a\cos 4t + \tfrac{1}{12}b\sin 2t + \tfrac{1}{120}b\sin 4t\right]$$

or

$$\ddot{u}_2 + u_2 = -\left(\delta_2 - \tfrac{63}{320}\right)a\cos t - \left(\delta_2 + \tfrac{7}{320}\right)b\sin t + \text{NST} \tag{173}$$

Eliminating the terms that lead to secular terms from (173) yields

$$\delta_2 = \tfrac{63}{320} \quad \text{and} \quad b = 0 \quad \text{or} \quad \delta_2 = -\tfrac{7}{320} \quad \text{and} \quad a = 0 \tag{174}$$

Hence one of the transition curves emanating from $\delta = 1$ is

$$\delta = 1 + \tfrac{63}{320}\varepsilon^2 + \cdots \tag{175}$$

along which

$$u = a\cos t + \varepsilon a\left(-\tfrac{3}{8} + \tfrac{1}{6}\cos 2t + \tfrac{1}{120}\cos 4t\right) + \cdots \tag{176}$$

The second transition curve emanating from $\delta = 1$ is

$$\delta = 1 - \tfrac{7}{320}\varepsilon^2 + \cdots \tag{177}$$

along which

$$u = b\sin t + \tfrac{1}{12}\varepsilon b\left(\sin 2t + \tfrac{1}{10}\sin 4t\right) + \cdots \tag{178}$$

When $n = 1$, the solution of (163) can be written as

$$u_0 = a\cos\tfrac{1}{2}t + b\sin\tfrac{1}{2}t \tag{179}$$

where a and b are constants. Then (164) becomes

$$\ddot{u}_1 + \tfrac{1}{4}u_1 = -\delta_1\left(a\cos\tfrac{1}{2}t + b\sin\tfrac{1}{2}t\right) - \tfrac{1}{4}(3\cos t + \cos 3t)$$
$$\times\left(a\cos\tfrac{1}{2}t + b\sin\tfrac{1}{2}t\right)$$

or

$$\ddot{u}_1 + \tfrac{1}{4}u_1 = -\left(\delta_1 + \tfrac{3}{8}\right) a \cos \tfrac{1}{2}t - \left(\delta_1 - \tfrac{3}{8}\right) b \sin \tfrac{1}{2}t - \tfrac{3}{8} a \cos \tfrac{3}{2}t$$

$$- \tfrac{1}{8} a \cos \tfrac{5}{2}t - \tfrac{1}{8} a \cos \tfrac{7}{2}t - \tfrac{3}{8} b \sin \tfrac{3}{2}t + \tfrac{1}{8} b \sin \tfrac{5}{2}t - \tfrac{1}{8} b \sin \tfrac{7}{2}t \tag{180}$$

Eliminating the terms that lead to secular terms from (180) demands that either

$$\delta_1 = -\tfrac{3}{8} \quad \text{and} \quad b = 0 \tag{181}$$

or

$$\delta_1 = \tfrac{3}{8} \quad \text{and} \quad a = 0 \tag{182}$$

Using (181), we write the solution of (180) as

$$u_1 = \tfrac{3}{16} a \cos \tfrac{3}{2}t + \tfrac{1}{48} a \cos \tfrac{5}{2}t + \tfrac{1}{96} a \cos \tfrac{7}{2}t \tag{183}$$

Using (182), we write the solution of (180) as

$$u_1 = \tfrac{3}{16} b \sin \tfrac{3}{2}t - \tfrac{1}{48} b \sin \tfrac{5}{2}t + \tfrac{1}{96} b \sin \tfrac{7}{2}t \tag{184}$$

Substituting (181) and (183) into (165) yields

$$\ddot{u}_2 + \tfrac{1}{4}u_2 = -\delta_2 a \cos \tfrac{1}{2}t + \tfrac{3}{8} a\left(\tfrac{3}{16} \cos \tfrac{3}{2}t + \tfrac{1}{48} \cos \tfrac{5}{2}t + \tfrac{1}{96} \cos \tfrac{7}{2}t\right)$$

$$- \tfrac{1}{4} a (3 \cos t + \cos 3t)\left(\tfrac{3}{16} \cos \tfrac{3}{2}t + \tfrac{1}{48} \cos \tfrac{5}{2}t + \tfrac{1}{96} \cos \tfrac{7}{2}t\right)$$

or

$$\ddot{u}_2 + \tfrac{1}{4}u_2 = -\left(\delta_2 + \tfrac{19}{256}\right) a \cos \tfrac{1}{2}t + \text{NST} \tag{185}$$

Eliminating the terms that lead to secular terms from (185) demands that $\delta_2 = -\frac{19}{256}$. Hence one of the transition curves emanating from $\delta = \frac{1}{4}$ is

$$\delta = \tfrac{1}{4} - \tfrac{3}{8}\varepsilon - \tfrac{19}{256}\varepsilon^2 + \cdots \tag{186}$$

along which

$$u = a \cos \tfrac{1}{2}t + \tfrac{1}{96} a\varepsilon\left(18 \cos \tfrac{3}{2}t + 2 \cos \tfrac{5}{2}t + \cos \tfrac{7}{2}t\right) + \cdots \tag{187}$$

Substituting (182) and (184) into (165) yields

$$\ddot{u}_2 + \tfrac{1}{4}u_2 = -\left(\delta_2 + \tfrac{19}{256}\right) b \sin \tfrac{1}{2}t + \text{NST} \tag{188}$$

Eliminating the terms that lead to secular terms from (188) demands that $\delta_2 = -\frac{19}{256}$. Hence the other transition curve that emanates from $\delta = \frac{1}{4}$ is

$$\delta = \tfrac{1}{4} + \tfrac{3}{8}\varepsilon - \tfrac{19}{256}\varepsilon^2 + \cdots \tag{189}$$

along which

$$u = b\sin\tfrac{1}{2}t + \tfrac{1}{96}b\varepsilon\left(18\sin\tfrac{3}{2}t - 2\sin\tfrac{5}{2}t + \sin\tfrac{7}{2}t\right) + \cdots \tag{190}$$

Whittaker's Technique. Substituting the normal form (75) into (161), we have

$$\ddot{\phi} + 2\gamma\dot{\phi} + \left(\delta + \gamma^2 + \varepsilon\cos^3 t\right)\phi = 0 \tag{191}$$

Substituting (77)–(79) into (191) and equating coefficients of like powers of ε, we obtain

$$\ddot{\phi}_0 + \delta_0\phi_0 = 0 \tag{192}$$

$$\ddot{\phi}_1 + \delta_0\phi_1 = -2\gamma_1\dot{\phi}_0 - \delta_1\phi_0 - \tfrac{1}{4}(3\cos t + \cos 3t)\phi_0 \tag{193}$$

$$\ddot{\phi}_2 + \delta_0\phi_2 = -2\gamma_2\dot{\phi}_0 - 2\gamma_1\dot{\phi}_1 - \gamma_1^2\phi_0 - \delta_2\phi_0 - \delta_1\phi_1 - \tfrac{1}{4}(3\cos t + \cos 3t)\phi_1 \tag{194}$$

When $\delta_0 = 0$, the bounded solution of (192) can be written as

$$\phi_0 = a \tag{195}$$

where a is a constant. Then (193) becomes

$$\ddot{\phi}_1 = -\delta_1 a - \tfrac{1}{4}a(3\cos t + \cos 3t) \tag{196}$$

In order that ϕ_1 be periodic, $\delta_1 = 0$. Then the solution of (196) can be written as

$$\phi_1 = \tfrac{3}{4}a\cos t + \tfrac{1}{36}a\cos 3t \tag{197}$$

Substituting ϕ_0 and ϕ_1 into (194) yields

$$\ddot{\phi}_2 = -\left(\gamma_1^2 + \delta_2 + \tfrac{41}{144}\right)a + \text{NST} \tag{198}$$

In order that ϕ_2 be periodic,

$$\gamma_1^2 + \delta_2 + \tfrac{41}{144} = 0$$

or

$$\gamma_1 = \pm i\sqrt{\delta_2 + \tfrac{41}{144}} \tag{199}$$

It follows from (75), (79), and (199) that u is bounded if $\delta_2 \geq -\frac{41}{144}$ and unbounded if $\delta_2 < -\frac{41}{144}$. Hence $\delta_2 = -\frac{41}{144}$ corresponds to the transition between stable and unstable motions and the transition curve separating stability from instability that emanates from $\delta = 0$ is

$$\delta = -\tfrac{41}{144}\varepsilon^2 + \cdots \tag{200}$$

When $\delta_0 = \frac{1}{4}$, the solution of (192) can be written as

$$\phi_0 = a\cos\tfrac{1}{2}t + b\sin\tfrac{1}{2}t \tag{201}$$

where a and b are constants. Then (193) becomes

$$\begin{aligned}\ddot{\phi}_1 + \tfrac{1}{4}\phi_1 = &-\left[\gamma_1 b + \left(\delta_1 + \tfrac{3}{8}\right)a\right]\cos\tfrac{1}{2}t + \left[\gamma_1 a - \left(\delta_1 - \tfrac{3}{8}\right)b\right]\sin\tfrac{1}{2}t \\ &-\tfrac{3}{8}a\cos\tfrac{3}{2}t - \tfrac{1}{8}a\cos\tfrac{5}{2}t - \tfrac{1}{8}a\cos\tfrac{7}{2}t - \tfrac{3}{8}b\sin\tfrac{3}{2}t \\ &+\tfrac{1}{8}b\sin\tfrac{5}{2}t - \tfrac{1}{8}b\sin\tfrac{7}{2}t\end{aligned} \tag{202}$$

In order that ϕ_1 be periodic,

$$\begin{aligned}\gamma_1 a - \left(\delta_1 - \tfrac{3}{8}\right)b &= 0 \\ \left(\delta_1 + \tfrac{3}{8}\right)a + \gamma_1 b &= 0\end{aligned} \tag{203}$$

Then the solution of (202) can be written as

$$\begin{aligned}\phi_1 = &A\cos\tfrac{1}{2}t + B\sin\tfrac{1}{2}t + \tfrac{3}{16}a\cos\tfrac{3}{2}t + \tfrac{1}{48}a\cos\tfrac{5}{2}t \\ &+\tfrac{1}{96}a\cos\tfrac{7}{2}t + \tfrac{3}{16}b\sin\tfrac{3}{2}t - \tfrac{1}{48}b\sin\tfrac{5}{2}t + \tfrac{1}{96}b\sin\tfrac{7}{2}t\end{aligned} \tag{204}$$

As discussed before, the homogeneous solution is included because a and b are related and the solvability condition at the next order would otherwise lead to a contradictory result. For a nontrivial solution, the determinant of the coefficient matrix in (203) must vanish; that is,

$$\gamma_1^2 = \tfrac{9}{64} - \delta_1^2$$

or

$$\gamma_1 = \pm\sqrt{\tfrac{9}{64} - \delta_1^2} \tag{205}$$

Then it follows from (203) that

$$\begin{aligned} b &= -\frac{\gamma_1}{3/8 - \delta_1}a = \mp\sqrt{\frac{3/8+\delta_1}{3/8-\delta_1}}\,a \quad \text{if } \delta_1 \neq \tfrac{3}{8} \\ a &= 0 \quad \text{if } \delta_1 = \tfrac{3}{8}\end{aligned} \tag{206}$$

Substituting ϕ_0 and ϕ_1 into (194) yields

$$\begin{aligned}\ddot{\phi}_2 + \tfrac{1}{4}\phi_2 = &-\left[\gamma_1 B + \left(\delta_1 + \tfrac{3}{8}\right)A + \left(\delta_2 + \tfrac{19}{256} + \gamma_1^2\right)a + \gamma_2 b\right]\cos\tfrac{1}{2}t \\ &+\left[\gamma_1 A + \left(\tfrac{3}{8} - \delta_1\right)B + \gamma_2 a - \left(\delta_2 + \tfrac{19}{256} + \gamma_1^2\right)b\right]\sin\tfrac{1}{2}t + \text{NST}\end{aligned} \tag{207}$$

In order that ϕ_2 be periodic,

$$\begin{aligned}\gamma_1 A + \left(\tfrac{3}{8} - \delta_1\right) B &= -\gamma_2 a + \left(\delta_2 + \gamma_1^2 + \tfrac{19}{256}\right) b \\ \left(\tfrac{3}{8} + \delta_1\right) A + \gamma_1 B &= -\left(\delta_2 + \gamma_1^2 + \tfrac{19}{256}\right) a - \gamma_2 b\end{aligned} \tag{208}$$

Since the homogeneous part of (208) has a nontrivial solution because of (205), the inhomogeneous equations have a solution if, and only if,

$$\gamma_1\left[-\gamma_2 a + \left(\delta_2 + \gamma_1^2 + \tfrac{19}{256}\right) b\right] = -\left(\tfrac{3}{8} - \delta_1\right)\left[\left(\delta_2 + \gamma_1^2 + \tfrac{19}{256}\right) a + \gamma_2 b\right] \tag{209}$$

Then, it follows from (205), (206), and (209) that

$$\gamma_2 = \pm \frac{1}{\sqrt{9/64 - \delta_1^2}} \delta_1\left(\delta_2 - \delta_1^2 + \tfrac{55}{256}\right) \quad \text{if } \delta_1 \neq \pm\tfrac{3}{8}$$

$$\gamma_2 = 0 \quad \text{and} \quad \delta_2 = -\tfrac{19}{256} \quad \text{if } \delta_1 = \pm\tfrac{3}{8} \tag{210}$$

Since the transition curves correspond to $\gamma = 0$, they correspond to $\delta_1 = \pm\frac{3}{8}$ and $\delta_2 = -\frac{19}{256}$. Therefore, the transition curves emanating from $\delta = \frac{1}{4}$ are

$$\delta = \tfrac{1}{4} \pm \tfrac{3}{8}\varepsilon - \tfrac{19}{256}\varepsilon^2 + \cdots \tag{211}$$

in agreement with (186) and (189), obtained using the method of strained parameters.

When $\delta_0 = 1$, the solution of (192) can be written as

$$\phi_0 = a\cos t + b\sin t \tag{212}$$

where a and b are constants. Then (193) becomes

$$\begin{aligned}\ddot{\phi}_1 + \phi_1 = &-(2\gamma_1 b + \delta_1 a)\cos t + (2\gamma_1 a - \delta_1 b)\sin t - \tfrac{3}{8}a - \tfrac{1}{2}a\cos 2t \\ &- \tfrac{1}{8}a\cos 4t - \tfrac{1}{4}b\sin 2t - \tfrac{1}{8}b\sin 4t\end{aligned} \tag{213}$$

In order that ϕ_1 be periodic,

$$\begin{aligned}2\gamma_1 a - \delta_1 b &= 0 \\ \delta_1 a + 2\gamma_1 b &= 0\end{aligned} \tag{214}$$

Then the solution of (213) can be written as

$$\phi_1 = -\tfrac{3}{8}a + \tfrac{1}{6}a\cos 2t + \tfrac{1}{120}a\cos 4t + \tfrac{1}{12}b\sin 2t + \tfrac{1}{120}b\sin 4t \tag{215}$$

For a nontrivial solution, the determinant of the coefficient matrix in (214) must vanish; that is,

$$\gamma_1^2 = -\tfrac{1}{4}\delta_1^2$$

or

$$\gamma_1 = \pm \tfrac{1}{2} i \delta_1$$

Since δ_1 is real, the variation of the solution with γ_1 is always bounded. Hence, without loss of generality, we put $\delta_1 = 0$ and $\gamma_1 = 0$. Substituting ϕ_0 and ϕ_1 into (194) yields

$$\ddot{\phi}_2 + \phi_2 = -\left[\left(\delta_2 - \tfrac{63}{320}\right) a + 2\gamma_2 b\right] \cos t + \left[2\gamma_2 a - \left(\delta_2 + \tfrac{7}{320}\right) b\right] \sin t + \text{NST} \tag{216}$$

For a periodic solution,

$$\begin{aligned} \left(\delta_2 - \tfrac{63}{320}\right) a + 2\gamma_2 b &= 0 \\ 2\gamma_2 a - \left(\delta_2 + \tfrac{7}{320}\right) b &= 0 \end{aligned} \tag{217}$$

For a nontrivial solution, the determinant of the coefficient matrix in (217) must vanish; that is,

$$4\gamma_2^2 = \left(\tfrac{7}{320} + \delta_2\right)\left(\tfrac{63}{320} - \delta_2\right)$$

or

$$\gamma_2 = \pm \tfrac{1}{2} \sqrt{\left(\tfrac{7}{320} + \delta_2\right)\left(\tfrac{63}{320} - \delta_2\right)} \tag{218}$$

It follows from (75), (79), and (218) that the solution is unbounded if $\frac{-7}{320} < \delta_2 < \frac{63}{320}$; otherwise, it is bounded. Therefore, the transition from bounded to unbounded solutions corresponds to $\delta_2 = -\frac{7}{320}$ or $\delta_2 = \frac{63}{320}$ and the transition curves emanating from $\delta = 1$ are

$$\delta = 1 - \tfrac{7}{320} \varepsilon^2 + \cdots \tag{219a}$$

$$\delta = 1 + \tfrac{63}{320} \varepsilon^2 + \cdots \tag{219b}$$

in agreement with (175) and (177), obtained using the method of strained parameters.

11.6. Consider the equation

$$\ddot{u} + \left(\delta + \varepsilon \cos 2t - \tfrac{1}{2} \varepsilon^2 \alpha \sin 2t + \tfrac{1}{8} \varepsilon^2 \cos 4t\right) u = 0 \tag{220}$$

Determine three terms for the transition curves when $\delta \approx 1$ and $\delta \approx 4$.

We use the method of strained parameters and let

$$\begin{aligned} u &= u_0(t) + \varepsilon u_1(t) + \varepsilon^2 u_2(t) + \cdots \\ \delta &= \delta_0 + \varepsilon \delta_1 + \varepsilon^2 \delta_2 + \cdots \end{aligned} \tag{221}$$

Substituting (221) into (220) and equating coefficients of like powers of ε, we obtain

$$\ddot{u}_0 + \delta_0 u_0 = 0 \tag{222}$$

$$\ddot{u}_1 + \delta_0 u_1 = -(\delta_1 + \cos 2t) u_0 \tag{223}$$

$$\ddot{u}_2 + \delta_0 u_2 = -\left(\delta_2 - \tfrac{1}{2}\alpha \sin 2t + \tfrac{1}{8}\cos 4t\right) u_0 - (\delta_1 + \cos 2t) u_1 \tag{224}$$

When $\delta_0 = 1$, the solution of (222) can be written as

$$u_0 = a \cos t + b \sin t \tag{225}$$

where a and b are constants. Then (223) becomes

$$\ddot{u}_1 + u_1 = -\left(\delta_1 + \tfrac{1}{2}\right) a \cos t - \left(\delta_1 - \tfrac{1}{2}\right) b \sin t - \tfrac{1}{2} a \cos 3t - \tfrac{1}{2} b \sin 3t \tag{226}$$

Eliminating the terms that lead to secular terms from (226) demands that

$$\left(\delta_1 + \tfrac{1}{2}\right) a = 0 \quad \text{and} \quad \left(\delta_1 - \tfrac{1}{2}\right) b = 0$$

Hence either

$$\delta_1 = -\tfrac{1}{2} \quad \text{and} \quad b = 0 \tag{227}$$

or

$$\delta_1 = \tfrac{1}{2} \quad \text{and} \quad a = 0 \tag{228}$$

Using (227), we write the solution of (226) as

$$u_1 = A \cos t + B \sin t + \tfrac{1}{16} a \cos 3t \tag{229}$$

where the solution of the homogeneous problem is included to avoid inconsistent results, as will be shown later. Then (224) becomes

$$\ddot{u}_2 + u_2 = -\left(\delta_2 + \tfrac{1}{32}\right) a \cos t + \left(B + \tfrac{1}{4}\alpha a\right) \sin t + \text{NST} \tag{230}$$

Eliminating the terms that lead to secular terms from (230) demands that $\delta_2 = -\frac{1}{32}$ and $B = -\frac{1}{4}\alpha a$. Had we put $B = 0$, we would have concluded that $\alpha = 0$ for a nontrivial solution, which is not right. Hence one of the transition curves emanating from $\delta = 1$ is

$$\delta = 1 - \tfrac{1}{2}\varepsilon - \tfrac{1}{32}\varepsilon^2 + \cdots \tag{231}$$

along which

$$u = a \cos t + \varepsilon\left[A \cos t - \tfrac{1}{4}\alpha a \sin t + \tfrac{1}{16} a \cos 3t\right] + \cdots \tag{232}$$

where A can be taken as zero because it can be absorbed into a.

Using (228), we write the solution of (226) as

$$u_1 = A\cos t + B\sin t + \tfrac{1}{16} b \sin 3t \tag{233}$$

where, as before, the complimentary solution is included. Then (224) becomes

$$\ddot{u}_2 + u_2 = -\left(A - \tfrac{1}{4}\alpha b\right)\cos t - \left(\delta_2 + \tfrac{1}{32}\right) b \sin t + \text{NST} \tag{234}$$

Eliminating the terms that lead to secular terms from (234) demands that $\delta_2 = -\frac{1}{32}$ and $A = \frac{1}{4}\alpha b$. Hence the other transition curve that emanates from $\delta = 1$ is

$$\delta = 1 + \tfrac{1}{2}\varepsilon - \tfrac{1}{32}\varepsilon^2 + \cdots \tag{235}$$

along which

$$u = b\sin t + \varepsilon\left[\tfrac{1}{4}\alpha b\cos t + B\sin t + \tfrac{1}{16} b \sin 3t\right] + \cdots \tag{236}$$

where B can be put at zero because it can be absorbed into b.

When $\delta_0 = 4$, the solution of (222) can be written as

$$u_0 = a\cos 2t + b\sin 2t \tag{237}$$

where a and b are constants. Then (223) becomes

$$\ddot{u}_1 + 4u_1 = -\delta_1(a\cos 2t + b\sin 2t) - \tfrac{1}{2}a - \tfrac{1}{2}a\cos 4t - \tfrac{1}{2}b\sin 4t \tag{238}$$

Eliminating the terms that lead to secular terms from (238) demands that $\delta_1 = 0$. Then the solution of (238) can be written as

$$u_1 = -\tfrac{1}{8}a + \tfrac{1}{24}a\cos 4t + \tfrac{1}{24}b\sin 4t \tag{239}$$

Substituting u_0 and u_1 into (224) yields

$$\ddot{u}_2 + 4u_2 = -\left(\delta_2 - \tfrac{1}{24}\right) a\cos 2t - \left(\delta_2 - \tfrac{1}{24}\right) b\sin 2t + \text{NST} \tag{240}$$

Eliminating the terms that lead to secular terms from (240) demands that $\delta_2 = \frac{1}{24}$. Hence, to second order, there is only one transition curve that emanates from $\delta = 4$. It is given by

$$\delta = 4 + \tfrac{1}{24}\varepsilon^2 + \cdots \tag{241}$$

11.7. Consider the equation

$$\ddot{u} + \tfrac{1}{4}(1 - \varepsilon\cos t)^{-2} a^{-2}\left[2(1 - \varepsilon\cos t)(2 - \varepsilon a^2 \cos t) + \varepsilon^2 a^2 \sin^2 t\right] u = 0 \tag{242}$$

Show that three of the transition curves are given by

$$a = 2 \pm \tfrac{1}{2}\varepsilon + \cdots \tag{243}$$

$$a = 1 + \tfrac{1}{6}\varepsilon^2 + \cdots \tag{244}$$

We first expand the coefficient of u in (242) for small ε and obtain

$$\ddot{u}+\frac{1}{a^2}\left[1+\varepsilon\cos t-\tfrac{1}{2}\varepsilon a^2\cos t+\varepsilon^2\cos^2 t+\tfrac{1}{4}\varepsilon^2 a^2-\tfrac{3}{4}\varepsilon^2 a^2\cos^2 t+\cdots\right]u=0 \tag{245}$$

When $a \approx 2$, we let

$$\begin{aligned} u &= u_0(t)+\varepsilon u_1(t)+\cdots \\ a &= 2-\varepsilon a_1+\cdots \end{aligned} \tag{246}$$

in (245), expand the result for small ε, equate coefficients of like powers of ε, and obtain

$$\ddot{u}_0+\tfrac{1}{4}u_0=0 \tag{247}$$

$$\ddot{u}_1+\tfrac{1}{4}u_1=-\tfrac{1}{4}(a_1+\cos t)\,u_0 \tag{248}$$

The solution of (247) can be written as

$$u_0=A\cos\tfrac{1}{2}t+B\sin\tfrac{1}{2}t \tag{249}$$

where A and B are constants. Then (248) becomes

$$\ddot{u}_1+\tfrac{1}{4}u_1=-\tfrac{1}{4}\left(a_1+\tfrac{1}{2}\right)A\cos\tfrac{1}{2}t-\tfrac{1}{4}\left(a_1-\tfrac{1}{2}\right)B\sin\tfrac{1}{2}t+\text{NST} \tag{250}$$

Eliminating the terms that lead to secular terms from (250) demands that $a_1=\pm\frac{1}{2}$. Hence the transition curves that emanate from $a=2$ are given to first order by (243).

When $a \approx 1$, we let

$$\begin{aligned} u &= u_0(t)+\varepsilon u_1(t)+\varepsilon^2 u_2(t)+\cdots \\ a &= 1+\varepsilon a_1+\varepsilon^2 a_2+\cdots \end{aligned} \tag{251}$$

in (245), expand the result for small ε, equate coefficients of like powers of ε, and obtain

$$\ddot{u}_0+u_0=0 \tag{252}$$

$$\ddot{u}_1+u_1=\left(2a_1-\tfrac{1}{2}\cos t\right)u_0 \tag{253}$$

$$\ddot{u}_2+u_2=\left(2a_1-\tfrac{1}{2}\cos t\right)u_1-\left[\tfrac{1}{4}(1+\cos^2 t)+3a_1^2-2a_2-2a_1\cos t\right]u_0 \tag{254}$$

The solution of (252) can be written as

$$u_0=A\cos t+B\sin t \tag{255}$$

where A and B are constants. Then (253) becomes

$$\ddot{u}_1+u_1=2a_1(A\cos t+B\sin t)-\tfrac{1}{4}A-\tfrac{1}{4}A\cos 2t-\tfrac{1}{4}B\sin 2t \tag{256}$$

Eliminating the terms that lead to secular terms from (256) demands that $a_1 = 0$. Then the solution of (256) can be written as

$$u_1 = -\tfrac{1}{4}A + \tfrac{1}{12}A\cos 2t + \tfrac{1}{12}B\sin 2t \tag{257}$$

Substituting u_0 and u_1 into (254) yields

$$\ddot{u}_2 + u_2 = \left(2a_2 - \tfrac{1}{3}\right)A\cos t + \left(2a_2 - \tfrac{1}{3}\right)B\cos t + \text{NST} \tag{258}$$

Eliminating the terms that lead to secular terms from (258) demands that $a_2 = \frac{1}{6}$. Hence the transition curve emanating from $a = 1$ is, to second order, given by (244).

11.8. Consider the problem

$$\ddot{u} + (\omega^2 + 2\varepsilon\cos 3t)u = 0, \qquad \varepsilon \ll 1 \tag{259}$$

(a) Use the method of variation of parameters to determine the equations describing the amplitude and the phase.

When $\varepsilon = 0$, the solution of (259) can be expressed as

$$u = a\cos(\omega t + \beta) \tag{260}$$

where a and β are constants. Differentiating (260) once with respect to t yields

$$\dot{u} = -a\omega\sin(\omega t + \beta) \tag{261}$$

When $\varepsilon \neq 0$, we still express the solution of (259) as in (260), subject to the constraint (261), but with time-varying a and β. Differentiating (260) once with respect to t and using (261), we conclude that

$$\dot{a}\cos\phi - a\dot{\beta}\sin\phi = 0 \tag{262}$$

where $\phi = \omega t + \beta$. Substituting (260) and (261) into (259) yields

$$\dot{a}\sin\phi + a\dot{\beta}\cos\phi = \frac{2\varepsilon a}{\omega}\cos\phi\cos 3t \tag{263}$$

Solving (262) and (263) for $\dot{a}$ and $\dot{\beta}$, we have

$$\begin{aligned}\dot{a} &= \frac{2\varepsilon a}{\omega}\sin\phi\cos\phi\cos 3t \\ &= \frac{\varepsilon a}{2\omega}\{\sin[(2\omega - 3)t + 2\beta] + \sin[(2\omega + 3)t + 2\beta]\}\end{aligned} \tag{264}$$

$$\begin{aligned}a\dot{\beta} &= \frac{2\varepsilon a}{\omega}\cos^2\phi\cos 3t \\ &= \frac{\varepsilon a}{2\omega}\{2\cos 3t + \cos[(2\omega - 3)t + 2\beta] + \cos[(2\omega + 3)t + 2\beta]\}\end{aligned} \tag{265}$$

(b) Use the method of averaging to determine the equations describing the slow variations in the amplitude and phase. Consider all cases.

All the terms on the right-hand side of (264) and (265) are fast-varying, unless $\omega \approx \frac{3}{2}$. Thus, when ω is away from $\frac{3}{2}$, keeping only the slowly varying parts in (264) and (265) yields

$$\begin{aligned} \dot{a} &= 0 \quad \text{or} \quad a = \text{constant} \\ \dot{\beta} &= 0 \quad \text{or} \quad \beta = \text{constant} \end{aligned} \tag{266}$$

When $\omega \approx \frac{3}{2}$, keeping the slowly varying parts in (264) and (265) yields

$$\begin{aligned} \dot{a} &= \frac{\varepsilon a}{2\omega} \sin[(2\omega - 3)t + 2\beta] \\ \dot{\beta} &= \frac{\varepsilon}{2\omega} \cos[(2\omega - 3)t + 2\beta] \end{aligned} \tag{267}$$

11.9. Use the methods of multiple scales and averaging to determine the equations describing the amplitudes and the phases to first order for a system governed by

$$\ddot{u} + \omega_0^2 u = \varepsilon u^2 \cos \Omega t \tag{268}$$

where Ω is away from zero when

(a) ω_0 is away from Ω and $\frac{1}{3}\Omega$

(b) $\omega_0 \approx \Omega$

(c) $3\omega_0 \approx \Omega$

The Method of Multiple Scales. We seek a first-order uniform expansion in the form

$$u = u_0(T_0, T_1) + \varepsilon u_1(T_0, T_1) + \cdots \tag{269}$$

where $T_0 = t$ and $T_1 = \varepsilon t$. Since Ω is away from zero, we express $\cos \Omega t$ as $\cos \Omega T_0$. Then, substituting (269) into (268) and equating coefficients of like powers of ε, we obtain

$$D_0^2 u_0 + \omega_0^2 u_0 = 0 \tag{270}$$

$$D_0^2 u_1 + \omega_0^2 u_1 = -2 D_0 D_1 u_0 + u_0^2 \cos \Omega T_0 \tag{271}$$

The solution of (270) can be expressed as

$$u_0 = A(T_1) e^{i\omega_0 T_0} + \bar{A}(T_1) e^{-i\omega_0 T_0} \tag{272}$$

Then (271) becomes

$$D_0^2 u_0 + \omega_0^2 u_1 = -2i\omega_0 A' e^{i\omega_0 T_0} + \tfrac{1}{2} e^{i\Omega T_0}\left(A^2 e^{2i\omega_0 T_0} + 2A\bar{A} + \bar{A}^2 e^{-2i\omega_0 T_0}\right) + \text{cc}$$

or

$$D_0^2 u_0 + \omega_0^2 u_1 = -2i\omega_0 A' e^{i\omega_0 T_0} + \tfrac{1}{2}A^2 e^{i(2\omega_0+\Omega)T_0} + A\bar{A}e^{i\Omega T_0} + \tfrac{1}{2}\bar{A}^2 e^{-i(2\omega_0-\Omega)T_0} + \text{cc} \tag{273}$$

(a) *When* ω_0 *is away from* Ω *and* $\frac{1}{3}\Omega$, only the first term on the right-hand side of (273) produces a secular term. Eliminating this secular term demands that

$$A' = 0 \tag{274}$$

Expressing A in the polar form

$$A = \tfrac{1}{2}ae^{i\beta} \tag{275}$$

and separating (274) into real and imaginary parts, we obtain

$$a' = 0, \qquad \beta' = 0 \tag{276}$$

(b) *When* $\omega_0 \approx \Omega$, we introduce a detuning parameter σ defined by

$$\Omega = \omega_0 + \varepsilon\sigma$$

and write

$$\Omega T_0 = \omega_0 T_0 + \sigma T_1$$

Eliminating the terms that lead to secular terms from (273) demands that

$$2i\omega_0 A' = A\bar{A}e^{i\sigma T_1} + \tfrac{1}{2}A^2 e^{-i\sigma T_1} \tag{277}$$

Substituting (275) into (277) and separating the result into real and imaginary parts, we obtain

$$a' = \frac{a^2}{8\omega_0}\sin(\sigma T_1 - \beta)$$
$$\beta' = -\frac{3a}{8\omega_0}\cos(\sigma T_1 - \beta) \tag{278}$$

(c) *When* $\omega_0 \approx \frac{1}{3}\Omega$, we introduce a detuning parameter σ defined by

$$\Omega = 3\omega_0 + \varepsilon\sigma$$

and write

$$\Omega T_0 = 3\omega_0 T_0 + \sigma T_1$$

Eliminating the terms that lead to secular terms from (273) demands that

$$2i\omega_0 A' = \tfrac{1}{2}\bar{A}^2 e^{i\sigma T_1} \tag{279}$$

Expressing A in the polar form (275) and separating (279) into real and imaginary parts, we obtain

$$a' = \frac{a^2}{8\omega_0}\sin(\sigma T_1 - 3\beta)$$

$$\beta' = -\frac{a}{8\omega_0}\cos(\sigma T_1 - 3\beta) \tag{280}$$

The Method of Averaging. When $\varepsilon = 0$, the solution of (268) can be expressed as

$$u = a\cos(\omega_0 t + \beta) \tag{281}$$

where a and β are constants. Differentiating (281) once with respect to t yields

$$\dot{u} = -\omega_0 a \sin(\omega_0 t + \beta) \tag{282}$$

When $\varepsilon \neq 0$, we still express the solution of (268) as in (281), subject to the constraint (282), but with time-varying rather than constant a and β. Then, differentiating (281) once with respect to t and using (282), we obtain

$$\dot{a}\cos(\omega_0 t + \beta) - a\dot{\beta}\sin(\omega_0 t + \beta) = 0 \tag{283}$$

Substituting (281) and (282) into (268) yields

$$\omega_0 \dot{a}\sin(\omega_0 t + \beta) + \omega_0 a\dot{\beta}\cos(\omega_0 t + \beta) = -\varepsilon a^2\cos^2(\omega_0 t + \beta)\cos\Omega t \tag{284}$$

Solving (283) and (284) for $\dot{a}$ and $\dot{\beta}$ yields

$$\begin{aligned}\dot{a} &= -\frac{\varepsilon a^2}{\omega_0}\sin(\omega_0 t + \beta)\cos^2(\omega_0 t + \beta)\cos\Omega t \\ &= -\frac{\varepsilon a^2}{8\omega_0}\{\sin[(\omega_0 - \Omega)t + \beta] + \sin[(\omega_0 + \Omega)t + \beta] + \sin[(3\omega_0 - \Omega)t + 3\beta] \\ &\quad + \sin[(3\omega_0 + \Omega)t + 3\beta]\}\end{aligned} \tag{285}$$

$$\begin{aligned}a\dot{\beta} &= -\frac{\varepsilon a^2}{\omega_0}\cos^3(\omega_0 t + \beta)\cos\Omega t \\ &= -\frac{\varepsilon a^2}{8\omega_0}\{3\cos[(\omega_0 - \Omega)t + \beta] + 3\cos[(\omega_0 + \Omega)t + \beta] + \cos[(3\omega_0 - \Omega)t + 3\beta] \\ &\quad + \cos[(3\omega_0 + \Omega)t + 3\beta]\}\end{aligned} \tag{286}$$

(a) *When ω_0 is away from Ω and $\frac{1}{3}\Omega$*, all the terms on the right-hand sides of (285) and (286) are fast varying. Hence, keeping only the slowly-varying parts in (285) and

(286), we have

$$\dot{a} = 0, \qquad \dot{\beta} = 0 \tag{287}$$

in agreement with (276).

(b) *When* $\omega_0 \approx \Omega$, keeping only the slowly-varying parts in (285) and (286), we obtain

$$\dot{a} = -\frac{\varepsilon a^2}{8\omega_0}\sin[(\omega_0 - \Omega)t + \beta]$$
$$\dot{\beta} = -\frac{3\varepsilon a}{8\omega_0}\cos[(\omega_0 - \Omega)t + \beta] \tag{288}$$

in agreement with (278), since $\Omega = \omega_0 + \varepsilon\sigma$ and $T_1 = \varepsilon t$.

(c) *When* $\omega_0 \approx \frac{1}{3}\Omega$, keeping only the slowly-varying parts, we obtain

$$\dot{a} = -\frac{\varepsilon a^2}{8\omega_0}\sin[(3\omega_0 - \Omega)t + 3\beta]$$
$$\dot{\beta} = -\frac{\varepsilon a}{8\omega_0}\cos[(3\omega_0 - \Omega)t + 3\beta] \tag{289}$$

in agreement with (280), because $\Omega = 3\omega_0 + \varepsilon\sigma$ and $T_1 = \varepsilon t$.

11.10. Consider the equation

$$\ddot{u} + \omega_0^2 u + 2\varepsilon u^3 \cos 2t = 0, \qquad \varepsilon \ll 1 \tag{290}$$

Use the methods of multiple scales and averaging to determine the equations describing the amplitude and the phase to first order when

(a) ω_0 is away from 1 and $\frac{1}{2}$
(b) $\omega_0 \approx 1$
(c) $\omega_0 \approx \frac{1}{2}$

The Method of Multiple Scales. Substituting (269) into (290), expressing $\cos 2t$ as $\cos 2T_0$, and equating coefficients of like powers of ε, we obtain

$$D_0^2 u_0 + \omega_0^2 u_0 = 0 \tag{291}$$

$$D_0^2 u_1 + \omega_0^2 u_1 = -2D_0 D_1 u_0 - 2u_0^3 \cos 2T_0 \tag{292}$$

The solution of (291) can be expressed as in (272). Then (292) becomes

$$D_0^2 u_1 + \omega_0^2 u_1 = -2i\omega_0 A' e^{i\omega_0 T_0} - A^3 e^{i(3\omega_0 + 2)T_0} - 3A^2 \bar{A} e^{i(\omega_0 + 2)T_0}$$
$$- 3A\bar{A}^2 e^{i(2-\omega_0)T_0} - \bar{A}^3 e^{i(2-3\omega_0)T_0} + cc \tag{293}$$

(a) *When* ω_0 *is away from* 1 *and* $\frac{1}{2}$, only the first term on the right-hand side of (293) produces a secular term. Hence eliminating the terms that lead to secular terms from (293) demands that $A' = 0$ which, upon using (275), yields (276).

(b) *When* $\omega_0 \approx 1$, we introduce a detuning parameter σ defined by

$$1 = \omega_0 + \varepsilon\sigma$$

and write

$$T_0 = \omega_0 T_0 + \sigma T_1$$

Then eliminating the terms that lead to secular terms from (293), we have

$$2i\omega_0 A' = -3A\bar{A}^2 e^{2i\sigma T_1} - A^3 e^{-2i\sigma T_1} \tag{294}$$

Expressing A in the polar form (275) and separating real and imaginary parts, we obtain

$$\begin{aligned} a' &= -\frac{a^3}{4\omega_0}\sin(2\sigma T_1 - 2\beta) \\ \beta' &= \frac{a^3}{2\omega_0}\cos(2\sigma T_1 - 2\beta) \end{aligned} \tag{295}$$

(c) *When* $\omega_0 \approx \frac{1}{2}$, we introduce a detuning parameter σ defined by

$$1 = 2\omega_0 + \varepsilon\sigma$$

and write

$$T_0 = 2\omega_0 T_0 + \sigma T_1$$

Eliminating the terms that lead to secular terms from (293) demands that

$$2i\omega_0 A' = -\bar{A}^3 e^{2i\sigma T_1} \tag{296}$$

Expressing A in the polar form (275) and separating (296) into real and imaginary parts, we obtain

$$\begin{aligned} a' &= -\frac{a^3}{8\omega_0}\sin(2\sigma T_1 - 4\beta) \\ \beta' &= \frac{a^2}{8\omega_0}\cos(2\sigma T_1 - 4\beta) \end{aligned} \tag{297}$$

The Method of Averaging. When $\varepsilon = 0$, the solution of (290) can be expressed as

$$u = a\cos(\omega_0 t + \beta) \tag{298}$$

where a and β are constants. Differentiating (298) once with respect to t yields

$$\dot{u} = -\omega_0 a \sin(\omega_0 t + \beta) \tag{299}$$

When $\varepsilon \neq 0$, we still express the solution of (290) as in (298), subject to the constraint (299), but with time-varying a and β. It follows from (298) and (299) that

$$\dot{a}\cos(\omega_0 t + \beta) - a\dot{\beta}\sin(\omega_0 t + \beta) = 0 \tag{300}$$

Substituting (298) and (299) into (290) yields

$$\dot{a}\sin(\omega_0 t + \beta) + a\dot{\beta}\cos(\omega_0 t + \beta) = \frac{2\varepsilon a^3}{\omega_0}\cos^3(\omega_0 t + \beta)\cos 2t \tag{301}$$

Solving (300) and (301) for $\dot{a}$ and $\dot{\beta}$ yields

$$\begin{aligned}\dot{a} &= \frac{2\varepsilon a^3}{\omega_0}\sin(\omega_0 t + \beta)\cos^3(\omega_0 t + \beta)\cos 2t \\ &= \frac{\varepsilon a^3}{8\omega_0}\{2\sin[(2\omega_0 - 2)t + 2\beta] + 2\sin[(2\omega_0 + 2)t + 2\beta] + \sin[(4\omega_0 - 2)t + 4\beta] \\ &\quad + \sin[(4\omega_0 + 2)t + 4\beta]\}\end{aligned} \tag{302}$$

$$\begin{aligned}\dot{\beta} &= \frac{2\varepsilon a^2}{\omega_0}\cos^4(\omega_0 t + \beta)\cos 2t \\ &= \frac{\varepsilon a^2}{8\omega_0}\{3\cos 2t + 4\cos[(2\omega_0 - 2)t + 2\beta] + 4\cos[(2\omega_0 + 2)t + 2\beta] \\ &\quad + \cos[(4\omega_0 - 2)t + 4\beta] + \cos[(4\omega_0 + 2)t + 4\beta]\}\end{aligned} \tag{303}$$

(a) *When ω_0 is away from* 1 *and* $\frac{1}{2}$, keeping only the slowly-varying parts on the right-hand sides of (302) and (303), we obtain

$$\dot{a} = 0 \quad \text{and} \quad \dot{\beta} = 0 \tag{304}$$

in agreement with (276).

(b) *When $\omega_0 \approx 1$*, keeping only the slowly-varying parts on the right-hand sides of (302) and (303), we obtain

$$\begin{aligned}\dot{a} &= \frac{\varepsilon a^3}{4\omega_0}\sin[(2\omega_0 - 2)t + 2\beta] \\ \dot{\beta} &= \frac{\varepsilon a^2}{2\omega_0}\cos[(2\omega_0 - 2)t + 2\beta]\end{aligned} \tag{305}$$

in agreement with (295), since $1 = \omega_0 + \varepsilon\sigma$ and $T_1 = \varepsilon t$.

(c) *When* $\omega_0 \approx \frac{1}{2}$, keeping only the slowly-varying parts on the right-hand sides of (302) and (303), we have

$$\dot{a} = \frac{\varepsilon a^3}{8\omega_0}\sin[(4\omega_0 - 2)t + 4\beta]$$
$$\dot{\beta} = \frac{\varepsilon a^2}{8\omega_0}\cos[(4\omega_0 - 2)t + 4\beta] \tag{306}$$

in agreement with (297), since $1 = 2\omega_0 + \varepsilon\sigma$ and $T_1 = \varepsilon t$.

11.11. The parametric excitation of a two-degree-of-freedom system is governed by

$$\ddot{u}_1 + \omega_1^2 u_1 + \varepsilon \cos\Omega t(f_{11}u_1 + f_{12}u_2) = 0$$
$$\ddot{u}_2 + \omega_2^2 u_2 + \varepsilon \cos\Omega t(f_{21}u_1 + f_{22}u_2) = 0 \tag{307}$$

Use the methods of multiple scales and averaging to determine the equations describing the amplitudes and the phases when $\Omega \approx \omega_2 \pm \omega_1$.

(a) ***The Method of Multiple Scales***

We seek a first-order uniform expansion using the method of multiple scales in the form

$$u_n = u_{n_0}(T_0, T_1) + \varepsilon u_{n_1}(T_0, T_1) + \cdots \tag{308}$$

where $T_0 = t$ and $T_1 = \varepsilon t$. Substituting (308) into (307), expressing $\cos\Omega t$ as $\cos\Omega T_0$, and equating coefficients of like powers of ε, we obtain

$$D_0^2 u_{10} + \omega_1^2 u_{10} = 0$$
$$D_0^2 u_{20} + \omega_2^2 u_{20} = 0 \tag{309}$$

$$D_0^2 u_{11} + \omega_1^2 u_{11} = -2D_0 D_1 u_{10} - \cos\Omega T_0(f_{11}u_{10} + f_{12}u_{20}) \tag{310}$$

$$D_0^2 u_{21} + \omega_2^2 u_{21} = -2D_0 D_1 u_{20} - \cos\Omega T_0(f_{21}u_{10} + f_{22}u_{20}) \tag{311}$$

The solution of (309) can be expressed as

$$u_{n_0} = A_n(T_1)e^{i\omega_n T_0} + cc \tag{312}$$

Then (310) and (311) become

$$D_0^2 u_{11} + \omega_1^2 u_{11} = -2i\omega_1 A_1' e^{i\omega_1 T_0} - \tfrac{1}{2}e^{i\Omega T_0}\big(f_{11}A_1 e^{i\omega_1 T_0} + f_{11}\bar{A}_1 e^{-i\omega_1 T_0} + f_{12}A_2 e^{i\omega_2 T_0} + f_{12}\bar{A}_2 e^{-i\omega_2 T_0}\big) + cc \tag{313}$$

$$D_0^2 u_{21} + \omega_2^2 u_{21} = -2i\omega_2 A_2' e^{i\omega_2 T_0} - \tfrac{1}{2}e^{i\Omega T_0}\big[f_{21}A_1 e^{i\omega_1 T_0} + f_{21}\bar{A}_1 e^{-i\omega_1 T_0} + f_{22}A_2 e^{i\omega_2 T_0} + f_{22}\bar{A}_2 e^{-i\omega_2 T_0}\big] + cc \tag{314}$$

When $\Omega \approx \omega_2 + \omega_1$ and there are no other resonances, we introduce a detuning parameter σ defined by

$$\Omega = \omega_2 + \omega_1 + \varepsilon\sigma$$

and write

$$\Omega T_0 = \omega_2 T_0 + \omega_1 T_0 + \sigma T_1$$

Then eliminating the terms that lead to secular terms from (313) and (314) demands that

$$\begin{aligned} 2i\omega_1 A_1' &= -\tfrac{1}{2} f_{12} \bar{A}_2 e^{i\sigma T_1} \\ 2i\omega_2 A_2' &= -\tfrac{1}{2} f_{21} \bar{A}_1 e^{i\sigma T_1} \end{aligned} \tag{315}$$

Expressing the A_n in the polar form

$$A_n = \tfrac{1}{2} a_n e^{i\beta_n} \tag{316}$$

and separating (315) into real and imaginary parts, we obtain

$$a_1' = -\frac{f_{12}}{4\omega_1} a_2 \sin\gamma \tag{317}$$

$$a_2' = -\frac{f_{21}}{4\omega_2} a_1 \sin\gamma \tag{318}$$

$$a_1 \beta_1' = \frac{f_{12}}{4\omega_1} a_2 \cos\gamma \tag{319}$$

$$a_2 \beta_2' = \frac{f_{21}}{4\omega_2} a_1 \cos\gamma \tag{320}$$

where

$$\gamma = \sigma T_1 - \beta_1 - \beta_2 \tag{321}$$

When $\Omega \approx \omega_2 - \omega_1$ and there are no other resonances, we introduce a detuning parameter according to

$$\Omega = \omega_2 - \omega_1 + \varepsilon\sigma$$

and write

$$\Omega T_0 = \omega_2 T_0 - \omega_1 T_0 + \sigma T_1$$

Eliminating the terms that lead to secular terms from (313) and (314) demands that

$$\begin{aligned} 2i\omega_1 A_1' &= -\tfrac{1}{2} f_{12} A_2 e^{-i\sigma T_1} \\ 2i\omega_2 A_2' &= -\tfrac{1}{2} f_{21} A_1 e^{i\sigma T_1} \end{aligned} \tag{322}$$

Substituting the polar form (316) into (322) and separating real and imaginary parts, we obtain

$$a_1' = \frac{f_{12}}{4\omega_1} a_2 \sin\gamma \tag{323}$$

$$a_2' = -\frac{f_{21}}{4\omega_2} a_1 \sin\gamma \tag{324}$$

$$a_1 \beta' = \frac{f_{12}}{4\omega_1} a_2 \cos\gamma \tag{325}$$

$$a_2 \beta_2' = \frac{f_{21}}{4\omega_2} a_1 \cos\gamma \tag{326}$$

where

$$\gamma = \sigma T_1 - \beta_2 + \beta_1 \tag{327}$$

(b) ***The Method of Averaging***

When $\varepsilon = 0$, the solutions of (307) can be expressed as

$$\begin{aligned} u_1 &= a_1 \cos(\omega_1 t + \beta_1) \\ u_2 &= a_2 \cos(\omega_2 t + \beta_2) \end{aligned} \tag{328}$$

where the a_n and β_n are constants. Differentiating (328) once with respect to t yields

$$\begin{aligned} \dot{u}_1 &= -\omega_1 a_1 \sin(\omega_1 t + \beta_1) \\ \dot{u}_2 &= -\omega_2 a_2 \sin(\omega_2 t + \beta_2) \end{aligned} \tag{329}$$

When $\varepsilon \neq 0$, we still represent the solutions of (307) as in (328), subject to the constraints (329), but with time-varying a_n and β_n. Then differentiating (328) once with respect to t yields

$$\begin{aligned} \dot{u}_1 &= -\omega_1 a_1 \sin(\omega_1 t + \beta_1) + \dot{a}_1 \cos(\omega_1 t + \beta_1) - a_1 \dot{\beta}_1 \sin(\omega_1 t + \beta_1) \\ \dot{u}_2 &= -\omega_2 a_2 \sin(\omega_2 t + \beta_2) + \dot{a}_2 \cos(\omega_2 t + \beta_2) - a_2 \dot{\beta}_2 \sin(\omega_2 t + \beta_2) \end{aligned} \tag{330}$$

Comparing (329) and (330), we conclude that

$$\begin{aligned}&\dot{a}_1\cos(\omega_1 t+\beta_1)-a_1\dot{\beta}_1\sin(\omega_1 t+\beta_1)=0\\&\dot{a}_2\cos(\omega_2 t+\beta_2)-a_2\dot{\beta}_2\sin(\omega_2 t+\beta_2)=0\end{aligned}\tag{331}$$

Differentiating (329) once with respect to t and substituting the result together with (328) into (307), we have

$$\begin{aligned}&\dot{a}_1\sin(\omega_1 t+\beta_1)+a_1\dot{\beta}_1\cos(\omega_1 t+\beta_1)\\&\qquad=\frac{\varepsilon}{\omega_1}\cos\Omega t[f_{11}a_1\cos(\omega_1 t+\beta_1)+f_{12}a_2\cos(\omega_2 t+\beta_2)]\end{aligned}\tag{332}$$

$$\begin{aligned}&\dot{a}_2\sin(\omega_2 t+\beta_2)+a_2\dot{\beta}_2\cos(\omega_2 t+\beta_2)\\&\qquad=\frac{\varepsilon}{\omega_2}\cos\Omega t[f_{21}a_1\cos(\omega_1 t+\beta_1)+f_{22}a_2\cos(\omega_2 t+\beta_2)]\end{aligned}\tag{333}$$

Solving (331)–(333) for the $\dot{a}_n$ and the $\dot{\beta}_n$ yields

$$\dot{a}_1=\frac{\varepsilon}{\omega_1}\cos\Omega t\sin(\omega_1 t+\beta_1)[f_{11}a_1\cos(\omega_1 t+\beta_1)+f_{12}a_2\cos(\omega_2 t+\beta_2)]\tag{334}$$

$$a_1\dot{\beta}_1=\frac{\varepsilon}{\omega_1}\cos\Omega t\cos(\omega_1 t+\beta_1)[f_{11}a_1\cos(\omega_1 t+\beta_1)+f_{12}a_2\cos(\omega_2 t+\beta_2)]\tag{335}$$

$$\dot{a}_2=\frac{\varepsilon}{\omega_2}\cos\Omega t\sin(\omega_2 t+\beta_2)[f_{21}a_1\cos(\omega_1 t+\beta_1)+f_{22}a_2\cos(\omega_2 t+\beta_2)]\tag{336}$$

$$a_2\dot{\beta}_2=\frac{\varepsilon}{\omega_2}\cos\Omega t\cos(\omega_2 t+\beta_2)[f_{21}a_1\cos(\omega_1 t+\beta_1)+f_{22}a_2\cos(\omega_2 t+\beta_2)]\tag{337}$$

When $\Omega\approx\omega_1+\omega_2$ and there are no other resonances, keeping only the slowly varying parts in (334)–(337), we have

$$\dot{a}_1=\frac{\varepsilon f_{12}}{4\omega_1}a_2\sin[(\omega_1+\omega_2-\Omega)t+\beta_1+\beta_2]\tag{338}$$

$$a_1\dot{\beta}_1=\frac{\varepsilon f_{12}}{4\omega_1}a_2\cos[(\omega_1+\omega_2-\Omega)t+\beta_1+\beta_2]\tag{339}$$

$$\dot{a}_2=\frac{\varepsilon f_{21}}{4\omega_2}a_1\sin[(\omega_1+\omega_2-\Omega)t+\beta_1+\beta_2]\tag{340}$$

$$a_2\dot{\beta}_2=\frac{\varepsilon f_{21}}{4\omega_2}a_1\cos[(\omega_1+\omega_2-\Omega)t+\beta_1+\beta_2]\tag{341}$$

in agreement with (317)–(321), because $\Omega=\omega_1+\omega_2+\varepsilon\sigma$ and $T_1=\varepsilon t$.

When $\Omega \approx \omega_2 - \omega_1$ and there are no other resonances, keeping the slowly varying parts in (334)–(337), we have

$$\dot{a}_1 = -\frac{\varepsilon f_{12}}{4\omega_1} a_2 \sin[(\omega_2 - \omega_1 - \Omega)t + \beta_2 - \beta_1] \tag{342}$$

$$a_2 \dot{\beta}_1 = \frac{\varepsilon f_{12}}{4\omega_1} a_2 \cos[(\omega_2 - \omega_1 - \Omega)t + \beta_2 - \beta_1] \tag{343}$$

$$\dot{a}_2 = \frac{\varepsilon f_{21}}{4\omega_2} a_1 \sin[(\omega_2 - \omega_1 - \Omega)t + \beta_2 - \beta_1] \tag{344}$$

$$a_2 \dot{\beta}_2 = \frac{\varepsilon f_{21}}{4\omega_2} a_1 \cos[(\omega_2 - \omega_1 - \Omega)t + \beta_2 - \beta_1] \tag{345}$$

in agreement with (323)–(327), because $\Omega = \omega_2 - \omega_1 + \varepsilon\sigma$ and $T_1 = \varepsilon t$.

Supplementary Exercises

11.12. Use the method of strained parameters and Whittaker's technique to determine second-order uniform expansions of the solutions of

$$\ddot{u} + (\delta + \varepsilon\cos 2t + \varepsilon\alpha\sin 2t)u = 0, \qquad \varepsilon \ll 1$$

when $\delta \approx 0$, 1, and 4.

11.13. Use the method of strained parameters and Whittaker's technique to determine second-order uniform expansions of the solutions of

$$\ddot{u} + (\delta + \varepsilon\cos t + \varepsilon\alpha\cos 2t)u = 0, \qquad \varepsilon \ll 1$$

when $\delta \approx 0$, 1, and 4.

11.14. Use the method of strained parameters and Whittaker's technique to determine second-order uniform expansions of the solutions of

$$\ddot{u} + (\delta + \varepsilon\cos^3 2t)u = 0, \qquad \varepsilon \ll 1$$

when $\delta \approx 0$, 1, and 4.

11.15. Determine second-order uniform expansions of the transition curves separating stable from unstable solutions of

$$\ddot{u} + 2\varepsilon\mu\dot{u} + (\delta + \varepsilon\cos 2t)u = 0, \qquad \varepsilon \ll 1$$

when $\delta \approx 0$, 1, and 4.

11.16. Determine second-order uniform expansions of the transition curves separating stable from unstable solutions of

$$\ddot{u} + \left(\delta + \varepsilon\cos 2t + \varepsilon^2\alpha_1\sin 2t + \varepsilon^2\alpha_2\cos 4t\right)u = 0, \qquad \varepsilon \ll 1$$

when $\delta \approx 0$, 1, and 4.

11.17. Use the methods of multiple scales and averaging to determine the equations describing the amplitude and the phase to first order for a system governed by

$$\ddot{u} + \omega_0^2 u = \varepsilon u^4 \cos \Omega t, \qquad \varepsilon \ll 1$$

11.18. Use the methods of multiple scales and averaging to determine the equations describing the amplitude and the phase to first order for a system governed by

$$\ddot{u} + \omega_0^2 u + \varepsilon \dot{u}|\dot{u}| + 2\varepsilon u \cos \Omega t = 0, \qquad \varepsilon \ll 1$$

when $\Omega \approx 2\omega_0$.

11.19. Use the methods of multiple scales and averaging to determine the equations describing the amplitude and the phase to first order for a system governed by

$$\ddot{u} + \omega_0^2 u + 2\varepsilon u \cos \Omega_1 t = F \cos \Omega_2 t, \qquad \varepsilon \ll 1$$

when $\omega_0 \approx \Omega_2 \pm \Omega_1$.

11.20. Consider the system of equations

$$\ddot{u}_1 + \omega_1^2 u_1 + \varepsilon \cos \Omega_1 t [f_{11} u_1 + f_{12} u_2] + \varepsilon \cos \Omega_2 t [g_{11} u_1 + g_{12} u_2] = 0$$

$$\ddot{u}_2 + \omega_2^2 u_2 + \varepsilon \cos \Omega_1 t [f_{21} u_1 + f_{22} u_2] + \varepsilon \cos \Omega_2 t [g_{21} u_1 + g_{22} u_2] = 0$$

where $\varepsilon \ll 1$. Determine the equations describing the amplitudes and the phases to first order when

(a) $\Omega_1 \approx 2\omega_1$ and $\Omega_2 \approx 2\omega_2$
(b) $\Omega_1 \approx 2\omega_1$ and $\Omega_2 \approx \omega_2 + \omega_1$
(c) $\Omega_1 \approx 2\omega_2$ and $\Omega_2 \approx \omega_2 - \omega_1$
(d) $\Omega_1 \approx \omega_2 - \omega_1$ and $\Omega_2 \approx \omega_2 + \omega_1$

11.21. Consider the system of equations

$$\ddot{u}_1 + \omega_1^2 u_1 + \alpha_1 u_1 u_2 + \varepsilon \cos \Omega t (f_{11} u_1 + f_{12} u_2) = 0$$

$$\ddot{u}_2 + \omega_2^2 u_2 + \alpha_2 u_1^2 + \varepsilon \cos \Omega t (f_{21} u_1 + f_{22} u_2) = 0$$

where $\varepsilon \ll 1$ and $\omega_2 = 2\omega_1 + \varepsilon\sigma_1$. Determine the equations describing the amplitudes and the phase to first order when

(a) $\Omega \approx 2\omega_1$,
(b) $\Omega \approx 2\omega_2$
(c) $\Omega \approx \omega_2 + \omega_1$
(d) $\Omega \approx \omega_2 - \omega_1$

CHAPTER 12

Boundary-Layer Problems

In Chapters 4–11, we considered methods of determining uniform approximate solutions of equations of the form

$$\ddot{u} + \omega_0^2 u = \varepsilon f(u, \dot{u}, t), \qquad \varepsilon \ll 1 \tag{1}$$

In this case, the small perturbation εf has a small effect only if it acts over short periods of times. However, it can have a large effect if it acts over a long period of time (i.e., its effects are cumulative). Consequently, the perturbation results in a slow modulation of the amplitudes and phases, which can be handled by slow scales. On the other hand, the small perturbation in problems of the form

$$\varepsilon y'' + p_1(x) y' + p_0(x) y = f(x), \qquad \varepsilon \ll 1 \tag{2a}$$

$$y(0) = \alpha, \qquad y(1) = \beta \tag{2b}$$

is operative over a narrow region across which the dependent variable undergoes very rapid changes. Frequently, these narrow regions adjoin the boundaries of the domain of interest, especially when the small parameter multiplies the highest derivative. Consequently, they are usually called boundary layers. However, there are many physical situations in which the sharp changes occur inside the domain of interest, such as shock waves. In contrast with the oscillation problems discussed in Chapters 4–11, boundary-layer problems cannot be handled by slow scales, but they can be handled by fast, magnified, or stretched scales.

There are a number of methods for treating boundary-layer problems. They include the method of matched asymptotic expansions, the method of composite expansions, the method of multiple scales, the WKBJ method, and the Langer and Olver transformations. The last two techniques are applicable to linear problems involving a large parameter; they are discussed in Chapter 14. In this chapter, we concentrate on the method of matched asymptotic expansions because it is the most versatile and effective technique for treating nonlinear as well as linear problems involving partial and ordinary-differential equations.

Seeking a straightforward expansion for the boundary-layer problem (2), let

$$y = y_0(x) + \varepsilon y_1(x) + \varepsilon^2 y_2(x) + \cdots \tag{3}$$

and obtain the perturbation problems

Order ε^0

$$p_1(x) y_0' + p_0(x) y_0 = f(x) \tag{4a}$$

$$y_0(0) = \alpha, \qquad y_0(1) = \beta \tag{4b}$$

Order ε^n

$$p_1(x) y_n' + p_0(x) y_n = -y_{n-1}'', \qquad n \geq 1 \tag{5a}$$

$$y_n(0) = 0, \qquad y_n(1) = 0 \tag{5b}$$

It is clear that all perturbation equations are of first order and hence none of them in general can satisfy both boundary conditions, and one of them must be dropped. Consequently, the straightforward expansion breaks down near the boundary where the boundary condition has been dropped. Determination of the boundary condition that must be dropped can be accomplished using either physical or mathematical arguments. The mathematical argument usually takes the form of the matching procedure discussed below.

Using the method of matched asymptotic expansions, we call the straightforward expansion the outer expansion and denote it y^o. Moreover, we call the independent variable x the outer variable. Thus the outer expansion is obtained by using the outer limiting process, that is,

$$y_0(x) = \lim_{\substack{\varepsilon \to 0 \\ x \text{ fixed}}} y(x; \varepsilon) \tag{6a}$$

$$y_n(x) = \lim_{\substack{\varepsilon \to 0 \\ x \text{ fixed}}} \frac{y(x; \varepsilon) - \sum_{m=0}^{n-1} \varepsilon^m y_m(x)}{\varepsilon^n} \tag{6b}$$

The outer expansion is supplemented by an inner expansion using the magnified or stretched independent variable

$$\xi = \frac{x}{\varepsilon^\lambda}, \qquad \lambda > 0 \tag{7a}$$

if the nonuniformity is at $x=0$ or

$$\xi = \frac{1-x}{\varepsilon^{\lambda}}, \qquad \lambda > 0 \tag{7b}$$

if the nonuniformity is at $x=1$. The independent variable ξ is called the inner variable. In terms of ξ, an inner expansion y^i is sought in the form

$$y^i(\xi;\varepsilon) = \sum_{m=0}^{N-1} \delta_m(\varepsilon) Y_m(\xi) + O[\delta_N(\varepsilon)] \tag{8}$$

where $\delta_m(\varepsilon)$ is an asymptotic sequence, using the inner limit processes in which $\varepsilon \to 0$ with ξ being fixed. Then, λ is chosen in such a way that the equation for $Y_0(\xi)$ contains $Y_0''(\xi)$ and has the least-degenerate form. This yields the so-called distinguished limits. If only one value of λ yields a distinguished limit, only one inner expansion exists. If two or more values of λ yield distinguished limits, then two or more inner expansions that are nested exist, and one speaks of multiple decks.

So far, if λ has only one distinguished limit, an approximate solution of the boundary-layer problem (2) has been given as two separate expansions in terms of the outer variable x and the inner variable ξ. The basic idea underlying the method of matched asymptotic expansions is that the domains of validity of the two expansions overlap and hence they can be matched (blended). Van Dyke has proposed the matching principle: the m-term inner expansion of the n-term outer expansion equals the n-term outer expansion of the m-term inner expansion. If the guess for the location of the boundary-layer is correct, the resulting expansions can usually be matched and all constants in the expansions can be determined. Finally, a single uniformly valid expansion called a composite expansion and denoted by y^c can be formed as follows:

$$y^c = y^o + y^i - (y^o)^i = y^o + y^i - (y^i)^o \tag{9}$$

Solved Exercises

12.1. Consider the problem

$$\varepsilon y'' + y' + y = 0 \tag{10a}$$

$$y(0) = \alpha, \qquad y(1) = \beta \tag{10b}$$

(a) Determine the exact solution.

(b) Use the method of matched asymptotic expansions to determine a first-order uniform expansion. Compare your answer with the exact solution.

(c) Use the method of multiple scales to determine a first-order uniform expansion. Compare your answer with those in (a) and (b).

(a) *The Exact Solution*

We seek solutions to the homogeneous equation (10a) in the form $y = \exp(sx)$. Then (10a) becomes

$$\varepsilon s^2 + s + 1 = 0$$

whose solution is

$$s_1 = \frac{-1+\sqrt{1-4\varepsilon}}{2\varepsilon}, \qquad s_2 = \frac{-1-\sqrt{1-4\varepsilon}}{2}$$

Hence the general solution of (10a) is

$$y = c_1 e^{s_1 x} + c_2 e^{s_2 x} \tag{11a}$$

Substituting (11a) into (10b) yields

$$c_1 + c_2 = \alpha$$

$$c_1 e^{s_1} + c_2 e^{s_2} = \beta$$

whose solution is

$$c_1 = \frac{\alpha e^{s_2} - \beta}{e^{s_2} - e^{s_1}}, \qquad c_2 = \frac{\beta - \alpha e^{s_1}}{e^{s_2} - e^{s_1}}$$

Hence the solution of (10) is

$$y = \frac{(\alpha e^{s_2} - \beta) e^{s_1 x} + (\beta - \alpha e^{s_1}) e^{s_2 x}}{e^{s_2} - e^{s_1}} \tag{11b}$$

(b) *The Method of Matched Asymptotic Expansions*

We seek a first-order outer expansion in the form

$$y^o = y_0(x) + \cdots \tag{12}$$

Since the coefficient of y' is 1 and it is positive, the boundary layer is at $x = 0$ and the outer expansion must satisfy the second boundary condition in (10b). Substituting (12) into (10a) with $y(1) = \beta$ and equating coefficients of like powers of ε, we obtain

$$y_0' + y_0 = 0, \qquad y_0(1) = \beta$$

whose solution is

$$y_0 = \beta e^{1-x}$$

Hence

$$y^o = \beta e^{1-x} + \cdots \tag{13}$$

To determine an inner expansion, we introduce the stretching transformation

$$\xi = \frac{x}{\varepsilon^\lambda}, \qquad \lambda > 0$$

in (10a) and obtain

$$\varepsilon^{1-2\lambda}\frac{d^2y}{d\xi^2} + \varepsilon^{-\lambda}\frac{dy}{d\xi} + y = 0 \tag{14}$$

As $\varepsilon \to 0$, the least degenerate form of (14) occurs when $\lambda = 1$. Thus we seek an inner expansion in the form

$$y^i = Y_0(\xi) + \cdots, \qquad \xi = \frac{x}{\varepsilon} \tag{15}$$

The inner expansion must satisfy the first boundary condition in (10b). Since $x = 0$ corresponds to $\xi = 0$,

$$y^i(0) = \alpha \tag{16}$$

Substituting (15) into (14) with $\lambda = 1$ and (16) and equating coefficients of like powers of ε, we obtain

$$Y_0'' + Y_0' = 0, \qquad Y_0(0) = \alpha$$

whose solution is

$$Y_0 = \alpha - c_1 + c_1 e^{-\xi}$$

Hence

$$y^i = \alpha - c_1 + c_1 e^{-\xi} + \cdots \tag{17}$$

Next, we match one term outer expansion with one term inner expansion as follows:

One-term outer expansion	$y \sim \beta e^{1-x}$	
Rewritten in inner variable	$= \beta e^{1-\varepsilon\xi}$	
Expanded for small ε	$= \beta e(1 - \varepsilon\xi + \cdots)$	
One-term inner expansion	$= \beta e$	(18a)
One-term inner expansion	$y \sim \alpha - c_1 + c_1 e^{-\xi}$	
Rewritten in outer variable	$= \alpha - c_1 + c_1 e^{-x/\varepsilon}$	
Expanded for small ε	$= \alpha - c_1 + \text{EST}$	
One-term outer expansion	$= \alpha - c_1$	(18b)

Equating (18a) and (18b) according to the matching principle yields

$$\alpha - c_1 = \beta e$$

Hence

$$c_1 = \alpha - \beta e$$

and

$$y^i = \beta e + (\alpha - \beta e) e^{-\xi} + \cdots$$

To form a composite expansion, we put

$$y^c = y^o + y^i - (y^o)^i = \beta e^{1-x} + \beta e + (\alpha - \beta e) e^{-\xi} - \beta e + \cdots$$

or

$$y^c = \beta e^{1-x} + (\alpha - \beta e) e^{-\xi} + \cdots \tag{19a}$$

To compare (19) with the exact solution (11b), we note that

$$s_1 = \frac{-1 + 1 - 2\varepsilon + \cdots}{2\varepsilon} = -1 + \cdots$$

$$s_2 = \frac{-1 - 1 + 2\varepsilon + \cdots}{2\varepsilon} = -\frac{1}{\varepsilon} + 1 + \cdots$$

Hence $\exp(s_2)$ is exponentially small and (11b) becomes

$$y^e = \beta e^{1-x} + (\alpha - \beta e) e^{x-(x/\varepsilon)} + \cdots \tag{19b}$$

Therefore

$$y^e - y^c = (\alpha - \beta e) e^{-x/\varepsilon}(e^x - 1) + \cdots \tag{19c}$$

which is $O(\varepsilon)$ in the inner region $x = O(\varepsilon)$ and exponentially small in the outer region.

(c) *The Method of Multiple Scales*

We seek a first-order expansion in the form

$$y(x; \varepsilon) = y_0(\xi, \zeta) + \varepsilon y_1(\xi, \zeta) + \cdots \tag{20a}$$

where

$$\zeta = x \quad \text{and} \quad \xi = \frac{x}{\varepsilon}$$

Then

$$\begin{aligned} \frac{dy}{dx} &= \frac{1}{\varepsilon}\frac{\partial y}{\partial \xi} + \frac{\partial y}{\partial \zeta} \\ \frac{d^2 y}{dx^2} &= \frac{1}{\varepsilon^2}\frac{\partial^2 y}{\partial \xi^2} + \frac{2}{\varepsilon}\frac{\partial^2 y}{\partial \xi \partial \zeta} + \frac{\partial^2 y}{\partial \zeta^2} \end{aligned} \tag{20b}$$

and (10a) becomes

$$\frac{1}{\varepsilon}\frac{\partial^2 y}{\partial \xi^2} + 2\frac{\partial^2 y}{\partial \xi \partial \zeta} + \varepsilon \frac{\partial^2 y}{\partial \zeta^2} + \frac{1}{\varepsilon}\frac{\partial y}{\partial \xi} + \frac{\partial y}{\partial \zeta} + y = 0 \tag{21}$$

Substituting (20a) into (21) and equating coefficients of like powers of ε, we obtain

$$\frac{\partial^2 y_0}{\partial \xi^2} + \frac{\partial y_0}{\partial \xi} = 0 \tag{22}$$

$$\frac{\partial^2 y_1}{\partial \xi^2} + \frac{\partial y_1}{\partial \xi} = -2\frac{\partial^2 y_0}{\partial \xi \partial \zeta} - \frac{\partial y_0}{\partial \zeta} - y_0 \tag{23}$$

The general solution of (22) can be written as

$$y_0 = A(\zeta) + B(\zeta)\, e^{-\xi} \tag{24}$$

where A and B are functions to be determined at the next level of approximation. Then (23) becomes

$$\frac{\partial^2 y_1}{\partial \xi^2} + \frac{\partial y_1}{\partial \xi} = -(A' + A) + (B' - B)\, e^{-\xi}$$

In order that y_1/y_0 be bounded for all ξ,

$$A' + A = 0 \tag{25a}$$

$$B' - B = 0 \tag{25b}$$

whose general solutions are

$$A = ae^{-\zeta}, \qquad B = be^{\zeta} \tag{26}$$

where a and b are constants. Substituting (26) into (24) and substituting the result into (20a), we have

$$y = ae^{-\zeta} + be^{\zeta - \xi} + \cdots$$

or

$$y = ae^{-x} + be^{x - x/\varepsilon} + \cdots \tag{27}$$

Substituting (27) into (10b) yields

$$a + b = \alpha$$

$$ae^{-1} + be^{1-(1/\varepsilon)} = \beta$$

whose solutions are

$$a = \beta e + \text{EST} \quad \text{and} \quad b = \alpha - \beta e$$

Then (27) becomes

$$y = \beta e^{1-x} + (\alpha - \beta e) e^{x - x/\varepsilon} + \cdots \tag{28}$$

in agreement with (19b).

12.2. Consider the problem

$$\varepsilon y'' - y' + y = 0 \tag{29}$$

$$y(0) = \alpha, \qquad y(1) = \beta \tag{30}$$

(a) Determine the exact solution.

(b) Use the method of matched asymptotic expansions to determine a first-order uniform expansion. Compare your answer with the exact solution.

(c) Use the method of multiple scales to determine a first-order uniform expansion. Compare your answer with those in (a) and (b).

(a) ***The Exact Solution***

We seek solutions of (29) in the form $y = \exp(sx)$ and obtain

$$\varepsilon s^2 - s + 1 = 0$$

whose general solution is

$$s_1 = \frac{1 + \sqrt{1 - 4\varepsilon}}{2\varepsilon}, \qquad s_2 = \frac{1 - \sqrt{1 - 4\varepsilon}}{2\varepsilon}$$

Hence the general solution of (29) has the form (11a). Imposing the boundary conditions (30) leads to the solution (11b).

(b) ***The Method of Matched Asymptotic Expansions***

Since the coefficient of y' is negative, the boundary layer is at the right end and the outer expansion must satisfy the left boundary condition, but it may not satisfy the right boundary condition.

We seek an outer expansion in the form

$$y^o = y_0(x) + \varepsilon y_1(x) + \cdots \tag{31}$$

Substituting (31) into (29) and the first boundary condition in (30) and equating the coefficients of ε^0 to zero, we obtain

$$y_0' - y_0 = 0, \qquad y_0(0) = \alpha \tag{32}$$

whose solution is

$$y_0 = \alpha e^x$$

Hence

$$y^o = \alpha e^x + \cdots \tag{33}$$

Since $y^o(1) = \alpha e$ is different, in general, from β, y^o is not valid near $x = 1$ and a boundary layer exists there.

To determine an inner expansion near $x = 1$, we introduce the stretching transformation

$$\xi = \frac{1-x}{\varepsilon^\lambda}, \qquad \lambda > 0$$

in (29) and obtain

$$\varepsilon^{1-2\lambda}\frac{d^2y}{d\xi^2} + \varepsilon^{-\lambda}\frac{dy}{d\xi} + y = 0 \tag{34}$$

At $\varepsilon \to 0$, the least degenerate form of (34) is

$$\frac{d^2y}{d\xi^2} + \frac{dy}{d\xi} = 0$$

which occurs when $\lambda = 1$. Then (34) becomes

$$\frac{d^2y}{d\xi^2} + \frac{dy}{d\xi} + \varepsilon y = 0 \tag{35}$$

We seek an inner expansion in the form

$$y^i = Y_0(\xi) + \varepsilon Y_1(\xi) + \cdots \tag{36}$$

Since the boundary layer is at $x = 1$, y^i must satisfy the right boundary condition. Moreover, since $x = 1$ corresponds to $\xi = 0$,

$$y(1) = \beta \Rightarrow y^i(0) = \beta \tag{37}$$

Substituting (36) into (35) and (37) and equating the coefficients of ε^0 on both sides, we obtain

$$Y_0'' + Y_0' = 0, \qquad Y_0(0) = \beta \tag{38}$$

whose solution is

$$Y_0 = \beta - c_1 + c_1 e^{-\xi} \tag{39}$$

where c_1 is a constant that needs to be determined by matching y^o and y^i.

We match one term outer expansion with one term inner expansion as follows:

One-term outer expansion	$y \sim \alpha e^x$	
Rewritten in inner variable	$= \alpha e^{1-\varepsilon\xi}$	
Expanded for small ε	$= \alpha e(1-\varepsilon\xi+\cdots)$	
One-term inner expansion	$= \alpha e$	(40a)
One-term inner expansion	$y \sim \beta - c_1 + c_1 e^{-\xi}$	
Rewritten in outer variable	$= \beta - c_1 + c_1 e^{-(1-x)/\varepsilon}$	
Expanded for small ε	$= \beta - c_1 + \text{EST}$	
One-term outer expansion	$= \beta - c_1$	(40b)

Equating (40a) and (40b) according to the matching principle yields

$$\alpha e = \beta - c_1 \quad \text{or} \quad c_1 = \beta - \alpha e$$

Hence

$$y^i = \alpha e + (\beta - \alpha e)\, e^{-\xi} + \cdots \tag{41}$$

Since

$$y^c = y^o + y^i - (y^o)^i$$

$$y^c = \alpha e^x + (\beta - \alpha e)\, e^{-\xi} + \cdots \tag{42}$$

To compare y^c with the exact solution, we note that as $\varepsilon \to 0$,

$$s_1 = \frac{1+1-2\varepsilon+\cdots}{2\varepsilon} = \frac{1}{\varepsilon} - 1 + \cdots$$

$$s_2 = \frac{1-1+2\varepsilon+\cdots}{2\varepsilon} = 1 + \cdots$$

Hence $\exp(s_2)$ is negligible compared with $\exp(s_1)$ and (11b) tends to

$$y^e = \alpha e^x + (\beta - \alpha e)\, e^{(1-x)(-1/\varepsilon+1)} + \cdots \tag{43}$$

Although y^e is different from y^c, their difference

$$y^e - y^c = (\beta - \alpha e)\, e^{(1-x)/\varepsilon}[e^{1-x} - 1] + \cdots$$

is $O(\varepsilon)$ in the inner region $1 - x = O(\varepsilon)$ and exponentially small in the outer region.

(c) *The Method of Multiple Scales*

We seek a first-order uniform expansion by using the outer variable $\zeta = x$ and the inner variable $\xi = (1-x)/\varepsilon$ as the appropriate scales. Thus we let

$$y(x;\varepsilon) = y_0(\zeta,\xi) + \varepsilon y_1(\zeta,\xi) + \cdots \tag{44}$$

In terms of ζ and ξ, the derivatives become

$$\frac{d}{dx} = -\frac{1}{\varepsilon}\frac{\partial}{\partial \xi} + \frac{\partial}{\partial \zeta}$$
$$\frac{d^2}{dx^2} = \frac{1}{\varepsilon^2}\frac{\partial^2}{\partial \xi^2} - \frac{2}{\varepsilon}\frac{\partial^2}{\partial \xi \partial \zeta} + \frac{\partial^2}{\partial \zeta^2} \tag{45a}$$

and (29) transforms into

$$\frac{1}{\varepsilon}\frac{\partial^2 y}{\partial \xi^2} - 2\frac{\partial^2 y}{\partial \xi \partial \zeta} + \varepsilon\frac{\partial^2}{\partial \zeta^2} + \frac{1}{\varepsilon}\frac{\partial y}{\partial \xi} - \frac{\partial y}{\partial \zeta} + y = 0 \tag{45b}$$

Substituting (44) into (45b) and equating coefficients of like powers of ε, we obtain

$$\frac{\partial^2 y_0}{\partial \xi^2} + \frac{\partial y_0}{\partial \xi} = 0 \tag{46}$$

$$\frac{\partial^2 y_1}{\partial \xi^2} + \frac{\partial y_1}{\partial \xi} = 2\frac{\partial^2 y_0}{\partial \xi \partial \zeta} + \frac{\partial y_0}{\partial \zeta} - y_0 \tag{47}$$

The general solution of (46) is

$$y_0 = A(\zeta) + B(\zeta)\, e^{-\xi} \tag{48}$$

Then (47) becomes

$$\frac{\partial^2 y_1}{\partial \xi^2} + \frac{\partial y_1}{\partial \xi} = -(B' + B)\, e^{-\xi} + A' - A \tag{49}$$

In order that y_1/y_0 be bounded for all ξ,

$$A' - A = 0$$
$$B' + B = 0 \tag{50}$$

whose solutions are

$$A = ae^{\zeta}, \qquad B = be^{-\zeta} \tag{51}$$

where a and b are constants. Substituting (51) into (48) and then substituting the result into (44), we have

$$y = ae^{\zeta} + be^{-(\zeta+\xi)} + \cdots$$

or

$$y = ae^{x} + be^{-x-(1-x)/\varepsilon} + \cdots \tag{52}$$

Substituting (52) into (30) yields

$$a = \alpha + \text{EST}, \qquad ae + be^{-1} = \beta$$

Hence

$$a = \alpha \quad \text{and} \quad b = \beta e - \alpha e^2$$

and (52) becomes

$$y = \alpha e^x + (\beta - \alpha e)\, e^{-(1-x)(1/\varepsilon - 1)} + \cdots \tag{53}$$

in agreement with (43).

12.3. Consider the problem

$$\varepsilon y'' - y' = 1 \tag{54}$$

$$y(0) = \alpha, \qquad y(1) = \beta \tag{55}$$

(a) Determine the exact solution.

(b) Use the method of matched asymptotic expansions to determine a first-order uniform expansion. Compare your answer with the exact solution.

(c) Use the method of multiple scales to determine a first-order uniform expansion. Compare your answer with those in (a) and (b).

(a) ***The Exact Solution***

The general solution of the homogeneous equation (54) can be written as

$$y_h = c_1 + c_2 e^{x/\varepsilon}$$

where c_1 and c_2 are constants, whereas a particular solution of (54) is

$$y_p = -x$$

Therefore, the general solution of (54) is

$$y = c_1 + c_2 e^{x/\varepsilon} - x$$

Imposing the boundary conditions (55) yields

$$c_1 + c_2 = \alpha$$

$$c_1 + c_2 e^{1/\varepsilon} = \beta + 1$$

whose solutions are

$$c_1 = \frac{\beta + 1 - \alpha e^{1/\varepsilon}}{1 - e^{1/\varepsilon}}, \qquad c_2 = \frac{\alpha - \beta - 1}{1 - e^{1/\varepsilon}}$$

Therefore

$$y = \frac{\beta + 1 - \alpha e^{1/\varepsilon} + (\alpha - \beta - 1)\, e^{x/\varepsilon}}{1 - e^{1/\varepsilon}} - x \tag{56}$$

(b) ***The Method of Matched Asymptotic Expansions***

As $\varepsilon \to 0$, (54) reduces to a first-order equation, which cannot satisfy, in general, the two boundary conditions (55), and one of them must be dropped and a boundary layer must be introduced there. Since the coefficient of y' is negative, the boundary layer exists at the right end, and the outer expansion must satisfy the left boundary condition. Thus, to determine a uniform expansion, we need to determine an outer expansion, an inner expansion, match them, and then form a composite expansion.

First, we seek an outer expansion in the form

$$y^o = y_0(x) + \varepsilon y_1(x) + \cdots \tag{57}$$

Substituting (57) into (54) and $y(0) = \alpha$ we obtain

$$-y_0' = 1, \qquad y_0(0) = \alpha$$

whose solution is

$$y_0 = \alpha - x$$

so that

$$y^o = \alpha - x + \cdots \tag{58}$$

Since $y(1) = \alpha - 1 \neq \beta$, in general, a boundary layer needs to be introduced at $x = 1$. To this end, we introduce the stretching transformation

$$\xi = \frac{1 - x}{\varepsilon^\lambda}, \qquad \lambda > 0$$

into (54) and obtain

$$\varepsilon^{1-2\lambda} \frac{d^2 y}{d\xi^2} + \varepsilon^{-\lambda} \frac{dy}{d\xi} = 1 \tag{59}$$

As $\varepsilon \to 0$, the least degenerate form of (59) occurs when $\lambda = 1$ and hence (59) can be rewritten as

$$\frac{d^2 y}{d\xi^2} + \frac{dy}{d\xi} = \varepsilon \tag{60}$$

Since $x = 1$ corresponds to $\xi = 0$, the boundary condition $y(1) = \beta$ that must be satisfied by the inner expansion y^i transforms into

$$y^i(0) = \beta \tag{61}$$

Next, we seek an inner expansion in the form

$$y^i = Y_0(\xi) + \varepsilon Y_1(\xi) + \cdots \tag{62}$$

Substituting (62) into (60) and (61) and equating the coefficients of ε^0 on both sides, we obtain

$$Y_0'' + Y_0' = 0, \qquad Y_0(0) = \beta$$

whose solution is

$$Y_0 = \beta - c_1 + c_1 e^{-\xi}$$

so that

$$y^i = \beta - c_1 + c_1 e^{-\xi} \tag{63}$$

To determine c_1, we match (58) and (63) as follows:

One-term outer expansion	$y \sim \alpha - x$	
Rewritten in inner variable	$= \alpha - (1 - \varepsilon\xi)$	
Expanded for small ε	$= \alpha - 1 + \varepsilon\xi$	
One-term inner expansion	$= \alpha - 1$	(64a)
One-term inner expansion	$y \sim \beta - c_1 + c_1 e^{-\xi}$	
Rewritten in outer variable	$= \beta - c_1 + c_1 e^{-(1-x)/\varepsilon}$	
Expanded for small ε	$= \beta - c_1 + \text{EST}$	
One-term outer expansion	$= \beta - c_1$	(64b)

Equating (64a) and (64b) according to the matching principle yields

$$\beta - c_1 = \alpha - 1$$

or

$$c_1 = \beta - \alpha + 1$$

Hence

$$y^i = \alpha - 1 + (\beta - \alpha + 1)\, e^{-\xi} + \cdots \tag{65}$$

To determine a single uniformly valid expansion, we form a composite expansion as follows:

$$y^c = y^o + y^i - (y^o)^i$$

or

$$y^c = \alpha - x + (\beta - \alpha + 1)\, e^{-\xi} + \cdots \tag{66}$$

To compare y^c with the exact solution (56), we let $\varepsilon \to 0$ in (56) and obtain

$$y = \alpha - x + (\beta - \alpha + 1) e^{-(1-x)/\varepsilon} + \cdots \tag{67}$$

in agreement with (66).

(c) ***The Method of Multiple Scales***

To determine a first-order uniform expansion using the method of multiple scales, we use the scales

$$\zeta = x \quad \text{and} \quad \xi = \frac{1-x}{\varepsilon}$$

and let

$$y = y_0(\zeta, \xi) + \varepsilon y_1(\zeta, \xi) + \cdots \tag{68}$$

The derivatives transform as in (45a) and (54) becomes

$$\frac{1}{\varepsilon}\frac{\partial^2 y}{\partial \xi^2} - 2\frac{\partial^2 y}{\partial \zeta \partial \xi} + \varepsilon \frac{\partial^2 y}{\partial \zeta^2} + \frac{1}{\varepsilon}\frac{\partial y}{\partial \xi} - \frac{\partial y}{\partial} = 1 \tag{69}$$

Substituting (68) into (69) and equating coefficients of like powers of ε, we obtain

$$\frac{\partial^2 y_0}{\partial \xi^2} + \frac{\partial y_0}{\partial \xi} = 0 \tag{70}$$

$$\frac{\partial^2 y_1}{\partial \xi^2} + \frac{\partial y_1}{\partial \xi} = 2\frac{\partial^2 y_0}{\partial \zeta \partial \xi} + \frac{\partial y_0}{\partial \zeta} + 1 \tag{71}$$

The general solution of (70) can be written as

$$y_0 = A(\zeta) + B(\zeta) e^{-\xi} \tag{72}$$

where A and B are to be determined by imposing the solvability condition at the next level of approximation. Then (71) becomes

$$\frac{\partial^2 y_1}{\partial \xi^2} + \frac{\partial y_1}{\partial \xi} = -B' e^{-\xi} + A' + 1 \tag{73}$$

In order that y_1/y_0 be bounded for all ξ,

$$B' = 0 \quad \text{and} \quad A' + 1 = 0 \tag{74}$$

so that

$$B = b \quad \text{and} \quad A = a - x \tag{75}$$

where a and b are constants. Substituting (75) into (72) and then substituting the result into (68), we obtain

$$y = a - x + be^{-(1-x)/\varepsilon} + \cdots \tag{76}$$

where ζ and ξ were expressed in terms of x. Imposing the boundary conditions (55) yields

$$a = \alpha \quad \text{and} \quad b = \beta - \alpha + 1$$

so that

$$y = \alpha - x + (\beta - \alpha + 1)\, e^{-(1-x)/\varepsilon} + \cdots \tag{77}$$

in agreement with (67).

12.4. Consider the problem

$$\varepsilon y'' + y' = 1 \tag{78}$$

$$y(0) = \alpha, \qquad y(1) = \beta \tag{79}$$

(a) Determine the exact solution.

(b) Use the method of matched asymptotic expansions to determine a first-order uniform expansion. Compare your answer with the exact solution.

(c) Use the method of multiple scales to determine a first-order uniform expansion. Compare your answer with those in (a) and (b).

(a) *The Exact Solution*

The solution of the homogeneous equation (78) can be written as

$$y_h = c_1 + c_2 e^{-x/\varepsilon}$$

where c_1 and c_2 are constants, whereas a particular solution of (78) is

$$y_p = x$$

Hence the general solution of (78) can be written as

$$y = x + c_1 + c_2 e^{-x/\varepsilon} \tag{80}$$

Imposing the boundary conditions (79) yields

$$c_1 + c_2 = \alpha$$

$$c_1 + c_2 e^{-1/\varepsilon} = \beta - 1$$

whose solutions are

$$c_1 = \frac{\beta - 1 - \alpha e^{-1/\varepsilon}}{1 - e^{-1/\varepsilon}}, \qquad c_2 = \frac{\alpha - \beta + 1}{1 - e^{-1/\varepsilon}}$$

Therefore, the solution of (78) and (79) is

$$y = x + \frac{\beta - 1 - \alpha e^{-1/\varepsilon} + (\alpha - \beta + 1) e^{-x/\varepsilon}}{1 - e^{-1/\varepsilon}} \tag{81}$$

As $\varepsilon \to 0$, $\exp(-1/\varepsilon)$ is exponentially small and (81) becomes

$$y^e = x + \beta - 1 + (\alpha - \beta + 1) e^{-x/\varepsilon} + \text{EST} \tag{82}$$

(b) *The Method of Matched Asymptotic Expansions*

As $\varepsilon \to 0$, (78) reduces to a first-order equation that cannot satisfy, in general, both boundary conditions in (79) and one of them must be dropped. Since the coefficient of y' is positive, a boundary layer exists at the left end and the boundary condition $y(0) = \alpha$ must be dropped and the outer expansion must satisfy $y(1) = \beta$. Thus, to determine a first-order uniform expansion using the method of matched expansions, we seek an outer expansion in the form

$$y^o = y_0(x) + \varepsilon y_1(x) + \cdots \tag{83}$$

Substituting (83) into (78) and $y(1) = \beta$ and equating the coefficients of ε^0 on both sides, we obtain

$$y_0' = 1, \qquad y(1) = \beta$$

Hence

$$y_0 = \beta - 1 + x$$

so that

$$y^o = \beta - 1 + x + \cdots \tag{84}$$

To determine an inner expansion, we introduce the stretching transformation

$$\xi = \frac{x}{\varepsilon^\lambda}, \qquad \lambda > 0$$

in (78) and obtain

$$\varepsilon^{1-2\lambda} \frac{d^2 y^i}{d\xi^2} + \varepsilon^{-\lambda} \frac{dy^i}{d\xi} = 1 \tag{85}$$

As $\varepsilon \to 0$, the distinguished limit of (85) occurs for $\lambda = 1$ so that it becomes

$$\frac{d^2 y^i}{d\xi^2} + \frac{dy^i}{d\xi} = \varepsilon \tag{86}$$

Since the inner expansion is valid in the neighborhood of the origin, it must satisfy the condition $y(0)=\alpha$. Since $x=0$ corresponds to $\xi=0$,

$$y^i(0)=\alpha \tag{87}$$

Next, we seek an inner expansion in the form

$$y^i = Y_0(\xi)+\varepsilon Y_1(\xi)+\cdots \tag{88}$$

Substituting (88) into (86) and (87) and equating the coefficients of ε^0 on both sides, we obtain

$$Y_0''+Y_0'=0, \qquad Y_0(0)=\alpha \tag{89}$$

whose solution is

$$Y_0=\alpha-c_1+c_1e^{-\xi}$$

where c_1 is a constant. Hence

$$y^i=\alpha-c_1+c_1e^{-\xi}+\cdots \tag{90}$$

To determine c_1, we match y^o and y^i as follows:

One-term outer expansion	$y\sim\beta-1+x$	
Rewritten in inner variable	$=\beta-1+\varepsilon\xi$	
Expanded for small ε	$=\beta-1+\varepsilon\xi$	
One-term inner expansion	$=\beta-1$	(91a)
One-term inner expansion	$y\sim\alpha-c_1+c_1e^{-\xi}$	
Rewritten in outer variable	$=\alpha-c_1+c_1e^{-x/\varepsilon}$	
Expanded for small ε	$=\alpha-c_1+\text{EST}$	
One-term outer expansion	$=\alpha-c_1$	(91b)

Equating (91a) and (91b) according to the matching principle, we have

$$\beta-1=\alpha-c_1$$

or

$$c_1=\alpha-\beta+1$$

Hence

$$Y_0=\beta-1+(\alpha-\beta+1)e^{-\xi}$$

and

$$y^i=\beta-1+(\alpha-\beta+1)e^{-\xi}+\cdots \tag{92}$$

To determine a single uniform expansion, we form a composite expansion as follows:

$$y^c=y^o+y^i-(y^o)^i$$

or

$$y^c = \beta - 1 + x + (\alpha - \beta + 1) e^{-\xi} + \cdots \tag{93}$$

in agreement with (82).

(c) ***The Method of Multiple Scales***

To determine a first-order uniform expansion using the method of multiple scales, we use the outer and inner variables as the appropriate scales and let

$$y = y_0(\zeta, \xi) + \varepsilon y_1(\zeta, \xi) + \cdots \tag{94}$$

where

$$\zeta = x \quad \text{and} \quad \xi = \frac{x}{\varepsilon}$$

Then the derivatives transform as in (20b) and (78) becomes

$$\frac{1}{\varepsilon}\frac{\partial^2 y}{\partial \xi^2} + 2\frac{\partial^2 y}{\partial \xi \partial \zeta} + \varepsilon \frac{\partial^2 y}{\partial \zeta^2} + \frac{1}{\varepsilon}\frac{\partial y}{\partial \xi} + \frac{\partial y}{\partial \zeta} = 1 \tag{95}$$

Substituting (94) into (95) and equating coefficients of like powers of ε, we obtain

$$\frac{\partial^2 y_0}{\partial \xi^2} + \frac{\partial y_0}{\partial \xi} = 0 \tag{96}$$

$$\frac{\partial^2 y_1}{\partial \xi^2} + \frac{\partial y_1}{\partial \xi} = -2\frac{\partial^2 y_0}{\partial \xi \partial \zeta} - \frac{\partial y_0}{\partial \zeta} + 1 \tag{97}$$

The general solution of (96) can be written as

$$y_0 = A(\zeta) + B(\zeta) e^{-\xi} \tag{98}$$

where A and B need to be determined by imposing the solvability condition at the next order. Substituting (98) into (97) yields

$$\frac{\partial^2 y_1}{\partial \xi^2} + \frac{\partial y_1}{\partial \xi} = B' e^{-\xi} - A' + 1 \tag{99}$$

In order that y_1/y_0 be bounded for all ξ,

$$B' = 0 \quad \text{and} \quad A' - 1 = 0$$

Hence

$$B = b \quad \text{and} \quad A = a + x \tag{100}$$

where a and b are constants. Substituting (98) into (94), using (100), and putting $\zeta = x$

and $\xi = x/\varepsilon$, we have

$$y = a + x + be^{-x/\varepsilon} + \cdots \tag{101}$$

Imposing the boundary conditions (79) yields

$$a + b = \alpha$$

$$a + be^{-1/\varepsilon} = \beta - 1$$

Hence

$$a = \beta - 1 + \text{EST}, \qquad b = \alpha - \beta + 1 + \text{EST}$$

and

$$y = x + \beta - 1 + (\alpha - \beta + 1)e^{-x/\varepsilon} + \cdots \tag{102}$$

in agreement with (82).

12.5. Determine first-order (one-term) uniform expansions for

$$\varepsilon y'' \pm (3x + 1)y' = 1 \tag{103}$$

$$y(0) = \alpha, \qquad y(1) = \beta \tag{104}$$

As $\varepsilon \to 0$, (103) reduces to a first-order equation that cannot satisfy, in general, both boundary conditions (104) and one of them must be dropped. In the case of the positive sign, the coefficient of y' is positive in $[0,1]$ and the boundary layer exists at the left end. In the case of the negative sign, the boundary layer exists at the right end. In both cases, we use the method of matched asymptotic expansions to determine a first-order uniform expansion.

(a) ***The Case of the Positive Sign***

We seek an outer expansion in the form

$$y^o = y_0(x) + \varepsilon y_1(x) + \cdots \tag{105}$$

In this case, the boundary layer is at the left end and the outer expansion must satisfy the second boundary condition in (104). Thus, substituting (105) into (103) with the positive sign and $y(1) = \beta$ and equating the coefficients of ε^0 on both sides, we obtain

$$(3x + 1)y_0' = 1, \qquad y_0(1) = \beta \tag{106}$$

whose solution is

$$y^o = \beta + \tfrac{1}{3}\ln\frac{3x + 1}{4} + \cdots \tag{107}$$

To determine an inner expansion, we introduce the stretching transformation

$$\xi = \frac{x}{\varepsilon^{\lambda}}, \qquad \lambda > 0$$

in (103) with the positive sign and obtain

$$\varepsilon^{1-2\lambda}\frac{d^2y^i}{d\xi^2} + \varepsilon^{-\lambda}(3\varepsilon^{\lambda}\xi + 1)\frac{dy^i}{d\xi} = 1 \tag{108}$$

As $\varepsilon \to 0$, the distinguished limit of (108) occurs when $\lambda = 1$ and (108) becomes

$$\frac{d^2y^i}{d\xi^2} + (3\varepsilon\xi + 1)\frac{dy^i}{d\xi} = \varepsilon \tag{109}$$

The inner expansion is valid at the origin and hence it must satisfy the boundary condition $y(0) = \alpha$. But $x = 0$ corresponds to $\xi = 0$, and hence

$$y^i(0) = \alpha \tag{110}$$

Then we seek an inner expansion in the form

$$y^i = Y_0(\xi) + \varepsilon Y_1(\xi) + \cdots \tag{111}$$

Substituting (111) into (109) and (110) and equating the coefficients of ε^0 on both sides, we obtain

$$Y_0'' + Y_0' = 0, \qquad Y_0(0) = \alpha \tag{112}$$

whose solution is

$$Y_0 = \alpha - c_1 + c_1 e^{-\xi}$$

where c_1 is a constant. Hence

$$y^i = \alpha - c_1 + c_1 e^{-\xi} + \cdots \tag{113}$$

To determine c_1, we match y^o and y^i as follows:

One-term outer expansion	$y \sim \beta + \frac{1}{3}\ln\frac{3x+1}{4}$	
Rewritten in inner variable	$= \beta + \frac{1}{3}\ln\frac{3\varepsilon\xi+1}{4}$	
Expanded for small ε	$= \beta - \frac{1}{3}\ln 4 + \frac{1}{3}(3\varepsilon\xi + \cdots)$	
One-term inner expansion	$= \beta - \frac{1}{3}\ln 4$	(114a)
One-term inner expansion	$y \sim \alpha - c_1 + c_1 e^{-\xi}$	
Rewritten in outer variable	$= \alpha - c_1 + c_1 e^{-x/\varepsilon}$	
Expanded for small ε	$= \alpha - c_1 + \text{EST}$	
One-term outer expansion	$= \alpha - c_1$	(114b)

Equating (114a) and (114b) according to the matching principle, we have

$$\alpha - c_1 = \beta - \tfrac{1}{3}\ln 4$$

or

$$c_1 = \alpha - \beta + \tfrac{1}{3}\ln 4$$

and

$$y^i = \beta - \tfrac{1}{3}\ln 4 + \left(\alpha - \beta + \tfrac{1}{3}\ln 4\right) e^{-\xi} + \cdots \tag{115}$$

To determine a single uniform expansion, we form a composite expansion as follows:

$$y^c = y^o + y^i - (y^o)^i$$

or

$$y^c = \beta + \tfrac{1}{3}\ln\frac{3x+1}{4} + \left(\alpha - \beta + \tfrac{1}{3}\ln 4\right) e^{-\xi} + \cdots \tag{116}$$

(b) *The Case of the Negative Sign*

In this case, the boundary layer is at the right end and the outer expansion satisfies $y(0) = \alpha$. Thus, substituting (105) into (103) with the negative sign and $y(0) = \alpha$ and equating the coefficients of ε^0 on both sides, we obtain

$$-(3x+1)\,y_0' = 1, \qquad y_0(0) = \alpha \tag{117}$$

whose solution is

$$y_0 = \alpha - \tfrac{1}{3}\ln(3x+1) + \cdots \tag{118}$$

To determine an inner expansion, we introduce the stretching transformation

$$\xi = \frac{1-x}{\varepsilon^\lambda}, \qquad \lambda > 0$$

in (103) with the negative sign and obtain

$$\varepsilon^{1-2\lambda}\frac{d^2y^i}{d\xi^2}+\varepsilon^{-\lambda}(4-3\varepsilon^{\lambda}\xi)\frac{dy^i}{d\xi}=1 \tag{119}$$

As $\varepsilon \to 0$, the distinguished limit occurs when $\lambda=1$ and (119) becomes

$$\frac{d^2y^i}{d\xi^2}+(4-3\varepsilon\xi)\frac{dy^i}{d\xi}=\varepsilon \tag{120}$$

The inner expansion must satisfy the condition $y(1)=\beta$. But $x=1$ corresponds to $\xi=0$. Hence

$$y^i(0)=\beta \tag{121}$$

Then we let

$$y^i=Y_0(\xi)+\varepsilon Y_1(\xi)+\cdots \tag{122}$$

in (120) and (121), equate the coefficients of ε^0 on both sides, and obtain

$$Y_0''+4Y_0'=0, \qquad Y_0(0)=\beta \tag{123}$$

Thus

$$Y_0=\beta-c_1+c_1e^{-4\xi}$$

and

$$y^i=\beta-c_1+c_1e^{-4\xi}+\cdots \tag{124}$$

where c_1 is a constant.

To determine c_1, we match y^o and y^i and proceed as follows:

One-term outer expansion	$y\sim\alpha-\frac{1}{3}\ln(3x+1)$	
Rewritten in inner variable	$=\alpha-\frac{1}{3}\ln(4-3\varepsilon\xi)$	
Expanded for small ε	$=\alpha-\frac{1}{3}\ln 4+\frac{1}{3}(\frac{3}{4}\varepsilon\xi+\cdots)$	
One-term inner expansion	$=\alpha-\frac{1}{3}\ln 4$	(125a)
One-term inner expansion	$y\sim\beta-c_1+c_1e^{-4\xi}$	
Rewritten in outer variable	$=\beta-c_1+c_1e^{-4(1-x)/\varepsilon}$	
Expanded for small ε	$=\beta-c_1+\text{EST}$	
One-term outer expansion	$=\beta-c_1$	(125b)

Equating (125a) and (125b) according to the matching principle, we have

$$\beta-c_1=\alpha-\frac{1}{3}\ln 4$$

or

$$c_1 = \beta - \alpha + \tfrac{1}{3}\ln 4$$

Hence

$$y^i = \alpha - \tfrac{1}{3}\ln 4 + (\beta - \alpha + \tfrac{1}{3}\ln 4)\,e^{-4\xi} + \cdots \tag{126}$$

To determine a single uniform expansion, we form a composite expansion as follows:

$$y^c = y^o + y^i - (y^o)^i$$

or

$$y^c = \alpha - \tfrac{1}{3}\ln(3x+1) + (\beta - \alpha + \tfrac{1}{3}\ln 4)\,e^{-4\xi} + \cdots \tag{127}$$

12.6. Determine first-order uniform expansions for

$$\varepsilon y'' \pm y' = 2x \tag{128}$$

$$y(0) = \alpha, \qquad y(1) = \beta \tag{129}$$

In each case, compare your answer with the exact solution.

As $\varepsilon \to 0$, (128) reduces to a first-order equation that cannot satisfy both boundary conditions in (129) and one of them must be dropped and a boundary layer exists there.

(a) ***The Case with the Positive Sign***

In this case, the boundary layer is at the left end and the outer expansion must satisfy the right boundary condition. To determine a uniform expansion using the method of matched asymptotic expansions, first we determine an outer expansion in the form

$$y^o = y_0(x) + \varepsilon y_1(x) + \cdots \tag{130}$$

Substituting (130) into (128) and $y(1) = \beta$ and equating the coefficients of ε^0 on both sides, we obtain

$$y_0' = 2x, \qquad y_0(1) = \beta \tag{131}$$

whose solution is

$$y_0 = x^2 - 1 + \beta$$

so that

$$y^o = x^2 - 1 + \beta + \cdots \tag{132}$$

Since $y^o(0) = \beta - 1 \neq \alpha$, in general, y^o is not valid near the origin, and hence we seek an inner expansion by introducing the stretching transformation

$$\xi = \frac{x}{\varepsilon^\lambda}, \qquad \lambda > 0$$

Then (128) becomes

$$\varepsilon^{1-2\lambda}\frac{d^2y^i}{d\xi^2}+\varepsilon^{-\lambda}\frac{dy^i}{d\xi}=2\varepsilon^{\lambda}\xi$$

As $\varepsilon \to 0$, the distinguished limit occurs when $\lambda=1$ and hence this equation can be rewritten as

$$\frac{d^2y^i}{d\xi^2}+\frac{dy^i}{d\xi}=2\varepsilon^2\xi \tag{133}$$

Next, we seek an inner expansion in the form

$$y^i=Y_0(\xi)+\varepsilon Y_1(\xi)+\cdots \tag{134}$$

The inner expansion must satisfy the left boundary condition. Since $x=0$ corresponds to $\xi=0$,

$$y^i(0)=\alpha \tag{135}$$

Then, substituting (134) into (133) and (135) and equating the coefficients of ε^0 on both sides, we obtain

$$Y_0''+Y_0'=0, \qquad Y_0(0)=\alpha \tag{136}$$

whose solution is

$$Y_0=\alpha-c_1+c_1e^{-\xi}$$

where c_1 is a constant. Thus

$$y^i=\alpha-c_1+c_1e^{-\xi}+\cdots \tag{137}$$

To determine c_1, we match y^o and y^i. To this end, we note that

$$(y^o)^i=\beta-1 \quad \text{and} \quad (y^i)^o=\alpha-c_1$$

Therefore

$$\alpha-c_1=\beta-1$$

or

$$c_1=\alpha-\beta+1$$

and

$$y^i=\beta-1+(\alpha-\beta+1)e^{-\xi}+\cdots \tag{138}$$

To determine a single uniform expansion, we form a composite expansion as follows:

$$y^c = y^o + y^i - (y^o)^i$$

Hence

$$y^c = x^2 - 1 + \beta + (\alpha - \beta + 1)e^{-\xi} + \cdots \tag{139}$$

Next we compare (139) with the exact solution of (128) and (129). A particular solution of (128) is

$$y_p = x^2 - 2\varepsilon x$$

whereas the complimentary solution of (128) is

$$y^c = c_1 + c_2 e^{-x/\varepsilon}$$

Hence

$$y = x^2 - 2\varepsilon x + c_1 + c_2 e^{-x/\varepsilon} \tag{140}$$

is the general solution of (128). Imposing the boundary conditions (129) yields

$$c_1 + c_2 = \alpha$$

$$c_1 + c_2 e^{-1/\varepsilon} = \beta - 1 + 2\varepsilon$$

whose solution is

$$c_1 = \frac{\beta - 1 + 2\varepsilon - \alpha e^{-1/\varepsilon}}{1 - e^{-1/\varepsilon}}, \qquad c_2 = \frac{\alpha - \beta + 1 - 2\varepsilon}{1 - e^{-1/\varepsilon}}$$

Therefore

$$y = x^2 - 2\varepsilon x + \frac{\beta - 1 + 2\varepsilon - \alpha e^{-1/\varepsilon} + (\alpha - \beta + 1 - 2\varepsilon)e^{-x/\varepsilon}}{1 - e^{-1/\varepsilon}} \tag{141}$$

in the exact solution. As $\varepsilon \to 0$, $\exp(-1/\varepsilon)$ is exponentially small, and hence (141) can be rewritten as

$$y = x^2 + \beta - 1 + (\alpha - \beta + 1)e^{-x/\varepsilon} + O(\varepsilon) \tag{142}$$

in agreement with (139).

(b) ***The Case with the Negative Sign***

In this case, the boundary layer is at the right end, the outer expansion satisfies the left boundary condition, and it has the form (130). Substituting (130) into (128) and $y(0) = \alpha$ and equating the coefficients of ε^0 on both sides, we obtain

$$y_0' = -2x, \qquad y_0(0) = \alpha \tag{143}$$

whose solution is

$$y_0 = \alpha - x^2$$

so that

$$y^o = \alpha - x^2 + \cdots \tag{144}$$

Since $y^o(1) = \alpha - 1 \neq \beta$, in general, y^o is not valid near the right end and an inner expansion must be introduced there. To this end, we introduce the stretching transformation

$$\xi = \frac{1-x}{\varepsilon^\lambda}, \qquad \lambda > 0$$

in (128) and obtain

$$\varepsilon^{1-2\lambda} \frac{d^2 y^i}{d\xi^2} + \varepsilon^{-\lambda} \frac{dy^i}{d\xi} = 2(1 - \varepsilon^\lambda \xi) \tag{145}$$

As $\varepsilon \to 0$, the distinguished limit of (145) occurs when $\lambda = 1$. Thus (145) can be rewritten as

$$\frac{d^2 y^i}{d\xi^2} + \frac{dy^i}{d\xi} = 2\varepsilon(1 - \varepsilon\xi) \tag{146}$$

Since $x = 1$ corresponds to $\xi = 0$,

$$y^i(0) = \beta \tag{147}$$

Seeking an inner expansion as in (134) and equating the coefficients of ε^0 on both sides of each of (146) and (147), we obtain

$$Y_0'' + Y_0' = 0, \qquad Y_0(0) = \beta \tag{148}$$

The solution of (148) can be expressed as

$$Y_0 = \beta - c_1 + c_1 e^{-\xi}$$

so that

$$y^i = \beta - c_1 + c_1 e^{-\xi} + \cdots \tag{149}$$

where c_1 is a constant.

To determine c_1, we match y^o and y^i. To this end, we note that

$$(y^o)^i = \alpha - 1, \qquad (y^i)^o = \beta - c_1$$

Hence

$$\alpha - 1 = \beta - c_1$$

or

$$c_1 = \beta - \alpha + 1$$

and

$$y^i = \alpha - 1 + (\beta - \alpha + 1)\, e^{-\xi} + \cdots \tag{150}$$

To determine a single uniform expansion, we form a composite expansion as follows:

$$y^c = y^o + y^i - (y^o)^i$$

or

$$y^c = \alpha - x^2 + (\beta - \alpha + 1)\, e^{-\xi} + \cdots \tag{151}$$

Next we compare (151) with the exact solution. To this end, we note that a particular solution of (128) is

$$y_p = -x^2 - 2\varepsilon x$$

whereas the complimentary solution of (128) is

$$y_c = c_1 + c_2 e^{x/\varepsilon}$$

Hence

$$y = -x^2 - 2\varepsilon x + c_1 + c_2 e^{x/\varepsilon} \tag{152}$$

is the general solution of (128). Imposing the boundary conditions (129) yields

$$c_1 + c_2 = \alpha$$

$$c_1 + c_2 e^{1/\varepsilon} = \beta + 1 + 2\varepsilon$$

whose solution is

$$c_1 = \frac{\alpha e^{1/\varepsilon} - \beta - 1 - 2\varepsilon}{e^{1/\varepsilon} - 1}, \qquad c_2 = \frac{\beta + 1 + 2\varepsilon - \alpha}{e^{1/\varepsilon} - 1}$$

Hence

$$y = -x^2 - 2\varepsilon x + \frac{\alpha e^{1/\varepsilon} - \beta - 1 - 2\varepsilon + (\beta + 1 + 2\varepsilon - \alpha)\, e^{x/\varepsilon}}{e^{1/\varepsilon} - 1} \tag{153}$$

is the exact solution of (128) and (129). As $\varepsilon \to 0$, $\exp(1/\varepsilon)$ is exponentially large and

(153) can be expressed as

$$y = -x^2 + \alpha + (\beta + 1 - \alpha) e^{-(1-x)/\varepsilon} + O(\varepsilon) \tag{154}$$

in agreement with (151).

12.7. Determine first-order uniform expansions for

$$\varepsilon y'' \pm (2x^2 + x + 1) y' = 4x + 1 \tag{155}$$

$$y(0) = \alpha, \qquad y(1) = \beta \tag{156}$$

As $\varepsilon \to 0$, (155) reduces to a first-order equation that cannot satisfy both boundary conditions in (156), and a boundary layer exists at one of the ends. The location of the boundary layer depends on the sign of y'.

(a) ***The Case with the Positive Sign***

In this case, the coefficient of y' is positive in the interval $[0,1]$, and hence the boundary layer is at the origin and the outer expansion satisfies $y(1) = \beta$. To determine a first-order uniform expansion using the method of matched asymptotic expansions, we determine first an outer expansion. It has the same form as in (130). Thus substituting (130) into (155) and $y(1) = \beta$ and equating the coefficients of ε^0 on both sides, we obtain

$$(2x^2 + x + 1) y_0' = 4x + 1, \qquad y_0(1) = \beta \tag{157}$$

whose solution is

$$y_0 = \beta + \ln \frac{2x^2 + x + 1}{4}$$

so that

$$y^o = \beta + \ln \frac{2x^2 + x + 1}{4} + \cdots \tag{158}$$

Since $y^o(0) = \beta - \ln 4 \neq \alpha$, in general, an inner expansion needs to be introduced at the origin. To this end, we introduce the stretching transformation

$$\xi = \frac{x}{\varepsilon^\lambda}, \qquad \lambda > 0$$

in (155) and obtain

$$\varepsilon^{1-2\lambda} \frac{d^2 y^i}{d\xi^2} + \varepsilon^{-\lambda} (2\varepsilon^{2\lambda} \xi^2 + \varepsilon^\lambda \xi + 1) \frac{dy^i}{d\xi} = 4\varepsilon^\lambda \xi + 1 \tag{159}$$

As $\varepsilon \to 0$, the distinguished limit of (159) corresponds to $\lambda = 1$ and (159) can be rewritten as

$$\frac{d^2y^i}{d\xi^2}+(2\varepsilon^2\xi^2+\varepsilon\xi+1)\frac{dy^i}{d\xi}=\varepsilon(4\varepsilon\xi+1) \tag{160}$$

Since the inner expansion is valid at the origin, it must satisfy the boundary condition there. Since $x=0$ corresponds to $\xi=0$,

$$y^i(0)=\alpha \tag{161}$$

Moreover, the inner expansion has the form (134). Substituting (134) into (160) and (161) and equating the coefficients of ε^0 on both sides, we obtain

$$Y_0''+Y_0'=0, \qquad Y_0(0)=\alpha \tag{162}$$

whose solution is

$$Y_0=\alpha-c_1+c_1e^{-\xi}$$

so that

$$y^i=\alpha-c_1+c_1e^{-\xi}+\cdots \tag{163}$$

where c_1 is a constant.

To determine c_1, we match y^o and y^i. To this end, we note that

$$(y^o)^i=\beta-\ln 4, \qquad (y^i)^o=\alpha-c_1$$

Therefore

$$\beta-\ln 4=\alpha-c_1$$

or

$$c_1=\alpha-\beta+\ln 4$$

and

$$y^i=\beta-\ln 4+(\alpha-\beta+\ln 4)e^{-\xi}+\cdots$$

To determine a single uniform expansion, we form a composite expansion as follows:

$$y^c=y^o+y^i-(y^o)^i$$

or

$$y^c=\beta+\ln\frac{2x^2+x+1}{4}+(\alpha-\beta+\ln 4)e^{-\xi}+\cdots \tag{164}$$

(b) ***The Case with the Negative Sign***

In this case, the boundary layer is at the right end and the outer expansion, which has the form (130), satisfies the left boundary condition. Substituting (130) into (155) and $y(0)=\alpha$ and equating the coefficients of ε^0 on both sides, we obtain

$$(2x^2+x+1)\,y_0' = -(4x+1), \qquad y_0(0)=\alpha \tag{165}$$

whose solution is

$$y_0 = \alpha - \ln(2x^2+x+1)$$

so that

$$y^o = \alpha - \ln(2x^2+x+1) + \cdots \tag{166}$$

Since $y^o(1) = \alpha - \ln 4 \neq \beta$, in general, y^o is not valid at the right end and a boundary layer must be introduced there. To this end, we introduce the stretching transformation

$$\xi = \frac{1-x}{\varepsilon^\lambda}, \qquad \lambda > 0$$

in (155) and obtain

$$\varepsilon^{1-2\lambda}\frac{d^2y^i}{d\xi^2} + \varepsilon^{-\lambda}\left(4-5\varepsilon^\lambda\xi+2\varepsilon^{2\lambda}\xi^2\right)\frac{dy^i}{d\xi} = 5-4\varepsilon^\lambda\xi \tag{167}$$

The distinguished limit of (167) corresponds to $\lambda=1$. Thus (167) can be rewritten as

$$\frac{d^2y^i}{d\xi^2} + \left(4-5\varepsilon\xi+2\varepsilon^2\xi^2\right)\frac{dy^i}{d\xi} = \varepsilon(5-4\varepsilon\xi) \tag{168}$$

Since the inner expansion must satisfy the right boundary condition and since $x=1$ corresponds to $\xi=0$,

$$y^i(0) = \beta \tag{169}$$

We seek an inner expansion in the form (134). Substituting (134) into (168) and (169) and equating the coefficients of ε^0 on both sides, we obtain

$$Y_0'' + 4Y_0' = 0, \qquad Y_0(0)=\beta \tag{170}$$

whose solution is

$$Y_0 = \beta - c_1 + c_1 e^{-4\xi}$$

so that

$$y^i = \beta - c_1 + c_1 e^{-4\xi} + \cdots \tag{171}$$

where c_1 is a constant.

To determine c_1, we match y^o and y^i. To this end, we note that

$$(y^o)^i = \alpha - \ln 4, \qquad (y^i)^o = \beta - c_1$$

Hence

$$\alpha - \ln 4 = \beta - c_1$$

or

$$c_1 = \beta - \alpha + \ln 4$$

so that

$$y^i = \alpha - \ln 4 - (\alpha - \beta - \ln 4)\, e^{-4\xi} + \cdots \tag{172}$$

To determine a single uniform expansion, we form a composite expansion as follows:

$$y^c = y^o + y^i - (y^o)^i$$

or

$$y^c = \alpha - \ln(2x^2 + x + 1) - (\alpha - \beta - \ln 4)\, e^{-4\xi} + \cdots \tag{173}$$

12.8. Determine first-order uniform expansions for

(a)
$$\varepsilon y'' + xy' + xy = 0 \tag{174}$$

$$y(0) = \alpha, \qquad y(1) = \beta \tag{175}$$

(b)
$$\varepsilon y'' - (1-x)\, y' - (1-x)\, y = 0 \tag{176}$$

$$y(0) = \alpha, \qquad y(1) = \beta \tag{177}$$

The First Case. As $\varepsilon \to 0$, (174) reduces to a first-order equation. Since the coefficient of y' is positive everywhere in the interval of interest except at the origin where it vanishes, a boundary layer is expected at the origin. We seek an outer expansion in the form (130). It must satisfy the right boundary condition. Thus, substituting (130) into (174) and $y(1) = \beta$ and equating the coefficients of ε^0 on both sides, we obtain

$$xy_0' + xy_0 = 0, \qquad y_0(1) = \beta$$

whose solution is

$$y_0 = \beta e^{1-x}$$

so that

$$y^o = \beta e^{1-x} + \cdots \tag{178}$$

Since $y^o(0) = \beta e \neq \alpha$, in general, a boundary layer is expected at the origin, and hence we seek an inner expansion by introducing the stretching transformation

$$\xi = \frac{x}{\varepsilon^\lambda}, \qquad \lambda > 0$$

Then (174) becomes

$$\varepsilon^{1-2\lambda}\frac{d^2y^i}{d\xi^2} + \xi\frac{dy^i}{d\xi} + \varepsilon^\lambda \xi y^i = 0$$

As $\varepsilon \to 0$, the distinguished limit of this equation corresponds to $\lambda = \frac{1}{2}$. Thus it can be rewritten as

$$\frac{d^2y^i}{d\xi^2} + \xi\frac{dy^i}{d\xi} + \varepsilon^{1/2}\xi y^i = 0 \tag{179}$$

Since the inner expansion is expected to satisfy the left boundary condition and since $x = 0$ corresponds to $\xi = 0$,

$$y^i(0) = \alpha \tag{180}$$

Seeking an inner expansion in the form (134) and equating the coefficients of ε^0 on both sides of (179) and (180), we obtain

$$Y_0'' + \xi Y_0' = 0, \qquad Y_0(0) = \alpha \tag{181}$$

whose solution is

$$Y_0 = c_1 \int_0^\xi e^{-\tau^2/2}\, d\tau + \alpha$$

so that

$$y^i = c_1 \int_0^\xi e^{-\tau^2/2}\, d\tau + \alpha + \cdots \tag{182}$$

where c_1 is a constant.

To determine c_1, we match y^o and y^i. To this end, we note that

$$(y^o)^i = \beta e, \qquad (y^i)^o = \alpha + c_1\sqrt{\frac{\pi}{2}} \tag{183}$$

Hence

$$\beta e = \alpha + c_1\sqrt{\frac{\pi}{2}}$$

or

$$c_1 = \sqrt{\frac{2}{\pi}}\,(\beta e - \alpha)$$

and

$$y^i = \sqrt{\frac{2}{\pi}}\,(\beta e - \alpha)\int_0^{\xi} e^{-\tau^2/2}\,d\tau + \alpha + \cdots \tag{184}$$

To determine a single uniform expansion, we form a composite expansion by adding (178) and (184) and then substracting (183). The result is

$$y^c = \beta e^{1-x} + \sqrt{\frac{2}{\pi}}\,(\beta e - \alpha)\int_0^{\xi} e^{-\tau^2/2}\,d\tau + \alpha - \beta e + \cdots \tag{185}$$

The Second Case. In this case, the coefficient of y' is negative everywhere in the interval of interest except at $x=1$, where it is zero. Thus a boundary layer is expected at $x=1$. Moreover, the outer expansion, which has the form (130), satisfies the left boundary condition. Hence, substituting (130) into (176) and $y(0)=\alpha$ and equating the coefficients of ε^0 on both sides, we obtain

$$y_0' + y_0 = 0, \qquad y_0(0) = \alpha \tag{186}$$

whose solution is

$$y_0 = \alpha e^{-x}$$

so that

$$y^o = \alpha e^{-x} + \cdots \tag{187}$$

Since $y^o(1) = \alpha e^{-1} \neq \beta$, in general, a boundary layer needs to be introduced at the right end. To this end, we introduce the stretching transformation

$$\xi = \frac{1-x}{\varepsilon^{\lambda}}, \qquad \lambda > 0$$

in (176) and obtain

$$\varepsilon^{1-2\lambda}\frac{d^2 y^i}{d\xi^2} + \xi\frac{dy^i}{d\xi} - \varepsilon^{\lambda}\xi y^i = 0 \tag{188}$$

whose distinguished limit corresponds to $\lambda = \frac{1}{2}$. Thus we rewrite (188) as

$$\frac{d^2 y^i}{d\xi^2} + \xi\frac{dy^i}{d\xi} - \varepsilon^{1/2}\xi y^i = 0 \tag{189}$$

Since the inner expansion satisfies the right boundary condition and since $x=1$

corresponds to $\xi = 0$,

$$y^i(0) = \beta \tag{190}$$

Seeking an inner expansion in the form (134) and equating the coefficients of ε^0 on both sides of (189) and (190), we obtain

$$Y_0'' + \xi Y_0' = 0, \qquad Y_0(0) = \beta \tag{191}$$

whose solution is

$$Y_0 = c_1 \int_0^{\xi} e^{-\tau^2/2}\, d\tau + \beta$$

so that

$$y^i = c_1 \int_0^{\xi} e^{-\tau^2/2}\, d\tau + \beta + \cdots \tag{192}$$

where c_1 is a constant.

To determine c_1, we match y^o and y^i. To this end, we note that

$$(y^o)^i = \alpha e^{-1}, \qquad (y^i)^o = -c_1\sqrt{\frac{\pi}{2}} + \beta \tag{193}$$

Hence

$$\alpha e^{-1} = -c_1\sqrt{\frac{\pi}{2}} + \beta$$

or

$$c_1 = \sqrt{\frac{2}{\pi}}\,(\beta - \alpha e^{-1})$$

and

$$y^i = \sqrt{\frac{2}{\pi}}\,(\beta - \alpha e^{-1}) \int_0^{\xi} e^{-\tau^2/2}\, d\tau + \beta + \cdots \tag{194}$$

To determine a single uniform expansion, we form a composite expansion by adding (187) and (194) and then subtracting (193). The result is

$$y^c = \alpha e^{-x} + \sqrt{\frac{2}{\pi}}\,(\beta - \alpha e^{-1}) \int_0^{\xi} e^{-\tau^2/2}\, d\tau + \beta - \alpha e^{-1} + \cdots \tag{195}$$

12.9. Determine a first-order uniform expansion for

$$\varepsilon y'' + x^2 y' - x^3 y = 0 \tag{196}$$

$$y(0) = \alpha, \qquad y(1) = \beta \tag{197}$$

As $\varepsilon \to 0$, (196) reduces to a first-order equation that cannot support both boundary conditions in (197). The coefficient of y' is positive everywhere in the interval of interest except at the origin where it vanishes. This suggests that a boundary layer may exist at the origin. This suggestion will be checked a posteriori.

Assuming the existence of a boundary layer at the origin demands that the outer expansion must satisfy the right boundary condition. Substituting the outer expansion (130) into (196) and $y(1) = \beta$ and equating the coefficients of ε^0 on both sides, we obtain

$$x^2 y_0' - x^3 y_0 = 0, \qquad y_0(1) = \beta \tag{198}$$

whose solution is

$$y_0 = \beta e^{(x^2-1)/2} \tag{199}$$

so that

$$y^o = \beta e^{(x^2-1)/2} + \cdots \tag{200}$$

Since $y^o(0) = \beta e^{-1/2} \neq \alpha$, in general, we introduce an inner expansion near the origin by using the stretching transformation

$$\xi = \frac{x}{\varepsilon^\lambda}, \qquad \lambda > 0$$

Then (196) becomes

$$\varepsilon^{1-2\lambda} \frac{d^2 y^i}{d\xi^2} + \varepsilon^\lambda \xi^2 \frac{dy^i}{d\xi} - \varepsilon^{3\lambda} \xi^3 y^i = 0 \tag{201}$$

As $\varepsilon \to 0$, the distinguished limit of (201) corresponds to $\lambda = \frac{1}{3}$. Thus we rewrite (201) as

$$\frac{d^2 y^i}{d\xi^2} + \xi^2 \frac{dy^i}{d\xi} - \varepsilon^{2/3} \xi^3 y^i = 0 \tag{202}$$

Since the inner expansion must be valid at the origin and since $x = 0$ corresponds to $\xi = 0$,

$$y^i(0) = \alpha \tag{203}$$

Substituting the inner expansion (134) into (202) and (203) and equating the coefficients of ε^0 on both sides, we obtain

$$Y_0'' + \xi^2 Y_0' = 0 \tag{204}$$

$$Y_0(0) = \alpha \tag{205}$$

Considering Y_0' as the dependent variable, one realizes that (204) is a first-order linear equation whose general solution is

$$Y_0' = c_1 e^{-\xi^3/3}$$

A further integration yields

$$Y_0 = c_1 \int_0^{\xi} e^{-\tau^3/3}\, d\tau + c_2$$

where c_1 and c_2 are constants. But $Y_0(0) = \alpha$, hence $c_2 = \alpha$ and

$$Y_0 = c_1 \int_0^{\xi} e^{-\tau^3/3}\, d\tau + \alpha$$

Thus

$$y^i = c_1 \int_0^{\xi} e^{-\tau^3/3}\, d\tau + \alpha + \cdots \tag{206}$$

To determine c_1, we match y^o and y^i. To this end, we note that

$$(y^o)^i = \beta e^{-1/2} \tag{207}$$

$$(y^i)^o = c_1 \int_0^{\infty} e^{-\tau^3/3}\, d\tau + \alpha \tag{208}$$

Letting $\frac{1}{3}\tau^3 = t$ in the integrand, we rewrite the integral in (208) as

$$\int_0^{\infty} e^{-\tau^3/3}\, d\tau = 3^{-2/3} \int_0^{\infty} t^{-2/3} e^{-t}\, dt = 3^{-2/3}\Gamma\left(\tfrac{1}{3}\right)$$

Hence

$$(y^i)^o = 3^{-2/3}\Gamma\left(\tfrac{1}{3}\right) c_1 + \alpha$$

Then, the matching principle demands that

$$\beta e^{-1/2} = 3^{-2/3}\Gamma\left(\tfrac{1}{3}\right) c_1 + \alpha$$

or

$$c_1 = \frac{3^{2/3}}{\Gamma\left(\frac{1}{3}\right)}\left(\beta e^{-1/2} - \alpha\right)$$

and

$$y^i = \frac{3^{2/3}}{\Gamma\left(\frac{1}{3}\right)}\left(\beta e^{-1/2} - \alpha\right) \int_0^{\xi} e^{-\tau^3/3}\, d\tau + \alpha + \cdots \tag{209}$$

Since the outer and inner expansions were successfully matched, our assumption about the location of the boundary layer is correct. To determine a single uniform expansion, we form a composite expansion by adding (200) and (209) and then substracting (207). The result is

$$y^c = \beta e^{(x^2-1)/2} + \frac{3^{2/3}}{\Gamma(\frac{1}{3})}(\beta e^{-1/2} - \alpha)\int_0^\xi e^{-\tau^3/3}\,d\tau + \alpha - \beta e^{-1/2} + \cdots \tag{210}$$

12.10. Consider the problem

$$\varepsilon y'' + x^n y' - x^m y = 0 \tag{211}$$

$$y(0) = \alpha, \qquad y(1) = \beta \tag{212}$$

Under what conditions will there be a boundary layer at the origin? Determine a first-order uniform expansion when a boundary layer exists there.

We assume that a boundary layer exists at the origin. Then the outer expansion must satisfy the right boundary condition, the inner expansion must satisfy the left boundary condition, and if there is only one distinguished limit the outer and inner expansions must match. To start we seek an outer expansion in the form (130). Substituting (130) into (211) and $y(1) = \beta$ and equating the coefficients of ε^0 on both sides, we obtain

$$x^n y_0' - x^m y_0 = 0, \qquad y_0(1) = \beta \tag{213}$$

whose solution is

$$y_0 = \beta x \quad \text{if } n - m = 1 \tag{214}$$

$$y_0 = \beta \exp\left(\frac{x^{m-n+1} - 1}{m - n + 1}\right) \quad \text{if } n - m \neq 1 \tag{215}$$

To determine an inner expansion, we introduce the stretching transformation

$$\xi = \frac{x}{\varepsilon^\lambda}, \quad \lambda > 0$$

in (211) and obtain

$$\varepsilon^{1-2\lambda}\frac{d^2y^i}{d\xi^2} + \varepsilon^{(n-1)\lambda}\xi^n\frac{dy^i}{d\xi} - \varepsilon^{m\lambda}\xi^m y^i = 0 \tag{216}$$

As $\varepsilon \to 0$, the distinguished limits of (216) are

(i) $$\frac{d^2y^i}{d\xi^2} + \xi^{m+1}\frac{dy^i}{d\xi} - \xi^m y^i = 0 \quad \text{if } n - m = 1 \text{ and } m \neq -2 \tag{217}$$

corresponding to $\lambda = 1/(m+2)$,

(ii)
$$\frac{d^2 y^i}{d\xi^2} + \xi^n \frac{dy^i}{d\xi} = 0 \quad \text{if } n - m < 1 \text{ and } n \neq -1 \tag{218}$$

corresponding to $\lambda = 1/(n+1)$, and

(iii)
$$\frac{d^2 y^i}{d\xi^2} - \xi^m y^i = 0 \quad \text{if } n - m > 1 \text{ and } m \neq -2 \tag{219}$$

corresponding to $\lambda/(m+2)$. There are no distinguished limits and there is no boundary layer at the origin when $n = -1$ and $m = -2$, $n = -1$ and $m > -2$, or $m = -2$ and $n > -1$.

In (i) the distinguished limit is the same as the original problem and no simplification was achieved. One needs to solve the original problem for the case $n = m + 1$.

In (ii) the general solution of (218) is

$$y^i = c_1 \int_0^\xi \exp\left(-\frac{\tau^{n+1}}{n+1}\right) d\tau + c_2 + \cdots \tag{220}$$

for $n > -1$. For $n < -1$, the integral does not exist, and $c_1 = 0$ for a finite $y^i(0)$. Thus, in the latter case, no boundary layer exists at the origin. For the former case, since the inner expansion must satisfy the boundary condition at the origin and because $x = 0$ corresponds to $\xi = 0$, $y^i(0) = \alpha$ and hence $c_2 = \alpha$. Therefore

$$y^i = c_1 \int_0^\xi \exp\left(-\frac{\tau^{n+1}}{n+1}\right) d\tau + \alpha + \cdots \tag{221}$$

To determine c_1 we match (215) and (221). To this end, we note that

$$(y^o)^i = \beta \exp\left(\frac{-1}{m-n+1}\right) \tag{222}$$

$$(y^i)^o = c_1 \int_0^\infty \exp\left(-\frac{\tau^{n+1}}{n+1}\right) d\tau + \alpha \tag{223}$$

Letting

$$\frac{\tau^{n+1}}{n+1} = t \quad \text{so that } \tau^n\, d\tau = dt$$

we have

$$\int_0^\infty \exp\left(-\frac{\tau^{n+1}}{n+1}\right) d\tau = \frac{1}{(n+1)^{n/(n+1)}} \int_0^\infty t^{-n/(n+1)} e^{-t}\, dt$$

$$= (n+1)^{-n/(n+1)} \Gamma[1/(n+1)]$$

Equating (222) and (223), we have

$$c_1 = \frac{(n+1)^{n/(n+1)}}{\Gamma[1/(n+1)]}\left[\beta \exp\left(-\frac{1}{m-n+1}\right) - \alpha\right]$$

and

$$y^i = \frac{(n+1)^{n/(n+1)}}{\Gamma[1/(n+1)]}\left[\beta \exp\left(-\frac{1}{m-n+1}\right) - \alpha\right]\int_0^\xi \exp\left(-\frac{\tau^{n+1}}{n+1}\right) d\tau + \alpha + \cdots \tag{224}$$

Adding (215) and (224) and then subtracting (222), we obtain the composite expansion

$$y^c = \beta \exp\left(\frac{x^{m-n+1}-1}{m-n+1}\right) + \frac{(n+1)^{n/(n+1)}}{\Gamma[1/(n+1)]}\left[\beta \exp\left(-\frac{1}{m-n+1}\right) - \alpha\right]$$

$$\times \int_0^\xi \exp\left(-\frac{\tau^{n+1}}{n+1}\right) d\tau + \alpha - \beta \exp\left(-\frac{1}{m-n+1}\right) + \cdots \tag{225}$$

for the case $n - m < 1$ and $n \neq -1$.

In (iii) the solution of (219) is

$$y^i = \sqrt{\xi}\left\{c_1 I_\nu\left[\frac{2}{m+2}\xi^{(m+2)/2}\right] + c_2 K_\nu\left[\frac{2}{m+2}\xi^{(m+2)/2}\right]\right\} + \cdots \tag{226}$$

where $\nu = 1/(m+2)$. Since

$$I_\nu \sim \frac{e^\xi}{\sqrt{2\pi\xi}}, \qquad K_\nu \sim \sqrt{\frac{\pi}{2\xi}}\, e^{-\xi} \quad \text{as } \xi \to \infty$$

matching y^i and y^o demands that $c_1 = 0$. Since the inner expansion must satisfy the boundary condition at the origin and because $x = 0$ corresponds to $\xi = 0$, $y^i(0) = \alpha$. Moreover, since

$$K_\nu(z) \sim \tfrac{1}{2}\Gamma(\nu)\left(\tfrac{1}{2}z\right)^{-\nu} \quad \text{as } z \to 0$$

it follows that

$$\sqrt{\xi}\, K_\nu\left[\frac{2}{m+2}\xi^{(m+2)/2}\right] \sim \tfrac{1}{2}\sqrt{\xi}\,\Gamma(\nu)\left[\frac{1}{m+2}\xi^{(m+2)/2}\right]^{-\nu}$$

$$= \tfrac{1}{2}\left(\frac{1}{m+2}\right)^{-\nu}\Gamma(\nu)$$

Hence imposing the condition $y^i(0)=\alpha$ yields

$$c_2 = \frac{2\alpha}{(m+2)^{\nu}\Gamma(\nu)}$$

Then

$$y^i = \frac{2\alpha}{(m+2)^{\nu}\Gamma(\nu)}\sqrt{\xi}\,K_{\nu}\left[\frac{2}{m+2}\xi^{(m+2)/2}\right]+\cdots \tag{227}$$

Since

$$(y^o)^i = 0 \quad \text{and} \quad (y^i)^o = 0$$

the inner and outer expansions match each other. Adding (215) and (227) yields the composite expansion

$$y^c = \beta\exp\left(\frac{x^{m-n+1}-1}{m-n+1}\right)+\frac{2\alpha}{(m+2)^{\nu}\Gamma(\nu)}\sqrt{\xi}\,K_{\nu}\left[\frac{2}{m+2}\xi^{(m+2)/2}\right]+\cdots \tag{228}$$

for the case $n-m>1$ and $m\neq -2$.

12.11. Determine a first-order uniform expansion for

$$\varepsilon y'' + xy' - xy = 0, \qquad \varepsilon \ll 1 \tag{229}$$

$$y(-1)=\alpha, \qquad y(1)=\beta \tag{230}$$

In this problem, the equation is linear and the coefficient of y' changes sign at $x=0$ in the middle of the interval $[-1,1]$. Thus we expect a boundary layer (transition layer) at $x=0$.

Outer Expansion. We seek an outer expansion in the form

$$y^o = y_0(x)+\varepsilon y_1(x)+\cdots \tag{231}$$

Substituting (231) into (229) and equating the coefficients of ε^0 to zero, we obtain

$$xy_0' - xy_0 = 0 \tag{232}$$

When $x\neq 0$, (232) becomes

$$y_0' - y_0 = 0 \tag{233}$$

whose general solution is

$$y_0 = ce^x \tag{234}$$

The outer expansion is expected to be valid everywhere, except in a small neighborhood around $x=0$. Thus there are two outer expansions: a y^l that satisfies the left boundary condition and a y^r that satisfies the right boundary condition.

Substituting (231) into $y(-1)=\alpha$ and equating the coefficients of ε^0 on both sides, we have

$$y_0(-1)=\alpha \tag{235}$$

Putting (235) in (234) yields $c=\alpha e$ so that

$$y^l=\alpha e^{1+x}+\cdots \tag{236}$$

Substituting (231) into $y(1)=\beta$ and equating the coefficients of ε^0 on both sides, we have

$$y_0(1)=\beta \tag{237}$$

Putting (237) in (234) yields $c=\beta e^{-1}$ so that

$$y^r=\beta e^{x-1}+\cdots \tag{238}$$

Inner Expansion. To investigate the interval near $x=0$, we introduce the stretching transformation

$$\xi=\frac{x}{\varepsilon^\lambda}, \qquad \lambda>0 \tag{239}$$

Then (229) becomes

$$\varepsilon^{1-2\lambda}\frac{d^2y^i}{d\xi^2}+\xi\frac{dy^i}{d\xi}-\varepsilon^\lambda\xi y^i=0 \tag{240}$$

As $\varepsilon\to 0$ with ξ being fixed, the distinguished limit of (240) corresponds to $\lambda=\frac{1}{2}$. Putting $\lambda=\frac{1}{2}$ in (240) gives

$$\frac{d^2y^i}{d\xi^2}+\xi\frac{dy^i}{d\xi}-\varepsilon^{1/2}\xi y^i=0 \tag{241}$$

Next we seek an expansion of the solutions of (241) in the form

$$y^i=Y_0(\xi)+\varepsilon^{1/2}Y_1(\xi)+\cdots \tag{242}$$

Substituting (242) into (241) and equating the coefficients of ε^0 to zero, we obtain

$$Y_0''+\xi Y_0'=0 \tag{243}$$

To solve (243), we let $Y_0'=u$ and obtain

$$u'+\xi u=0 \tag{244}$$

Using separation of variables or an integrating factor, we find that the general solution of (244) is

$$u = a_0 e^{-\xi^2/2} \tag{245}$$

But $u = Y_0'$, hence

$$Y_0' = a_0 e^{-\xi^2/2} \tag{246}$$

Integrating (246) yields

$$Y_0 = a_0 \int_0^\xi e^{-\tau^2/2}\, d\tau + a_1 \tag{247}$$

where the dummy variable of integration is denoted by τ and the lower limit of integration is taken to be $\xi = 0$, so that it coincides with the location of the transition layer $x = 0$.

Putting (247) into (242), we have

$$y^i = a_0 \int_0^\xi e^{-\tau^2/2}\, d\tau + a_1 + \cdots \tag{248}$$

The constants of integration a_0 and a_1 need to be determined by matching, because y^i is not valid near either end where the boundary conditions are specified.

Matching. To match y^i as given by (248) with y^r as given by (238), we let $m = n = 1$ in van Dyke's matching principle and proceed as follows:

$$\begin{aligned}
&\text{One-Term outer expansion} && = \beta e^{x-1} \\
&\quad\text{Rewritten in inner variable} && = \beta e^{\sqrt{\varepsilon}\xi - 1} \\
&\quad\text{Expanded for small } \varepsilon && = \beta e^{-1}(1 + \sqrt{\varepsilon}\,\xi + \cdots) \\
&\quad\text{One-term inner expansion} && = \beta e^{-1}
\end{aligned} \tag{249}$$

$$\begin{aligned}
&\text{One-term inner expansion} && = a_0 \int_0^\xi e^{-\tau^2/2}\, d\tau + a_1 \\
&\quad\text{Rewritten in outer variable} && = a_0 \int_0^{x/\sqrt{\varepsilon}} e^{-\tau^2/2}\, d\tau + a_1 \\
&\quad\text{Expanded for small } \varepsilon && = a_0 \int_0^\infty e^{-\tau^2/2}\, d\tau + a_1 + \cdots \\
& && = \frac{a_0\sqrt{\pi}}{\sqrt{2}} + a_1 + \cdots \\
&\quad\text{One-term outer expansion} && = \frac{a_0\sqrt{\pi}}{\sqrt{2}} + a_1
\end{aligned} \tag{250}$$

Equating (249) and (250), we obtain

$$\frac{a_0\sqrt{\pi}}{\sqrt{2}} + a_1 = \beta e^{-1} \tag{251}$$

Similarly, we match y^l and y^i as follows:

$$
\begin{aligned}
&\text{One-term outer expansion} && = \alpha e^{x+1} \\
&\quad\text{Rewritten in inner variable} && = \alpha e^{\sqrt{\varepsilon}\xi+1} \\
&\quad\text{Expanded for small } \varepsilon && = \alpha e(1+\sqrt{\varepsilon}\,\xi+\cdots) \\
&\quad\text{One-term inner expansion} && = \alpha e \qquad (252)
\end{aligned}
$$

$$
\begin{aligned}
&\text{One-term inner expansion} && = a_0\int_0^{\xi} e^{-\tau^2/2}\,d\tau + a_1 \\
&\quad\text{Rewritten in outer variable} && = a_0\int_0^{x/\sqrt{\varepsilon}} e^{-\tau^2/2}\,d\tau + a_1 \\
&\quad\text{Expanded for small } \varepsilon && = a_0\int_0^{-\infty} e^{-\tau^2/2}\,d\tau + a_1 + \cdots \\
&\quad\text{One-term outer expansion} && = -\frac{a_0\sqrt{\pi}}{\sqrt{2}} + a_1 \qquad (253)
\end{aligned}
$$

Equating (252) and (253) gives

$$
-\frac{a_0\sqrt{\pi}}{\sqrt{2}} + a_1 = \alpha e \tag{254}
$$

Solving (251) and (254), we obtain

$$
a_0 = \frac{1}{\sqrt{2\pi}}\left(\beta e^{-1} - \alpha e\right), \qquad a_1 = \tfrac{1}{2}\left(\beta e^{-1} + \alpha e\right) \tag{255}
$$

Hence

$$
y^i = \frac{1}{\sqrt{2\pi}}\left(\beta e^{-1} - \alpha e\right)\int_0^{\xi} e^{-\tau^2/2}\,d\tau + \tfrac{1}{2}\left(\beta e^{-1} + \alpha e\right) + \cdots \tag{256}
$$

Then

$$
\begin{aligned}
y_r^c &= \beta e^{x-1} + a_0\int_0^{\xi} e^{-\tau^2/2}\,d\tau + a_1 - \frac{a_0\sqrt{\pi}}{\sqrt{2}} - a_1 + \cdots \\
&= \beta e^{x-1} + \tfrac{1}{2}\left(\beta e^{-1} - \alpha e\right)\left(\frac{\sqrt{2}}{\sqrt{\pi}}\int_0^{\xi} e^{-\tau^2/2}\,d\tau - 1\right) + \cdots
\end{aligned} \tag{257}
$$

$$
\begin{aligned}
y_l^c &= \alpha e^{x+1} + a_0\int_0^{\xi} e^{-\tau^2/2}\,d\tau + a_1 + \frac{a_0\sqrt{\pi}}{\sqrt{2}} - a_1 + \cdots \\
&= \alpha e^{x+1} + \tfrac{1}{2}\left(\beta e^{-1} - \alpha e\right)\left(\frac{\sqrt{2}}{\sqrt{\pi}}\int_0^{\xi} e^{-\tau^2/2}\,d\tau + 1\right) + \cdots
\end{aligned} \tag{258}
$$

12.12. Consider the problem

$$
\varepsilon y''' - y' = 1 \tag{259}
$$

$$
y(0) = \alpha, \qquad y'(0) = \beta, \qquad y(1) = \gamma \tag{260}
$$

(a) Determine the exact solution and use it to show that there is in general a boundary layer at each end.

(b) Determine a two-term uniform expansion and compare your answer with the exact solution.

(a) ***The Exact Solution***

A particular solution of (259) is

$$y_p = -x \tag{261}$$

The homogeneous solution is

$$y_h = a_0 + a_1 e^{x/\sqrt{\varepsilon}} + a_2 e^{-x/\sqrt{\varepsilon}} \tag{262}$$

Therefore, the general solution of (259) is

$$y = -x + a_0 + a_1 e^{x/\sqrt{\varepsilon}} + a_2 e^{-x/\sqrt{\varepsilon}} \tag{263}$$

Substituting (263) into (260) gives

$$a_0 + a_1 + a_2 = \alpha \tag{264}$$

$$a_1 - a_2 = (\beta + 1)\sqrt{\varepsilon} \tag{265}$$

$$a_0 + a_1 e^{1/\sqrt{\varepsilon}} + a_2 e^{-1/\sqrt{\varepsilon}} = \gamma + 1 \tag{266}$$

Subtracting (264) from (266), we have

$$a_1(e^{1/\sqrt{\varepsilon}} - 1) + a_2(e^{-1/\sqrt{\varepsilon}} - 1) = \gamma - \alpha + 1 \tag{267}$$

Solving (265) and (267) yields

$$\begin{aligned} a_1 &= \frac{\gamma - \alpha + 1 + (\beta + 1)\sqrt{\varepsilon}(e^{-1/\sqrt{\varepsilon}} - 1)}{e^{-1/\sqrt{\varepsilon}} + e^{1/\sqrt{\varepsilon}} - 2} \\ a_2 &= \frac{\gamma - \alpha + 1 - (\beta + 1)\sqrt{\varepsilon}(e^{1/\sqrt{\varepsilon}} - 1)}{e^{-1/\sqrt{\varepsilon}} + e^{1/\sqrt{\varepsilon}} - 2} \end{aligned} \tag{268}$$

Therefore, the solution of (259) and (260) is

$$\begin{aligned} y = -x + \alpha &+ \frac{\gamma - \alpha + 1 + (\beta + 1)\sqrt{\varepsilon}(e^{-1/\sqrt{\varepsilon}} - 1)}{e^{1/\sqrt{\varepsilon}} + e^{-1/\sqrt{\varepsilon}} - 2}(e^{x/\sqrt{\varepsilon}} - 1) \\ &+ \frac{\gamma - \alpha + 1 - (\beta + 1)\sqrt{\varepsilon}(e^{1/\sqrt{\varepsilon}} - 1)}{e^{-1/\sqrt{\varepsilon}} + e^{1/\sqrt{\varepsilon}} - 2}(e^{-x/\sqrt{\varepsilon}} - 1) \end{aligned} \tag{269}$$

Expanding (269) for small ε, we obtain

$$y = -x + \alpha + (\gamma - \alpha + 1)\, e^{(x-1)/\sqrt{\varepsilon}} - (\beta + 1)\sqrt{\varepsilon}\, e^{(x-1)/\sqrt{\varepsilon}}$$

$$- (\beta + 1)\sqrt{\varepsilon}\,(e^{-x/\sqrt{\varepsilon}} - 1) + \cdots \tag{270}$$

Hence there are two boundary layers, one at $x = 0$ and the other at $x = 1$. Each of them is of width $\sqrt{\varepsilon}$. More precisely

$$\lim_{\varepsilon \to 0} y = -x + \alpha \quad \text{for } x < 1 \tag{271}$$

$$\lim_{x \to 0} \lim_{\varepsilon \to 0} y = \alpha \tag{272}$$

$$\lim_{x \to 1} \lim_{\varepsilon \to 0} y = -1 + \alpha \tag{273}$$

$$\lim_{\varepsilon \to 0} \lim_{x \to 0} y = \lim_{\varepsilon \to 0} \left[\alpha + (\gamma - \alpha + 1)\, e^{-1/\sqrt{\varepsilon}} - (\beta + 1)\sqrt{\varepsilon}\, e^{-1/\sqrt{\varepsilon}}\right] = \alpha \tag{274}$$

$$\lim_{\varepsilon \to 0} \lim_{x \to 1} y = \lim_{\varepsilon \to 0} \left[-1 + \alpha + \gamma - \alpha + 1 - (\beta + 1)\sqrt{\varepsilon} - (\beta + 1)\sqrt{\varepsilon}\,(e^{-1/\sqrt{\varepsilon}} - 1)\right] = \gamma \tag{275}$$

$$y' = -1 + \frac{\gamma - \alpha + 1}{\sqrt{\varepsilon}}\, e^{(x-1)/\sqrt{\varepsilon}} - (\beta + 1)\, e^{(x-1)/\sqrt{\varepsilon}} + (\beta + 1)\, e^{-x/\sqrt{\varepsilon}} + \cdots$$

$$\lim_{x \to 0} \lim_{\varepsilon \to 0} y' = \lim_{x \to 0} [-1] = -1 \tag{276}$$

$$\lim_{\varepsilon \to 0} \lim_{x \to 0} y' = \lim_{\varepsilon \to 0} [\beta] = \beta \tag{277}$$

Thus

$$\lim_{\varepsilon \to 0} \lim_{x \to 0} y' \neq \lim_{x \to 0} \lim_{\varepsilon \to 0} y'$$

and a boundary layer exists at $x = 0$. Moreover

$$\lim_{\varepsilon \to 0} \lim_{x \to 1} y \neq \lim_{x \to 1} \lim_{\varepsilon \to 0} y$$

and a boundary layer exists at $x = 1$.

(b) ***The Method of Matched Asymptotic Expansions***

Outer Expansion. We seek a first-order outer expansion in the form

$$y^o = y_0(x) + \delta(\varepsilon)\, y_1 + \cdots \tag{278}$$

where $\delta(\varepsilon) \to 0$ as $\varepsilon \to 0$ and will be determined from matching to be $\sqrt{\varepsilon}$. Substituting

(278) into (259) and equating coefficients of like powers of ε, we obtain

$$-y_0' = 1 \Rightarrow y_0 = -x + c_0 \tag{279}$$

$$-y_1' = 0 \Rightarrow y_1 = c_1 \tag{280}$$

where c_0 and c_1 are constants. Since the difference between the derivatives of the term multiplying ε and the second term is 2, we expect to have two boundary layers, one at each end. Hence the outer expansion is not expected to satisfy any of the boundary conditions. Therefore

$$y^o = c_0 - x + \varepsilon^{1/2} c_1 + \cdots \tag{281}$$

where c_0 and c_1 need to be determined from matching.

Inner Expansion at $x = 0$. We introduce the stretching transformation

$$\xi = x/\varepsilon^{\lambda}, \qquad \lambda > 0$$

in (259) and obtain

$$\varepsilon^{1-3\lambda} \frac{d^3 y^i}{d\xi^3} - \varepsilon^{-\lambda} \frac{dy^i}{d\xi} = 1 \tag{282}$$

As $\varepsilon \to 0$, the least degenerate limit is

$$\frac{d^3 y^i}{d\xi^3} - \frac{dy^i}{d\xi} = 0 \tag{283}$$

corresponding to $\lambda = \frac{1}{2}$. Putting $\lambda = \frac{1}{2}$ in (282) gives

$$\frac{d^3 y^i}{d\xi^3} - \frac{dy^i}{d\xi} = \varepsilon^{1/2} \tag{284}$$

We seek a two-term inner expansion in the form

$$y^i = Y_0(\xi) + \sqrt{\varepsilon}\, Y_1(\xi) + \cdots \tag{285}$$

Substituting (285) into (284) and equating coefficients of like powers of ε, we have

$$Y_0''' - Y_0' = 0 \tag{286}$$

$$Y_1''' - Y_1' = 1 \tag{287}$$

This inner expansion is expected to satisfy both boundary conditions at $x = 0$. Since

$x = 0$ corresponds to $\xi = 0$, the boundary conditions at $x = 0$ transform into

$$y^i(0) = \alpha, \qquad \frac{dy^i}{d\xi}(0) = \varepsilon^{1/2}\beta \tag{288}$$

Substituting (285) into (288) and equating coefficients of like powers of ε, we obtain

$$Y_0(0) = \alpha, \qquad Y_0'(0) = 0 \tag{289}$$

$$Y_1(0) = 0, \qquad Y_1'(0) = \beta \tag{290}$$

The general solution of (286) is

$$Y_0 = a_0 + a_1 e^{-\xi} + a_2 e^{\xi} \tag{291}$$

where a_2 must be zero to enable the matching of y^i and y^o. Using (289) in (291) gives

$$a_0 + a_1 = \alpha, \qquad a_1 = 0 \tag{292}$$

so that $a_0 = \alpha$ and

$$Y_0 = \alpha \tag{293}$$

The general solution of (287) is

$$Y_1 = -\xi + b_0 + b_1 e^{-\xi} + b_2 e^{\xi} \tag{294}$$

where again b_2 must be zero from matching considerations. Using (290) in (294) gives

$$b_0 + b_1 = 0, \qquad b_1 = -1 - \beta \tag{295}$$

so that $b_0 = 1 + \beta$ and

$$Y_1 = -\xi + (1+\beta)(1 - e^{-\xi}) \tag{296}$$

Therefore

$$y^i = \alpha + \varepsilon^{1/2}\left[-\xi + (1+\beta)(1 - e^{-\xi})\right] + \cdots \tag{297}$$

Matching y^o and y^i. Letting $m = n = 2$ in van Dyke's matching principle, we proceed as follows:

Two-term outer expansion	$= c_0 - x + \varepsilon^{1/2} c_1$	
Rewritten in inner variable	$= c_0 - \varepsilon^{1/2}\xi + \varepsilon^{1/2} c_1$	
Expanded for small ε	$= c_0 - \varepsilon^{1/2}\xi + \varepsilon^{1/2} c_1$	
Two-term inner expansion	$= c_0 + \varepsilon^{1/2}(c_1 - \xi)$	(298)
Two-term inner expansion	$= \alpha + \varepsilon^{1/2}[-\xi + (1+\beta)(1-e^{-\xi})]$	
Rewritten in outer variable	$= \alpha + \varepsilon^{1/2}\left[(1+\beta) - \dfrac{x}{\varepsilon^{1/2}} - (1+\beta)\, e^{-x/\sqrt{\varepsilon}}\right]$	
Expanded for small ε	$= \alpha - x + (1+\beta)\varepsilon^{1/2}$	
Two-term outer expansion	$= \alpha - x + \varepsilon^{1/2}(1+\beta)$	(299)

Equating (298) and (299) yields

$$c_0 + \varepsilon^{1/2}(c_1 - \xi) = \alpha - x + \varepsilon^{1/2}(1+\beta)$$

which when expressed in terms of x becomes

$$c_0 - x + \varepsilon^{1/2} c_1 = \alpha - x + \varepsilon^{1/2}(1+\beta) \tag{300}$$

It follows from (300) that $c_0 = \alpha$ and $c_1 = 1 + \beta$, so that (281) becomes

$$y^o = \alpha - x + \varepsilon^{1/2}(1+\beta) + \cdots \tag{301}$$

Inner Expansion at $x = 1$. We introduce the stretching transformation

$$\zeta = \frac{1-x}{\varepsilon^\lambda}, \qquad \lambda > 0 \tag{302}$$

or

$$x = 1 - \varepsilon^\lambda \zeta$$

in (259) and obtain

$$-\varepsilon^{1-3\lambda}\frac{d^3 y^I}{d\zeta^3} + \varepsilon^{-\lambda}\frac{dy^I}{d\zeta} = 1 \tag{303}$$

The least degenerate limit of (303) as $\varepsilon \to 0$ is

$$\frac{d^3 y^I}{d\zeta^3} - \frac{dy^I}{d\zeta} = 0$$

corresponding to $\lambda = \frac{1}{2}$. Putting $\lambda = \frac{1}{2}$ in (303), we have

$$\frac{d^3 y^I}{d\zeta^3} - \frac{dy^I}{d\zeta} = -\varepsilon^{1/2} \tag{304}$$

This inner expansion is expected to satisfy the boundary condition at $x=1$. Since $x=1$ corresponds to $\zeta=0$, $y(1)=\gamma$ is transformed into

$$y^I(0)=\gamma \tag{305}$$

We let

$$y^I = Y_0(\zeta)+\varepsilon^{1/2}Y_1(\zeta)+\cdots \tag{306}$$

in (304) and (305), equate coefficients of like powers of ε, and obtain

$$Y_0''' - Y_0' = 0 \tag{307}$$

$$Y_1''' - Y_1' = -1 \tag{308}$$

$$Y_0(0)=\gamma, \qquad Y_1(0)=0 \tag{309}$$

The general solution of (307) is

$$Y_0 = d_0 + d_1 e^{-\zeta} \tag{310}$$

where the exponentially growing solution was omitted from matching considerations. Putting (310) in (309) gives

$$d_0 + d_1 = \gamma \quad \text{or} \quad d_0 = \gamma - d_1$$

so that (310) becomes

$$Y_0 = \gamma + d_1(e^{-\zeta}-1) \tag{311}$$

The general solution of (308) is

$$Y_1 = \zeta + d_3 + d_4 e^{-\zeta} \tag{312}$$

where the exponentially growing solution was omitted from matching considerations. Putting (312) in (309) yields

$$d_3 + d_4 = 0 \quad \text{or} \quad d_4 = -d_3$$

so that (312) becomes

$$Y_1 = \zeta + d_3(1-e^{-\zeta}) \tag{313}$$

Therefore

$$y^I = \gamma + d_1(e^{-\zeta}-1)+\varepsilon^{1/2}\left[\zeta + d_3(1-e^{-\zeta})\right]+\cdots \tag{314}$$

Matching y^o and y^I. We let $m=n=2$ in van Dyke's matching principle and proceed as follows:

Two-term outer expansion	$=\alpha - x + \varepsilon^{1/2}(1+\beta)$	
Rewritten in inner variable	$=\alpha - (1-\varepsilon^{1/2}\zeta) + \varepsilon^{1/2}(1+\beta)$	
Expanded for small ε	$=\alpha - 1 + \varepsilon^{1/2}\zeta + \varepsilon^{1/2}(1+\beta)$	
Two-term inner expansion	$=\alpha - 1 + \varepsilon^{1/2}(\zeta + 1 + \beta)$	(315)
Two-term inner expansion	$=\gamma + d_1(e^{-\zeta}-1) + \varepsilon^{1/2}[\zeta + d_3(1-e^{-\zeta})]$	
Rewritten in outer variable	$=\gamma + d_1\left(e^{-(1-x)/\varepsilon^{1/2}} - 1\right) + \varepsilon^{1/2}\left[\dfrac{1-x}{\varepsilon^{1/2}} + d_3\left(1 - e^{-(1-x)/\varepsilon^{1/2}}\right)\right]$	
Expanded for small ε	$=\gamma - d_1 + 1 - x + \varepsilon^{1/2}d_3$	
Two-term outer expansion	$=\gamma - d_1 + 1 - x + \varepsilon^{1/2}d_3$	(316)

Equating (315) and (316) yields

$$\alpha - 1 + \varepsilon^{1/2}(\zeta + 1 + \beta) = \gamma - d_1 + 1 - x + \varepsilon^{1/2} d_3$$

which when written in terms of x becomes

$$\alpha - 1 + 1 - x + \varepsilon^{1/2}(1+\beta) = \gamma - d_1 + 1 - x + \varepsilon^{1/2} d_3$$

Hence

$$\alpha = \gamma - d_1 + 1, \qquad d_3 = 1 + \beta$$

or

$$d_1 = \gamma - \alpha + 1$$

Thus

$$y^I = \gamma + (\gamma - \alpha + 1)(e^{-\zeta} - 1) + \varepsilon^{1/2}\left[\zeta + (1+\beta)(1-e^{-\zeta})\right] + \cdots \tag{317}$$

Composite Expansion

$$\begin{aligned}
y^c &= y^o + y^i + y^I - (y^i)^o - (y^I)^o \\
&= \alpha - x + \varepsilon^{1/2}(1+\beta) + \alpha + \varepsilon^{1/2}\left[-\xi + (1+\beta)(1-\varepsilon^{-\xi})\right] + \gamma \\
&\quad + (\gamma - \alpha + 1)(e^{-\zeta} - 1) + \varepsilon^{1/2}\left[\zeta + (1+\beta)(1-e^{-\zeta})\right] - \alpha + x \\
&\quad - \varepsilon^{1/2}(1+\beta) - (\alpha - 1) - \varepsilon^{1/2}(\zeta + 1 + \beta) \\
&= \alpha - x + (\gamma - \alpha + 1)e^{-\zeta} + \varepsilon^{1/2}(1+\beta)(1 - e^{-\xi} - e^{-\zeta}) + \cdots
\end{aligned} \tag{318}$$

in agreement with (270).

12.13. Consider the problem

$$\varepsilon y''' - y' + y = 0 \tag{319}$$

$$y(0) = \alpha, \qquad y'(0) = \beta, \qquad y(1) = \gamma \tag{320}$$

(a) Determine the exact solution and use it to show that there is in general a boundary layer at each end.

(b) Determine a two-term uniform expansion and compare your answer with the exact solution.

(a) ***The Exact Solution***

Since (319) is a linear homogeneous equation with constant coefficients, its solutions have the form $\exp(sx)$, where

$$\varepsilon s^3 - s + 1 = 0 \tag{321}$$

Let us denote the three roots of (321) by s_1, s_2, and s_3. Using the methods of Chapter 2, we find that

$$s_1 = 1 + \varepsilon + \cdots, \qquad s_2 = \frac{1}{\sqrt{\varepsilon}} - \tfrac{1}{2} + \cdots, \qquad s_3 = -\frac{1}{\sqrt{\varepsilon}} - \tfrac{1}{2} + \cdots \tag{322}$$

Then the general solution of (319) can be written as

$$y = c_1 e^{s_1 x} + c_2 e^{s_2 x} + c_3 e^{s_3 x} \tag{323}$$

where the c_n are constants. Substituting (323) into (320) gives

$$c_1 + c_2 + c_3 = \alpha$$

$$s_1 c_1 + s_2 c_2 + s_3 c_3 = \beta$$

$$c_1 e^{s_1} + c_2 e^{s_2} + c_3 e^{s_3} = \gamma$$

whose solution is

$$c_1 = \frac{1}{\Delta}\left[\beta(e^{s_2} - e^{s_3}) + \alpha s_2 e^{s_3} - \alpha s_3 e^{s_2} + (s_3 - s_2)\gamma\right]$$

$$c_2 = \frac{1}{\Delta}\left[(\beta - \alpha s_1)(e^{s_3} - e^{s_1}) - (s_3 - s_1)(\gamma - \alpha e^{s_1})\right]$$

$$c_3 = \frac{1}{\Delta}\left[-(\beta - \alpha s_1)(e^{s_2} - e^{s_1}) + (s_2 - s_1)(\gamma - \alpha e^{s_1})\right]$$

where

$$\Delta = (s_2 - s_1)(e^{s_3} - e^{s_1}) - (s_3 - s_1)(e^{s_2} - e^{s_1})$$

Using (322) we find that as $\varepsilon \to 0$,

$$c_1 = \alpha + \sqrt{\varepsilon}(\beta - \alpha) + \cdots$$

$$c_2 = \left[\gamma - \alpha e - \sqrt{\varepsilon}(\beta - \alpha)e\right] e^{-1/\sqrt{\varepsilon} + 1/2}$$

$$c_3 = -\sqrt{\varepsilon}(\beta - \alpha) + \cdots$$

Hence, as $\varepsilon \to 0$, (323) tends to

$$y = \left[\alpha + \sqrt{\varepsilon}(\beta - \alpha)\right] e^x - \sqrt{\varepsilon}(\beta - \alpha) e^{-(1-x)/\sqrt{\varepsilon} - x/2 + 3/2}$$
$$+ (\gamma - \alpha e) e^{-(1-x)/\sqrt{\varepsilon} + (1-x)/2} - \sqrt{\varepsilon}(\beta - \alpha) e^{-x/\sqrt{\varepsilon} - x/2} + \cdots \tag{324}$$

It follows from (324) that

$$\lim_{\varepsilon \to 0} \lim_{x \to 0} y' = \beta \neq \lim_{x \to 0} \lim_{\varepsilon \to 0} y' = \alpha$$

$$\lim_{\varepsilon \to 0} \lim_{x \to 1} y = \gamma \neq \lim_{x \to 1} \lim_{\varepsilon \to 0} y = \alpha e$$

Thus there are two boundary layers, one at $x = 0$ and one at $x = 1$. It follows from (324) that each of them is of width $\sqrt{\varepsilon}$.

(b) *The Method of Matched Asymptotic Expansions*

Outer Expansion. We seek a first-order outer expansion in the form

$$y^o = y_0(x) + \delta(\varepsilon) y_1(x) + \cdots \tag{325}$$

where $\delta(\varepsilon) \to 0$ as $\varepsilon \to 0$ and can be shown to be $\sqrt{\varepsilon}$ by matching. Substituting (325) into (319) and equating coefficients of like powers of ε, we obtain

$$y_0' - y_0 = 0 \tag{326}$$

$$y_1' - y_1 = 0 \tag{327}$$

The general solutions of (326) and (327) are

$$y_0 = c_0 e^x, \qquad y_1 = c_1 e^x$$

where c_0 and c_1 are constants. Hence (325) becomes

$$y^o = c_0 e^x + \sqrt{\varepsilon}\, c_1 e^x + \cdots \tag{328}$$

The constants c_0 and c_1 need to be determined from matching because the outer expansion is not expected to be valid near any of the ends where the boundary conditions are imposed.

Inner Expansion at $x = 0$. We introduce the stretching transformation

$$\xi = x/\varepsilon^{\lambda}, \qquad \lambda > 0$$

in (319) and obtain

$$\varepsilon^{1-3\lambda}\frac{d^3y^i}{d\xi^3} - \varepsilon^{-\lambda}\frac{dy^i}{d\xi} + y^i = 0 \tag{329}$$

As $\varepsilon \to 0$, the distinguished limit of (329) corresponds to $\lambda = \frac{1}{2}$. Then (329) becomes

$$\frac{d^3y^i}{d\xi^3} - \frac{dy^i}{d\xi} + \sqrt{\varepsilon}\, y^i = 0 \tag{330}$$

We seek a two-term inner expansion in the form

$$y^i = Y_0(\xi) + \sqrt{\varepsilon}\, Y_1(\xi) + \cdots \tag{331}$$

Since y^i is expected to be valid at $x = 0$ and since $x = 0$ corresponds to $\xi = 0$,

$$y^i(0) = \alpha, \qquad \frac{dy^i}{d\xi} = \sqrt{\varepsilon}\,\beta \tag{332}$$

Substituting (331) into (330) and (332) and equating coefficients of like powers of ε, we obtain

$$Y_0''' - Y_0' = 0, \qquad Y_0(0) = \alpha, \qquad Y_0'(0) = 0 \tag{333}$$

$$Y_1''' - Y_1' = -Y_0, \qquad Y_1(0) = 0, \qquad Y_1'(0) = \beta \tag{334}$$

The general solution of the equation in (333) is

$$Y_0 = a_0 + a_1 e^{-\xi} \tag{335}$$

where the exponentially growing solution was excluded to enable the matching of y^i and y^o. Using the conditions in (333), we conclude that $a_1 = 0$ and $a_0 = \alpha$. Then the equation in (334) becomes

$$Y_1''' - Y_1' = -\alpha \tag{336}$$

whose general solution is

$$Y_1 = b_0 + b_1 e^{-\xi} + \alpha\xi \tag{337}$$

Again, the exponentially growing solution was excluded using matching considerations. Using the conditions in (334), we find that

$$b_0 = -b_1 = \beta - \alpha$$

Hence

$$Y_1 = (\beta - \alpha)(1 - e^{-\xi}) + \alpha\xi \tag{338}$$

Therefore

$$y^i = \alpha + \sqrt{\varepsilon}\left[(\beta - \alpha)(1 - e^{-\xi}) + \alpha\xi\right] + \cdots \tag{339}$$

Matching y^o and y^i. Letting $m = n = 2$ in van Dyke's matching principle, we proceed as follows:

$$\begin{aligned}
&\text{Two-term outer expansion} && = c_0 e^x + \sqrt{\varepsilon}\, c_1 e^x \\
&\quad\text{Rewritten in inner variable} && = c_0 e^{\sqrt{\varepsilon}\xi} + \sqrt{\varepsilon}\, c_1 e^{\sqrt{\varepsilon}\xi} \\
&\quad\text{Expanded for small } \varepsilon && = c_0(1 + \sqrt{\varepsilon}\,\xi + \cdots) + \sqrt{\varepsilon}\, c_1(1 + \cdots) \\
&\quad\text{Two-term inner expansion} && = c_0 + \sqrt{\varepsilon}\,(c_0\xi + c_1)
\end{aligned} \tag{340}$$

$$\begin{aligned}
&\text{Two-term inner expansion} && = \alpha + \sqrt{\varepsilon}[(\beta - \alpha)(1 - e^{-\xi}) + \alpha\xi] \\
&\quad\text{Rewritten in outer variable} && = \alpha + \sqrt{\varepsilon}\left[(\beta - \alpha)(1 - e^{-x/\sqrt{\varepsilon}}) + \frac{\alpha x}{\sqrt{\varepsilon}}\right] \\
&\quad\text{Expanded for small } \varepsilon && = \alpha + \alpha x + \sqrt{\varepsilon}\,(\beta - \alpha) + \text{EST} \\
&\quad\text{Two-term outer expansion} && = \alpha + \alpha x + \sqrt{\varepsilon}\,(\beta - \alpha)
\end{aligned} \tag{341}$$

Equating (340) and (341) yields

$$c_0 + \sqrt{\varepsilon}\,(c_0\xi + c_1) = \alpha + \alpha x + \sqrt{\varepsilon}\,(\beta - \alpha)$$

which, when expressed in terms of x, becomes

$$c_0(1 + x) + \sqrt{\varepsilon}\, c_1 = \alpha(1 + x) + \sqrt{\varepsilon}\,(\beta - \alpha)$$

Hence $c_0 = \alpha$ and $c_1 = \beta - \alpha$ so that (328) becomes

$$y^o = \alpha e^x + \sqrt{\varepsilon}\,(\beta - \alpha)\, e^x + \cdots \tag{342}$$

Inner Expansion at $x = 1$. We introduce the stretching transformation

$$\zeta = \frac{(1 - x)}{\varepsilon^\lambda}, \qquad \lambda > 0$$

or

$$x = 1 - \varepsilon^\lambda \zeta$$

in (319) and obtain

$$\varepsilon^{1 - 3\lambda}\frac{d^3 y^I}{d\zeta^3} - \varepsilon^{-\lambda}\frac{dy^I}{d\zeta} - y^I = 0 \tag{343}$$

The distinguished limit of (343) as $\varepsilon \to 0$ corresponds to $\lambda = \frac{1}{2}$. Then (343) becomes

$$\frac{d^3 y^I}{d\zeta^3} - \frac{dy^I}{d\zeta} - \sqrt{\varepsilon}\, y^I = 0 \tag{344}$$

Since y^I is expected to be valid at $x = 1$ and since $x = 1$ corresponds to $\zeta = 0$,

$$y^I(0) = \gamma \tag{345}$$

Next we seek a two-term inner expansion in the form

$$y^I = Y_0(\zeta) + \sqrt{\varepsilon}\, Y_1(\zeta) + \cdots \tag{346}$$

Substituting (346) into (344) and (345) and equating coefficients of like powers of ε, we obtain

$$Y_0''' - Y_0' = 0, \qquad Y_0(0) = \gamma \tag{347}$$

$$Y_1''' - Y_1' = Y_0, \qquad Y_1(0) = 0 \tag{348}$$

The general solution of the equation in (347) is

$$Y_0 = b_0 + b_1 e^{-\zeta}$$

where the exponentially growing solution was excluded because of matching considerations. Using the boundary condition in (347), we find that

$$b_0 + b_1 = \gamma \quad \text{or} \quad b_0 = \gamma - b_1$$

and hence

$$Y_0 = \gamma + b_1(e^{-\zeta} - 1) \tag{349}$$

Then the equation in (348) becomes

$$Y_1''' - Y_1' = \gamma - b_1 + b_1 e^{-\zeta} \tag{350}$$

The complementary solution of (350) is

$$Y_{1h} = d_0 + d_1 e^{-\zeta}$$

where, as before, the exponentially growing solution was excluded. A particular solution of (350) is

$$Y_{1p} = -(\gamma - b_1)\zeta + \tfrac{1}{2} b_1 \zeta e^{-\zeta}$$

Hence

$$Y_1 = d_0 + d_1 e^{-\zeta} - (\gamma - b_1)\zeta + \tfrac{1}{2} b_1 \zeta e^{-\zeta}$$

Using the boundary condition in (348), we conclude that $d_0 = -d_1$. Therefore

$$y^I = \gamma + b_1(e^{-\zeta} - 1) + \sqrt{\varepsilon}\left[d_1(e^{-\zeta} - 1) - (\gamma - b_1)\zeta + \tfrac{1}{2}b_1\zeta e^{-\zeta}\right] + \cdots \tag{351}$$

Matching y^o and y^I. Letting $m = n = 2$ in van Dyke's matching principle, we proceed as follows:

$$\begin{aligned}
&\text{Two-term outer expansion} &&= \alpha e^x + \sqrt{\varepsilon}(\beta - \alpha)e^x \\
&\quad\text{Rewritten in inner variable} &&= \alpha e^{1-\sqrt{\varepsilon}\zeta} + \sqrt{\varepsilon}(\beta - \alpha)e^{1-\sqrt{\varepsilon}\zeta} \\
&\quad\text{Expanded for small } \varepsilon &&= \alpha e(1 - \sqrt{\varepsilon}\zeta + \cdots) + \sqrt{\varepsilon}(\beta - \alpha)e(1 + \cdots) \\
&\quad\text{Two-term inner expansion} &&= \alpha e + \sqrt{\varepsilon}\, e(\beta - \alpha - \alpha\zeta)
\end{aligned} \tag{352}$$

$$\begin{aligned}
&\text{Two-term inner expansion} &&= \gamma + b_1(e^{-\zeta} - 1) + \sqrt{\varepsilon}[d_1(e^{-\zeta} - 1) \\
& && \quad - (\gamma - b_1)\zeta + \tfrac{1}{2}b_1\zeta e^{-\zeta}] \\
&\quad\text{Rewritten in outer variable} &&= \gamma + b_1(e^{-(1-x)/\sqrt{\varepsilon}} - 1) \\
& && \quad + \sqrt{\varepsilon}\Big[d_1(e^{-(1-x)/\sqrt{\varepsilon}} - 1) \\
& && \quad - (\gamma - b_1)\frac{(1-x)}{\sqrt{\varepsilon}} + \tfrac{1}{2}b_1\frac{(1-x)}{\sqrt{\varepsilon}}e^{-(1-x)/\sqrt{\varepsilon}}\Big] \\
&\quad\text{Expanded for small } \varepsilon &&= \gamma - b_1 + \sqrt{\varepsilon}\left[-d_1 - \frac{(\gamma - b_1)(1-x)}{\sqrt{\varepsilon}}\right] + \text{EST} \\
&\quad\text{Two-term outer expansion} &&= (\gamma - b_1)x - \sqrt{\varepsilon}\, d_1
\end{aligned} \tag{353}$$

Equating (352) and (353) yields

$$\alpha e + \sqrt{\varepsilon}\, e(\beta - \alpha - \alpha\zeta) = (\gamma - b_1)x - \sqrt{\varepsilon}\, d_1$$

which, when written in terms of x, becomes

$$\alpha e x + \sqrt{\varepsilon}\, e(\beta - \alpha) = (\gamma - b_1)x - \sqrt{\varepsilon}\, d_1$$

Hence

$$\gamma - b_1 = \alpha e \quad \text{or} \quad b_1 = \gamma - \alpha e$$

and

$$d_1 = -(\beta - \alpha)e$$

Therefore, (351) becomes

$$y^I = \alpha e + (\gamma - \alpha e)e^{-\zeta} + \sqrt{\varepsilon}\left[-(\beta - \alpha)e(e^{-\zeta} - 1) - \alpha e\zeta + \tfrac{1}{2}(\gamma - \alpha e)\zeta e^{-\zeta}\right] + \cdots \tag{354}$$

Composite Expansion

$$\begin{aligned}
y^c &= y^o + y^i + y^I - (y^i)^o - (y^I)^o \\
&= \alpha e^x + \sqrt{\varepsilon}(\beta-\alpha)e^x + \alpha + \sqrt{\varepsilon}\left[(\beta-\alpha)(1-e^{-\xi}) + \alpha\xi\right] + \alpha e \\
&\quad + (\gamma - \alpha e)e^{-\zeta} + \sqrt{\varepsilon}\left[-(\beta-\alpha)e(e^{-\zeta}-1) - \alpha e\zeta + \tfrac{1}{2}(\gamma-\alpha e)\zeta e^{-\zeta}\right] \\
&\quad - \alpha - \alpha x - \sqrt{\varepsilon}(\beta-\alpha) - \alpha e - \sqrt{\varepsilon}\, e(\beta - \alpha - \alpha\zeta) + \cdots \\
&= \alpha e^x + \sqrt{\varepsilon}(\beta-\alpha)e^x - \sqrt{\varepsilon}(\beta-\alpha)e^{-\xi} + (\gamma-\alpha e)e^{-\zeta} \\
&\quad - \sqrt{\varepsilon}(\beta-\alpha)e^{1-\zeta} + \tfrac{1}{2}\sqrt{\varepsilon}(\gamma-\alpha e)\zeta e^{-\zeta} + \cdots
\end{aligned} \tag{355}$$

in agreement with (324) to $O(\sqrt{\varepsilon})$.

12.14. Consider the problem

$$\varepsilon y''' - (2x+1)y' = 1 \tag{356}$$

$$y(0) = \alpha, \qquad y'(0) = \beta, \qquad y(1) = \gamma \tag{357}$$

Determine a two-term uniform expansion.

Outer Expansion. We seek an outer expansion in the form

$$y^o = y_0(x) + \delta(\varepsilon)y_1(x) + \cdots \tag{358}$$

where $\delta(\varepsilon) \to 0$ as $\varepsilon \to 0$ and must be $\sqrt{\varepsilon}$ in order that the outer and inner expansions match. Since the difference between the derivatives of the term multiplying ε and the second term is 2, we expect to have two boundary layers, one at each end. Hence the outer expansion is not expected to satisfy in general any of the boundary conditions. Substituting (358) into (356) and equating coefficients of like powers of ε, we obtain

$$(2x+1)y_0' = -1, \qquad y_1' = 0$$

whose solutions are

$$y_0 = -\tfrac{1}{2}\ln(1+2x) + c_0, \qquad y_1 = c_1$$

where c_0 and c_1 are constants. Therefore

$$y^o = c_0 - \tfrac{1}{2}\ln(1+2x) + \sqrt{\varepsilon}\,c_1 + \cdots \tag{359}$$

Inner Expansion at $x = 0$. We introduce the stretching transformation

$$\xi = \frac{x}{\varepsilon^\lambda}, \qquad \lambda > 0$$

in (356) and obtain

$$\varepsilon^{1-3\lambda}\frac{d^3y^i}{d\xi^3}-(1+2\varepsilon^\lambda\xi)\,\varepsilon^{-\lambda}\frac{dy^i}{d\xi}=1 \tag{360}$$

The distinguished limit of (360) as $\varepsilon\to 0$ corresponds to $\lambda=\frac{1}{2}$. Then (360) becomes

$$\frac{d^3y^i}{d\xi^3}-(1+2\sqrt{\varepsilon}\,\xi)\frac{dy^i}{d\xi}=\sqrt{\varepsilon} \tag{361}$$

Since y^i is expected to be valid at $x=0$ and since $x=0$ corresponds to $\xi=0$,

$$y^i(0)=\alpha, \qquad \frac{dy^i}{d\xi}(0)=\sqrt{\varepsilon}\,\beta \tag{362}$$

Next we seek a two-term inner expansion in the form

$$y^i=Y_0(\xi)+\sqrt{\varepsilon}\,Y_1(\xi)+\cdots \tag{363}$$

Substituting (363) into (361) and (362) and equating coefficients of like powers of ε, we obtain

$$Y_0'''-Y_0'=0, \qquad Y_0(0)=\alpha, \qquad Y_0'(0)=0 \tag{364}$$

$$Y_1'''-Y_1'=1+2\xi Y_0', \qquad Y_1(0)=0, \qquad Y_1'(0)=\beta \tag{365}$$

Excluding the exponentially growing part, we find that the solution of (364) is

$$Y_0=\alpha$$

Then the equation in (365) becomes

$$Y_1'''-Y_1'=1$$

whose solution, subject to the boundary conditions in (365), is

$$Y_1=-\xi+(1+\beta)(1-e^{-\xi})$$

Therefore

$$y^i=\alpha+\sqrt{\varepsilon}\left[-\xi+(1+\beta)(1-e^{-\xi})\right]+\cdots \tag{366}$$

Matching y^o and y^i

Two-term inner expansion of (two-term outer expansion) $=c_0+\sqrt{\varepsilon}\,(c_1-\xi)$

Two-term outer expansion of (two-term inner expansion) $=\alpha-x+\sqrt{\varepsilon}\,(1+\beta)$

Hence

$$c_0 + \sqrt{\varepsilon}\,(c_1 - \xi) = \alpha - x + \sqrt{\varepsilon}\,(1+\beta)$$

which, upon using $\xi = x/\sqrt{\varepsilon}$, yields

$$c_0 = \alpha \quad \text{and} \quad c_1 = 1+\beta$$

Therefore

$$y^o = \alpha - \tfrac{1}{2}\ln(1+2x) + \sqrt{\varepsilon}\,(1+\beta) + \cdots \tag{367}$$

Inner Expansion at $x = 1$. We introduce the stretching transformation

$$\zeta = \frac{(1-x)}{\varepsilon^{\lambda}}, \qquad \lambda > 0$$

in (356) and obtain

$$\varepsilon^{1-3\lambda}\frac{d^3y^I}{d\zeta^3} - \varepsilon^{-\lambda}(3 - 2\varepsilon^{\lambda}\zeta)\frac{dy^I}{d\zeta} = -1 \tag{368}$$

As $\varepsilon \to 0$, the distinguished limit of (368) corresponds to $\lambda = \frac{1}{2}$. Then (368) becomes

$$\frac{d^3y^I}{d\zeta^3} - (3 - 2\sqrt{\varepsilon}\,\zeta)\frac{dy^I}{d\zeta} = -\sqrt{\varepsilon} \tag{369}$$

Since y^I is expected to be valid at $x = 1$ and since $x = 1$ corresponds to $\zeta = 0$,

$$y^I(0) = \gamma \tag{370}$$

We seek a two-term inner expansion in the form

$$y^I = Y_0(\zeta) + \sqrt{\varepsilon}\,Y_1(\zeta) + \cdots \tag{371}$$

Substituting (371) into (369) and (370) and equating coefficients of like powers of ε, we obtain

$$Y_0''' - 3Y_0' = 0, \qquad Y_0(0) = \gamma \tag{372}$$

$$Y_1''' - 3Y_1' = -1 - 2\zeta Y_0', \qquad Y_1(0) = 0 \tag{373}$$

Excluding the exponentially growing part, we find that the solution of (372) is

$$Y_0 = \gamma - b_1 + b_1 e^{-\sqrt{3}\,\zeta}$$

Then the equation in (373) becomes

$$Y_1''' - 3Y_1' = -1 + 2\sqrt{3}\,b_1\zeta e^{-\sqrt{3}\,\zeta} \tag{374}$$

Excluding the exponentially growing part, we find that the solution of (374) that satisfies $Y_1(0) = 0$ is

$$Y_1 = d_1(1 - e^{-\sqrt{3}\zeta}) + \tfrac{1}{3}\zeta + \frac{\sqrt{3}}{6}(\zeta^2 + \sqrt{3}\zeta)\, b_1 e^{-\sqrt{3}\zeta}$$

Therefore

$$y^I = \gamma - b_1 + b_1 e^{-\sqrt{3}\zeta} + \sqrt{\varepsilon}\left[d_1(1 - e^{-\sqrt{3}\zeta}) + \tfrac{1}{3}\zeta + \frac{\sqrt{3}}{6}(\zeta^2 + \sqrt{3}\zeta)\, b_1 e^{-\sqrt{3}\zeta}\right] + \cdots \tag{375}$$

Matching y^o and y^I

Two-term inner expansion of (two-term outer expansion) $= \alpha - \tfrac{1}{2}\ln 3 + \sqrt{\varepsilon}(1 + \beta + \tfrac{1}{3}\zeta)$ (376)

Two-term outer expansion of (two-term inner expansion) $= \gamma - b_1 + \tfrac{1}{3}(1 - x) + \sqrt{\varepsilon}\, d_1$ (377)

Equating (376) and (377) and using $\zeta = (1 - x)/\sqrt{\varepsilon}$, we have

$$b_1 = \gamma - \alpha + \tfrac{1}{2}\ln 3, \qquad d_1 = 1 + \beta$$

Therefore

$$y^I = \alpha - \tfrac{1}{2}\ln 3 + (\gamma - \alpha + \tfrac{1}{2}\ln 3)\, e^{-\sqrt{3}\zeta} + \sqrt{\varepsilon}\Big[(1 + \beta)(1 - e^{-\sqrt{3}\zeta})$$
$$+ \tfrac{1}{3}\zeta + \frac{\sqrt{3}}{6}(\zeta^2 + \sqrt{3}\zeta)(\gamma - \alpha + \tfrac{1}{2}\ln 3)\, e^{-\sqrt{3}\zeta}\Big] + \cdots \tag{378}$$

Composite Expansion

$$y^c = y^o + y^i + y^I - (y^i)^o - (y^I)^o = \alpha - \tfrac{1}{2}\ln(1 + 2x) + \sqrt{\varepsilon}(1 + \beta) + \alpha$$
$$+ \sqrt{\varepsilon}\left[-\xi + (1 + \beta)(1 - e^{-\xi})\right] + \alpha - \tfrac{1}{2}\ln 3 + (\gamma - \alpha + \tfrac{1}{2}\ln 3)\, e^{-\sqrt{3}\zeta}$$
$$+ \sqrt{\varepsilon}\left[(1 + \beta)(1 - e^{-\sqrt{3}\zeta}) + \tfrac{1}{3}\zeta + \frac{\sqrt{3}}{6}(\zeta^2 + \sqrt{3}\zeta)(\gamma - \alpha + \tfrac{1}{2}\ln 3)\, e^{-\sqrt{3}\zeta}\right]$$
$$- \alpha + x - \sqrt{\varepsilon}(1 + \beta) - \alpha + \tfrac{1}{2}\ln 3 - \sqrt{\varepsilon}(1 + \beta + \tfrac{1}{3}\zeta) + \cdots$$
$$= \alpha - \tfrac{1}{2}\ln(1 + 2x) + \sqrt{\varepsilon}(1 + \beta) - \sqrt{\varepsilon}(1 + \beta)\, e^{-\xi}$$
$$+ (\gamma - \alpha + \tfrac{1}{2}\ln 3)\, e^{-\sqrt{3}\zeta} - \sqrt{\varepsilon}(1 + \beta)\, e^{-\sqrt{3}\zeta}$$
$$+ \frac{\sqrt{3}}{6}\sqrt{\varepsilon}(\zeta^2 + \sqrt{3}\zeta)(\gamma - \alpha + \tfrac{1}{2}\ln 3)\, e^{-\sqrt{3}\zeta} + \cdots \tag{379}$$

12.15. Consider the problem

$$\varepsilon y^{iv} - y'' = 1 \tag{380}$$

$$y(0) = \alpha, \qquad y'(0) = \beta, \qquad y(1) = \gamma, \qquad y'(1) = \delta \tag{381}$$

(a) Determine the exact solution and use it to show that there is in general a boundary layer at each end.

(b) Determine a two-term uniform expansion and compare your answer with the exact solution.

(a) ***The Exact Solution***

A particular solution of (380) is

$$y_p = -\tfrac{1}{2}x^2$$

whereas the complementary function is

$$y_h = c_0 + c_1 x + c_2 e^{x/\sqrt{\varepsilon}} + c_3 e^{-x/\sqrt{\varepsilon}}$$

Hence the general solution of (380) is

$$y = -\tfrac{1}{2}x^2 + c_0 + c_1 x + c_2 e^{x/\sqrt{\varepsilon}} + c_3 e^{-x/\sqrt{\varepsilon}} \tag{382}$$

Imposing the boundary conditions (381) yields

$$c_0 + c_2 + c_3 = \alpha$$

$$c_1 + \frac{c_2}{\sqrt{\varepsilon}} - \frac{c_3}{\sqrt{\varepsilon}} = \beta$$

$$c_0 + c_1 + c_2 e^{1/\sqrt{\varepsilon}} + c_3 e^{-1/\sqrt{\varepsilon}} = \gamma + \tfrac{1}{2}$$

$$c_1 + \frac{c_2}{\sqrt{\varepsilon}} e^{1/\sqrt{\varepsilon}} - \frac{c_3}{\sqrt{\varepsilon}} e^{-1/\sqrt{\varepsilon}} = \delta + 1$$

whose solution is

$$c_0 = \alpha - \sqrt{\varepsilon}\left(\gamma - \alpha - \beta + \tfrac{1}{2}\right) + \cdots$$

$$c_1 = \gamma - \alpha + \tfrac{1}{2} + \sqrt{\varepsilon}\left(2\gamma - 2\alpha - \beta - \delta\right) + \cdots$$

$$c_2 = -\left(\gamma - \alpha - \delta - \tfrac{1}{2}\right)\sqrt{\varepsilon}\, e^{-1/\sqrt{\varepsilon}} + \cdots$$

$$c_3 = \sqrt{\varepsilon}\left(\gamma - \alpha - \beta + \tfrac{1}{2}\right) + \cdots$$

Therefore

$$y = -\tfrac{1}{2}x^2 + \alpha - \sqrt{\varepsilon}\left(\gamma - \alpha - \beta + \tfrac{1}{2}\right)$$
$$+\left[\gamma - \alpha + \tfrac{1}{2} + \sqrt{\varepsilon}\left(2\gamma - 2\alpha - \beta - \delta\right)\right]x - \sqrt{\varepsilon}\left(\gamma - \alpha - \delta - \tfrac{1}{2}\right)e^{-(1-x)/\sqrt{\varepsilon}}$$
$$+\sqrt{\varepsilon}\left(\gamma - \alpha - \beta + \tfrac{1}{2}\right)e^{-x/\sqrt{\varepsilon}} + \cdots \tag{383}$$

It can be easily shown that

$$\lim_{\varepsilon \to 0}\lim_{x \to 0} y' = \beta \neq \lim_{x \to 0}\lim_{\varepsilon \to 0} y' = \gamma - \alpha + \tfrac{1}{2}$$

$$\lim_{\varepsilon \to 0}\lim_{x \to 1} y' = \delta \neq \lim_{x \to 1}\lim_{\varepsilon \to 0} y' = \gamma - \alpha - \tfrac{1}{2}$$

Therefore, two boundary layers exist, one at each end.

(b) *The Method of Matched Asymptotic Expansions*

Outer Expansion. We seek an outer expansion in the form

$$y^o = y_0(x) + \delta(\varepsilon)\, y_1(x) + \cdots \tag{384}$$

where $\delta(\varepsilon) \to 0$ as $\varepsilon \to 0$ and must be $\sqrt{\varepsilon}$ in order that the outer expansion match with the inner expansions. Since the difference between the derivatives of the term multiplying ε and the second term is 2, one expects the presence of a boundary layer at each end. Hence the outer expansion is not expected to satisfy in general any of the boundary conditions. Substituting (384) into (380) and equating coefficients of like powers of ε, we obtain

$$y_0'' = -1, \qquad y_1'' = 0$$

whose solutions are

$$y_0 = -\tfrac{1}{2}x^2 + c_0 + c_1 x, \qquad y_1 = c_2 + c_3 x$$

where the c_n are constants. Hence

$$y^o = -\tfrac{1}{2}x^2 + c_0 + c_1 x + \sqrt{\varepsilon}\left(c_2 + c_3 x\right) + \cdots \tag{385}$$

Inner Expansion at $x = 0$. We introduce the stretching transformation

$$\xi = \frac{x}{\varepsilon^\lambda}, \qquad \lambda > 0$$

in (380) and obtain

$$\varepsilon^{1-4\lambda}\frac{d^4 y^i}{d\xi^4} - \varepsilon^{-2\lambda}\frac{d^2 y^i}{d\xi^2} = 1 \tag{386}$$

As $\varepsilon \to 0$, the distinguished limit of (386) corresponds to $\lambda = \frac{1}{2}$. Then (386) becomes

$$\frac{d^4 y^i}{d\xi^4} - \frac{d^2 y^i}{d\xi^2} = \varepsilon \tag{387}$$

Since the inner expansion is valid at $x = 0$ and since $x = 0$ corresponds to $\xi = 0$,

$$y^i(0) = \alpha, \qquad \frac{dy^i}{d\xi}(0) = \sqrt{\varepsilon}\,\beta \tag{388}$$

Next, we seek a two-term expansion in the form

$$y^i = Y_0(\xi) + \sqrt{\varepsilon}\, Y_1(\xi) + \cdots \tag{389}$$

Substituting (389) into (387) and (388) and equating coefficients of like powers of ε, we obtain

$$Y_0^{iv} - Y_0'' = 0, \qquad Y_0(0) = \alpha, \qquad Y_0'(0) = 0 \tag{390}$$

$$Y_1^{iv} - Y_1'' = 0, \qquad Y_1(0) = 0, \qquad Y_1'(0) = \beta \tag{391}$$

The general solution of (390) is

$$Y_0 = a_0 + a_1 \xi + a_2 e^{-\xi} + a_3 e^{\xi}$$

where the a_n are constants and a_3 must be zero to enable the matching of y^i and y^o. Using the boundary conditions in (390), we conclude that $a_0 = \alpha - a_2$ and $a_1 = a_2$. Hence

$$Y_0 = \alpha - a_2(1 - \xi - e^{-\xi})$$

Similarly, the solution of (391) excluding the exponentially growing term is

$$Y_1 = \beta\xi - b_2(1 - \xi - e^{-\xi})$$

Hence

$$y^i = \alpha - a_2(1 - \xi - e^{-\xi}) + \sqrt{\varepsilon}\left[\beta\xi - b_2(1 - \xi - e^{-\xi})\right] + \cdots \tag{392}$$

Matching y^o and y^i

Two-term inner expansion of (two-term outer expansion)

$$= c_0 + \sqrt{\varepsilon}\,(c_1 \xi + c_2) \tag{393}$$

Two-term outer expansion of (two-term inner expansion)

$$= \begin{cases} \dfrac{a_2 x}{\sqrt{\varepsilon}} + \alpha - a_2 & \text{if } a_2 \neq 0 \\ \alpha + (\beta + b_2)x - \sqrt{\varepsilon}\, b_2 & \text{if } a_2 = 0 \end{cases} \tag{394}$$

Equating (393) and (394) demands that $a_2 = 0$ and

$$c_0 + \sqrt{\varepsilon}(c_1 \xi + c_2) = \alpha + (\beta + b_2)x - \sqrt{\varepsilon}\, b_2$$

Since $x = \sqrt{\varepsilon}\, \xi$,

$$c_0 = \alpha, \qquad c_2 = -b_2, \qquad c_1 = \beta + b_2 \tag{395}$$

Inner Expansion at $x = 1$. We introduce the stretching transformation

$$\zeta = \frac{(1-x)}{\varepsilon^{\lambda}}, \qquad \lambda > 0$$

or

$$x = 1 - \varepsilon^{\lambda} \zeta$$

in (380) and obtain

$$\varepsilon^{1-4\lambda} \frac{d^4 y^I}{d\zeta^4} - \varepsilon^{-2\lambda} \frac{d^2 y^I}{d\zeta^2} = 1 \tag{396}$$

As $\varepsilon \to 0$, the distinguished limit of (396) corresponds to $\lambda = \frac{1}{2}$. Then (396) becomes

$$\frac{d^4 y^I}{d\zeta^4} - \frac{d^2 y^I}{d\zeta^2} = \varepsilon \tag{397}$$

Since y^I is valid at $x = 1$ and since $x = 1$ corresponds to $\zeta = 0$,

$$y^I(0) = \gamma, \qquad \frac{dy^I}{d\zeta}(0) = -\sqrt{\varepsilon}\, \delta \tag{398}$$

Next, we seek a two-term inner expansion in the form

$$y^I = Y_0(\zeta) + \sqrt{\varepsilon}\, Y_1(\zeta) + \cdots \tag{399}$$

Substituting (399) into (397) and (398) and equating coefficients of like powers of ε, we obtain

$$Y_0^{iv} - Y_0'' = 0, \qquad Y_0(0) = \gamma, \qquad Y_0'(0) = 0 \tag{400}$$

$$Y_1^{iv} - Y_1'' = 0, \qquad Y_1(0) = 0, \qquad Y_1'(0) = -\delta \tag{401}$$

Excluding the exponentially growing part, we find, as before, that the solution of (400) is

$$Y_0 = \gamma - d_1(1 - \zeta - e^{-\zeta})$$

and the solution of (401) is

$$Y_1 = -\delta\zeta - d_2(1 - \zeta - e^{-\zeta})$$

Hence

$$y^I = \gamma - d_1(1-\zeta - e^{-\zeta}) + \sqrt{\varepsilon}\left[-\delta\zeta - d_2(1-\zeta-e^{-\zeta})\right] + \cdots \tag{402}$$

Matching y^o and y^I

Two-term inner expansion of (two-term outer expansion)

$$= -\tfrac{1}{2} + c_0 + c_1 + \sqrt{\varepsilon}\,(\zeta - c_1\zeta + c_2 + c_3) \tag{403}$$

Two-term outer expansion of (two-term inner expansion)

$$= \begin{cases} \dfrac{d_1(1-x)}{\sqrt{\varepsilon}} + \gamma - d_1 & \text{if } d_1 \neq 0 \\ \gamma + (d_2 - \delta)(1-x) - \sqrt{\varepsilon}\, d_2 & \text{if } d_1 = 0 \end{cases} \tag{404}$$

Equating (403) and (404), we conclude that $d_1 = 0$ and

$$-\tfrac{1}{2} + c_0 + c_1 + \sqrt{\varepsilon}\,(\zeta - c_1\zeta + c_2 + c_3) = \gamma + (d_2 - \delta)(1-x) - \sqrt{\varepsilon}\, d_2$$

Hence

$$-\tfrac{1}{2} + c_0 + c_1 = \gamma$$

$$1 - c_1 = d_2 - \delta$$

$$c_2 + c_3 = -d_2$$

Recalling (395) and solving these equations, we obtain

$$c_1 = \gamma - \alpha + \tfrac{1}{2}$$

$$d_2 = \alpha + \delta - \gamma + \tfrac{1}{2}$$

$$b_2 = -c_2 = \gamma - \alpha - \beta + \tfrac{1}{2}$$

$$c_3 = 2\gamma - 2\alpha - \beta - \delta$$

Therefore

$$y^o = -\tfrac{1}{2}x^2 + \alpha + (\gamma - \alpha + \tfrac{1}{2})x + \sqrt{\varepsilon}[-\gamma + \alpha + \beta - \tfrac{1}{2}$$

$$+(2\gamma - 2\alpha - \beta - \delta)x] + \cdots \tag{405}$$

$$y^i = \alpha + \sqrt{\varepsilon}[\beta\xi - (\gamma - \alpha - \beta + \tfrac{1}{2})(1 - \xi - e^{-\xi})] + \cdots \tag{406}$$

$$y^I = \gamma + \sqrt{\varepsilon}[-\delta\zeta - (\alpha + \delta - \gamma + \tfrac{1}{2})(1 - \zeta - e^{-\zeta})] + \cdots \tag{407}$$

Composite Expansion

$$y^c = y^o + y^i + y^I - (y^i)^o - (y^I)^o = -\tfrac{1}{2}x^2 + c_0 + c_1 x$$

$$+\sqrt{\varepsilon}(c_2 + c_3 x) + \alpha + \sqrt{\varepsilon}[\beta\xi - b_2(1 - \xi - e^{-\xi})] + \gamma$$

$$+\sqrt{\varepsilon}[-\delta\zeta - d_2(1 - \zeta - e^{-\zeta})] - \alpha - (\beta + b_2)x + \sqrt{\varepsilon}\, b_2 - \gamma$$

$$-(d_2 - \delta)(1 - x) + \sqrt{\varepsilon}\, d_2$$

$$= -\tfrac{1}{2}x^2 + \alpha + (\gamma - \alpha + \tfrac{1}{2})x + \sqrt{\varepsilon}[-\gamma + \alpha + \beta - \tfrac{1}{2} + (2\gamma - 2\alpha - \beta - \delta)x]$$

$$+\sqrt{\varepsilon}(\gamma - \alpha - \beta + \tfrac{1}{2})e^{-\xi} + \sqrt{\varepsilon}(\alpha + \delta - \gamma + \tfrac{1}{2})e^{-\zeta} + \cdots \tag{408}$$

in agreement with (383).

12.16. Consider the problem

$$\varepsilon y^{iv} - (2x+1)y'' = 1 \tag{409}$$

$$y(0) = \alpha, \qquad y'(0) = \beta, \qquad y(1) = \gamma, \qquad y'(1) = \delta \tag{410}$$

Determine a two-term uniform expansion.

Outer Expansion. We seek a two-term outer expansion in the form

$$y^o = y_0(x) + \delta(\varepsilon)y_1(x) + \cdots \tag{411}$$

where $\delta(\varepsilon) \to 0$ as $\varepsilon \to 0$ and must be $\sqrt{\varepsilon}$ to enable the matching of outer and inner expansions. Since the difference between the derivatives of the term multiplying ε and the second term is 2, a boundary layer is expected at each end and the outer expansion is not expected to satisfy in general any of the boundary conditions. Substituting (411) into (409) and equating coefficients of like powers of ε, we obtain

$$(2x+1)y_0'' = -1, \qquad y_1'' = 0$$

whose solutions are

$$y_0 = -\tfrac{1}{2}x\ln(1+2x) + \tfrac{1}{2}x - \tfrac{1}{4}\ln(1+2x) + A_0 x + B_0$$

$$y_1 = A_1 x + B_1$$

where the A_n and B_n are constants. Hence,

$$y^o = -\tfrac{1}{2}x\ln(1+2x) + \tfrac{1}{2}x - \tfrac{1}{4}\ln(1+2x) + A_0 x + B_0 + \sqrt{\varepsilon}(A_1 x + B_1) + \cdots \tag{412}$$

Inner Expansion at $x = 0$. We introduce the stretching transformation

$$\xi = \frac{x}{\varepsilon^\lambda}, \qquad \lambda > 0$$

in (409) and obtain

$$\varepsilon^{1-4\lambda}\frac{d^4 y^i}{d\xi^4} - \varepsilon^{-2\lambda}(1+2\varepsilon^\lambda \xi)\frac{d^2 y^i}{d\xi^2} = 1 \tag{413}$$

As $\varepsilon \to 0$, the distinguished limit of (413) corresponds to $\lambda = \frac{1}{2}$. Then (413) becomes

$$\frac{d^4 y^i}{d\xi^4} - (1+2\sqrt{\varepsilon}\,\xi)\frac{d^2 y^i}{d\xi^2} = \varepsilon \tag{414}$$

Since y^i is valid at $x = 0$ and since $x = 0$ corresponds to $\xi = 0$,

$$y^i(0) = \alpha, \qquad \frac{dy^i}{d\xi}(0) = \sqrt{\varepsilon}\,\beta \tag{415}$$

We seek a two-term inner expansion in the form

$$y^i = Y_0(\xi) + \sqrt{\varepsilon}\, Y_1(\xi) + \cdots \tag{416}$$

Substituting (416) into (414) and (415) and equating coefficients of like powers of ε, we obtain

$$Y_0^{iv} - Y_0'' = 0, \qquad Y_0(0) = \alpha, \qquad Y_0'(0) = 0 \tag{417}$$

$$Y_1^{iv} - Y_1'' = 2\xi Y_0'', \qquad Y_1(0) = 0, \qquad Y_1'(0) = \beta \tag{418}$$

The solution of the equation in (417) that excludes the exponentially growing solution is

$$Y_0 = a_0 + a_1\xi + a_2 e^{-\xi}$$

As in the preceding exercise, matching y^o and y^i demands that $a_1 = 0$. Then using the boundary conditions in (417) yields $a_2 = 0$ and $a_0 = \alpha$. Thus

$$Y_0 = \alpha$$

Then the equation in (418) becomes

$$Y_1^{iv} - Y_1'' = 0$$

Excluding the exponentially growing term, we find, as in the preceding exercise, that its solution that satisfies the boundary conditions in (418) is

$$Y_1 = \beta\xi - b_2(1 - \xi - e^{-\xi})$$

Hence

$$y^i = \alpha + \sqrt{\varepsilon}\left[\beta\xi - b_2(1 - \xi - e^{-\xi})\right] + \cdots \tag{419}$$

Matching y^o and y^i

$$\text{Two-term inner expansion of (two-term outer expansion)} = B_0 + \sqrt{\varepsilon}\left[A_0\xi + B_1\right] \tag{420}$$

$$\text{Two-term outer expansion of (two-term inner expansion)} = \alpha + (b_2 + \beta)x - \sqrt{\varepsilon}\, b_2 \tag{421}$$

Equating (420) and (421) and recalling that $\sqrt{\varepsilon}\,\xi = x$, we have

$$B_0 = \alpha, \qquad A_0 = b_2 + \beta, \qquad B_1 = -b_2 \tag{422}$$

Inner Expansion at $x = 1$. We introduce the stretching transformation

$$\zeta = \frac{(1-x)}{\varepsilon^\lambda}, \qquad \lambda > 0$$

or

$$x = 1 - \varepsilon^\lambda \zeta$$

in (409) and obtain

$$\varepsilon^{1-4\lambda}\frac{d^4y^I}{d\zeta^4} - \varepsilon^{-2\lambda}(3 - 2\varepsilon^\lambda\zeta)\frac{d^2y^I}{d\zeta^2} = 1 \tag{423}$$

As $\varepsilon \to 0$, the distinguished limit of (423) corresponds to $\lambda = \frac{1}{2}$. Then (423) becomes

$$\frac{d^4y^I}{d\zeta^4} - (3 - 2\sqrt{\varepsilon}\,\zeta)\frac{d^2y^I}{d\zeta^2} = \varepsilon \tag{424}$$

Since y^I is valid at $x=1$ and since $x=1$ corresponds to $\zeta=0$,

$$y^I(0)=\gamma, \qquad \frac{dy^I}{d\zeta}=-\sqrt{\varepsilon}\,\delta \tag{425}$$

We seek a two-term inner expansion in the form

$$y^I=Y_0(\zeta)+\sqrt{\varepsilon}\,Y_1(\zeta)+\cdots \tag{426}$$

Substituting (426) into (424) and (425) and equating coefficients of like powers of ε, we obtain

$$Y_0^{iv}-3Y_0''=0, \qquad Y_0(0)=\gamma, \qquad Y_0'(0)=0 \tag{427}$$

$$Y_1^{iv}-3Y_1''=-2\zeta Y_0'', \qquad Y_1(0)=0, \qquad Y_0'(0)=-\delta \tag{428}$$

The general solution of the equation in (427) is

$$Y_0=d_0+d_1\zeta+d_2e^{-\sqrt{3}\zeta}+d_3e^{\sqrt{3}\zeta}$$

As in the preceding exercise, matching y^o and y^I demands that $d_1=d_3=0$. Using the boundary conditions in (427) yields $d_0=\gamma$ and $d_2=0$ so that

$$Y_0=\gamma$$

Then the equation in (428) becomes

$$Y_1^{iv}-3Y_1''=0$$

whose general solution is

$$Y_1=a_0+a_1\zeta+a_2e^{-\sqrt{3}\zeta}$$

where the exponentially growing term was excluded using matching considerations. Using the boundary conditions in (428), we have

$$Y_1=-\delta\zeta-a_2\left(1-\sqrt{3}\,\zeta-e^{-\sqrt{3}\zeta}\right)$$

Hence

$$y^I=\gamma+\sqrt{\varepsilon}\left[-\delta\zeta-a_2\left(1-\sqrt{3}\,\zeta-e^{-\sqrt{3}\zeta}\right)\right]+\cdots \tag{429}$$

Matching y^o and y^I

Two-term inner expansion of (two-term outer expansion)

$$=-\tfrac{3}{4}\ln 3+\tfrac{1}{2}+A_0+B_0+\sqrt{\varepsilon}\left[\left(\tfrac{1}{2}\ln 3-A_0\right)\zeta+A_1+B_1\right] \tag{430}$$

Two-term outer expansion of (two-term inner expansion)

$$= \gamma + \left(a_2\sqrt{3} - \delta\right)(1-x) - a_2\sqrt{\varepsilon} \tag{431}$$

Equating (430) and (431) and recalling that $(1-x) = \sqrt{\varepsilon}\,\zeta$, we find that

$$-\tfrac{3}{4}\ln 3 + \tfrac{1}{2} + A_0 + B_0 = \gamma$$

$$a_2\sqrt{3} - \delta = \tfrac{1}{2}\ln 3 - A_0$$

$$A_1 + B_1 = -a_2$$

Using (422), we solve for A_n, B_n, a_2, and b_2 and obtain

$$A_0 = \gamma - \alpha + \tfrac{3}{4}\ln 3 - \tfrac{1}{2}$$

$$a_2 = \tfrac{1}{3}\sqrt{3}\left(\alpha + \delta - \gamma - \tfrac{1}{4}\ln 3 + \tfrac{1}{2}\right)$$

$$b_2 = -B_1 = \gamma - \alpha - \beta + \tfrac{3}{4}\ln 3 - \tfrac{1}{2}$$

$$A_1 = \left(1 + \tfrac{1}{3}\sqrt{3}\right)(\gamma - \alpha) - \beta - \tfrac{1}{3}\sqrt{3}\,\delta + \tfrac{1}{4}\ln 3\left(3 + \tfrac{1}{3}\sqrt{3}\right) - \tfrac{1}{2}\left(1 + \tfrac{1}{3}\sqrt{3}\right)$$

Composite Expansion

$$\begin{aligned}
y^c &= y^o + y^i + y^I - (y^i)^o - (y^I)^o = -\tfrac{1}{2}x\ln(1+2x) + \tfrac{1}{2}x \\
&\quad - \tfrac{1}{4}\ln(1+2x) + A_0 x + B_0 + \sqrt{\varepsilon}\left(A_1 x + B_1\right) + \alpha \\
&\quad + \sqrt{\varepsilon}\left[\beta\xi - b_2\left(1 - \xi - e^{-\xi}\right)\right] + \gamma + \sqrt{\varepsilon}\left[-\delta\zeta - a_2\left(1 - \sqrt{3}\,\zeta - e^{-\sqrt{3}\zeta}\right)\right] \\
&\quad - \alpha - (b_2 + \beta)x + \sqrt{\varepsilon}\,b_2 - \gamma - \left(a_2\sqrt{3} - \delta\right)(1-x) + a_2\sqrt{\varepsilon} \\
&= -\tfrac{1}{2}x\ln(1+2x) + \tfrac{1}{2}x - \tfrac{1}{4}\ln(1+2x) + \left(\gamma - \alpha + \tfrac{3}{4}\ln 3 - \tfrac{1}{2}\right)x \\
&\quad + \alpha + \sqrt{\varepsilon}\left[\left(1 + \tfrac{1}{3}\sqrt{3}\right)(\gamma - \alpha) - \beta - \tfrac{1}{3}\sqrt{3}\,\delta + \tfrac{1}{4}\ln 3\left(3 + \tfrac{1}{3}\sqrt{3}\right) - \tfrac{1}{2}\left(1 + \tfrac{1}{3}\sqrt{3}\right)\right]x \\
&\quad - \sqrt{\varepsilon}\left(\gamma - \alpha - \beta + \tfrac{3}{4}\ln 3 - \tfrac{1}{2}\right)\left(1 - e^{-\xi}\right) \\
&\quad + \tfrac{1}{3}\sqrt{3\varepsilon}\left(\alpha + \delta - \gamma - \tfrac{1}{4}\ln 3 + \tfrac{1}{2}\right)e^{-\sqrt{3}\zeta} + \cdots
\end{aligned} \tag{432}$$

12.17. Consider the problem

$$\varepsilon^2 y'' + x^2 y' - \left(x^2 + \varepsilon^{1/2}\right)y = 0 \tag{433}$$

$$y(0) = \alpha, \qquad y(1) = \beta \tag{434}$$

Show that there are two distinguished limits. Then develop a first-order triple-deck solution.

Since the small parameter multiplies the highest derivative, and since the coefficient of y' is greater than or equal to zero in the interval $[0,1]$, there is at least one boundary layer at the origin. To investigate the behavior of y in the neighborhood of the origin, we introduce the stretching transformation

$$\xi = \frac{x}{\varepsilon^{\lambda}}, \qquad \lambda > 0$$

in (433) and obtain

$$\varepsilon^{2-2\lambda}\frac{d^2y}{d\xi^2} + \varepsilon^{\lambda}\xi^2\frac{dy}{d\xi} - \left(\varepsilon^{2\lambda}\xi^2 + \varepsilon^{1/2}\right)y = 0 \tag{435}$$

As $\varepsilon \to 0$, the term $O(\varepsilon^{2\lambda})$ is small compared with ε^{λ} since $\lambda > 0$. Hence the dominant part of (435) is

$$\varepsilon^{2-2\lambda}\frac{d^2y}{d\xi^2} + \varepsilon^{\lambda}\xi^2\frac{dy}{d\xi} - \varepsilon^{1/2}y + \cdots = 0 \tag{436}$$

The distinguished limits can be obtained by balancing any two terms in (436). Balancing the first and second terms demands that $\lambda = \frac{2}{3}$. Then (436) becomes

$$\varepsilon^{2/3}\frac{d^2y}{d\xi^2} + \varepsilon^{2/3}\xi^2\frac{dy}{d\xi} - \varepsilon^{1/2}y + \cdots = 0$$

whose dominant part is the trivial case $y = 0$. Hence this case must be discarded. Balancing the first and third terms in (436) demands that $\lambda = \frac{3}{4}$. Then (435) becomes

$$\varepsilon^{1/2}\frac{d^2y}{d\xi^2} + \varepsilon^{3/4}\xi^2\frac{dy}{d\xi} - \left(\varepsilon^{3/2}\xi^2 + \varepsilon^{1/2}\right)y = 0 \tag{437}$$

whose dominant part is nontrivial, and hence this case must be included. Balancing the second and third terms demands that $\lambda = \frac{1}{2}$. To distinguish this case from the preceding case, we use the variable ζ instead of ξ so that $\xi = x/\varepsilon^{3/4}$ and $\zeta = x/\varepsilon^{1/2}$. Then (435) becomes

$$\varepsilon\frac{d^2y}{d\zeta^2} + \varepsilon^{1/2}\zeta^2\frac{dy}{d\zeta} - \left(\varepsilon\zeta^2 + \varepsilon^{1/2}\right)y = 0 \tag{438}$$

whose dominant part is nontrivial, and hence this case must be included. Therefore, there are two distinguished limits, one corresponding to the stretching transformation $\xi = x/\varepsilon^{3/4}$ and one corresponding to the stretching transformation $\zeta = x/\varepsilon^{1/2}$.

Outer Expansion. We seek a first-order expansion in the form

$$y^o = y_0(x) + \cdots \tag{439}$$

Since the boundary layers are at the origin, y^o must satisfy $y(1)=\beta$. Hence

$$y^o(1)=\beta \tag{440}$$

Substituting (439) into (433) and (440) and equating the coefficients of ε^0 on both sides, we obtain

$$y_0'-y_0=0, \qquad y_0(1)=\beta$$

Hence

$$y_0=\beta e^{x-1}$$

and

$$y^o=\beta e^{x-1}+\cdots \tag{441}$$

Middle Deck. We seek a first-order middle-deck expansion in the form

$$y^m=Y_0(\zeta)+\cdots$$

which, when substituted into (438) and the coefficient of $\varepsilon^{1/2}$ is set equal to zero, yields

$$\zeta^2 Y_0'-Y_0=0$$

Separating variables, we have

$$\frac{dY_0}{Y_0}=\frac{d\zeta}{\zeta^2}$$

which, upon integration, gives

$$\ln Y_0=-\frac{1}{\zeta}+\ln c$$

where c is a constant. Hence

$$Y_0=c\exp\left(-\frac{1}{\zeta}\right)$$

so that

$$y^m=c\exp\left(-\frac{1}{\zeta}\right)+\cdots \tag{442}$$

Since $x=0$ corresponds to $\zeta=0$, $y^m\to 0$ as $x\to 0$ and hence it cannot satisfy the boundary condition $y(0)=\alpha$. Therefore, c needs to be determined from matching y^m with either y^o or the left deck y^i.

Left Deck. We seek a first-order left-deck expansion in the form

$$y^l = Y_0(\xi) + \cdots$$

which, when substituted into (437) and the coefficient of $\varepsilon^{1/2}$ is equated to zero, yields

$$Y_0'' - Y_0 = 0$$

Since $x = 0$ corresponds to $\xi = 0$ and since the left deck must satisfy the boundary condition at the origin,

$$y^l(0) = \alpha$$

so that

$$Y_0(0) = \alpha$$

Consequently, excluding the exponentially growing part, we find that the solution of Y_0 is

$$Y_0 = \alpha e^{-\xi}$$

so that

$$y^l = \alpha e^{-\xi} + \cdots \tag{443}$$

Matching. To match y^l and y^m, we note that

$$(y^l)^m = 0, \qquad (y^m)^l = 0$$

Hence y^l and y^m match. To match y^m and y^o, we note that

$$(y^o)^m = \beta e^{-1}, \qquad (y^m)^o = c$$

Hence $c = \beta e^{-1}$ so that

$$y^m = \beta \exp\left(-1 - \frac{1}{\zeta}\right) + \cdots$$

Composite Expansion. We form a composite expansion in the form

$$y^c = y^o + y^m + y^l - (y^o)^m - (y^l)^m$$

Hence

$$y^c = \beta e^{x-1} + \beta \exp\left(-1 - \frac{1}{\zeta}\right) + \alpha e^{-\xi} - \beta e^{-1} + \cdots \tag{444}$$

One can easily show that

$$(y^c)^o = y^o, \qquad (y^c)^m = y^m, \qquad (y^c)^l = y^l$$

Hence y^c is valid everywhere in the interval [0,1].

12.18. Consider the problem

$$u'' + \frac{3}{r}u' + \varepsilon u u' = 0 \tag{445}$$

$$u(1) = 0, \qquad u(\infty) = 1 \tag{446}$$

Determine a two-term uniform expansion.

We seek a two-term outer expansion in the form

$$u^o = u_0(r) + \varepsilon u_1(r) + \cdots \tag{447}$$

Substituting (447) into (445) and (446) and equating coefficients of like powers of ε, we obtain

$$u_0'' + \frac{3}{r}u_0' = 0$$
$$u_0(1) = 0, \qquad u_0(\infty) = 1 \tag{448}$$

$$u_1'' + \frac{3}{r}u_1' = -u_0 u_0'$$
$$u_1(1) = 0, \qquad u_1(\infty) = 0 \tag{449}$$

Since (448) is of the Euler type, its solutions have the form $u_0 = r^s$. Hence

$$s(s-1) + 3s = 0 \quad \text{or} \quad s^2 + 2s = 0$$

so that $s = 0$ or $s = -2$. Therefore, the general solution of u_0 is

$$u_0 = c_0 + \frac{c_1}{r^2}$$

Imposing the boundary conditions in (448), we obtain

$$c_0 + c_1 = 0, \qquad c_0 = 1$$

Hence

$$u_0 = 1 - \frac{1}{r^2}$$

which is uniform because it satisfies the reduced equation and both boundary conditions. Then the equation in (449) becomes

$$u_1'' + \frac{3}{r} u_1' = -\frac{2}{r^3} + \frac{2}{r^5}$$

which, upon multiplication by r^3, becomes

$$r^3 u_1'' + 3r^2 u_1' = -2 + \frac{2}{r^2}$$

or

$$\left(r^3 u_1'\right)' = -2 + \frac{2}{r^2}$$

Integrating once yields

$$r^3 u_1' = -2r - \frac{2}{r} + a_0$$

or

$$u_1' = -\frac{2}{r^2} - \frac{2}{r^4} + \frac{a_0}{r^3}$$

Integrating once again yields

$$u_1 = \frac{2}{r} + \frac{2}{3r^3} - \frac{a_0}{2r^2} + a_1$$

Imposing the boundary conditions in (449), we have

$$a_1 - \tfrac{1}{2} a_0 = -2\tfrac{2}{3}, \qquad a_1 = 0$$

Hence $a_0 = \frac{16}{3}$ and

$$u_1 = \frac{2}{r} + \frac{2}{3r^3} - \frac{8}{3r^2}$$

Therefore

$$u = 1 - \frac{1}{r^2} + \frac{2\varepsilon}{3}\left(\frac{3}{r} - \frac{4}{r^2} + \frac{1}{r^3}\right) + \cdots \tag{450}$$

which is uniformly valid. Continuing the expansion to second order, a nonuniformity at large r appears.

12.19. Consider the problem

$$u'' + \frac{1}{r} u' + \varepsilon u u' = 0 \tag{451}$$

$$u(1) = 0, \qquad u(\infty) = 1 \tag{452}$$

Determine a two-term uniform expansion.

As $\varepsilon \to 0$, (451) reduces to

$$u'' + \frac{1}{r}u' = 0$$

which, upon multiplication by r, can be rewritten as

$$(ru')' = 0$$

Hence

$$u' = \frac{c_0}{r}$$

where c_0 is a constant. A further integration yields

$$u = c_0 \ln r + c_1$$

where c_1 is a constant. Imposing the boundary condition $u(1) = 0$ gives $c_1 = 0$ and u becomes

$$u = c_0 \ln r \tag{453}$$

Since $\ln r \to \infty$ as $r \to \infty$, (453) cannot satisfy the boundary condition $u(\infty) = 1$. Therefore, (453) is not valid for large r where a nonuniformity exists.

To determine a uniform expansion, we consider (453) as the first term in an outer expansion and seek an inner expansion using the contracting transformation

$$\xi = r\varepsilon^{\lambda}, \qquad \lambda > 0$$

In terms of ξ, (451) becomes

$$\frac{d^2u^i}{d\xi^2} + \frac{1}{\xi}\frac{du^i}{d\xi} + \varepsilon^{1-\lambda}u^i\frac{du^i}{d\xi} = 0$$

whose distinguished limit as $\varepsilon \to 0$ is

$$\frac{d^2u^i}{d\xi^2} + \frac{1}{\xi}\frac{du^i}{d\xi} + u^i\frac{du^i}{d\xi} = 0 \tag{454}$$

corresponding to $\lambda = 1$.

To determine the form of the inner expansion, we use the matching principle and note that

$$(u^i)^o = (u^o)^i = c_0 \ln\left(\frac{\xi}{\varepsilon}\right) = c_0 \ln\left(\frac{1}{\varepsilon}\right) + c_0 \ln \xi$$

But the boundary condition $u(\infty)=1$ suggests that

$$u^i = 1 + \delta(\varepsilon) U_1(\xi) + \cdots \tag{455}$$

where $\delta(\varepsilon) \to 0$ as $\varepsilon \to 0$, hence

$$(u^i)^o = 1 + \delta(\varepsilon) \lim_{\xi \to 0} U_1(\xi) + \cdots$$

Thus

$$1 + \delta(\varepsilon) \lim_{\xi \to 0} U_1(\xi) = c_0 \ln\left(\frac{1}{\varepsilon}\right) + c_0 \ln \xi$$

Therefore

$$c_0 = \delta(\varepsilon) = \left(\ln \frac{1}{\varepsilon}\right)^{-1}$$

and

$$\lim_{\xi \to 0} U_1(\xi) = \ln \xi \tag{456}$$

Substituting (455) into (454) and equating the coefficient of $\delta(\varepsilon)$ to zero, we have

$$\frac{d^2 U_1}{d\xi^2} + \frac{1}{\xi} \frac{dU_1}{d\xi} + \frac{dU_1}{d\xi} = 0 \tag{457a}$$

To solve (457a), we let $U_1' = v$ and obtain

$$v' + \left(\frac{1}{\xi} + 1\right) v = 0$$

which, upon separating variables, can be rewritten as

$$\frac{dv}{v} = -\left(\frac{1}{\xi} + 1\right) d\xi$$

Hence

$$\ln v = -(\ln \xi + \xi) + \ln a_0$$

where a_0 is a constant. Then

$$v = \frac{dU_1}{d\xi} = \frac{a_0}{\xi} e^{-\xi}$$

A further integration gives

$$U_1 = a_0 \int_{\infty}^{\xi} \frac{1}{\tau} e^{-\tau}\, d\tau + a_1$$

where a_1 is a constant. Since u^i is expected to be valid at $r = \infty$ and since $r = \infty$ corresponds to $\xi = \infty$, $u^i(\infty) = 1$ and hence $U_1(\infty) = 0$. Therefore, $a_1 = 0$ and

$$U_1 = a_0 \int_{\infty}^{\xi} \frac{1}{\tau} e^{-\tau}\, d\tau$$

To impose the matching condition on U_1 in (456), we need the expansion of U_1 for small ξ. To this end, we note that

$$\frac{dU_1}{d\xi} = a_0 \left[\frac{1}{\xi} e^{-\xi} \right] = a_0 \left[\frac{1}{\xi} - 1 + \tfrac{1}{2}\xi + \cdots \right]$$

Hence

$$U_1 = a_0 \left[\ln \xi + c - \xi + \tfrac{1}{4}\xi^2 + \cdots \right]$$

where c is a constant. To determine c, we note that

$$\int_{\infty}^{\xi} \frac{1}{\tau} e^{-\tau}\, d\tau = \ln \xi + c + \cdots$$

so that

$$c = \lim_{\xi \to 0} \left[\int_{\infty}^{\xi} \frac{1}{\tau} e^{-\tau}\, d\tau - \ln \xi \right]$$

Using integration by parts, we have

$$c = \lim_{\xi \to 0} \left[\ln \xi e^{-\xi} + \int_{\infty}^{\xi} \ln \tau e^{-\tau}\, d\tau - \ln \xi \right] = - \int_0^{\infty} \ln \tau e^{-\tau}\, d\tau = \gamma \approx 0.5772$$

is Euler's constant. Consequently,

$$U_1 = a_0[\ln \xi + \gamma + O(\xi)] \quad \text{as } \xi \to 0 \tag{457b}$$

and

$$(u^i)^o = 1 + a_0 \left(\ln \frac{1}{\varepsilon} \right)^{-1} [\ln \xi + \gamma] + \cdots \tag{457c}$$

Comparing (456) and (457b), we see that there is an extra term in (457), preventing the matching of (453) and (455). To effect the matching, we need to add another term to the outer expansion (453) whose inner expansion is $a_0(\ln(1/\varepsilon)^{-1}\gamma$. Since $c_0 =$

$(\ln 1/\varepsilon)^{-1}$, we try

$$u^o = \left(\ln\frac{1}{\varepsilon}\right)^{-1}\ln r + \left(\ln\frac{1}{\varepsilon}\right)^{-2} u_1(r) + \cdots \tag{458}$$

Substituting (458) into (451) and $u(1) = 0$, we find that

$$u_1'' + \frac{1}{r}u_1' = 0, \qquad u_1(1) = 0$$

Hence $u_1 = b\ln r$ and

$$u^o = \left(\ln\frac{1}{\varepsilon}\right)^{-1}\ln r + b\left(\ln\frac{1}{\varepsilon}\right)^{-2}\ln r + \cdots$$

Then

$$(u^o)^i = \left(\ln\frac{1}{\varepsilon}\right)^{-1}\ln\left(\frac{\xi}{\varepsilon}\right) + b\left(\ln\frac{1}{\varepsilon}\right)^{-2}\ln\left(\frac{\xi}{\varepsilon}\right) = 1 + \left(\ln\frac{1}{\varepsilon}\right)^{-1}[\ln\xi + b]$$
$$+ O\left[\left(\ln\frac{1}{\varepsilon}\right)^{-2}\right] \tag{459}$$

Comparing (457c) and (459), we conclude that $a_0 = 1$ and $b = \gamma$. Hence

$$u^i = 1 + \left(\ln\frac{1}{\varepsilon}\right)^{-1}\int_\infty^\xi \frac{1}{\tau}e^{-\tau}\,d\tau + \cdots$$

$$u^c = u^o + u^i - (u^o)^i = 1 + \left(\ln\frac{1}{\varepsilon}\right)^{-1}\int_\infty^\xi \frac{1}{\tau}e^{-\tau}\,d\tau + \gamma\left(\ln\frac{1}{\varepsilon}\right)^{-2}\ln\xi + \cdots$$

12.20. Consider the problem

$$\tfrac{1}{2}\dot{x}^2 = \frac{1-\varepsilon}{x} + \frac{\varepsilon}{1-x}, \qquad t(0) = 0 \tag{460}$$

Show that

$$\sqrt{2}\,t^c = \tfrac{2}{3}x^{3/2} + \varepsilon\left[\tfrac{1}{2} - \ln 2 + \tfrac{1}{2}\ln\varepsilon + \tfrac{2}{3}x^{3/2} + x^{1/2} - \ln(1 + x^{1/2})\right.$$
$$\left. + \xi - \sqrt{\xi(\xi+1)} + \sinh^{-1}\sqrt{\xi}\right] + \cdots \tag{461}$$

where $\xi = (1-x)/\varepsilon$.

We interchange the roles of x and t and rewrite (460) as

$$\sqrt{2}\,\frac{dt}{dx} = \left(\frac{1-\varepsilon}{x} + \frac{\varepsilon}{1-x}\right)^{-1/2} \tag{462}$$

where the positive sign was taken so that t increases with x. Then we seek an outer expansion in the form

$$\sqrt{2}\, t^o = t_0(x) + \varepsilon t_1(x) + \cdots \tag{463}$$

Substituting (463) into (462) and $t(0) = 0$ and equating coefficients of like powers of ε, we obtain

$$\sqrt{2}\, t_0' = x^{1/2}, \qquad t_0(0) = 0 \tag{464}$$

$$\sqrt{2}\, t_1' = \tfrac{1}{2} x^{1/2} - \frac{x^{3/2}}{2(1-x)}, \qquad t_1(0) = 0 \tag{465}$$

The solution of (464) is

$$\sqrt{2}\, t_0 = \tfrac{2}{3} x^{3/2}$$

Then the solution of (465) can be written as

$$\sqrt{2}\, t_1 = \int_0^x \left[\tfrac{1}{2} u^{1/2} - \frac{u^{3/2}}{2(1-u)} \right] du = \int_0^x \left[\tfrac{1}{2} u^{1/2} + \tfrac{1}{2} u^{1/2} - \frac{u^{1/2}}{2(1-u)} \right] du$$

$$= \tfrac{2}{3} x^{3/2} - \tfrac{1}{2} \int_0^x \frac{u^{1/2}}{1-u}\, du$$

We let $u^{1/2} = v$ so that $du = 2v\,dv$. Hence

$$\tfrac{1}{2} \int_0^x \frac{u^{1/2}}{1-u}\, du = \int_0^{x^{1/2}} \frac{v^2\, dv}{1-v^2} = -\int_0^{x^{1/2}} dv + \int_0^{x^{1/2}} \frac{1}{1-v^2}\, dv$$

$$= -x^{1/2} + \tfrac{1}{2} \int_0^{x^{1/2}} \frac{dv}{1+v} + \tfrac{1}{2} \int_0^{x^{1/2}} \frac{dv}{1-v}$$

$$= -x^{1/2} + \tfrac{1}{2} \ln(1 + x^{1/2}) - \tfrac{1}{2} \ln(1 - x^{1/2}) = -x^{1/2} + \tfrac{1}{2} \ln \frac{1 + x^{1/2}}{1 - x^{1/2}}$$

Thus

$$\sqrt{2}\, t_1 = \tfrac{2}{3} x^{3/2} + x^{1/2} - \tfrac{1}{2} \ln \frac{1 + x^{1/2}}{1 - x^{1/2}}$$

and

$$\sqrt{2}\, t^0 = \tfrac{2}{3} x^{3/2} + \varepsilon \left[\tfrac{2}{3} x^{3/2} + x^{1/2} - \tfrac{1}{2} \ln \frac{1 + x^{1/2}}{1 - x^{1/2}} \right] + \cdots \tag{466}$$

The expansion (466) breaks down as $x \to 1$ because of the presence of the logarithmic term. To determine an expansion valid near $x = 1$, we introduce the stretching transformation

$$\xi = \frac{1-x}{\varepsilon^\lambda}, \quad \lambda > 0$$

in (462) and obtain

$$\sqrt{2}\,\frac{dt^i}{d\xi} = -\varepsilon^\lambda\left[\frac{1-\varepsilon}{1-\varepsilon^\lambda\xi} + \frac{\varepsilon^{1-\lambda}}{\xi}\right]^{-1/2} \tag{467}$$

As $\varepsilon \to 0$, the distinguished limit of (467) corresponds to $\lambda = 1$. Hence (467) becomes

$$\sqrt{2}\,\frac{dt^i}{d\xi} = -\varepsilon\left[\frac{1-\varepsilon}{1-\varepsilon\xi} + \frac{1}{\xi}\right]^{-1/2} \tag{468}$$

We seek a two-term inner expansion in the form

$$\sqrt{2}\,t^i = T_0(\xi) + \varepsilon T_1(\xi) + \cdots \tag{469}$$

Substituting (469) into (468) and equating coefficients of like powers ε, we obtain

$$T_0'(\xi) = 0$$

$$\sqrt{2}\,T_1'(\xi) = -\left[1 + \frac{1}{\xi}\right]^{-1/2}$$

Hence $T_0 = \tau_0$ is a constant and

$$\sqrt{2}\,T_1 = -\int \frac{\xi^{1/2}}{(1+\xi)^{1/2}}\,d\xi + \tau_1$$

where τ_1 is a constant. To evaluate the above integral, we let

$$\xi = \sinh^2\theta$$

so that

$$d\xi = 2\sinh\theta\cosh\theta\,d\theta$$

Hence

$$\begin{aligned}
\sqrt{2}\,T_1 &= -2\int \sinh^2\theta\,d\theta + \tau_1 \\
&= \int (1 - \cosh 2\theta)\,d\theta + \tau_1 \\
&= \theta - \tfrac{1}{2}\sinh 2\theta + \tau_1 \\
&= \sinh^{-1}\left(\sqrt{\xi}\right) - \sqrt{\xi(\xi+1)} + \tau_1
\end{aligned}$$

Therefore

$$\sqrt{2}\,t^i = \sqrt{2}\,\tau_0 + \varepsilon\left[\sinh^{-1}\left(\sqrt{\xi}\right) - \sqrt{\xi(\xi+1)} + \tau_1\right] + \cdots \tag{470}$$

where τ_0 and τ_1 need to be determined from matching because t^i is not expected to be valid at $x = 0$.

To match the outer and inner expansions, we let $m = n = 2$ in van Dyke's matching principle and proceed as follows:

Two-term outer expansion

$$= \tfrac{2}{3}x^{3/2} + \varepsilon\left[\tfrac{2}{3}x^{3/2} + x^{1/2} - \tfrac{1}{2}\ln\frac{1+x^{1/2}}{1-x^{1/2}}\right]$$

Rewritten in inner variable

$$= \tfrac{2}{3}(1-\varepsilon\xi)^{3/2} + \varepsilon\left[\tfrac{2}{3}(1-\varepsilon\xi)^{3/2} + (1-\varepsilon\xi)^{1/2} - \tfrac{1}{2}\ln\frac{1+(1-\varepsilon\xi)^{1/2}}{1-(1-\varepsilon\xi)^{1/2}}\right]$$

Expanded for small ε

$$= \tfrac{2}{3}(1 - \tfrac{3}{2}\varepsilon\xi + \cdots) + \varepsilon\left[\tfrac{2}{3} + 1 - \tfrac{1}{2}\ln\frac{1+1-(1/2)\,\varepsilon\xi}{(1/2)\,\varepsilon\xi} + \cdots\right]$$

Two-term inner expansion

$$= \tfrac{2}{3} + \varepsilon\left[\tfrac{5}{3} - \xi - \ln 2 + \tfrac{1}{2}\ln\varepsilon\xi\right] \tag{471}$$

Two-term inner expansion

$$= \sqrt{2}\,\tau_0 + \varepsilon\left[\sinh^{-1}\sqrt{\xi} - \sqrt{\xi(\xi+1)} + \tau_1\right]$$

Rewritten in outer variable

$$= \sqrt{2}\,\tau_0 + \varepsilon\left[\sinh^{-1}\left(\sqrt{\frac{1-x}{\varepsilon}}\right) - \sqrt{\frac{1-x}{\varepsilon}\left(1+\frac{1-x}{\varepsilon}\right)} + \tau_1\right]$$

Expanded for small ε

$$= \sqrt{2}\,\tau_0 + \varepsilon\left\{\tfrac{1}{2}\ln\frac{1-x}{\varepsilon} + \ln 2 - \frac{1-x}{\varepsilon}\left[1 + \frac{\varepsilon}{2(1-x)} + \cdots\right] + \tau_1\right\}$$

Two-term inner expansion

$$= \sqrt{2}\,\tau_0 + \varepsilon\left[\tfrac{1}{2}\ln\frac{1-x}{\varepsilon} + \ln 2 - \frac{1-x}{\varepsilon} - \tfrac{1}{2} + \tau_1\right] \tag{472}$$

Equating (471) and (472) and putting $(1-x)/\varepsilon = \xi$, we have

$$\tfrac{2}{3} + \varepsilon\left[\tfrac{5}{3} - \ln 2 + \tfrac{1}{2}\ln\varepsilon - \xi + \tfrac{1}{2}\ln\xi\right] = \sqrt{2}\,\tau_0 + \varepsilon\left[\tau_1 - \tfrac{1}{2} + \ln 2 + \tfrac{1}{2}\ln\xi - \xi\right]$$

Hence

$$\sqrt{2}\,\tau_0 = \tfrac{2}{3}, \qquad \tau_1 = \tfrac{13}{6} - 2\ln 2 + \tfrac{1}{2}\ln\varepsilon$$

To determine a single uniform expansion, we form a composite expansion according to

$$t^c = t^o + t^i - (t^o)^i$$

which, upon using t^o, t^i, and $(t^o)^i$, yields (461).

12.21. Consider the problem

$$\varepsilon y'' - yy' - y = 0 \tag{473}$$

$$y(0) = \alpha, \qquad y(1) = \beta \tag{474}$$

Determine a first-order uniform solution for the case $y = O(1)$.

In this case, the small parameter multiplies the highest derivative, and hence a boundary layer is expected. However, the location of the boundary layer depends on the sign of the coefficient y of y'. But the value of y is a function of its values α and β at the boundaries. Hence the location of the boundary layer depends on the values of α and β.

Outer Expansion. We seek an outer expansion in the form

$$y^o = y_0(x) + \cdots$$

and obtain from (473) and (474) that

$$y_0 y_0' + y_0 = 0 \tag{475}$$

$$y_0(0) = \alpha, \qquad y_0(1) = \beta \tag{476}$$

Equation (475) provides two branches for the outer expansion, namely

$$y_0 = 0$$

and

$$y_0 = -x + c_0$$

where c_0 is a constant. The first branch must be discarded because it cannot satisfy general boundary conditions. The second branch yields the two special outer solutions

$$\begin{aligned} y^r &= -x + \beta + 1 + \cdots \\ y^l &= -x + \alpha + \cdots \end{aligned} \tag{477}$$

where y^r satisfies the right boundary condition and y^l satisfies the left boundary condition. It follows from (477) that

$$y^r(0) = \beta + 1, \qquad y^l(1) = \alpha - 1$$

Hence, if $\alpha \neq \beta + 1$, y^r is not valid near $x = 0$ and y^l is not valid near $x = 1$. Thus a boundary layer is needed. If the boundary layer is at the left end, y^l is discarded. If the boundary layer is at the right end, y^r is discarded. If the boundary layer is in the interior of the interval, both y^r and y^l are needed. If $\alpha = \beta + 1$, then $y^r = y^l = -x + \alpha$ satisfies the differential equation and boundary conditions and is the exact solution.

Inner Expansion. When $\alpha \neq \beta + 1$, a boundary layer develops somewhere in [0,1]. To investigate the behavior of y in the boundary layer, we introduce the stretching transformation

$$\xi = \frac{x - x_b}{\varepsilon^\lambda} \quad \text{or} \quad x = x_b + \varepsilon^\lambda \xi, \qquad \lambda > 0$$

where x_b is the location of the boundary layer, which is not known a priori. Since the problem is nonlinear, y needs to be scaled. However, since the special case $y = O(1)$ is specified in the problem, there is no need to scale y. In terms of the stretched variable, (473) becomes

$$\varepsilon^{1-2\lambda} \frac{d^2 y^i}{d\xi^2} - \varepsilon^{-\lambda} y^i \frac{dy^i}{d\xi} - y^i = 0$$

or

$$\frac{d^2 y^i}{d\xi^2} - \varepsilon^{\lambda-1} y^i \frac{dy^i}{d\xi} - \varepsilon^{2\lambda-1} y^i = 0$$

whose distinguished limit is

$$\frac{d^2 y^i}{d\xi^2} - y^i \frac{dy^i}{d\xi} = 0$$

corresponding to $\lambda = 1$.

Hence if

$$y^i = Y(\xi) + \cdots$$

then

$$Y'' - YY' = 0$$

which, upon integration, gives

$$Y' - \tfrac{1}{2} Y^2 = \tfrac{1}{2} b \quad \text{or} \quad Y' = \tfrac{1}{2}(b + Y^2)$$

where b is a constant of integration. It must be negative; otherwise, $Y \to \infty$ as $\xi \to \infty$, making y^i unmatchable with the outer expansions. Separating variables, we have

$$\frac{2\,dY}{Y^2 - k^2} = d\xi$$

where b is replaced with $-k^2$, since it is negative. A further integration yields

$$Y = \begin{cases} -k \tanh[\tfrac{1}{2} k(\xi + d)] & \text{if } Y^2 \le k^2 \\ -k \coth[\tfrac{1}{2} k(\xi + d)] & \text{if } Y^2 \ge k^2 \end{cases} \tag{478}$$

We note that Y is independent of the sign of k. Hence, without loss of generality, we take it to be positive.

Matching. When the boundary layer is at the left end, $x_b = 0$, the outer expansion is given by y^r and y^l must be discarded. Expressing y^r in terms of ξ and expanding the result for small ε, we have

$$(y^o)^i = \beta + 1$$

Expressing y^i or to the first approximation Y in terms of x, expanding the result for small ε, and noting that x is positive, we obtain

$$(y^i)^o = -k < 0$$

Hence $k = -(\beta + 1)$ and

$$Y = \begin{cases} -(\beta+1)\tanh[\frac{1}{2}(\beta+1)(\xi+d)] & \text{if } Y^2 \le (\beta+1)^2 \\ -(\beta+1)\coth[\frac{1}{2}(\beta+1)(\xi+d)] & \text{if } Y^2 \ge (\beta+1)^2 \end{cases} \tag{479}$$

Since the boundary layer is assumed at $x = 0$ corresponding to $\xi = 0$, it must satisfy $y(0) = \alpha$ or $Y(0) = \alpha$. Hence

$$\alpha = -(\beta+1)\tanh[\tfrac{1}{2}(\beta+1)d]$$

or

$$\alpha = -(\beta+1)\coth[\tfrac{1}{2}(\beta+1)d]$$

Since k is positive, $\beta < -1$ so the inner solution descends or ascends to $(\beta + 1)$. If $\alpha > (\beta + 1)$, it must also be less than $-(\beta + 1)$ and the inner solution is given by a tanh that descends from α to $(\beta + 1)$. If $\alpha < \beta + 1$, the inner solution is given by a coth that ascends from α to $(\beta + 1)$.

When the boundary layer is at the right end, $x_b = 1$, y^r must be discarded, and the outer solution is given by y^l. To match y^l and y^i, we note that

$$(y^o)^i = \alpha - 1$$

and

$$(y^i)^o = k$$

because as $\varepsilon \to 0$ with x fixed, $\xi \to -\infty$. Hence $k = \alpha - 1$ so that α must be greater than or equal to 1. Moreover

$$Y = \begin{cases} -(\alpha-1)\tanh[\frac{1}{2}(\alpha-1)(\xi+d)] & \text{if } Y^2 \le (1-\alpha)^2 \\ -(\alpha-1)\coth[\frac{1}{2}(\alpha-1)(\xi+d)] & \text{if } Y^2 \ge (1-\alpha)^2 \end{cases} \tag{480}$$

Since the boundary layer is at $x = 1$ corresponding to $\xi = 0$, $Y(0) = \beta$, and

$$\beta = -(\alpha-1)\tanh[\tfrac{1}{2}(\alpha-1)d]$$

or

$$\beta = -(\alpha - 1)\coth[\tfrac{1}{2}(\alpha - 1)d]$$

When $\beta \geq \alpha - 1$, the inner solution is given by a coth that descends from β to $\alpha - 1$. When $\beta \leq \alpha - 1$, it must also be greater than $1 - \alpha$, and the inner solution is given by a tanh that ascends from β to $\alpha - 1$.

Finally, when the boundary layer is at an interior point, both y^r and y^l are needed. To match y^r with y^i, we note that

$$(y^r)^i = [-x + \beta + 1]^i = [-x_b - \varepsilon\xi + \beta + 1]^i = -x_b + \beta + 1$$

$$(y^i)^o = -k$$

because ξ is positive when $x > x_b$. Hence

$$-k = -x_b + \beta + 1$$

To match y^l with y^i, we note that

$$(y^l)^i = [-x + \alpha]^i = [-x_b - \varepsilon\xi + \alpha]^i = -x_b + \alpha$$

$$(y^i)^o = k$$

because ξ is negative when $x < x_b$. Hence

$$k = -x_b + \alpha$$

Solving for k and x_b, we have

$$x_b = \tfrac{1}{2}(\alpha + \beta + 1), \qquad k = \tfrac{1}{2}(\alpha - \beta - 1)$$

Since $0 < x_b < 1$ and $k \geq 0$,

$$0 < \tfrac{1}{2}(\alpha + \beta + 1) < 1 \quad \text{and} \quad \alpha - \beta \geq 1$$

We note that the inner solution must rise from $-x_b + \alpha$ at $\xi = -\infty$ to $-x_b + \beta + 1$ as $\xi \to \infty$. Therefore, the inner expansion is given by a tanh; namely it is given by

$$y^i = -\tfrac{1}{2}(\alpha - \beta - 1)\tanh[\tfrac{1}{4}(\alpha - \beta - 1)\xi] + \cdots \tag{481}$$

where d is taken to be zero because the boundary layer is assumed to be centered at $x = x_b$.

12.22. Consider the problem

$$\varepsilon y'' + yy' - xy = 0 \tag{482}$$

$$y(0) = \alpha, \qquad y(1) = \beta \tag{483}$$

Determine a first-order uniform expansion for the case $y = O(1)$.

Since the small parameter multiplies the highest derivative, the location of the boundary layer depends on the coefficient y of y' and since the value of y depends on the boundary values α and β, the location of the boundary layer depends on the values of α and β.

Outer Expansion. We seek a first-order outer expansion in the form

$$y^o = y_0(x) + \cdots$$

and obtain from (482) and (483) that

$$y_0 y_0' - x y_0 = 0 \tag{484}$$

$$y_0(0) = \alpha, \qquad y_0(1) = \beta \tag{485}$$

Equation (484) provides the two branches

$$y_0 = 0$$

and

$$y_0 = \tfrac{1}{2}x^2 + c_0$$

where c_0 is a constant. The first branch must be discarded because it cannot satisfy general boundary conditions. The second branch yields the two special outer expansions

$$\begin{aligned} y^r &= \tfrac{1}{2}x^2 + \beta - \tfrac{1}{2} + \cdots \\ y^l &= \tfrac{1}{2}x^2 + \alpha \end{aligned} \tag{486}$$

where y^r satisfies the right boundary condition and y^l satisfies the left boundary condition. We note that

$$y^r(0) = \beta - \tfrac{1}{2} \quad \text{and} \quad y^l(1) = \alpha + \tfrac{1}{2}$$

Hence, if $\alpha \neq \beta - \frac{1}{2}$, y^r is not valid near $x = 0$, y^l is not valid near $x = 1$, and a boundary layer needs to be introduced. If $\alpha = \beta - \frac{1}{2}$, a uniform first approximation to y is given by $y^r = y^l = \frac{1}{2}x^2 + \alpha$. In contrast with the preceding exercise, $y^r = y^l$ is only a first approximation rather than the exact solution, because it does not satisfy the differential equation (482).

Inner Expansion. When $\alpha \neq \beta - \frac{1}{2}$, a boundary layer develops somewhere in $[0,1]$. Let x_b be its location. To investigate the behavior of y in the boundary layer, we introduce the stretching transformation

$$\xi = \frac{x - x_b}{\varepsilon^\lambda} \quad \text{or} \quad x = x_b + \varepsilon^\lambda \xi, \qquad \lambda > 0$$

As in the preceding exercise, although the problem is nonlinear, y need not be scaled because only the case $y = O(1)$ is being considered. In terms of the stretched variable,

(482) becomes

$$\varepsilon^{1-2\lambda}\frac{d^2y^i}{d\xi^2}+\varepsilon^{-\lambda}y^i\frac{dy^i}{d\xi}-\left(x_b+\varepsilon^\lambda\xi\right)y^i=0$$

whose distinguished limit is

$$\frac{d^2y^i}{d\xi^2}+y^i\frac{dy^i}{d\xi}=0$$

corresponding to $\lambda=1$. If we let

$$y^i=Y(\xi)+\cdots$$

then to the first approximation,

$$Y''+YY'=0$$

Integration once yields

$$Y'+\tfrac{1}{2}Y^2=\tfrac{1}{2}b$$

where b is a constant of integration. It must be positive, or $Y\to-\infty$ as $\xi\to\infty$ and $Y\to\infty$ as $\xi\to-\infty$, making y^i unmatchable with the outer expansion. Thus we let $b=k^2$. A further integration (see Exercise 12.21) yields

$$Y=\begin{cases} k\tanh\left[\tfrac{1}{2}k(\xi+d)\right] & \text{if } Y^2\le k^2 \\ k\coth\left[\tfrac{1}{2}k(\xi+d)\right] & \text{if } Y^2\ge k^2 \end{cases} \tag{487}$$

We note that Y is independent of the sign of k. Hence, without loss of generality, we take it to be positive.

Matching. When the boundary layer is at the left end, $x_b=0$, y^l must be discarded and the outer expansion is given by y^r. To match y^r and y^i, we note that

$$\left(y^r\right)^i=\left[\tfrac{1}{2}x^2+\beta-\tfrac{1}{2}\right]^i=\left[\tfrac{1}{2}\varepsilon^2\xi^2+\beta-\tfrac{1}{2}\right]^i=\beta-\tfrac{1}{2}$$

$$\left(y^i\right)^o=k$$

because $\xi\to\infty$ as $\varepsilon\to0$ with x fixed. Hence $k=\beta-\frac{1}{2}$. Consequently, β must be greater than $\frac{1}{2}$ since k is positive. Moreover

$$Y=\begin{cases} \left(\beta-\tfrac{1}{2}\right)\tanh\left[\tfrac{1}{2}\left(\beta-\tfrac{1}{2}\right)(\xi+d)\right] & \text{if } Y^2\le\left(\beta-\tfrac{1}{2}\right)^2 \\ \left(\beta-\tfrac{1}{2}\right)\coth\left[\tfrac{1}{2}\left(\beta-\tfrac{1}{2}\right)(\xi+d)\right] & \text{if } Y^2\ge\left(\beta-\tfrac{1}{2}\right)^2 \end{cases} \tag{488}$$

Since the boundary layer is assumed at $x=0$ corresponding to $\xi=0$, y must satisfy

$y(0) = \alpha$ so that $Y(0) = \alpha$. Hence

$$\alpha = \left(\beta - \tfrac{1}{2}\right)\tanh\left[\tfrac{1}{2}\left(\beta - \tfrac{1}{2}\right)d\right]$$

or

$$\alpha = \left(\beta - \tfrac{1}{2}\right)\coth\left[\tfrac{1}{2}\left(\beta - \tfrac{1}{2}\right)d\right]$$

It follows from the behavior of tanh and coth shown in Figure 12.1 that if $\alpha \geq \beta - \frac{1}{2}$, the inner solution is given by a coth that descends from α to $\beta - \frac{1}{2}$. Moreover, if $\alpha < \beta - \frac{1}{2}$, α must also be greater than $\frac{1}{2} - \beta$ and the inner solution is given by a tanh that ascends from α to $\beta - \frac{1}{2}$.

When the boundary layer is at the right end, $x_b = 1$, y^r must be discarded, and the outer expansion is given by y^l. To match y^l and y^i, we note that

$$\left(y^l\right)^i = \left[\tfrac{1}{2}x^2 + \alpha\right]^i = \left[\tfrac{1}{2}(1 + \varepsilon\xi)^2 + \alpha\right]^i = \alpha + \tfrac{1}{2}$$

$$\left(y^i\right)^o = -k$$

because $\xi \to -\infty$ as $\varepsilon \to 0$ with x fixed. Hence $k = -\left(\alpha + \frac{1}{2}\right)$ and

$$Y = \begin{cases} -\left(\alpha + \frac{1}{2}\right)\tanh\left[-\frac{1}{2}\left(\alpha + \frac{1}{2}\right)(\xi + d)\right] & \text{if } Y^2 \leq \left(\alpha + \frac{1}{2}\right)^2 \\ -\left(\alpha + \frac{1}{2}\right)\coth\left[-\frac{1}{2}\left(\alpha + \frac{1}{2}\right)(\xi + d)\right] & \text{if } Y^2 \geq \left(\alpha + \frac{1}{2}\right)^2 \end{cases} \tag{489}$$

Since the inner expansion is expected to satisfy the right boundary condition and since $x = 1$ corresponds to $\xi = 0$, $Y(0) = \beta$. Consequently

$$\beta = -\left(\alpha + \tfrac{1}{2}\right)\tanh\left[-\tfrac{1}{2}\left(\alpha + \tfrac{1}{2}\right)d\right]$$

or

$$\beta = -\left(\alpha + \tfrac{1}{2}\right)\coth\left[-\tfrac{1}{2}\left(\alpha + \tfrac{1}{2}\right)d\right]$$

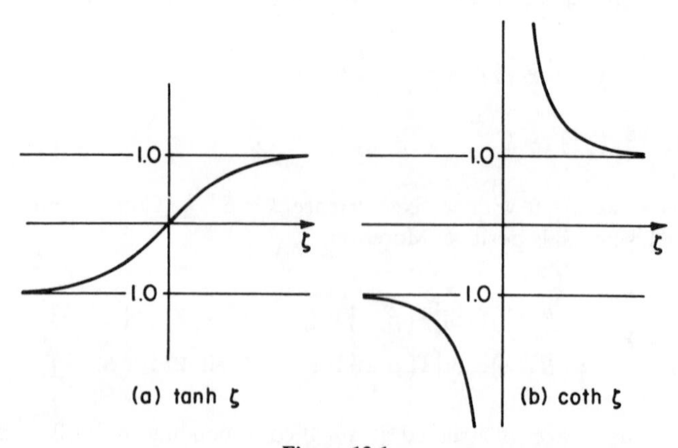

Figure 12.1

It follows from the behavior of tanh and coth that if $\beta < (\alpha + \frac{1}{2})$, the inner solution is given by a coth that ascends from β to $\alpha + \frac{1}{2}$. Moreover, if $\beta > (\alpha + \frac{1}{2})$, then β must also be less than $-(\alpha + \frac{1}{2})$ and the inner solution is given by a tanh that descends from β to $\alpha + \frac{1}{2}$.

When the boundary layer is at an interior point, both y^r and y^l are needed. To match y^r and y^i, we note that

$$(y^r)^i = \left[\tfrac{1}{2}x^2 + \beta - \tfrac{1}{2}\right]^i = \left[\tfrac{1}{2}(x_b + \varepsilon\xi)^2 + \beta - \tfrac{1}{2}\right]^i = \beta + \tfrac{1}{2}(x_b^2 - 1)$$

$$(y^i)^o = k$$

because $\xi \to \infty$ as $\varepsilon \to 0$ with x fixed. Hence

$$k = \beta + \tfrac{1}{2}(x_b^2 - 1)$$

To match y^l with y^i, we note that

$$(y^l)^i = \left[\tfrac{1}{2}x^2 + \alpha\right]^i = \left[\tfrac{1}{2}(x_b + \varepsilon\xi)^2 + \alpha\right]^i = \alpha + \tfrac{1}{2}x_b^2$$

$$(y^i)^o = -k$$

because $\xi \to -\infty$ as $\varepsilon \to 0$ with x fixed. Hence

$$-k = \alpha + \tfrac{1}{2}x_b^2$$

Solving for k and x_b, we obtain

$$x_b = \left(\tfrac{1}{2} - \alpha - \beta\right)^{1/2}, \qquad k = \tfrac{1}{2}\left(\beta - \alpha - \tfrac{1}{2}\right)$$

Hence $\alpha + \beta$ must be less than or equal to $\frac{1}{2}$ so that x_b is real. Since $0 < x_b < 1$ and k is positive, $0 < (\frac{1}{2} - \alpha - \beta)^{1/2} < 1$ and $\beta - \alpha \geq \frac{1}{2}$. For the matching to be effected, the inner solution must be given by a tanh that rises from $-k$ when $\xi = -\infty$ to k when $\xi = \infty$. Therefore

$$y^i = \tfrac{1}{2}\left(\beta - \alpha - \tfrac{1}{2}\right)\tanh\left[\tfrac{1}{2}\left(\beta - \alpha - \tfrac{1}{2}\right)\xi\right] + \cdots \tag{490}$$

where d is taken to be zero because the boundary layer is assumed to be centered at $x = x_b$.

12.23. Consider the problem

$$\varepsilon y'' - y^2 = 0 \tag{491}$$

$$y(0) = \alpha, \qquad y(1) = \beta \tag{492}$$

Determine a first-order uniform expansion.

Since the small parameter multiplies the highest derivative and the difference between the derivatives of the first two terms is 2, two boundary layers are expected, one at each end. Hence the outer expansion is not expected to satisfy any of the boundary conditions.

Outer Expansion. We seek an outer expansion in the form

$$y^o = y_0(x) + \cdots$$

and obtain from (491) that $y_0 \equiv 0$ so that

$$y^o \equiv 0 \tag{493}$$

Inner Expansion Near $x = 0$. To determine an inner expansion near $x = 0$, we introduce the stretching transformation

$$\xi = \frac{x}{\varepsilon^\lambda} \quad \text{or} \quad x = \varepsilon^\lambda \xi, \qquad \lambda > 0$$

in (491) and obtain

$$\varepsilon^{1-2\lambda} \frac{d^2 y^i}{d\xi^2} - (y^i)^2 = 0$$

Its distinguished limit is

$$\frac{d^2 y^i}{d\xi^2} - (y^i)^2 = 0 \tag{494}$$

corresponding to $\lambda = \frac{1}{2}$. We note that the distinguished limit, except for a transformation, is the same as the original differential equation. In other words, no simplification has been obtained using the inner limit process. The simplification occurs in the right boundary condition obtained by matching y^i with y^o, as shown below.

Since y^i is expected to be valid at $x = 0$ corresponding to $\xi = 0$,

$$y^i(0) = \alpha \tag{495}$$

To determine the second boundary condition, we match y^o and y^i. To this end, we note that

$$(y^o)^i \equiv 0$$

$$(y^i)^o = \left[y^i\left(\frac{x}{\varepsilon}\right) \right]^o = \lim_{\xi \to \infty} y^i$$

Hence

$$\lim_{\xi \to \infty} y^i \equiv 0 \tag{496}$$

We seek a first-order inner expansion in the form

$$y^i = Y_0(\xi) + \cdots$$

and obtain from (494)–(496) that

$$Y_0'' - Y_0^2 = 0 \tag{497}$$

$$Y_0(0) = \alpha, \qquad \lim_{\xi \to \infty} Y_0 \equiv 0 \tag{498}$$

To solve for Y_0, we note that Y_0' is absent from (497). Hence we let $u = Y_0'$ and observe that

$$Y_0'' = u' = \frac{du}{dY_0} \cdot Y_0' = u\frac{du}{dY_0}$$

Thus (497) can be rewritten as

$$u\frac{du}{dY_0} = Y_0^2$$

which can be integrated once to obtain

$$\tfrac{1}{2}u^2 = \tfrac{1}{3}Y_0^3 + c_0$$

or

$$\left(\frac{dY_0}{d\xi}\right)^2 = \tfrac{2}{3}Y_0^3 + 2c_0$$

where c_0 is a constant. Imposing the second boundary condition in (498), $Y_0 \to 0$ and $Y_0' \to 0$ as $\xi \to \infty$, yields $c_0 = 0$. Hence

$$\frac{dY_0}{d\xi} = \pm\sqrt{\tfrac{2}{3}}\, Y_0^{3/2}$$

which, upon separation of variables, becomes

$$Y_0^{-3/2}\, dY_0 = \pm\sqrt{\tfrac{2}{3}}\, d\xi$$

A further integration yields

$$-2Y_0^{-1/2} + c_1 = \pm\sqrt{\tfrac{2}{3}}\,\xi$$

where c_1 is a constant. Imposing the boundary condition $Y_0(0) = \alpha$, we have $c_1 = 2\alpha^{-1/2}$, which demands that α be positive. Then

$$-2Y_0^{-1/2} + 2\alpha^{-1/2} = \pm\sqrt{\tfrac{2}{3}}\,\xi$$

or

$$Y_0 = \frac{\alpha}{\left(1 \mp \sqrt{\alpha/6}\,\xi\right)^2}$$

Since $\sqrt{\alpha}\,\xi$ is positive, Y_0 tends to an infinite value at a finite value of ξ instead of Y_0 tending to zero as $\xi \to \infty$, unless the negative sign is discarded. Therefore, to the first approximation,

$$y^i = \frac{\alpha}{\left(1 + \sqrt{\alpha/6}\,\xi\right)^2} + \cdots \tag{499}$$

Inner Expansion at $x=1$. To determine an inner expression valid near $x=1$, we introduce the stretching transformation

$$\zeta = \frac{1-x}{\varepsilon^\lambda} \quad \text{or} \quad x = 1 - \varepsilon^\lambda \zeta, \qquad \lambda > 0$$

and obtain from (491) that

$$\varepsilon^{1-2\lambda} \frac{d^2 y^I}{d\zeta^2} - \left(y^I\right)^2 = 0$$

Its distinguished limit is

$$\frac{d^2 y^I}{d\zeta^2} - \left(y^I\right)^2 = 0$$

corresponding to $\lambda = \frac{1}{2}$. Since y^I is expected to be valid at $x=1$ corresponding to $\zeta = 0$,

$$y^I(0) = \beta$$

Moreover, matching y^o and y^I yields

$$\lim_{\zeta \to \infty} y^I \equiv 0$$

We note that the problem governing y^I is the same as that governing y^i if α is replaced with β. Therefore, β must be positive and

$$y^I = \frac{\beta}{\left(1 + \sqrt{\beta/6}\,\zeta\right)^2} + \cdots \tag{500}$$

The analysis could be easily extended to the case

$$\varepsilon y'' - f(x)\, y^2 = 0 \tag{501}$$

where $f(x)>0$ in [0,1]. The result is

$$y^o \equiv 0 \tag{502}$$

$$y^i = \frac{\alpha}{\left\{1+\sqrt{[\alpha f(0)/6]}\,\xi\right\}^2} + \cdots \tag{503}$$

$$y^I = \frac{\beta}{\left\{1+\sqrt{[\beta f(1)/6]}\,\zeta\right\}^2} + \cdots \tag{504}$$

The analysis could be easily extended to the case

$$\varepsilon y'' - y^n = 0 \tag{505}$$

where $n>1$. The result is

$$y^o \equiv 0 \tag{506}$$

$$y^i = \frac{\alpha}{\left\{1+\left[(n-1)\alpha^{(n-1)/2}/\sqrt{2(n+1)}\right]\xi\right\}^{2/(n-1)}} + \cdots \tag{507}$$

$$y^I = \frac{\beta}{\left\{1+\left[(n-1)\beta^{(n-1)/2}/\sqrt{2(n+1)}\right]\zeta\right\}^{2/(n-1)}} + \cdots \tag{508}$$

Finally, the analysis could be easily extended to the case

$$\varepsilon y'' - f(x)\,y^n = 0 \tag{509}$$

where $n>1$ and $f(x)>0$ in [0,1]. The result is

$$y^o \equiv 0 \tag{510}$$

$$y^i = \frac{\alpha}{\left\{1+(n-1)\sqrt{[\alpha^{n-1}f(0)/2(n+1)]}\,\xi\right\}^{2/(n-1)}} + \cdots \tag{511}$$

$$y^I = \frac{\beta}{\left\{1+(n-1)\sqrt{[\beta^{n-1}f(1)/2(n+1)]}\,\zeta\right\}^{2/(n-1)}} + \cdots \tag{512}$$

12.24. Consider the problem

$$\varepsilon y'' \pm (2x+1)\,y' + y^2 = 0 \tag{513}$$

$$y(0)=\alpha, \qquad y(1)=\beta \tag{514}$$

Determine a first-order uniform expansion.

Since the small parameter multiplies the highest derivative and the difference between the derivatives of the first two terms is one, a boundary layer is expected at the left end in the case of the positive sign and at the right end in the case of the negative sign. Since the problem in nonlinear, nonuniformities may also arise in the interior of the interval, as shown below.

(a) *The Case of the Positive Sign*

In this case, the boundary layer is expected at the left end and the outer expansion must satisfy the right boundary condition. Hence, if we seek a first-order outer expansion in the form

$$y^o = y_0(x) + \cdots$$

we obtain from (513) and $y(1) = \beta$ that

$$(2x+1)\, y_0' + y_0^2 = 0, \qquad y_0(1) = \beta$$

Separating variables, we have

$$\frac{dy_0}{y_0^2} + \frac{dx}{2x+1} = 0$$

which, upon integration, yields

$$-\frac{1}{y_0} + c_0 + \tfrac{1}{2}\ln(2x+1) = 0$$

where c_0 is a constant. Imposing the boundary condition $y_0(1) = \beta$ yields

$$c_0 = \frac{1}{\beta} - \tfrac{1}{2}\ln 3$$

Hence

$$y^o = y_0 + \cdots = \frac{\beta}{1 + \frac{1}{2}\beta \ln[(2x+1)/3]} + \cdots \tag{515}$$

We note that y^o is singular somewhere in $[0,1]$ if $\beta \leq 2/\ln 3$ and a special treatment near the singularity, where another boundary layer exists, is needed. This case is quite involved and it is not discussed here; that is, only the case $\beta > 2/\ln 3$ is discussed here.

Since

$$y^o(0) = y_0(0) + \cdots = \frac{\beta}{1 - \frac{1}{2}\beta \ln 3} + \cdots \neq \alpha$$

in general, y^o is not valid near $x = 0$ and a boundary layer needs to be introduced. To this end, we introduce the stretching transformation

$$\xi = \frac{x}{\varepsilon^\lambda} \quad \text{or} \quad x = \varepsilon^\lambda \xi, \qquad \lambda > 0$$

If $\alpha = O(1)$, then $y = O(1)$ in the boundary layer and (513) can be rewritten as

$$\varepsilon^{1-2\lambda}\frac{d^2y^i}{d\xi^2} + \varepsilon^{-\lambda}(1+2\varepsilon^{\lambda}\xi)\frac{dy^i}{d\xi} + (y^i)^2 = 0$$

Its distinguished limit is

$$\frac{d^2y^i}{d\xi^2} + \frac{dy^i}{d\xi} = 0$$

corresponding to $\lambda = 1$. Hence

$$y^i = a_0 + a_1 e^{-\xi} + \cdots$$

where a_0 and a_1 are constants. Since y^i is valid near $x = 0$ corresponding to $\xi = 0$, $y^i(0) = \alpha$ and hence $a_0 + a_1 = \alpha$. Then

$$y^i = a_0 + (\alpha - a_0)e^{-\xi} + \cdots$$

To determine a_0, we match y^o and y^i and proceed as follows:

$$(y^o)^i = \left\{\frac{\beta}{1+\frac{1}{2}\beta\ln[(2x+1)/3]}\right\}^i + \cdots = \left\{\frac{\beta}{1+\frac{1}{2}\beta\ln[(1+2\varepsilon\xi)/3]}\right\}^i + \cdots$$

$$= \frac{\beta}{1-\frac{1}{2}\beta\ln 3}$$

$$(y^i)^o = [a_0 + (\alpha - a_0)e^{-\xi}]^o = [a_0 + (\alpha - a_0)e^{-x/\varepsilon}]^o = a_0$$

Therefore

$$a_0 = \frac{\beta}{1-\frac{1}{2}\beta\ln 3}$$

and

$$y^i = \frac{\beta}{1-\frac{1}{2}\beta\ln 3} + \left(\alpha - \frac{\beta}{1-\frac{1}{2}\beta\ln 3}\right)e^{-\xi} + \cdots \tag{516}$$

A single uniformly valid expansion can be obtained by forming a composite expansion as follows:

$$y^c = y^o + y^i - (y^i)^o = \frac{\beta}{1+\frac{1}{2}\beta\ln[(2x+1)/3]} + \left(\alpha - \frac{\beta}{1-\frac{1}{2}\beta\ln 3}\right)e^{-\xi} + \cdots \tag{517}$$

(b) *The Case of the Negative Sign*

In this case, the outer expansion satisfies the left boundary condition and a boundary layer is expected at the right end.

We seek a first-order outer expansion in the form

$$y^o = y_0(x) + \cdots$$

and obtain from $y(0) = \alpha$ and (513) with the negative sign that

$$-(2x+1)\,y_0' + y_0^2 = 0, \qquad y_0(0) = \alpha$$

Separating variables, we have

$$\frac{dy_0}{y_0^2} - \frac{dx}{2x+1} = 0$$

which, upon integration, yields

$$-\frac{1}{y_0} + c_0 - \tfrac{1}{2}\ln(2x+1) = 0$$

Imposing the boundary condition $y_0(0) = \alpha$, we obtain

$$c_0 = \frac{1}{\alpha}$$

Hence

$$y^o = y_0 + \cdots = \frac{\alpha}{1 - \frac{1}{2}\alpha\ln(2x+1)} + \cdots \tag{518}$$

We note that y_0 and hence $y^o \to \infty$ at a point in the interval $[0,1]$ if $\alpha \geq 2/\ln 3$. In this case, another boundary layer needs to be introduced at

$$x = x_b = \tfrac{1}{2}(e^{2/\alpha} - 1)$$

and the analysis is quite involved. This case is not treated here.

When $\alpha < 2/\ln 3$, y_0 and hence y^o does not develop any singularities in $[0,1]$ and only one boundary layer needs to be introduced at the right end if

$$\frac{\alpha}{1 - \frac{1}{2}\alpha\ln 3} \neq \beta$$

To accomplish this, we introduce the stretching transformation

$$\xi = \frac{1-x}{\varepsilon^\lambda} \quad \text{or} \quad x = 1 - \varepsilon^\lambda\xi, \qquad \lambda > 0$$

in (513) with the negative sign and obtain

$$\varepsilon^{1-2\lambda}\frac{d^2y^i}{d\xi^2} + \varepsilon^{-\lambda}(3 - 2\varepsilon^\lambda\xi)\frac{dy^i}{d\xi} + (y^i)^2 = 0$$

Its distinguished limit is

$$\frac{d^2y^i}{d\xi^2}+3\frac{dy^i}{d\xi}=0$$

corresponding to $\lambda=1$. Hence

$$y^i=a_0+a_1e^{-3\xi}+\cdots$$

Since y^i is valid near $x=1$ corresponding to $\xi=0$, $y^i(0)=\beta$ and hence $a_1=\beta-a_0$. Then

$$y^i=a_0+(\beta-a_0)e^{-3\xi}+\cdots$$

To determine a_0, we match y^o and y^i and proceed as follows:

$$(y^o)^i=\left[\frac{\alpha}{1-\frac{1}{2}\alpha\ln(2x+1)}\right]^i+\cdots=\left[\frac{\alpha}{1-\frac{1}{2}\alpha\ln(3-2\varepsilon\xi)}\right]^i+\cdots$$

$$=\frac{\alpha}{1-\frac{1}{2}\alpha\ln 3}$$

$$(y^i)^o=\left[a_0+(\beta-a_0)e^{-3\xi}\right]^o=\left[a_0+(\beta-a_0)e^{-3(1-x)/\varepsilon}\right]^o=a_0$$

Therefore

$$a_0=\frac{\alpha}{1-\frac{1}{2}\alpha\ln 3}$$

and

$$y^i=\frac{\alpha}{1-\frac{1}{2}\alpha\ln 3}+\left(\beta-\frac{\alpha}{1-\frac{1}{2}\alpha\ln 3}\right)e^{-3\xi}+\cdots \tag{519}$$

A single uniformly valid expansion can be obtained by forming a composite expansion as follows:

$$y^c=y^o+y^i-(y^i)^o=\frac{\alpha}{1-\frac{1}{2}\alpha\ln(2x+1)}+\left(\beta-\frac{\alpha}{1-\frac{1}{2}\alpha\ln 3}\right)e^{-3\xi}+\cdots \tag{520}$$

Supplementary Exercises

12.25. Consider the problem

$$\varepsilon y''+2y'+y=0,\qquad \varepsilon\ll 1$$

$$y(0)=\alpha,\qquad y(1)=\beta$$

(a) Determine the exact solution.

(b) Use the method of matched asymptotic expansions to determine a first-order uniform expansion. Compare your answer with the exact solution.

(c) Use the method of multiple scales to determine a first-order uniform expansion. Compare your answer with those in (a) and (b).

12.26. Consider the problem

$$\varepsilon y'' - 2y' + y = 0, \qquad \varepsilon \ll 1$$

$$y(0) = \alpha, \qquad y(1) = \beta$$

(a) Determine the exact solution.

(b) Use the method of matched asymptotic expansions to determine a first-order uniform expansion. Compare your answer with the exact solution.

(c) Use the method of multiple scales to determine a first-order uniform expansion. Compare your answer with those in (a) and (b).

12.27. Consider the problem

$$\varepsilon y'' + y' + y = 2, \qquad \varepsilon \ll 1$$

$$y(0) = \alpha, \qquad y(1) = \beta$$

(a) Determine the exact solution.

(b) Use the method of matched asymptotic expansions to determine a first-order uniform expansion. Compare your answer with the exact solution.

(c) Use the method of multiple scales to determine a first-order uniform expansion. Compare your answer with those in (a) and (b).

12.28. Consider the problem

$$\varepsilon y'' - y' + y = 2x, \qquad \varepsilon \ll 1$$

$$y(0) = \alpha, \qquad y(1) = \beta$$

(a) Determine the exact solution.

(b) Use the method of matched asymptotic expansions to determine a first-order uniform expansion. Compare your answer with the exact solution.

(c) Use the method of multiple scales to determine a first-order uniform expansion. Compare your answer with those in (a) and (b).

12.29. Determine first-order uniform expansions for

$$\varepsilon y'' \pm (2x^2 + x + 1)\,y' + (4x + 1)\,y = 0, \qquad \varepsilon \ll 1$$

$$y(0) = \alpha, \qquad y(1) = \beta$$

12.30. Determine first-order uniform expansions for

$$\varepsilon y'' \pm (x^3 + 1)\,y' = 3x^2$$

$$y(0) = \alpha, \qquad y(1) = \beta$$

12.31. Determine first-order uniform expansions for

$$\varepsilon y'' \pm (x^2+1)y' + 2xy = 3x^2 + 2x + 3, \qquad \varepsilon \ll 1$$

$$y(0) = \alpha, \qquad y(1) = \beta$$

12.32. Determine first-order uniform expansions for

(a) $\varepsilon y'' - xy' + xy = 0, \qquad \varepsilon \ll 1$
$y(0) = \alpha, \qquad y(1) = \beta$

(b) $\varepsilon y'' + (1-x)y' + (1-x)y = 0, \qquad \varepsilon \ll 1$
$y(0) = \alpha, \qquad y(1) = \beta$

12.33. Determine first-order uniform expansions for

(a) $\varepsilon y'' + (x-1)y' + (x-1)y = 0, \qquad \varepsilon \ll 1$
$y(1) = \alpha, \qquad y(2) = \beta$

(b) $\varepsilon y'' - xy' - xy = 0, \qquad \varepsilon \ll 1$
$y(-1) = \alpha, \qquad y(0) = \beta$

12.34. Determine a first-order uniform expansion for

$$\varepsilon y'' - (1-x)^2 y' - (1-x)^3 y = 0, \qquad \varepsilon \ll 1$$

$$y(0) = \alpha, \qquad y(1) = \beta$$

12.35. Determine a first-order uniform expansion for

$$\varepsilon y'' + (x^2 + x)y' + xy = 0, \qquad \varepsilon \ll 1$$

$$y(0) = \alpha, \qquad y(1) = \beta$$

12.36. Determine a first-order uniform expansion for

$$\varepsilon y'' - (1-x^3)y' + (1+x-2x^2)y = 0, \qquad \varepsilon \ll 1$$

$$y(0) = \alpha, \qquad y(1) = \beta$$

12.37. Determine a first-order uniform expansion for

$$\varepsilon y'' + xy' - x^2 y = 0, \qquad \varepsilon \ll 1$$

$$y(0) = \alpha, \qquad y(1) = \beta$$

12.38. Determine a first-order uniform expansion for

$$\varepsilon y'' + x^3 y' - xy = 0, \qquad \varepsilon \ll 1$$

$$y(0) = \alpha, \qquad y(1) = \beta$$

12.39. Determine a first-order uniform expansion for

$$\varepsilon y'' + xy' - x^2 y = 0, \qquad \varepsilon \ll 1$$

$$y(-1) = \alpha, \qquad y(1) = \beta$$

12.40. Determine a second-order uniform expansion for

$$\varepsilon y''' - y' = 1$$

$$y(0) = \alpha, \qquad y(1) = \beta, \qquad y'(1) = \gamma$$

Compare your answer with the exact solution.

12.41. Determine a second-order uniform expansion for

$$\varepsilon y''' - y' + y = 0$$

$$y(0) = \alpha, \qquad y(1) = \beta, \qquad y'(1) = \gamma$$

Compare your answer with the exact solution.

12.42. Determine a second-order uniform expansion for

$$\varepsilon y''' - y' = 2x + 1$$

$$y(0) = \alpha, \qquad y'(0) = \beta, \qquad y(1) = \gamma$$

12.43. Determine a two-term uniform expansion of the solution of

$$\varepsilon y^{iv} - (3x+1)^2 y'' = 3, \qquad \varepsilon \ll 1$$

$$y(0) = \alpha, \qquad y'(0) = \beta, \qquad y(1) = \gamma, \qquad y'(1) = \delta$$

12.44. Determine a one-term uniform expansion of the solution of

$$\varepsilon^2 y'' + (x^2 + 2x^3) y' - (2x^2 + \varepsilon^{1/2}) y = 0, \qquad \varepsilon \ll 1$$

$$y(0) = \alpha, \qquad y(1) = \beta$$

12.45. Determine a three-term uniform expansion of the solution of

$$u'' + \frac{3}{r} u' + \varepsilon u u' = 0, \qquad \varepsilon \ll 1$$

$$u(1) = 0, \qquad u(\infty) = 1$$

12.46. Determine a three-term uniform expansion of the solution of

$$u'' + \frac{4}{r}u' + \varepsilon u u' = 0, \qquad \varepsilon \ll 1$$

$$u(1) = 0, \qquad u(\infty) = 1$$

12.47. Determine a first-order uniform expansion of the solution of

$$\varepsilon y'' - y y' - x y = 0, \qquad \varepsilon \ll 1$$

$$y(0) = \alpha, \qquad y(1) = \beta$$

for the case $y = O(1)$.

12.48. Determine a first-order uniform expansion of the solution of

$$\varepsilon y'' + y y' - (2x+1) y = 0, \qquad \varepsilon \ll 1$$

$$y(0) = \alpha, \qquad y(1) = \beta$$

for the case $y = O(1)$.

12.49. Determine a first-order uniform expansion of the solution of

$$\varepsilon y'' - y y' - f(x) y = 0, \qquad \varepsilon \ll 1$$

$$y(0) = \alpha, \qquad y(1) = \beta$$

for the case $y = O(1)$, where $f(x)$ is a known function of x.

12.50. Determine a first-order uniform expansion of the solution of

$$\varepsilon y'' - f(x) y^2 = 0, \qquad \varepsilon \ll 1$$

$$y(0) = \alpha, \qquad y(1) = \beta$$

where $f(x) > 0$ in $[0,1]$.

12.51. Determine a first-order uniform expansion of the solution of

$$\varepsilon y'' - y^n = 0, \qquad \varepsilon \ll 1$$

$$y(0) = \alpha, \qquad y(1) = \beta$$

where $n > 1$.

12.52. Determine a first-order uniform expansion of the solution of

$$\varepsilon y'' - f(x) y^n = 0, \qquad \varepsilon \ll 1$$

$$y(0) = \alpha, \qquad y(1) = \beta$$

where $n > 1$ and $f(x) > 0$ in $[0,1]$.

12.53. Determine first-order uniform expansions of the solutions of

$$\varepsilon y'' \pm (3x+1)\,y' + y^2 = 0, \qquad \varepsilon \ll 1$$

$$y(0) = \alpha, \qquad y(1) = \beta$$

12.54. Determine first-order uniform expansions of the solutions of

$$\varepsilon y'' \pm y' + y^3 = 0, \qquad \varepsilon \ll 1$$

$$y(0) = \alpha, \qquad y(1) = \beta$$

12.55. Determine first-order uniform expansions of the solutions of

$$\varepsilon y'' \pm y' + y^n = 0, \qquad \varepsilon \ll 1$$

$$y(0) = \alpha, \qquad y(1) = \beta$$

where $n \geq 2$.

12.56. Determine first-order uniform expansions of the solutions of

$$\varepsilon y'' \pm f(x)\,y' + f'(x)\,y^2 = 0, \qquad \varepsilon \ll 1$$

$$y(0) = \alpha, \qquad y(1) = \beta$$

where $f(x) > 0$ in $[0,1]$.

12.57. Determine first-order uniform expansions of the solutions of

$$\varepsilon y'' \pm f(x)\,y' + f'(x)\,y^n = 0, \qquad \varepsilon \ll 1$$

$$y(0) = \alpha, \qquad y(1) = \beta$$

where $n \geq 2$ and $f(x) > 0$ in $[0,1]$.

CHAPTER 13

Linear Equations with Variable Coefficients

In this chapter we determine the asymptotic solutions of linear ordinary-differential equations with variable coefficients in the neighborhood of a given finite point $x = x_0$. The asymptotic developments of the solutions depend on whether the point $x = x_0$ is an ordinary point, a regular singular point, or an irregular singular point. For the second-order equation

$$y'' + p(x)y' + q(x)y = 0 \tag{1}$$

the type of the point $x = x_0$ depends on the behavior of $p(x)$ and $q(x)$ at $x = x_0$. Let us assume that they can be developed in ascending power series of $(x - x_0)$ as

$$p(x) = p_0(x - x_0)^{\alpha}[1 + p_1(x - x_0) + \cdots]$$

$$q(x) = q_0(x - x_0)^{\beta}[1 + q_1(x - x_0) + \cdots]$$

where p_0 and q_0 are different from zero. The point $x = x_0$ is called an ordinary point if $\alpha \geq 0$ and $\beta \geq 0$; otherwise, it is called a singular point. A singular point is called a regular singular point if $\alpha \geq -1$ and $\beta \geq -2$; otherwise, it is called an irregular singular point. The point x_0 can always be transferred to the origin by letting $\xi = x - x_0$. The point at infinity can be investigated by transferring it first to the origin by letting $z = x^{-1}$. Therefore, the point x_0 is taken to be the origin, without loss of generality.

If the origin is an ordinary point of a differential equation, all linear independent solutions of the equation can be expressed as convergent Taylor series as

$$y(x) = \sum_{n=0}^{\infty} a_n x^n \tag{2}$$

If the origin is a regular singular point of a differential equation, at least one of the solutions of the equation can be expressed in the Frobenius form

$$y = \sum_{n=0}^{\infty} a_n x^{\sigma+n} \tag{3}$$

where σ is called an index and the series converges. Substituting this solution into the differential equation and extracting the dominant terms yield an algebraic equation for σ; it is called the indicial equation. The resulting solution depends on the roots σ_1 and σ_2 of the indicial equation. There are three cases: $\sigma_1 \neq \sigma_2$ and $\sigma_2 - \sigma_1$ is not an integer, $\sigma_1 \neq \sigma_2$ and $\sigma_2 - \sigma_1$ is an integer, and $\sigma_1 = \sigma_2$. When the indices are unequal and differ by a quantity that is not an integer, the differential equation possesses two linearly independent solutions of the Frobenius type corresponding to the two indices σ_1 and σ_2.

When the indices are equal, one of the solutions of the differential equation has the Frobenius form (3). Another linearly independent solution can be obtained by substituting the assumed form into the differential equation, determining a_n in terms of a_0 and σ, and obtaining

$$z(x;\sigma) = \sum_{n=0}^{\infty} a_n(\sigma) x^{n+\sigma} \tag{4}$$

Then two linearly independent solutions $y_1(x)$ and $y_2(x)$ of the differential equation are given by

$$y_1(x) = \lim_{\sigma \to \sigma_1} z(x;\sigma) \tag{5a}$$

$$y_2(x) = \lim_{\sigma \to \sigma_1} \left[\frac{\partial z(x;\sigma)}{\partial \sigma} \right] \tag{5b}$$

When the indices σ_2 and σ_1 are unequal and differ by an integer, there are three possibilities: the differential equation possesses two linearly independent solutions of the Frobenius form, one or more of the coefficients of $z(x;\sigma)$ is infinite at $\sigma = \sigma_1$, or one coefficient of $z(x;\sigma)$ is indeterminate at $\sigma = \sigma_1$. As an example of the first case, we consider

$$(1 - x^2) y'' + 2xy' + y = 0$$

The indices are $\sigma = 0$ and $\sigma = 1$ and two linearly independent solutions of this equation are

$$y_1(x) = 1 - \frac{x^2}{2} + \frac{x^4}{8} + \frac{x^6}{80} + \cdots$$

$$y_2(x) = x - \frac{x^3}{2} + \frac{x^5}{40} + \frac{3x^7}{560} + \cdots$$

As an example of the second case, we consider

$$x^2y'' + xy' + (x^2 - 1)y = 0$$

The indices are $\sigma = -1$ and $\sigma = 1$ and

$$z(x;\sigma) = a_0 x^\sigma \left[1 - \frac{x^2}{(\sigma+1)(\sigma+3)} + \frac{x^4}{(\sigma+1)(\sigma+3)^2(\sigma+5)} - \frac{x^6}{(\sigma+1)(\sigma+3)^2(\sigma+5)^2(\sigma+7)} + \cdots\right] \tag{6}$$

To obtain two linearly independent solutions from (6), we let $a_0 = (\sigma+1)$ and put

$$y_1(x) = \lim_{\sigma\to -1}[z(x;\sigma)] = -\frac{x}{2} + \frac{x^3}{16} - \frac{x^5}{384} + \cdots$$

$$y_2(x) = \lim_{\sigma\to -1}\left[\frac{\partial z(x;\sigma)}{\partial\sigma}\right] = y_1(x)\ln x + x^{-1} + \frac{x}{4} - \frac{5x^3}{64} + \cdots$$

As an example of the third case, we consider

$$x^2y'' - (4x + x^2)y' + (4 + 3x)y = 0 \tag{7}$$

The indices are $\sigma = 1$ and $\sigma = 4$ and

$$a_{m+1} = \frac{\sigma + m - 3}{(\sigma+m)(\sigma+m-3)} a_m$$

Thus a_3 is indeterminate when $\sigma = 1$. Consequently two linearly independent solutions of (7) are

$$y_1(x) = x\left(1 + x + \frac{x^2}{2}\right)$$

$$y_2(x) = x^4\left(1 + \frac{x}{4} + \frac{x^2}{4.5} + \frac{x^3}{4.5.6} + \cdots\right)$$

If the origin is an irregular singular point of a differential equation, its solutions in the neighborhood of the origin have the form

$$y(x) = e^{\Lambda(x)} x^\sigma u(x) \tag{8}$$

where $u(x)$ can be expressed as a power series in $x^{m/n}$, which need not be convergent, and $\Lambda(x)$ is a polynomial in $x^{-m/n}$, where m and n are prime integers.

Solved Exercises

13.1. Determine two linearly independent solutions of each of the following equations:

(a)
$$xy'' + y' = 0 \tag{9}$$

This equation is of the Euler type. We seek its solutions in the form

$$y = x^{\sigma} \tag{10}$$

Substituting (10) into (9) yields

$$xy'' + y' = \sigma^2 x^{\sigma-1} = 0$$

In this case, the roots of the indicial equation are $\sigma = 0$ with a multiplicity of 2. Thus two linearly independent solutions of (9) are

$$y_1 = x^{\sigma}\Big|_{\sigma=0} = 1$$

$$y_2 = \frac{\partial}{\partial\sigma}(x^{\sigma})\Big|_{\sigma=0} = x^{\sigma}\ln x\Big|_{\sigma=0} = \ln x$$

(b)
$$x^2y'' - y = 0 \tag{11}$$

Again (11) is of the Euler type. Substituting (10) into (11) yields

$$x^2y'' - y = [\sigma(\sigma-1) - 1]x^{\sigma} = 0$$

The indicial equation is

$$\sigma^2 - \sigma - 1 = 0$$

whose solutions are

$$\sigma = \tfrac{1}{2} \mp \sqrt{\tfrac{5}{4}} = \frac{1+\sqrt{5}}{2}, \frac{1-\sqrt{5}}{2}$$

Therefore, two linearly independent solutions of (11) are

$$y_1 = x^{(1+\sqrt{5})/2}, \qquad y_2 = x^{(1-\sqrt{5})/2}$$

(c)
$$x^2y'' + xy' - y = 0 \tag{12}$$

Equation (12) is of the Euler type. Substituting (10) into (12) yields

$$x^2y'' + xy' - y = (\sigma^2 - 1)x^\sigma = 0$$

The indicial equation is

$$\sigma^2 - 1 = 0$$

whose solutions are $\sigma = 1$ and $\sigma = -1$. Hence two linearly independent solutions of (12) are

$$y_1 = x, \qquad y_2 = x^{-1}$$

(d)

$$x^2y'' + 2xy' - 4y = 0 \tag{13}$$

Equation (13) is of the Euler type. Substituting (10) into (13) gives

$$x^2y'' + 2xy' - 4y = (\sigma^2 + \sigma - 4)x^\sigma = 0$$

The indicial equation is

$$\sigma^2 + \sigma - 4 = 0$$

whose solutions are

$$\sigma = \frac{-1+\sqrt{17}}{2}, \qquad \frac{-1-\sqrt{17}}{2}$$

Therefore, two linearly independent solutions of (13) are

$$y_1 = x^{(-1+\sqrt{17})/2}, \qquad y_2 = x^{-(1+\sqrt{17})/2}$$

(e)

$$x^2y'' - xy' + y = 0 \tag{14}$$

Equation (14) is of the Euler type. Substituting (10) into (14) gives

$$x^2y'' - xy' + y = (\sigma - 1)^2 x^\sigma = 0 \tag{15}$$

In this case, the roots of the indicial equation are $\sigma = 1$ with a multiplicity of 2. Thus two linearly independent solutions of (14) are

$$y_1 = x^\sigma \Big|_{\sigma=1} = x \tag{16a}$$

$$y_2 = \frac{\partial}{\partial \sigma}(x^\sigma)\Big|_{\sigma=1} = x^\sigma \ln x\Big|_{\sigma=1} = x \ln x \tag{16b}$$

13.2. Determine three linearly independent solutions of the following equations:

(a)
$$x^2 y''' + 2xy'' - 2y' = 0 \tag{17}$$

(b)
$$x^3 y''' - 3xy' + 3y = 0 \tag{18}$$

(c)
$$x^3 y''' + 2x^2 y'' - 3xy' + 3y = 0 \tag{19}$$

(d)
$$x^3 y''' - 6x^2 y'' + 7xy' - 7y = 0 \tag{20}$$

All of these equations are of the Euler type. Hence some of their solutions have the form (10).

(a) Substituting (10) into (17) yields

$$x^2 y''' + 2xy'' - 2y' = \sigma(\sigma+1)(\sigma-2)x^\sigma = 0 \tag{21}$$

The roots of the indicial equation are

$$\sigma = 0, -1, 2$$

and three linearly independent solutions of (17) are

$$y_1 = 1, \qquad y_2 = x^{-1}, \qquad y_3 = x^2$$

(b) Substituting (10) into (18) gives

$$x^3 y''' - 3xy' + 3y = (\sigma+1)(\sigma-1)(\sigma-3)x^\sigma = 0 \tag{22}$$

The roots of the indicial equation are

$$\sigma = -1, 1, 3$$

and three linearly independent solutions of (18) are

$$y_1 = x^{-1}, \qquad y_2 = x, \qquad y_3 = x^3$$

(c) Substituting (10) into (19) yields

$$x^3 y''' + 2x^2 y'' - 3xy' + 3y = (\sigma-1)(\sigma^2-3)x^\sigma = 0 \tag{23}$$

The roots of the indicial equation are

$$\sigma = 1, \sqrt{3}, -\sqrt{3}$$

and three linearly independent solutions of (19) are

$$y_1 = x, \qquad y_2 = x^{\sqrt{3}}, \qquad y_3 = x^{-\sqrt{3}}$$

(d) Substituting (10) into (20) yields

$$x^3 y''' - 6x^2 y'' + 7xy' - 7y = (\sigma-1)^2(\sigma-7)x^\sigma = 0 \tag{24}$$

The roots of the indicial equation are

$$\sigma = 1, 1, 7$$

and three linearly independent solutions of (20) are

$$y_1 = x, \qquad y_2 = \frac{\partial}{\partial \sigma}(x^\sigma)\bigg|_{\sigma=1} = x \ln x, \qquad y_3 = x^7$$

13.3. Determine two linearly independent solutions of each of the following equations near the origin and determine the radius of convergence for each series:

(a)
$$y'' - xy = 0 \tag{25}$$

(b)
$$y'' + xy' - 2y = 0 \tag{26}$$

(c)
$$y'' + y' - 2xy = 0 \tag{27}$$

(d)
$$y'' - xy' - y = 0 \tag{28}$$

In all of these equations, the origin is a regular point; hence their solutions can be expressed as convergent series of the form

$$y = \sum_{n=0}^{\infty} a_n x^n \tag{29}$$

(a) Substituting (29) into (25) yields

$$\sum_{n=2}^{\infty} n(n-1) a_n x^{n-2} - \sum_{n=0}^{\infty} a_n x^{n+1} = 0$$

or

$$2a_2 + \sum_{n=3}^{\infty} n(n-1) a_n x^{n-2} - \sum_{n=0}^{\infty} a_n x^{n+1} = 0$$

Letting $n = 3 + m$ in the first summation and $n = m$ in the second summation, we obtain

$$2a_2 + \sum_{m=0}^{\infty} (m+2)(m+3) a_{m+3} x^{m+1} - \sum_{m=0}^{\infty} a_m x^{m+1} = 0$$

Equating the coefficient of each power of x to zero, we have

$$a_2 = 0$$

$$a_{m+3} = \frac{a_m}{(m+2)(m+3)}, \qquad m = 0, 1, 2, \ldots \tag{30}$$

Hence

$$a_3 = \frac{a_0}{2\cdot 3}, \qquad a_4 = \frac{a_1}{3\cdot 4} = \frac{2}{4!}a_1, \qquad a_5 = 0$$

$$a_6 = \frac{a_3}{5\cdot 6} = \frac{1\cdot 4}{6!}a_0$$

$$a_7 = \frac{a_4}{6\cdot 7} = \frac{2\cdot 5}{7!}a_1$$

$$a_8 = 0, \qquad a_9 = \frac{a_6}{8\cdot 9} = \frac{1\cdot 4\cdot 7}{9!}a_0$$

$$a_{10} = \frac{a_7}{9\cdot 10} = \frac{2\cdot 5\cdot 8}{10!}a_1$$

and

$$y = a_0\left[1 + \frac{x^3}{3!} + \frac{1\cdot 4x^6}{6!} + \frac{1\cdot 4\cdot 7x^9}{9!} + \cdots\right]$$

$$+ a_1\left[x + \frac{2x^4}{4!} + \frac{2.5x^7}{7!} + \frac{2\cdot 5\cdot 8x^{10}}{10!} + \cdots\right] \tag{31}$$

Consequently, two linearly independent solutions of (25) are

$$y_1 = 1 + \frac{1}{3!}x^3 + \frac{1.4}{6!}x^6 + \frac{1\cdot 4\cdot 7}{9!}x^9 + \cdots$$

$$= 1 + \sum_{n=1}^{\infty} \frac{1\cdot 4\cdot 7\cdots(3n-2)}{(3n)!}x^{3n} \tag{32}$$

$$y_2 = x + \frac{2x^4}{4!} + \frac{2.5x^7}{7!} + \frac{2\cdot 5\cdot 8x^{10}}{10!} + \cdots$$

$$= x + \sum_{n=1}^{\infty} \frac{2\cdot 5\cdot 8\cdots(3n-1)}{(3n+1)!}x^{3n+1} \tag{33}$$

It follows from (32) that

$$\lim_{n\to\infty}\left|\frac{n\text{th term}}{(n-1)\text{th term}}\right| = \lim_{n\to\infty}\left|\frac{x^3}{3n(3n-1)}\right| = 0$$

Hence the radius of convergence of (32) is infinite. Also, it follows from (33) that

$$\lim_{n\to\infty}\left|\frac{n\text{th term}}{(n-1)\text{th term}}\right| = \lim_{n\to\infty}\left|\frac{x^3}{3n(3n+1)}\right| = 0$$

Thus the radius of convergence of (33) is also infinite.

(b) Substituting (29) into (26) yields

$$\sum_{n=2}^{\infty} n(n-1)a_n x^{n-2} + \sum_{n=0}^{\infty} na_n x^n - 2\sum_{n=0}^{\infty} a_n x^n = 0$$

Letting $n = m+2$ in the first summation and $n = m$ in the last two summations, we obtain

$$\sum_{m=0}^{\infty} [(m+1)(m+2)a_{m+2} + (m-2)a_m]x^m = 0$$

which, upon equating the coefficient of each power of x to zero, yields

$$a_{m+2} = -\frac{m-2}{(m+1)(m+2)}a_m, \qquad m = 0, 1, 2, \ldots \tag{34}$$

It follows from (34) that

$$a_2 = \frac{2}{1\cdot 2}a_0 = a_0, \qquad a_4 = 0, \qquad a_6 = -\frac{2}{5\cdot 6}a_4 = 0$$

Hence

$$a_{2n} = 0 \quad \text{for } n = 2, 3, 4, \ldots$$

Moreover, it follows from (34) that

$$a_3 = \frac{1}{2\cdot 3}a_1, \qquad a_5 = \frac{-1}{4\cdot 5} = -\frac{a_1}{5!}$$

$$a_7 = -\frac{3}{6\cdot 7} = \frac{3}{7!}a_1, \qquad a_9 = -\frac{5}{8\cdot 9}a_7 = -\frac{3\cdot 5}{9!}a_1$$

Thus

$$y = a_0(1+x^2) + a_1\left[x + \frac{x^3}{3!} - \frac{x^5}{5!} + \frac{3x^7}{7!} - \frac{3\cdot 5x^9}{9!} + \cdots\right]$$

and two linearly independent solutions of (26) are

$$y_1 = 1 + x^2 \tag{35}$$

$$y_2 = x\left[1 + \frac{x^2}{3!} - \frac{x^4}{5!} + \frac{3x^6}{7!} - \frac{3\cdot 5x^8}{9!} + \cdots\right] = x + \frac{x^3}{3!}$$

$$+ \sum_{n=1}^{\infty} \frac{1\cdot 3\cdot 5\cdot 7\cdots(2n-1)(-x)^{2n+3}}{(2n+3)!} \tag{36}$$

The series (35) terminates. To determine the radius of convergence of (36), we note that

$$\lim_{n\to\infty}\left|\frac{n\text{th term}}{(n-1)\text{th term}}\right| = \lim_{n\to\infty}\left|\frac{-(2n-1)x^2}{(2n+3)(2n+2)}\right| = 0$$

Hence the radius of convergence of (36) is infinite.

(c) Substituting (29) into (27) yields

$$\sum_{n=2}^{\infty} n(n-1)a_n x^{n-2} + \sum_{n=1}^{\infty} n a_n x^{n-1} - 2\sum_{n=0}^{\infty} a_n x^{n+1} = 0$$

or

$$2a_2 + a_1 + \sum_{n=3}^{\infty} n(n-1)a_n x^{n-2} + \sum_{n=2}^{\infty} n a_n x^{n-1} - 2\sum_{n=0}^{\infty} a_n x^{n+1} = 0$$

Letting $n = 3 + m$ in the first summation, $n = m + 2$ in the second summation, and $n = m$ in the third summation, we obtain

$$2a_2 + a_1 + \sum_{m=0}^{\infty} [(m+2)(m+3)a_{m+3} + (m+2)a_{m+2} - 2a_m] x^{m+1} = 0$$

which, upon equating the coefficient of each power of x to zero, yields

$$a_2 = -\tfrac{1}{2}a_1$$

$$a_{m+3} = \frac{2a_m}{(m+2)(m+3)} - \frac{a_{m+2}}{m+3}, \qquad m = 0,1,2,\ldots \tag{37}$$

It follows from (37) that

$$a_3 = \frac{2a_0}{2\cdot 3} - \frac{a_2}{3} = \frac{a_0}{3} + \frac{a_1}{6}$$

$$a_4 = \frac{2a_1}{3\cdot 4} - \frac{a_3}{4} = -\frac{a_0}{12} + \frac{a_1}{8}$$

$$a_5 = \frac{2a_2}{4\cdot 5} - \frac{a_4}{5} = \frac{a_0}{60} - \frac{3a_1}{40}$$

$$a_6 = \frac{2a_3}{5\cdot 6} - \frac{a_5}{6} = \frac{7a_0}{360} + \frac{17a_1}{720}$$

Hence

$$y = a_0\left[1 + \frac{x^3}{3} - \frac{x^4}{12} + \frac{x^5}{60} + \frac{7x^6}{360} + \cdots\right]$$

$$+ a_1\left[x - \frac{x^2}{2} + \frac{x^3}{6} + \frac{x^4}{8} - \frac{3x^5}{40} + \frac{17x^6}{720} + \cdots\right]$$

and two linearly independent solutions of (27) are

$$y_1 = 1 + \frac{x^3}{3} - \frac{x^4}{12} + \frac{x^5}{60} + \frac{7x^6}{360} + \cdots$$

$$y_2 = x - \frac{x^2}{2} + \frac{x^3}{6} + \frac{x^4}{8} - \frac{3x^5}{40} + \frac{17x^6}{720} + \cdots$$

It follows from (29) and (37) that

$$z = \lim_{n\to\infty}\left|\frac{n\text{th term}}{(n-1)\text{th term}}\right| = \lim_{n\to\infty}\left|\frac{xa_n}{a_{n-1}}\right| = \lim_{n\to\infty}\left|\frac{2xa_{n-3}}{n(n-1)a_{n-1}} - \frac{x}{n}\right|$$

$$= \lim_{n\to\infty}\left|\frac{2xa_{n-3}}{n(n-1)a_{n-1}}\right|$$

But

$$a_{n-1} = \frac{2a_{n-4}}{(n-2)(n-1)} - \frac{a_{n-2}}{(n-1)}$$

according to (37), hence

$$z = \lim_{n\to\infty}\left|\frac{1}{[na_{n-4}/(n-2)xa_{n-3}] - (na_{n-2}/2xa_{n-3})}\right| \tag{38}$$

We note that as $n \to \infty$

$$\lim_{n\to\infty}\left|\frac{n\text{th term}}{(n-1)\text{th term}}\right| = \lim_{n\to\infty}\left|\frac{(n-3)\text{th term}}{(n-4)\text{th term}}\right| = \lim_{n\to\infty}\left|\frac{(n-2)\text{th term}}{(n-3)\text{th term}}\right|$$

$$= \lim_{n\to\infty}\left|\frac{xa_n}{a_{n-1}}\right| = \lim_{n\to\infty}\left|\frac{xa_{n-3}}{a_{n-4}}\right| = \lim_{n\to\infty}\left|\frac{xa_{n-2}}{a_{n-3}}\right| = z$$

It follows from (38) that

$$z = \lim_{n\to\infty}\left|\frac{1}{1/z - \frac{1}{2}nz}\right| = 0$$

and consequently, the radii of convergence of the series in y_1 and y_2 are infinite.

(d) Substituting (29) into (28) gives

$$\sum_{n=2}^{\infty} n(n-1)a_n x^{n-2} - \sum_{n=0}^{\infty} na_n x^n - \sum_{n=0}^{\infty} a_n x^n = 0$$

Letting $n = 2 + m$ in the first summation and $n = m$ in the last two summations, we have

$$\sum_{m=0}^{\infty} [(m+1)(m+2)a_{m+2} - (m+1)a_m]x^m = 0$$

which, upon equating the coefficient of each power of x to zero, yields

$$a_{m+2} = \frac{a_m}{m+2}, \qquad m = 0, 1, 2, \ldots \tag{39}$$

Hence

$$a_2 = \frac{a_0}{2}, \qquad a_4 = \frac{a_2}{4} = \frac{a_0}{2\cdot 4}$$

$$a_6 = \frac{a_4}{6} = \frac{a_0}{2\cdot 4\cdot 6}, \qquad a_8 = \frac{a_6}{8} = \frac{a_0}{2\cdot 4\cdot 6\cdot 8}$$

$$a_3 = \frac{a_1}{3}, \qquad a_5 = \frac{a_3}{5} = \frac{a_1}{3\cdot 5}$$

$$a_7 = \frac{a_5}{7} = \frac{a_1}{3\cdot 5\cdot 7}, \qquad a_9 = \frac{a_7}{9} = \frac{a_1}{3\cdot 5\cdot 7\cdot 9}$$

Then

$$y = a_0\left[1 + \frac{x^2}{2} + \frac{x^4}{2\cdot 4} + \frac{x^6}{2\cdot 4\cdot 6} + \frac{x^8}{2\cdot 4\cdot 6\cdot 8} + \cdots\right]$$

$$+ a_1\left[x + \frac{x^3}{3} + \frac{x^5}{3\cdot 5} + \frac{x^7}{3\cdot 5\cdot 7} + \frac{x^9}{3\cdot 5\cdot 7\cdot 9} + \cdots\right]$$

and two linearly independent solutions of (28) are

$$y_1 = 1 + \frac{x^2}{2} + \frac{x^4}{2\cdot 4} + \frac{x^6}{2\cdot 4\cdot 6} + \frac{x^8}{2\cdot 4\cdot 6\cdot 8} + \cdots = \sum_{m=0}^{\infty} \frac{x^{2m}}{2^m m!} \tag{40}$$

$$y_2 = x + \frac{x^3}{3} + \frac{x^5}{3\cdot 5} + \frac{x^7}{3\cdot 5\cdot 7} + \frac{x^9}{3\cdot 5\cdot 7\cdot 9} + \cdots = \sum_{m=0}^{\infty} \frac{2^m m! x^{2m+1}}{(2m+1)!} \tag{41}$$

It follows from (40) that

$$\lim_{m\to\infty}\left|\frac{(m+1)\text{th term}}{m\text{th term}}\right| = \lim_{m\to\infty}\left|\frac{x^{2m}2^{m-1}(m-1)!}{2^m m! x^{2m-2}}\right| = 0$$

and hence the radius of convergence of (40) is infinite. It follows from (41) that

$$\lim_{m\to\infty}\left|\frac{(m+1)\text{th term}}{m\text{th term}}\right| = \lim_{m\to\infty}\left|\frac{2^m m! x^{2m+1}(2m-1)!}{(2m+1)!2^{m-1}(m-1)!x^{2m-1}}\right| = 0$$

and hence the radius of convergence of (41) is infinite.

13.4. Consider the Hermite equation

$$y'' - 2xy' + \gamma y = 0 \tag{42}$$

Determine two linearly independent solutions of this equation in power series near the origin and show that one of them terminates if $\gamma = 2n$, where $n = 0, 1, 2, \ldots$.

The origin is an ordinary point of (42) so that its solutions have the form (29). Substituting (29) into (42) gives

$$\sum_{n=2}^{\infty} n(n-1)a_n x^{n-2} - \sum_{n=0}^{\infty} (2n-\gamma)a_n x^n = 0$$

Letting $n = 2+m$ in the first summation and $n = m$ in the second summation, we obtain

$$\sum_{m=0}^{\infty} [(m+1)(m+2)a_{m+2} - (2m-\gamma)a_m]x^m = 0$$

which, upon equating the coefficient of each power of x to zero, yields

$$a_{m+2} = \frac{2m-\gamma}{(m+1)(m+2)}a_m \tag{43}$$

Hence two linearly independent solutions of (42) are

$$y_1 = 1 + \sum_{m=1}^{\infty} \frac{-\gamma(2\cdot 2-\gamma)(2\cdot 4-\gamma)\cdots[2(2m-2)-\gamma]}{(2m)!}x^{2m}$$

$$y_2 = x + \sum_{m=1}^{\infty} \frac{(2\cdot 1-\gamma)(2\cdot 3-\gamma)\cdots[2(2m-1)-\gamma]}{(2m+1)!}x^{2m+1}$$

It follows from (43) that $a_{s+2} = 0$ for all $s \geq n$ if $\gamma = 2n$.

13.5. Consider the Legendre equation

$$(1-x^2)y'' - 2xy' + \gamma y = 0 \tag{44}$$

Determine two linearly independent solutions of this equation in power series near the origin and show that one of them terminates if $\gamma = n(n+1)$, $n = 0, 1, 2, \ldots$.

Since the origin is an ordinary point of (44), its solutions can be expressed as in (29). Substituting (29) into (44) yields

$$\sum_{n=2}^{\infty} n(n-1)a_n x^{n-2} - \sum_{n=2}^{\infty} n(n-1)a_n x^n - 2\sum_{n=1}^{\infty} na_n x^n + \gamma \sum_{n=0}^{\infty} a_n x^n = 0$$

or

$$\sum_{n=2}^{\infty} n(n-1)a_n x^{n-2} + \sum_{n=0}^{\infty} [\gamma - n(n+1)]a_n x^n = 0$$

Putting $n = m+2$ in the first summation and $n = m$ in the second summation, we have

$$\sum_{m=0}^{\infty} (m+1)(m+2)a_{m+2}x^m + \sum_{m=0}^{\infty} [\gamma - m(m+1)]a_m x^m = 0$$

Equating the coefficient of each power of x to zero, we obtain

$$a_{m+2} = -\frac{\gamma - m(m+1)}{(m+1)(m+2)}a_m \tag{45}$$

It follows from (45) that

$$a_2 = -\frac{\gamma}{2!}a_0, \qquad a_4 = \frac{\gamma(\gamma - 2\cdot 3)}{4!}a_0, \qquad a_6 = -\frac{\gamma(\gamma-2\cdot 3)(\gamma - 4\cdot 5)}{6!}a_0$$

$$a_3 = -\frac{\gamma - 1\cdot 2}{3!}a_1, \qquad a_5 = \frac{(\gamma - 1\cdot 2)(\gamma - 3\cdot 4)}{5!}a_1$$

$$a_7 = -\frac{(\gamma-1\cdot 2)(\gamma - 3\cdot 4)(\gamma - 5\cdot 6)}{7!}a_1$$

Therefore, two linearly independent solutions of (44) are

$$y_1 = 1 - \frac{\gamma}{2!}x^2 + \frac{\gamma(\gamma-2\cdot 3)}{4!}x^4 - \frac{\gamma(\gamma - 2\cdot 3)(\gamma - 4\cdot 5)}{6!}x^6 + \cdots$$

$$y_2 = x - \frac{\gamma - 1\cdot 2}{3!}x^3 + \frac{(\gamma - 1\cdot 2)(\gamma - 3\cdot 4)}{5!}x^5 - \frac{(\gamma - 1\cdot 2)(\gamma-3\cdot 4)(\gamma - 5\cdot 6)}{7!}x^7 + \cdots$$

Moreover, it follows from (45) that $a_{s+2} = 0$ for $s \geq n$ if $\gamma = n(n+1)$, and hence one of the above series terminates.

13.6. Determine two linearly independent solutions in power series near the origin of the Tschebycheff equation

$$(1-x^2)y'' - xy' + \gamma y = 0 \tag{46}$$

and show that one of them terminates if $\gamma = n^2$, $n = 0, 1, 2, \ldots$.

Since the origin is an ordinary point of (46), the solutions of (46) can be expressed as in (29). Substituting (29) into (46) yields

$$\sum_{n=2}^{\infty} n(n-1)a_n x^{n-2} - \sum_{n=2}^{\infty} n(n-1)a_n x^n - \sum_{n=1}^{\infty} na_n x^n + \gamma\sum_{n=0}^{\infty} a_n x^n = 0$$

or

$$\sum_{n=2}^{\infty} n(n-1)a_n x^{n-2} + \sum_{n=0}^{\infty} (\gamma - n^2)a_n x^n = 0$$

Putting $n = m+2$ in the first summation and $n = m$ in the second summation, we obtain

$$\sum_{m=0}^{\infty} (m+1)(m+2)a_{m+2}x^m + \sum_{m=0}^{\infty} (\gamma - m^2)a_m x^m = 0$$

Equating the coefficient of each power of x to zero, we have

$$a_{m+2} = -\frac{\gamma - m^2}{(m+1)(m+2)}a_m \tag{47}$$

It follows from (47) that

$$a_2 = -\frac{\gamma}{2!}a_0, \qquad a_4 = \frac{\gamma(\gamma-4)}{4!}a_0, \qquad a_6 = -\frac{\gamma(\gamma-4)(\gamma-16)}{6!}a_0$$

$$a_3 = -\frac{\gamma-1}{3!}a_1, \qquad a_5 = \frac{(\gamma-1)(\gamma-9)}{5!}a_1$$

$$a_7 = -\frac{(\gamma-1)(\gamma-9)(\gamma-25)}{7!}a_1$$

Thus two linearly independent solutions of (46) are

$$y_1 = 1 - \frac{\gamma}{2!}x^2 + \frac{\gamma(\gamma-4)}{4!}x^4 - \frac{\gamma(\gamma-4)(\gamma-16)}{6!}x^6 + \cdots$$

$$y_2 = x - \frac{\gamma-1}{3!}x^3 + \frac{(\gamma-1)(\gamma-9)}{5!}x^5 - \frac{(\gamma-1)(\gamma-9)(\gamma-25)}{7!}x^7 + \cdots$$

Moreover, it follows from (47) that $a_{s+2} = 0$ for $s \geq n$ if $\gamma = n^2$, and hence one of the series terminates.

13.7. Determine two linearly independent solutions in power series near the origin for each of the following equations:

(a)
$$4xy'' + 2y' - y = 0 \tag{48}$$

(b)
$$(2x + x^2)y'' + y' - 6xy = 0 \tag{49}$$

(c)
$$9x(1-x)y'' - 12y' + 4y = 0 \tag{50}$$

(d)
$$2x(1-x)y'' + (1-x)y' + 3y = 0 \tag{51}$$

The origin is a regular singular point of each of the equations (48)–(51). Hence at least one solution of each of these equations has the Frobenius form

$$y = \sum_{n=0}^{\infty} a_n x^{\sigma+n} \tag{52}$$

(a) Substituting (52) into (48) yields

$$\sum_{n=0}^{\infty} 2(\sigma+n)(2\sigma+2n-1)\, a_n x^{\sigma+n-1} - \sum_{n=0}^{\infty} a_n x^{\sigma+n} = 0$$

or

$$2\sigma(2\sigma-1)\, a_0 x^{\sigma-1} + \sum_{n=1}^{\infty} 2(\sigma+n)(2\sigma+2n-1)\, a_n x^{\sigma+n-1} - \sum_{n=0}^{\infty} a_n x^{\sigma+n} = 0$$

Putting $n = m+1$ in the first summation and $n = m$ in the second summation, we have

$$2\sigma(2\sigma-1)\, a_0 x^{\sigma-1} + \sum_{m=0}^{\infty} \left[2(\sigma+m+1)(2\sigma+2m+1)\, a_{m+1} - a_m\right] x^{\sigma+m} = 0$$

Equating the coefficient of each power of x to zero yields

$$2\sigma(2\sigma-1)\, a_0 = 0 \tag{53}$$

$$a_{m+1} = \frac{a_m}{2(\sigma+m+1)(2\sigma+2m+1)} \tag{54}$$

It follows from (53) and (54) that, for a nontrivial solution, $a_0 \neq 0$ and hence

$$\sigma = 0 \quad \text{or} \quad \sigma = \tfrac{1}{2}$$

Since the indices are distinct and their difference is not an integer, (48) possesses two linearly independent solutions of the Frobenius type.

When $\sigma = 0$, it follows from (54) that

$$a_{m+1} = \frac{a_m}{2(m+1)(2m+1)}$$

and hence one of the solutions of (48) is

$$y_1 = 1 + \frac{x}{2!} + \frac{x^2}{4!} + \frac{x^3}{6!} + \cdots = \sum_{n=0}^{\infty} \frac{x^n}{(2n)!}$$

When $\sigma = \frac{1}{2}$, it follows from (54) that

$$a_{m+1} = \frac{a_m}{(2m+2)(2m+3)}$$

and a second solution of (48) that is linearly independent from y_1 is

$$y_2 = x^{1/2}\left[1 + \frac{x}{3!} + \frac{x^2}{5!} + \frac{x^3}{7!} + \cdots\right] = \sum_{n=0}^{\infty} \frac{x^{n+1/2}}{(2n+1)!}$$

Using the ratio test, one can show that y_1 and y_2 have infinite radii of convergence.

(b) Substituting (52) into (49) yields

$$\sum_{n=0}^{\infty} (\sigma+n)(2\sigma+2n-1)\,a_n x^{\sigma+n-1} + \sum_{n=0}^{\infty} (\sigma+n)(\sigma+n-1)\,a_n x^{\sigma+n}$$
$$-6\sum_{n=0}^{\infty} a_n x^{\sigma+n+1} = 0$$

or

$$\sigma(2\sigma-1)\,a_0 x^{\sigma-1} + [(\sigma+1)(2\sigma+1)\,a_1 + \sigma(\sigma-1)\,a_0]\,x^{\sigma}$$
$$+ \sum_{n=2}^{\infty} (\sigma+n)(2\sigma+2n-1)\,a_n x^{\sigma+n-1}$$
$$+ \sum_{n=1}^{\infty} (\sigma+n)(\sigma+n-1)\,a_n x^{\sigma+n} - 6\sum_{n=0}^{\infty} a_n x^{\sigma+n+1} = 0$$

Putting $n = m+2$ in the first summation, $n = m+1$ in the second summation, and $n = m$ in the third summation, we obtain

$$\sigma(2\sigma-1)\,a_0 x^{\sigma-1} + [(\sigma+1)(2\sigma+1)\,a_1 + \sigma(\sigma-1)\,a_0]\,x^{\sigma}$$
$$+ \sum_{m=0}^{\infty} [(\sigma+m+2)(2\sigma+2m+3)\,a_{m+2}$$
$$+(\sigma+m+1)(\sigma+m)\,a_{m+1} - 6a_m]\,x^{\sigma+m+1} = 0$$

Equating the coefficient of each power of x to zero yields

$$\sigma(2\sigma-1)\,a_0 = 0 \tag{55}$$

$$a_1 = -\frac{\sigma(\sigma-1)\,a_0}{(\sigma+1)(2\sigma+1)} \tag{56}$$

$$a_{m+2} = \frac{6a_m - (\sigma+m)(\sigma+m+1)\,a_{m+1}}{(\sigma+m+2)(2\sigma+2m+3)} \tag{57}$$

For a nontrivial solution, it follows from (55)–(57) that $a_0 \neq 0$ and the indices are

$$\sigma = 0 \quad \text{and} \quad \tfrac{1}{2}$$

Since they are distinct and their difference is not an integer, two linearly independent solutions of (49) are of the Frobenius type.

When $\sigma = 0$, it follows from (56) that $a_1 = 0$ and it follows from (57) that

$$a_{m+2} = \frac{6a_m - m(m+1)a_{m+1}}{(m+2)(2m+3)}$$

Hence one solution of (49) is

$$y_1 = 1 + x^2 - \tfrac{2}{15}x^3 + \tfrac{17}{70}x^4 + \cdots$$

When $\sigma = \frac{1}{2}$, it follows from (56) and (57) that $a_1 = \frac{1}{12}a_0$ and

$$a_{m+2} = \frac{24a_m - (2m+1)(2m+3)a_{m+1}}{4(2m+5)(m+2)}$$

Hence a second solution of (49) that is linearly independent of y_1 is

$$y_2 = x^{1/2}\left[1 + \tfrac{1}{12}x + \tfrac{19}{32}x^2 - \tfrac{221}{2688}x^3 + \cdots\right]$$

(c) Substituting (52) into (50) yields

$$\sum_{n=0}^{\infty} 3(\sigma+n)(3\sigma+3n-7)a_n x^{\sigma+n-1} - \sum_{n=0}^{\infty}[9(\sigma+n)(\sigma+n-1)-4]a_n x^{\sigma+n} = 0$$

or

$$3\sigma(3\sigma-7)a_0 x^{\sigma-1} + \sum_{n=1}^{\infty} 3(\sigma+n)(3\sigma+3n-7)a_n x^{\sigma+n-1}$$

$$-\sum_{n=0}^{\infty}[9(\sigma+n)(\sigma+n-1)-4]a_n x^{\sigma+n} = 0$$

Putting $n = m+1$ in the first summation and $n = m$ in the second summation, we have

$$3\sigma(3\sigma-7)a_0 x^{\sigma-1} + \sum_{m=0}^{\infty}\{3(\sigma+m+1)(3\sigma+3m-4)a_{m+1}$$

$$-[9(\sigma+m)(\sigma+m-1)-4]a_m\}x^{\sigma+m} = 0$$

Equating the coefficient of each power of x to zero yields

$$\sigma(3\sigma-7)a_0 = 0 \tag{58}$$

$$a_{m+1} = \frac{9(\sigma+m)(\sigma+m-1)-4}{3(\sigma+m+1)(3\sigma+3m-4)}a_m \tag{59}$$

For a nontrivial solution $a_0 \neq 0$ and $\sigma = 0$ or $\frac{7}{3}$. Since the indices are distinct and differ by a quantity that is not an integer, two linearly independent solutions of (50) are of the Frobenius type.

When $\sigma = 0$, it follows from (59) that

$$a_{m+1} = \frac{9m(m-1)-4}{3(m+1)(3m-4)} a_m = \frac{3m+1}{3(m+1)} a_m$$

Hence one of the solutions of (50) is

$$y_1 = 1 + \tfrac{1}{3}x + \frac{1\cdot 4}{3\cdot 6}x^2 + \frac{1\cdot 4\cdot 7}{3\cdot 6\cdot 9}x^3 + \cdots = (1-x)^{-1/3}$$

When $\sigma = \frac{7}{3}$, it follows from (59) that

$$a_{m+1} = \frac{(7+3m)(4+3m)-4}{(3m+10)(3m+3)} a_m = \frac{3m+8}{3m+10} a_m$$

and a second solution of (50) that is linearly independent of y_1 is

$$y_2 = x^{7/3}\left[1 + \tfrac{8}{10}x + \frac{8\cdot 11}{10\cdot 13}x^2 + \frac{8\cdot 11\cdot 14}{10\cdot 13\cdot 16}x^3 + \cdots\right]$$

(d) Substituting (52) into (51) yields

$$\sum_{n=0}^{\infty} (\sigma+n)(2\sigma+2n-1)a_n x^{\sigma+n-1} - \sum_{n=0}^{\infty} (2\sigma+2n-3)(\sigma+n+1)a_n x^{\sigma+n} = 0$$

or

$$\sigma(2\sigma-1)a_0 x^{\sigma-1} + \sum_{n=1}^{\infty} (\sigma+n)(2\sigma+2n-1)a_n x^{\sigma+n-1}$$

$$-\sum_{n=0}^{\infty} (2\sigma+2n-3)(\sigma+n+1)a_n x^{\sigma+n} = 0$$

Putting $n = m+1$ in the first summation and $n = m$ in the second summation, we have

$$\sigma(2\sigma-1)a_0 x^{\sigma-1} + \sum_{m=0}^{\infty} [(\sigma+m+1)(2\sigma+2m+1)a_{m+1}$$

$$-(2\sigma+2m-3)(\sigma+m+1)a_m]x^{\sigma+m} = 0$$

Equating the coefficient of each power of x to zero yields

$$\sigma(2\sigma-1)a_0 = 0 \tag{60}$$

$$a_{m+1} = \frac{2\sigma+2m-3}{2\sigma+2m+1} a_m \tag{61}$$

For a nontrivial solution, $a_0 \neq 0$ and the indices are $\sigma = 0$ and $\frac{1}{2}$. Since they are unequal and differ by a quantity that is not an integer, two linearly independent solutions of (51) are of the Frobenius type.

When $\sigma = 0$, it follows from (61) that

$$a_{m+1} = \frac{2m-3}{2m+1} a_m$$

and a solution of (51) is

$$y_1 = 1 - 3x + \frac{3x^2}{1 \cdot 3} + \frac{3x^3}{3 \cdot 5} + \frac{3x^4}{5 \cdot 7} + \cdots$$

When $\sigma = \frac{1}{2}$, it follows from (61) that

$$a_{m+1} = \frac{m-1}{m+1} a_m$$

and hence

$$a_1 = -a_0, \qquad a_2 = 0, \qquad a_3 = 0, \qquad \ldots, \qquad a_n = 0$$

Therefore, a second solution of (51) that is linearly independent of y_1 is

$$y_2 = x^{1/2}(1-x)$$

13.8. Determine two linearly independent solutions in power series near the origin for each of the following equations:

(a) $$x^2 y'' + x(x-1) y' - xy = 0 \tag{62}$$

(b) $$x^2 y'' + xy' + (x^2 - 4) y = 0 \tag{63}$$

(c) $$(1-x^2) y'' - 2xy' + 2y = 0 \tag{64}$$

(d) $$x(1-x) y'' - 3xy' - y = 0 \tag{65}$$

(e) $$y'' + x^2 y = 0 \tag{66}$$

(f) $$(2+x^2) y'' + xy' + (1+x) y = 0 \tag{67}$$

(a) The origin in a regular singular point of (62) and hence at least one of its solutions is of the Frobenius type. Substituting (52) into (62) yields

$$\sum_{n=0}^{\infty} (\sigma+n)(\sigma+n-2) a_n x^{\sigma+n} + \sum_{n=0}^{\infty} (\sigma+n-1) a_n x^{\sigma+n+1} = 0$$

or

$$\sigma(\sigma-2) a_0 + \sum_{n=1}^{\infty} (\sigma+n)(\sigma+n-2) a_n x^{\sigma+n} + \sum_{n=0}^{\infty} (\sigma+n-1) a_n x^{\sigma+n+1} = 0$$

Putting $n = m+1$ in the first summation and $n = m$ in the second summation, we have

$$\sigma(\sigma-2)a_0 + \sum_{m=0}^{\infty} [(\sigma+m+1)(\sigma+m-1)a_{m+1} + (\sigma+m-1)a_m] x^{\sigma+m+1} = 0$$

Equating the coefficient of each power of x to zero, we obtain

$$\sigma(\sigma-2)a_0 = 0 \tag{68}$$

$$(\sigma+m+1)(\sigma+m-1)a_{m+1} = -(\sigma+m-1)a_m \tag{69}$$

It follows from (68) that $\sigma = 0$ or 2. Thus the difference between the indices is an integer.

When $\sigma = 0$, it follows from (69) that a_2 is indeterminate and two linearly independent solutions of (62) are of the Frobenius type, one proportional to a_0 and the other proportional to a_2. Then it follows from (69) that

$$a_{m+1} = -\frac{a_m}{m+1}$$

provided that $m \neq 1$. Hence

$$a_1 = -a_0, \qquad a_3 = -\frac{a_2}{3}, \qquad a_4 = -\frac{a_3}{4} = \frac{a_2}{3\cdot 4},$$

$$a_5 = -\frac{a_4}{5} = -\frac{a_2}{3\cdot 4\cdot 5}, \qquad a_6 = -\frac{a_5}{6} = \frac{a_2}{3\cdot 4\cdot 5\cdot 6}$$

Putting $a_2 = \frac{1}{2}b$, we have

$$y = a_0(1-x) + b\left(\frac{x^2}{2!} - \frac{x^3}{3!} + \frac{x^4}{4!} - \frac{x^5}{5!} + \frac{x^6}{6!} - \cdots\right) = a_0(1-x) + b(e^{-x} - 1 + x)$$

Therefore, two linearly independent solutions of (62) are

$$y_1 = 1 - x, \qquad y_2 = e^{-x}$$

Thus the origin is an apparent singularity of (62). When $\sigma = 2$, it follows from (69) that

$$a_{m+1} = -\frac{a_m}{m+3}$$

Hence

$$y = a_0 x^2\left(1 - \frac{x}{3} + \frac{x^2}{3\cdot 4} - \frac{x^3}{3\cdot 4\cdot 5} + \frac{x^6}{3\cdot 4\cdot 5\cdot 6} - \cdots\right)$$

Putting $a_0 = \frac{1}{2}$, we have

$$y = \frac{x^2}{2!} - \frac{x^3}{3!} + \frac{x^4}{4!} - \frac{x^5}{5!} + \cdots = e^{-x} - 1 + x$$

which is a linear combination of y_1 and y_2.

(b) The origin is a regular singular point of (63) and has at least one solution of the Frobenius type. Substituting (52) into (63) yields

$$\sum_{n=0}^{\infty} \left[(\sigma+n)^2 - 4\right] a_n x^{\sigma+n} + \sum_{n=0}^{\infty} a_n x^{\sigma+n+2} = 0$$

or

$$(\sigma^2 - 4) a_0 x^\sigma + \left[(\sigma+1)^2 - 4\right] a_1 x^{\sigma+1}$$
$$+ \sum_{n=2}^{\infty} \left[(\sigma+n)^2 - 4\right] a_n x^{\sigma+n} + \sum_{n=0}^{\infty} a_n x^{\sigma+n+2} = 0$$

Putting $n = m+2$ in the first summation and $n = m$ in the second summation, we have

$$(\sigma^2 - 4) a_0 x^\sigma + \left[(\sigma+1)^2 - 4\right] a_1 x^{\sigma+1}$$
$$+ \sum_{m=0}^{\infty} \left\{\left[(\sigma+m+2)^2 - 4\right] a_{m+2} + a_m\right\} x^{\sigma+m+2} = 0$$

Hence

$$(\sigma^2 - 4) a_0 = 0 \tag{70}$$

$$\left[(\sigma+1)^2 - 4\right] a_1 = 0 \tag{71}$$

$$\left[(\sigma+m+2)^2 - 4\right] a_{m+2} = -a_m \tag{72}$$

It follows from (70) that if $a_0 \neq 0$, then $\sigma = 2$ or -2. When $\sigma = 2$, it follows from (71) that $a_1 = 0$ and from (72) that

$$a_{m+2} = -\frac{a_m}{(m+2)(m+6)}$$

and if $a_0 = \frac{1}{8}$, one solution of (63) is

$$y_1 = \frac{(x/2)^2}{2!} - \frac{(x/2)^4}{1!\,3!} + \frac{(x/2)^6}{2!\,4!} - \frac{(x/2)^8}{3!\,5!} + \cdots$$

When $\sigma = -2$, it follows from (72) that a_4 is infinite and the second solution of (63) is not of the Frobenius type. To determine the second solution, we first determine the series in terms of σ. Thus

$$y(x;\sigma) = a_0 x^\sigma \left[1 - \frac{x^2}{\sigma(\sigma+4)} + \frac{x^4}{\sigma(\sigma+2)(\sigma+4)(\sigma+6)} - \frac{x^6}{\sigma(\sigma+2)(\sigma+4)^2(\sigma+6)(\sigma+8)} + \frac{x^8}{\sigma(\sigma+2)(\sigma+4)^2(\sigma+6)^2(\sigma+8)(\sigma+10)} + \cdots\right] \tag{73}$$

It is clear that $y \to \infty$ as $\sigma \to -2$. To overcome this difficulty, we put $a_0 = b(\sigma+2)$ in (73) and obtain

$$y(x,\sigma) = bx^\sigma \left[\sigma + 2 - \frac{x^2(\sigma+2)}{\sigma(\sigma+4)} + \frac{x^4}{\sigma(\sigma+4)(\sigma+6)} - \frac{x^6}{\sigma(\sigma+4)^2(\sigma+6)(\sigma+8)} + \frac{x^8}{\sigma(\sigma+4)^2(\sigma+6)^2(\sigma+8)(\sigma+10)} + \cdots\right]$$

Putting $\sigma = -2$ yields

$$y = bx^{-2}\left[\frac{x^4}{-2\cdot 2\cdot 4} - \frac{x^6}{-2\cdot 2^2\cdot 4\cdot 6} + \frac{x^8}{-2\cdot 2^2\cdot 4^2\cdot 6\cdot 8} + \cdots\right]$$

which is the same as y_1 if $b = -2$. A second linearly independent solution is

$$y_2 = \left.\frac{\partial y}{\partial \sigma}\right|_{\sigma=-2} = y_1 \ln x - 2x^{-2}\left[1 + \frac{x^2}{4} + \frac{x^4}{64} + \cdots\right]$$

(c) The origin is an ordinary point of (64) and its solutions can be represented in convergent power series. Substituting (29) into (64) yields

$$\sum_{n=2}^{\infty} n(n-1)a_n x^{n-2} + \sum_{n=0}^{\infty} [2 - n(n+1)]a_n x^n = 0$$

Putting $n = m+2$ in the first summation and $n = m$ in the second summation, we obtain

$$\sum_{m=0}^{\infty} \{(m+1)(m+2)a_{m+2} + [2 - m(m+1)]a_m\} x^m = 0$$

Hence

$$a_{m+2} = \frac{m(m+1)-2}{(m+1)(m+2)} a_m \tag{74}$$

It follows from (29) and (74) that

$$y = a_0\left[1 - x^2 - \tfrac{1}{3}x^4 - \tfrac{1}{5}x^6 + \cdots\right] + a_1 x$$

and thus two linearly independent solutions of (64) are

$$y_1 = x$$

$$y_2 = 1 - x^2 - \tfrac{1}{3}x^4 - \tfrac{1}{5}x^6 + \cdots$$

(d) The origin is a regular singular point of (65) and at least one of its solutions is of the Frobenius type. Substituting (52) into (65) yields

$$\sum_{n=0}^{\infty} (\sigma+n)(\sigma+n-1) a_n x^{\sigma+n-1} - \sum_{n=0}^{\infty} (\sigma+n+1)^2 a_n x^{\sigma+n} = 0$$

or

$$\sigma(\sigma-1) a_0 x^{\sigma-1} + \sum_{n=1}^{\infty} (\sigma+n)(\sigma+n-1) a_n x^{\sigma+n-1} - \sum_{n=0}^{\infty} (\sigma+n+1)^2 a_n x^{\sigma+n} = 0$$

Putting $n = m+1$ in the first summation and $n = m$ in the second summation, we have

$$\sigma(\sigma-1) a_0 x^{\sigma-1} + \sum_{m=0}^{\infty} \left[(\sigma+m)(\sigma+m+1) a_{m+1} - (\sigma+m+1)^2 a_m\right] x^{\sigma+m} = 0$$

Equating the coefficient of each power of x to zero yields

$$\sigma(\sigma-1) a_0 = 0 \tag{75}$$

$$(\sigma+m)(\sigma+m+1) a_{m+1} = (\sigma+m+1)^2 a_m \tag{76}$$

It follows from (75) that the indices are $\sigma = 0$ and 1. Since their difference is an integer, both solutions of (65) may or may not be of the Frobenius type. When $\sigma = 0$, it follows from (76) that

$$m a_{m+1} = (m+1) a_m$$

and hence a_1 is infinite. Therefore, one solution is of the Frobenius type and the other solution involves a logarithm. Thus we determine the series for general σ first. It follows from (76) that

$$a_{m+1} = \frac{\sigma+m+1}{\sigma+m} a_m$$

Hence

$$y(x;\sigma) = a_0\left[1 + \frac{\sigma+1}{\sigma}x + \frac{(\sigma+2)}{\sigma}x^2 + \frac{\sigma+3}{\sigma}x^3 + \frac{\sigma+4}{\sigma}x^4 + \frac{\sigma+5}{\sigma}x^5 + \cdots + \frac{\sigma+n}{\sigma}x^n + \cdots\right]$$

To circumvent the difficulty at $\sigma = 0$, we put $a_0 = b\sigma$ and obtain

$$y(x;\sigma) = bx^\sigma\left[\sigma + (\sigma+1)x + (\sigma+2)x^2 + (\sigma+3)x^3 + (\sigma+4)x^4 + (\sigma+5)x^5 + \cdots + (\sigma+n)x^n + \cdots\right]$$

Then two linearly independent solutions of (65) are

$$y_1(x) = y(x;\sigma)\Big|_{\substack{\sigma=0\\ b=1}} = x + 2x^2 + 3x^3 + \cdots + nx^n + \cdots = \sum_{n=1}^{\infty} nx^n = \frac{x}{(1-x)^2}$$

$$y_2(x) = \frac{\partial y}{\partial\sigma}\Big|_{\substack{\sigma=0\\ b=1}} = y_1(x)\ln x + 1 + x + x^2 + \cdots + x^n + \cdots = \frac{x\ln x}{(1-x)^2} + \frac{1}{1-x}$$

(e) The origin is an ordinary point of (66) and hence its solutions can be expressed in power series of x. Substituting (29) into (66) yields

$$\sum_{n=2}^{\infty} n(n-1)a_n x^{n-2} + \sum_{n=0}^{\infty} a_n x^{n+2} = 0$$

or

$$2a_2 + 6a_3x + \sum_{n=4}^{\infty} n(n-1)a_n x^{n-2} + \sum_{n=0}^{\infty} a_n x^{n+2} = 0$$

Putting $n = 4 + m$ in the first summation and $n = m$ in the second summation, we obtain

$$2a_2 + 6a_3x + \sum_{m=0}^{\infty}\left[(m+3)(m+4)a_{m+4} + a_m\right]x^{m+2} = 0$$

Equating the coefficient of each power of x to zero yields

$$2a_2 = 0$$

$$6a_3 = 0$$

$$a_{m+4} = -\frac{a_m}{(m+3)(m+4)} \tag{77}$$

Hence two linearly independent solutions of (66) are

$$y_1 = 1 - \frac{x^4}{3\cdot4} + \frac{x^8}{3\cdot4\cdot7\cdot8} - \frac{x^{12}}{3\cdot4\cdot7\cdot8\cdot11\cdot12} + \cdots$$

$$y_2 = x - \frac{x^5}{4\cdot5} + \frac{x^9}{4\cdot5\cdot8\cdot9} - \frac{x^{13}}{4\cdot5\cdot8\cdot9\cdot12\cdot13} + \cdots$$

(f) The origin is an ordinary point of (67) and hence its solutions can be expressed in power series of x. Substituting (29) into (67) yields

$$\sum_{n=2}^{\infty} 2n(n-1)a_n x^{n-2} + \sum_{n=0}^{\infty} (n^2+1)a_n x^n + \sum_{n=0}^{\infty} a_n x^{n+1} = 0$$

Putting $n = m+2$ in the first summation, $n = m$ in the second summation, and $n = m-1$ in the third summation, we obtain

$$\sum_{m=0}^{\infty} \left[2(m+1)(m+2)a_{m+2} + (m^2+1)a_m\right] x^m + \sum_{m=1}^{\infty} a_{m-1}x^m = 0$$

Equating the coefficient of each power of x to zero yields

$$4a_2 + a_0 = 0$$

$$a_{m+2} = -\frac{(m^2+1)a_m + a_{m-1}}{2(m+1)(m+2)} \quad \text{for } m \geq 1 \tag{78}$$

Hence two linearly independent solutions of (67) are

$$y_1 = 1 - \tfrac{1}{4}x^2 - \tfrac{1}{12}x^3 + \tfrac{5}{96}x^4 + \cdots$$

$$y_2 = x - \tfrac{1}{6}x^3 - \tfrac{1}{24}x^4 + \tfrac{41}{960}x^5 + \cdots$$

13.9. Determine two linearly independent solutions in power series near the origin for each of the following equations:

(a)
$$xy'' + (1+x)y' + 2y = 0 \tag{79}$$

(b)
$$(x - x^2)y'' + (1-x)y' - y = 0 \tag{80}$$

(c)
$$(x - x^2)y'' + (1-5x)y' - 4y = 0 \tag{81}$$

(d)
$$4(x^4 - x^2)y'' + 8x^3y' - y = 0 \tag{82}$$

In Equations (79)–(82), the origin is a regular singular point, and hence at least one solution of each equation is of the Frobenius type.

(a) Substituting (52) into (79) yields

$$\sum_{n=0}^{\infty} (\sigma+n)^2 a_n x^{\sigma+n-1} + \sum_{n=0}^{\infty} (\sigma+n+2) a_n x^{\sigma+n} = 0$$

or

$$\sigma^2 a_0 x^{\sigma-1} + \sum_{n=1}^{\infty} (\sigma+n)^2 a_n x^{\sigma+n-1} + \sum_{n=0}^{\infty} (\sigma+n+2) a_n x^{\sigma+n} = 0$$

Putting $n = m+1$ in the first summation and $n = m$ in the second summation, we obtain

$$\sigma^2 a_0 x^{\sigma-1} + \sum_{m=0}^{\infty} \left[(\sigma+m+1)^2 a_{m+1} + (\sigma+m+2) a_m\right] x^{\sigma+m} = 0$$

Equating the coefficient of each power of x to zero yields

$$\sigma^2 a_0 = 0$$

$$a_{m+1} = -\frac{\sigma+m+2}{(\sigma+m+1)^2} a_m \tag{83}$$

In this case, the indices are equal. Hence one solution of (79) is of the Frobenius type whereas the other involves a logarithm. Using (83) and (52), we have

$$y(x,\sigma) = x^\sigma \left[1 - \frac{\sigma+2}{(\sigma+1)^2} x + \frac{\sigma+3}{(\sigma+1)^2(\sigma+2)} x^2 - \frac{\sigma+4}{(\sigma+1)^2(\sigma+2)(\sigma+3)} x^3 \right.$$
$$\left. + \frac{(\sigma+5)}{(\sigma+1)^2(\sigma+2)(\sigma+3)(\sigma+4)} x^4 + \cdots \right]$$

Hence two linearly independent solutions of (79) are

$$y_1 = y(x,\sigma)\Big|_{\sigma=0} = 1 - 2x + \frac{3}{2!}x^2 - \frac{4}{3!}x^3 + \frac{5}{4!}x^4 - \cdots$$

$$y_2 = \frac{\partial y}{\partial \sigma}\Big|_{\sigma=0}$$

$$= y_1 \ln x + \left[2(2-\tfrac{1}{1})x - \frac{3}{2!}(2+\tfrac{1}{2}-\tfrac{1}{1})x^2 + \frac{4}{3!}(2+\tfrac{1}{2}+\tfrac{1}{3}-\tfrac{1}{4})x^3 \right.$$
$$\left. - \frac{5}{4!}(2+\tfrac{1}{2}+\tfrac{1}{3}+\tfrac{1}{4}-\tfrac{1}{5})x^4 + \cdots\right]$$

(b) Substituting (52) into (80) yields

$$\sum_{n=0}^{\infty} (\sigma+n)^2 a_n x^{\sigma+n-1} - \sum_{n=0}^{\infty} \left[(\sigma+n)^2+1\right] a_n x^{\sigma+n} = 0$$

or

$$\sigma^2 a_0 x^{\sigma-1} + \sum_{n=1}^{\infty} (\sigma+n)^2 a_n x^{\sigma+n-1} - \sum_{n=0}^{\infty} \left[(\sigma+n)^2+1\right] a_n x^{\sigma+n} = 0$$

Putting $n = m+1$ in the first summation and $n = m$ in the second summation yields

$$\sigma^2 a_0 x^{\sigma-1} + \sum_{m=0}^{\infty} \left\{(\sigma+m+1)^2 a_{m+1} - \left[(\sigma+m)^2+1\right] a_m\right\} x^{\sigma+m} = 0$$

Equating the coefficient of each power of x to zero yields

$$\sigma^2 a_0 = 0$$

$$a_{m+1} = \frac{(\sigma+m)^2+1}{(\sigma+m+1)^2} a_m \tag{84}$$

In this case, the indices are equal. Hence one of the solutions of (80) is of the Frobenius type, whereas the other involves a logarithm. Using (52) and (84), we have

$$y(x;\sigma) = x^\sigma\left\{1 + \frac{\sigma^2+1}{(\sigma+1)^2}x + \frac{(\sigma^2+1)\left[(\sigma+1)^2+1\right]}{(\sigma+1)^2(\sigma+2)^2}x^2 + \frac{(\sigma^2+1)\left[(\sigma+1)^2+1\right]\left[(\sigma+2)^2+1\right]}{(\sigma+1)^2(\sigma+2)^2(\sigma+3)^2}x^3 + \cdots\right\}$$

Hence two linearly independent solutions of (80) are

$$y_1 = y\Big|_{\sigma=0} = 1 + x + \tfrac{2}{4}x^2 + \frac{2\cdot5}{4\cdot9}x^3 + \frac{2\cdot5\cdot10}{4\cdot9\cdot16}x^4 + \frac{2\cdot5\cdot10\cdot17}{4\cdot9\cdot16\cdot25}x^5 + \cdots$$

$$y_2 = \frac{\partial y}{\partial\sigma}\Big|_{\sigma=0} = y_1 \ln x + \left(-2x - x^2 - \tfrac{14}{27}x^3 + \cdots\right)$$

(c) Substituting (52) into (81) yields

$$\sum_{n=0}^{\infty} (\sigma+n)^2 a_n x^{\sigma+n-1} - \sum_{n=0}^{\infty} (\sigma+n+2)^2 a_n x^{\sigma+n} = 0$$

or

$$\sigma^2 a_0 x^{\sigma-1} + \sum_{n=1}^{\infty} (\sigma+n)^2 a_n x^{\sigma+n-1} - \sum_{n=0}^{\infty} (\sigma+n+2)^2 a_n x^{\sigma+n} = 0$$

Putting $n = m+1$ in the first summation and $n = m$ in the second summation, we have

$$\sigma^2 a_0 x^{\sigma-1} + \sum_{m=0}^{\infty} \left[(\sigma+m+1)^2 a_{m+1} - (\sigma+m+2)^2 a_m\right] x^{\sigma+m} = 0$$

Equating the coefficient of each power of x to zero yields

$$\sigma^2 a_0 = 0$$

$$a_{m+1} = \frac{(\sigma+m+2)^2}{(\sigma+m+1)^2} a_m \tag{85}$$

Since the indices are equal, one of the solutions of (81) is of the Frobenius type, whereas the other involves a logarithm. It follows from (52) and (85) that

$$y(x;\sigma) = x^\sigma \left[1 + \frac{(\sigma+2)^2}{(\sigma+1)^2} x + \frac{(\sigma+3)^2}{(\sigma+1)^2} x^2 + \frac{(\sigma+4)^2}{(\sigma+1)^2} x^3 + \cdots\right]$$

Hence two linearly independent solutions of (81) are

$$y_1 = y\bigg|_{\sigma=0} = 1 + 4x + 9x^2 + 16x^3 + 25x^4 + \cdots = \sum_{n=0}^{\infty} (n+1)^2 x^n$$

$$y_2 = \frac{\partial y}{\partial \sigma}\bigg|_{\sigma=0} = y_1 \ln x - 2(1\cdot 2x + 2\cdot 3x^2 + 3\cdot 4x^3 + 4\cdot 5x^4 + 5\cdot 6x^5 + \cdots)$$

$$= y_1 \ln x - 2\sum_{n=1}^{\infty} n(n+1)x^n$$

(d) Substituting (52) into (82) yields

$$-\sum_{n=0}^{\infty} (2\sigma+2n-1)^2 a_n x^{\sigma+n} + 4\sum_{n=0}^{\infty} (\sigma+n)(\sigma+n+1) a_n x^{\sigma+n+2} = 0$$

or

$$-(2\sigma-1)^2 a_0 x^\sigma - (2\sigma+1)^2 a_1 x^{\sigma+1} - \sum_{n=2}^{\infty} (2\sigma+2n-1)^2 a_n x^{\sigma+n}$$

$$+4\sum_{n=0}^{\infty} (\sigma+n)(\sigma+n+1) a_n x^{\sigma+n+2} = 0$$

Putting $n=m+2$ in the first summation and $n=m$ in the second summation, we have

$$-(2\sigma-1)^2 a_0 x^\sigma-(2\sigma+1)^2 a_1 x^{\sigma+1}$$

$$-\sum_{m=0}^{\infty}\left[(2\sigma+2m+3)^2 a_{m+2}-4(\sigma+m)(\sigma+m+1)a_m\right]x^{\sigma+m+2}=0$$

Equating the coefficient of each power of x to zero yields

$$(2\sigma-1)^2 a_0=0 \tag{86}$$

$$(2\sigma+1)^2 a_1=0 \tag{87}$$

$$a_{m+2}=\frac{4(\sigma+m)(\sigma+m+1)}{(2\sigma+2m+3)^2}a_m \tag{88}$$

It follows from (86) that if $a_0\neq 0$, $\sigma=\frac{1}{2}$, and from (87) that $a_1=0$. Since the indices are equal one of the solutions of (81) is of the Frobenius type, whereas the other solution involves a logarithm. Using (52) and (88) and the fact that $a_1=0$, we obtain

$$y(x;\sigma)=x^\sigma\left[1+\frac{4\sigma(\sigma+1)}{(2\sigma+3)^2}x^2+\frac{4^2\sigma(\sigma+1)(\sigma+2)(\sigma+3)}{(2\sigma+3)^2(2\sigma+7)^2}x^4\right.$$

$$\left.+\frac{4^3\sigma(\sigma+1)(\sigma+2)(\sigma+3)(\sigma+4)(\sigma+5)}{(2\sigma+3)^2(2\sigma+7)^2(2\sigma+11)^2}x^6+\cdots\right]$$

Hence two linearly independent solutions of (82) are

$$y_1=y\Big|_{\sigma=1/2}=x^{1/2}\left[1+\frac{1\cdot 3}{4^2}x^2+\frac{1\cdot 3\cdot 5\cdot 7}{4^2\cdot 8^2}x^4+\frac{1\cdot 3\cdot 5\cdot 7\cdot 9\cdot 11}{4^2\cdot 8^2\cdot 12^2}x^6+\cdots\right]$$

$$y_2=\frac{\partial y}{\partial\sigma}\Big|_{\sigma=1/2}=y_1\ln x+2x^{1/2}\left[\frac{1\cdot 3}{4^2}\left(1+\tfrac{1}{3}-\tfrac{1}{2}\right)x^2\right.$$

$$\left.+\frac{1\cdot 3\cdot 5\cdot 7}{4^2\cdot 8^2}\left(1+\tfrac{1}{3}-\tfrac{1}{2}+\tfrac{1}{5}+\tfrac{1}{7}-\tfrac{1}{4}\right)x^4+\cdots\right]$$

13.10. Determine two linearly independent solutions in power series near the origin for each of the following equations:

(a) $$x^2y''+x^2y'-2y=0 \tag{89}$$

(b) $$xy''-(1+x)y'+2(1-x)y=0 \tag{90}$$

The origin is a regular singular point of each of Equations (89) and (90). Hence at least one solution of each of these equations is of the Frobenius type.

(a) Substituting (52) into (89), we have

$$\sum_{n=0}^{\infty} (\sigma+n-2)(\sigma+n+1)\,a_n x^{\sigma+n} + \sum_{n=0}^{\infty} (\sigma+n)\,a_n x^{\sigma+n+1} = 0$$

or

$$(\sigma-2)(\sigma+1)\,a_0 x^{\sigma} + \sum_{n=1}^{\infty} (\sigma+n-2)(\sigma+n+1)\,a_n x^{\sigma+n} + \sum_{n=0}^{\infty} (\sigma+n)\,a_n x^{\sigma+n+1} = 0$$

Putting $n=m+1$ in the first summation and $n=m$ in the second summation, we obtain

$$(\sigma-2)(\sigma+1)\,a_0 x^{\sigma} + \sum_{m=0}^{\infty} \left[(\sigma+m-1)(\sigma+m+2)\,a_{m+1} + (\sigma+m)\,a_m\right] x^{\sigma+m+1} = 0$$

Equating the coefficient of each power of x to zero yields

$$(\sigma-2)(\sigma+1)\,a_0 = 0$$

$$(\sigma+m-1)(\sigma+m+2)\,a_{m+1} + (\sigma+m)\,a_m = 0 \tag{91}$$

In this case, the indices are $\sigma=-1$ and 2. Since they differ by an integer, both solutions of (89) may or may not be of the Frobenius type, depending on whether one of the a_n is indeterminate or infinite, respectively. When $\sigma=-1$, it follows from (91) that a_3 is indeterminate, and hence

$$a_{m+1} = -\frac{m-1}{(m+1)(m-2)}\,a_m \quad \text{for } m \neq 2 \tag{92}$$

Using (52) and (92), we obtain

$$y = a_0 x^{-1}\left(1-\tfrac{1}{2}x\right) + a_3 x^{-1}\left(x^3 - \frac{2}{1\cdot 4}x^4 + \frac{2\cdot 3}{1\cdot 4\cdot 2\cdot 5}x^5 - \frac{2\cdot 3\cdot 4}{1\cdot 4\cdot 2\cdot 5\cdot 3\cdot 6}x^6 + \cdots\right)$$

Thus two linearly independent solutions of (89) are

$$y_1 = \frac{1}{x} - \frac{1}{2}$$

$$y_2 = x^2 - \frac{2}{1\cdot 4}x^3 + \frac{2\cdot 3}{1\cdot 4\cdot 2\cdot 5}x^4 - \frac{2\cdot 3\cdot 4}{1\cdot 4\cdot 2\cdot 5\cdot 3\cdot 6}x^5 + \cdots$$

and $x=0$ is an apparent singularity of (89).

(b) Substituting (52) into (90) yields

$$\sum_{n=0}^{\infty}(\sigma+n)(\sigma+n-2)a_n x^{\sigma+n-1}-\sum_{n=0}^{\infty}(\sigma+n-2)a_n x^{\sigma+n}-2\sum_{n=0}^{\infty}a_n x^{\sigma+n+1}=0$$

or

$$\sigma(\sigma-2)a_0x^{\sigma-1}+\sum_{n=1}^{\infty}(\sigma+n)(\sigma+n-2)a_n x^{\sigma+n-1}-\sum_{n=0}^{\infty}(\sigma+n-2)a_n x^{\sigma+n}$$

$$-2\sum_{n=0}^{\infty}a_n x^{\sigma+n+1}=0$$

Putting $n=m+1$ in the first summation, $n=m$ in the second summation, and $n=m-1$ in the third summation, we obtain

$$\sigma(\sigma-2)a_0x^{\sigma-1}$$

$$+\sum_{m=0}^{\infty}[(\sigma+m+1)(\sigma+m-1)a_{m+1}-(\sigma+m-2)a_m-2a_{m-1}]x^{\sigma+m}=0$$

Equating the coefficient of each power of x to zero yields

$$\sigma(\sigma-2)a_0=0$$

$$(\sigma+m+1)(\sigma+m-1)a_{m+1}-(\sigma+m-2)a_m-2a_{m-1}=0 \tag{93}$$

In this case, the indices are $\sigma=0$ and 2. Since the difference between the indices is an integer, both solutions of (90) may or may not be of the Frobenius type, depending on whether one of the a_n is indeterminate or infinite, respectively. When $\sigma=0$, it follows from (93) that

$$-a_1+2a_0=0 \tag{94}$$

$$0\cdot a_2+a_1-2a_0=0 \tag{95}$$

$$a_{m+1}=\frac{(m-2)a_m+2a_{m-1}}{(m+1)(m-1)} \quad \text{for } m\geq 2 \tag{96}$$

It follows from (94) that $a_1=2a_0$ and from (95) that a_2 is indeterminate. Hence two linearly independent solutions of (90) of the Frobenius type can be found, one proportional to a_0 and the other proportional to a_2. It follows from (96) that

$$a_3=\tfrac{2}{3}a_1,\qquad a_4=\tfrac{1}{12}a_1+\tfrac{1}{4}a_2,\qquad a_5=\tfrac{1}{10}a_1+\tfrac{1}{30}a_2$$

Hence

$$y=a_0\left(1+2x+\tfrac{4}{3}x^3+\tfrac{1}{6}x^4+\tfrac{1}{5}x^5+\cdots\right)+a_2\left(x^2+\tfrac{1}{4}x^4+\tfrac{1}{30}x^5+\cdots\right)$$

and two linearly independent solutions of (90) are

$$y_1 = 1 + 2x + \tfrac{4}{3}x^3 + \tfrac{1}{6}x^4 + \tfrac{1}{5}x^5 + \cdots$$

$$y_2 = x^2 + \tfrac{1}{4}x^4 + \tfrac{1}{30}x^5 + \cdots$$

Therefore, the origin is an apparent singularity of (90).

13.11. Show that the origin is an apparent singularity for the equation

$$x^2y'' - (4x + \lambda_1 x^2)\,y' + (4 - \lambda_2 x)\,y = 0 \tag{97}$$

when (a) $\lambda_2 = -\lambda_1$, (b) $\lambda_2 = -2\lambda_1$, and (c) $\lambda_3 = -3\lambda_1$.

We seek a solution of the Frobenius type because the origin appears to be a regular singular point. Substituting (52) into (97) yields

$$\sum_{n=0}^{\infty} (\sigma + n - 1)(\sigma + n - 4)\,a_n x^{\sigma+n} - \sum_{n=0}^{\infty} [\lambda_1(\sigma + n) + \lambda_2]\,a_n x^{\sigma+n+1} = 0$$

or

$$(\sigma - 1)(\sigma - 4)\,a_0 x^{\sigma} + \sum_{n=1}^{\infty} (\sigma + n - 1)(\sigma + n - 4)\,a_n x^{\sigma+n}$$

$$- \sum_{n=0}^{\infty} [\lambda_1(\sigma + n) + \lambda_2]\,a_n x^{\sigma+n+1} = 0$$

Putting $n = m + 1$ in the first summation and $n = m$ in the second summation, we have

$$(\sigma - 1)(\sigma - 4)\,a_0 x^{\sigma}$$

$$+ \sum_{m=0}^{\infty} \{(\sigma + m)(\sigma + m - 3)\,a_{m+1} - [\lambda_1(\sigma + m) + \lambda_2]\,a_m\}\,x^{\sigma+m+1} = 0$$

Equating the coefficient of each power of x to zero yields

$$(\sigma - 1)(\sigma - 4)\,a_0 = 0$$

$$(\sigma + m)(\sigma + m - 3)\,a_{m+1} = [\lambda_1(\sigma + m) + \lambda_2]\,a_m \tag{98}$$

In this case, the indices are $\sigma = 1$ and 4. Since they are unequal and their difference is an integer, both solutions of (97) are of the Frobenius type and hence bounded if none of the a_m are infinite. When $\sigma = 4$, it follows from (98) that all a_m are finite if a_0 is finite. When $\sigma = 1$, it follows from (98) that

$$(m + 1)(m - 2)\,a_{m+1} = [\lambda_1(m + 1) + \lambda_2]\,a_m \tag{99}$$

Hence

$$a_1 = -\tfrac{1}{2}(\lambda_1 + \lambda_2) a_0 \tag{100}$$

$$a_2 = -\tfrac{1}{2}(2\lambda_1 + \lambda_2) a_1 = \tfrac{1}{4}(\lambda_1 + \lambda_2)(2\lambda_1 + \lambda_2) a_0 \tag{101}$$

$$0 \cdot a_3 = (3\lambda_1 + \lambda_2) a_2 = \tfrac{1}{4}(\lambda_1 + \lambda_2)(2\lambda_1 + \lambda_2)(3\lambda_1 + \lambda_2) a_0 \tag{102}$$

It follows from (99)–(102) that a_3 and hence all a_m for $m \geq 3$ are infinite unless the right-hand side of (102) vanishes, and one solution of (97) is of the Frobenius type and the other involves a logarithm, so that the origin is a real singularity. When the right-hand side of (102) vanishes (i.e., $\lambda_2 = -\lambda_1, -2\lambda_1$, or $-3\lambda_1$), a_3 is indeterminate and both solutions of (97) are of the Frobenius type (some of them may terminate), so that the origin is an apparent singularity.

When the right-hand side of (102) vanishes, it follows from (99) that

$$a_{m+1} = \frac{\lambda_1(m+1) + \lambda_2}{(m+1)(m-2)} a_m, \qquad m \neq 2$$

Hence

$$\begin{aligned} y = a_0 x\left[1 - \tfrac{1}{2}(\lambda_1 + \lambda_2)x + \tfrac{1}{4}(\lambda_1 + \lambda_2)(2\lambda_1 + \lambda_2)x^2\right] \\ + a_3 x\left[x^3 + \frac{\lambda_2 + 4\lambda_1}{1 \cdot 4}x^4 + \frac{(\lambda_2 + 4\lambda_1)(\lambda_2 + 5\lambda_1)}{1 \cdot 4 \cdot 2 \cdot 5}x^5\right. \\ \left. + \frac{(\lambda_2 + 4\lambda_1)(\lambda_2 + 5\lambda_1)(\lambda_2 + 6\lambda_1)}{1 \cdot 4 \cdot 2 \cdot 5 \cdot 3 \cdot 6}x^6 + \cdots\right] \end{aligned} \tag{103}$$

When $\lambda_2 = -\lambda_1$, it follows from (103) that two linearly independent solutions of (97) are

$$y_1 = x$$

$$y_2 = x^4\left(1 + \tfrac{3}{4}\lambda_1 x + \tfrac{3}{10}\lambda_1^2 x^2 + \tfrac{1}{12}\lambda_1^3 x^3 + \cdots\right)$$

When $\lambda_2 = -2\lambda_1$, it follows from (103) that two linearly independent solutions of (97) are

$$y_1 = x\left(1 + \tfrac{1}{2}\lambda_1 x\right)$$

$$y_2 = x^4\left(1 + \tfrac{1}{2}\lambda_1 x + \tfrac{3}{20}\lambda_1^2 x^2 + \tfrac{1}{30}\lambda_1^3 x^3 + \cdots\right)$$

When $\lambda_2 = -3\lambda_1$, two linearly independent solutions of (97) are

$$y_1 = x\left(1 + \lambda_1 x + \tfrac{1}{2}\lambda_1^2 x^2\right)$$

$$y_2 = x^4\left(1 + \tfrac{1}{4}\lambda_1 x + \tfrac{1}{20}\lambda_1^2 x^2 + \tfrac{1}{120}\lambda_1^3 x^3 + \cdots\right)$$

13.12. Show that

$$y_1 = x^{-1/3}\left[1 + \frac{3}{3!}x + \frac{9}{6!}x^2 + \frac{27}{9!}x^3 + \cdots\right] \tag{104}$$

$$y_2 = \frac{1}{1!} + \frac{3}{4!}x + \frac{9}{7!}x^2 + \frac{27}{10!}x^3 + \cdots \tag{105}$$

$$y_3 = x^{1/3}\left[\frac{1}{2!} + \frac{3}{5!}x + \frac{9}{8!}x^2 + \frac{27}{11!}x^3 + \cdots\right] \tag{106}$$

are linearly independent solutions of

$$9x^2y''' + 27xy'' + 8y' - y = 0 \tag{107}$$

Since the origin is a regular singular point of (107), it possesses solutions of the Frobenius type. Substituting (52) into (107) yields

$$\sum_{n=0}^{\infty} (\sigma + n)(3\sigma + 3n + 1)(3\sigma + 3n - 1)\, a_n x^{\sigma+n-1} - \sum_{n=0}^{\infty} a_n x^{\sigma+n} = 0$$

or

$$\sigma(3\sigma + 1)(3\sigma - 1)\, a_0 x^{\sigma-1} + \sum_{n=1}^{\infty} (\sigma + n)(3\sigma + 3n + 1)(3\sigma + 3n - 1)\, a_n x^{\sigma+n-1}$$

$$- \sum_{n=0}^{\infty} a_n x^{\sigma+n} = 0$$

Putting $n = m + 1$ in the first summation and $n = m$ in the second summation, we obtain

$$\sigma(3\sigma + 1)(3\sigma - 1)\, a_0 x^{\sigma-1}$$

$$+ \sum_{m=0}^{\infty} \left[(\sigma + m + 1)(3\sigma + 3m + 4)(3\sigma + 3m + 2)\, a_{m+1} - a_m\right] x^{\sigma+m} = 0$$

Equating the coefficient of each power of x to zero, we have

$$\sigma(3\sigma + 1)(3\sigma - 1)\, a_0 = 0$$

$$a_{m+1} = \frac{a_m}{(\sigma + m + 1)(3\sigma + 3m + 4)(3\sigma + 3m + 2)} \tag{108}$$

In this case, the indices are

$$\sigma = -\tfrac{1}{3}, 0, \tfrac{1}{3}$$

Since the difference between any two of the indices is not an integer, three linearly independent solutions of (107) are of the Frobenius type. When $\sigma = -\frac{1}{3}$, it follows from (108) that

$$a_{m+1} = \frac{a_m}{(m+1)(3m+1)(3m+2)}$$

Hence one solution of (107) is

$$y = x^{-1/3}\left(1 + \tfrac{1}{2}x + \frac{1}{2\cdot 2\cdot 4\cdot 5}x^2 + \frac{1}{2\cdot 2\cdot 4\cdot 5\cdot 3\cdot 7\cdot 8}x^3 + \cdots\right)$$

which can be expressed as in (104). When $\sigma = 0$, it follows from (108) that

$$a_{m+1} = \frac{a_m}{(m+1)(3m+2)(3m+4)}$$

Hence one solution of (107) is

$$y = 1 + \frac{1}{1\cdot 2\cdot 4}x + \frac{1}{1\cdot 2\cdot 4\cdot 2\cdot 5\cdot 7}x^2 + \frac{1}{1\cdot 2\cdot 4\cdot 2\cdot 5\cdot 7\cdot 3\cdot 8\cdot 10}x^3 + \cdots$$

which can be expressed as in (105). When $\sigma = \frac{1}{3}$, it follows from (108) that

$$a_{m+1} = \frac{a_m}{(m+1)(3m+4)(3m+5)}$$

Hence one solution of (107) is

$$y = x^{1/3}\left(1 + \frac{1}{1\cdot 4\cdot 5}x + \frac{1}{1\cdot 4\cdot 5\cdot 2\cdot 7\cdot 8}x^2 + \frac{1}{1\cdot 4\cdot 5\cdot 2\cdot 7\cdot 8\cdot 3\cdot 10\cdot 11}x^3 + \cdots\right)$$

which when multiplied by $\frac{1}{2}$ can be expressed as in (106).

13.13. Show that

$$y_1 = 1 + x + \frac{x^2}{2^2} + \frac{x^3}{2^2\cdot 3^2} + \cdots \tag{109}$$

$$y_2 = y_1 \ln x + 2\left[-x - \frac{1}{2^2}\left(1 - \tfrac{1}{2}\right)x^2 - \frac{1}{2^2\cdot 3^2}\left(1 + \tfrac{1}{2} + \tfrac{1}{3}\right)x^3 + \cdots\right] \tag{110}$$

$$y_3 = 2y_2 \ln x - y_1(\ln x)^2 + \left[6x + \left(\frac{6}{2^2} + \frac{8}{2^3} + \frac{6}{2^4}\right)x^2 + \cdots\right] \tag{111}$$

are three linearly independent solutions of

$$x^2y''' + 3xy'' + (1-x)y' - y = 0 \tag{112}$$

Since the origin is a regular singular point of (112), it possesses solutions of the Frobenius type. Substituting (52) into (112) yields

$$\sum_{n=0}^{\infty}(\sigma+n)^3 a_n x^{\sigma+n-1}-\sum_{n=0}^{\infty}(\sigma+n+1)a_n x^{\sigma+n}=0$$

or

$$\sigma^3 a_0 x^{\sigma-1}+\sum_{n=1}^{\infty}(\sigma+n)^3 a_n x^{\sigma+n-1}-\sum_{n=0}^{\infty}(\sigma+n+1)a_n x^{\sigma+n}=0$$

Putting $n=m+1$ in the first summation and $n=m$ in the second summation, we have

$$\sigma^3 a_0 x^{\sigma-1}+\sum_{m=0}^{\infty}\left[(\sigma+m+1)^3 a_{m+1}-(\sigma+m+1)a_m\right]x^{\sigma+m}=0$$

Equating the coefficient of each power of x to zero, we have

$$\sigma^3 a_0=0$$

$$a_{m+1}=\frac{a_m}{(\sigma+m+1)^2} \tag{113}$$

In this case, the three indices are equal to zero. Hence only one solution is of the Frobenius type, whereas the other two involve logarithms. Using (52) and (113), we obtain

$$y(x;\sigma)=x^{\sigma}\left[1+\frac{x}{(\sigma+1)^2}+\frac{x^2}{(\sigma+1)^2(\sigma+2)^2}+\frac{x^3}{(\sigma+1)^2(\sigma+2)^2(\sigma+3)^2}+\cdots\right] \tag{114}$$

Using (113), we find that

$$x^2y'''+3xy''+(1-x)y'-y=a_0\sigma^3x^{\sigma-1}$$

Hence

$$y(x;\sigma)|_{\sigma=0},\quad \left.\frac{\partial y}{\partial\sigma}\right|_{\sigma=0},\quad \left.\frac{\partial^2 y}{\partial\sigma^2}\right|_{\sigma=0}$$

yield three linearly independent solutions of (112). Putting $\sigma=0$ in (114) yields (109).

Differentiating (114) with respect to σ, we have

$$\frac{\partial y}{\partial \sigma} = x^{\sigma} \ln x \left[1 + \frac{x}{(\sigma+1)^2} + \frac{x^2}{(\sigma+1)^2(\sigma+2)^2} + \frac{x^3}{(\sigma+1)^2(\sigma+2)^2(\sigma+3)^2} + \cdots\right]$$

$$+ x^{\sigma}\left[-\frac{2x}{(\sigma+1)^3} - \frac{2x^2}{(\sigma+1)^2(\sigma+2)^2}\left(\frac{1}{\sigma+1} + \frac{1}{\sigma+2}\right)\right.$$

$$\left. - \frac{2x^3}{(\sigma+1)^2(\sigma+2)^2(\sigma+3)^2}\left(\frac{1}{\sigma+1} + \frac{1}{\sigma+2} + \frac{1}{\sigma+3}\right) + \cdots\right] \tag{115}$$

which reduces to (110) when $\sigma = 0$. Differentiating (115) with respect to σ yields

$$\frac{\partial^2 y}{\partial \sigma^2} = x^{\sigma}(\ln x)^2 \left[1 + \frac{x}{(\sigma+1)^2} + \frac{x^2}{(\sigma+1)^2(\sigma+2)^2} + \frac{x^3}{(\sigma+1)^2(\sigma+2)^2(\sigma+3)^2} + \cdots\right]$$

$$+ 2x^{\sigma}\ln x\left[-\frac{2x}{(\sigma+1)^3} - \frac{2x^2}{(\sigma+1)^2(\sigma+2)^2}\left(\frac{1}{\sigma+1} + \frac{1}{\sigma+2}\right)\right.$$

$$\left. - \frac{2x^3}{(\sigma+1)^2(\sigma+2)^2(\sigma+3)^2}\left(\frac{1}{\sigma+1} + \frac{1}{\sigma+2} + \frac{1}{\sigma+3}\right) + \cdots\right]$$

$$+ x^{\sigma}\left[\frac{6x}{(\sigma+1)^4} - \frac{2x^2}{(\sigma+1)^3(\sigma+2)^2}\left(\frac{-3}{\sigma+1} - \frac{2}{\sigma+2}\right)\right.$$

$$\left. - \frac{2x^2}{(\sigma+1)^2(\sigma+2)^3}\left(\frac{-2}{\sigma+1} - \frac{3}{\sigma+2}\right) + \cdots\right]$$

Letting $\sigma = 0$, we have

$$\left.\frac{\partial^2 y}{\partial \sigma^2}\right|_{\sigma=0} = y_1(\ln x)^2 + 4\ln x\left[-x - \frac{x^2}{2^2}\left(1+\tfrac{1}{2}\right) - \frac{x^3}{2^2 \cdot 3^2}\left(1 + \tfrac{1}{2} + \tfrac{1}{3}\right) + \cdots\right]$$

$$+ \left[6x + \left(\frac{6}{2^2} + \frac{8}{2^3} + \frac{6}{2^4}\right)x^2 + \cdots\right]$$

which can be expressed as in (111).

13.14. Show that

$$y = c_1 \exp\left(x + \frac{1}{x}\right) + c_2 \exp\left(-x - \frac{1}{x}\right) \tag{116}$$

is the general solution of

$$x^4(1-x^2)y'' + 2x^3y' - (1-x^2)^3 y = 0 \tag{117}$$

Show that both the origin and infinity are irregular singular points of this equation

Equation (117) can be rewritten as

$$y'' + \frac{2}{x(1-x^2)}y' - \frac{(1-x^2)^2}{x^4}y = 0 \tag{118}$$

Since the coefficient of y has a pole of order 4 at $x=0$, the origin is an irregular singular point of (117). To investigate the behavior of (117) at infinity, we let $x = z^{-1}$ and investigate the behavior of (117) at $z=0$. In terms of z, we rewrite (118) as

$$\frac{d^2y}{dz^2} + \left[\frac{2}{z} - \frac{1}{z^2}\cdot\frac{2z}{(1-z^{-2})}\right]\frac{dy}{dz} - \frac{1}{z^4}\cdot z^4(1-z^{-2})^2 y = 0$$

or

$$\frac{d^2y}{dz^2} + \frac{2}{z(1-z^2)}\frac{dy}{dz} - \frac{(1-z^2)^2}{z^4}y = 0 \tag{119}$$

Since the coefficient of y has a pole of order 4 at $z=0$, $x=\infty$ is an irregular singular point of (117). Moreover, since (119) is the same as (118),

$$y(x) = y(z) = y(x^{-1})$$

Next, we show by direct substitution that (116) is a solution of (117). To this end, we differentiate (116) twice and obtain

$$y' = c_1\left(1 - \frac{1}{x^2}\right)\exp\left(x + \frac{1}{x}\right) + c_2\left(\frac{1}{x^2} - 1\right)\exp\left(-x - \frac{1}{x}\right)$$

$$y'' = c_1\left[\left(1 - \frac{1}{x^2}\right)^2 + \frac{2}{x^3}\right]\exp\left(x + \frac{1}{x}\right) + c_2\left[\left(\frac{1}{x^2} - 1\right)^2 - \frac{2}{x^3}\right]\exp\left(-x - \frac{1}{x}\right)$$

Substituting y, y', and y'' into (117), one can easily show that it is satisfied. Since the function multiplying c_1 is linearly independent from that multiplying c_2 and since (117) is a second-order equation, (116) is its general solution.

As an alternative, we derive the solution (116) instead of just verifying that it is the solution of (117) by substitution. Since the origin is an irregular singular point of (117), its solutions have the form

$$y(x) \sim e^{\Lambda(x)}x^{\sigma}u(x) \tag{120a}$$

where $u(x)$ can be expressed as a power series in $x^{m/n}$ and $\Lambda(x)$ is a polynomial in $x^{-m/n}$, where m and n are prime integers. To determine the form of $\Lambda(x)$, we determine first its leading term. Assuming the leading term to be $\lambda x^{-\nu}$, we let

$$y \sim e^{\lambda x^{-\nu}} \tag{120b}$$

where $\nu > 0$. Then

$$y' \sim -\frac{\lambda \nu}{x^{\nu+1}} e^{\lambda x^{-\nu}}, \qquad y'' \sim \left[\frac{\lambda^2 \nu^2}{x^{2\nu+2}} + \frac{\lambda \nu(\nu+1)}{x^{\nu+2}}\right] e^{\lambda x^{-\nu}} \tag{120c}$$

Substituting y, y', and y'' into (117), we have

$$\frac{\lambda^2 \nu^2}{x^{2\nu-2}} + \frac{\lambda \nu(\nu+1)}{x^{\nu-2}} - \frac{2\lambda \nu}{x^{\nu-2}} - 1 + \cdots = 0$$

Extracting the dominant terms, we obtain

$$\frac{\lambda^2 \nu^2}{x^{2\nu-2}} - 1 + \cdots = 0$$

Hence $\nu = 1$ and $\lambda = \pm 1$. Therefore, (120a) becomes

$$y(x) \sim e^{\lambda x^{-1}} v(x), \qquad \lambda = \pm 1 \tag{121a}$$

Then

$$y' \sim \left(-\frac{\lambda}{x^2} v + v'\right) e^{\lambda x^{-1}} \tag{121b}$$

$$y'' \sim \left(\frac{\lambda^2}{x^4} v - \frac{2\lambda v'}{x^2} + \frac{2\lambda}{x^3} v + v''\right) e^{\lambda x^{-1}} \tag{121c}$$

and (117) becomes

$$x^2(1-x^2)v'' - (2\lambda - 2x - 2\lambda x^2)v' + (2 - 2\lambda x - 3x^2 + x^4)v = 0 \tag{122}$$

Guided by (120a) and (121a), we seek the solution of (122) in the form of a power series as

$$v = \sum_{n=0}^{\infty} a_n x^{n+\sigma}$$

Then (122) yields

$$\begin{aligned} &-2\lambda\sigma a_0 x^{\sigma-1} - 2\lambda \sum_{n=1}^{\infty} (n+\sigma) a_n x^{n+\sigma-1} + \sum_{n=0}^{\infty} \left[(n+\sigma)^2 + n + \sigma + 2\right] a_n x^{n+\sigma} \\ &+ \sum_{n=0}^{\infty} 2\lambda(n+\sigma-1) a_n x^{n+\sigma+1} - \sum_{n=0}^{\infty} \left[(n+\sigma)^2 - n - \sigma + 3\right] a_n x^{n+\sigma+2} \\ &+ \sum_{n=0}^{\infty} a_n x^{n+\sigma+4} = 0 \end{aligned} \tag{123}$$

Equating the coefficient of $x^{\sigma-1}$ to zero yields $\sigma = 0$. Then, equating each of the coefficients of 1, x, x^2, and x^3 to zero yields

$$-2\lambda a_1 + 2a_0 = 0$$

$$-4\lambda a_2 + 4a_1 - 2\lambda a_0 = 0$$

$$-6\lambda a_3 + 8a_2 - 3a_0 = 0$$

$$-8\lambda a_4 + 14a_3 + 2\lambda a_2 - 3a_1 = 0$$

Hence

$$a_1 = \lambda a_0, \qquad a_2 = \tfrac{1}{2!}a_0, \qquad a_3 = \tfrac{1}{3!}\lambda a_0, \qquad a_4 = \tfrac{1}{4!}a_0$$

because $\lambda^2 = 1$. Then, putting $n = m + 5$ in the first summation, $n = m + 4$ in the second summation, $n = m + 3$ in the third summation, $n = m + 2$ in the fourth summation, and $n = m$ in the fifth summation, we rewrite (123) as

$$\sum_{m=0}^{\infty} \left[-2\lambda(m+5)a_{m+5} + (m^2 + 9m + 22)a_{m+4} + 2\lambda(m+2)a_{m+3} \right.$$

$$\left. -(m^2 + 3m + 5)a_{m+2} + a_m \right] x^{m+4} = 0$$

Hence

$$a_{m+5} = \frac{(m^2 + 9m + 22)a_{m+4} + 2\lambda(m+2)a_{m+3} - (m^2 + 3m + 5)a_{m+2} + a_m}{2\lambda(m+5)}$$

which yields

$$a_n = \frac{\lambda^n a_0}{n!} \quad \text{because } \lambda^2 = 1$$

Therefore

$$v(x) = a_0 \sum_{n=0}^{\infty} \frac{(\lambda x)^n}{n!} = a_0 e^{\lambda x}$$

which when substituted into (121a) yields the following two linearly independent solutions:

$$y_1(x) = \exp\left(x + \frac{1}{x}\right), \qquad y_2(x) = \exp\left(-x - \frac{1}{x}\right)$$

13.15. Determine two linearly independent solutions near the origin for each of the following equations:

(a)
$$x^3 y'' + x(1-2x)\, y' - 2y = 0 \tag{124}$$

(b)
$$x^4 y'' + 2x^3 y' - y = 0 \tag{125}$$

(c)
$$x^2 y'' + 2(1-x)\, y' - y = 0 \tag{126}$$

The origin is an irregular singular point of each of equations (124)–(126) and hence their solutions have asymptotic expansions in the form (120a).

(a) Substituting (120b) and (120c) into (124) yields

$$\frac{\lambda^2 \nu^2}{x^{2\nu-1}} + \frac{\lambda\nu(\nu+1)}{x^{\nu-1}} - \frac{\lambda\nu}{x^\nu} - 2 + \cdots = 0$$

Balancing the dominant terms (the first and third), we have $\nu = 1$ and $\lambda = 0$ or 1.

When $\lambda = 0$,

$$y \sim \sum_{n=0}^{\infty} a_n x^{\sigma+n} \tag{127}$$

Substituting (127) into (124) yields

$$\sum_{n=0}^{\infty} (\sigma+n-2)\, a_n x^{\sigma+n} + \sum_{n=0}^{\infty} (\sigma+n)(\sigma+n-3)\, a_n x^{\sigma+n+1} = 0$$

or

$$(\sigma-2)\, a_0 x^\sigma + \sum_{n=1}^{\infty} (\sigma+n-2)\, a_n x^{\sigma+n} + \sum_{n=0}^{\infty} (\sigma+n)(\sigma+n-3)\, a_n x^{\sigma+n+1} = 0$$

or

$$(\sigma-2)\, a_0 x^\sigma + \sum_{m=0}^{\infty} [(\sigma+m-1)\, a_{m+1} + (\sigma+m)(\sigma+m-3)\, a_m]\, x^{\sigma+m+1} = 0$$

Equating the coefficient of each power of x to zero yields

$$(\sigma-2)\, a_0 = 0 \quad \text{or} \quad \sigma = 2 \quad \text{when } a_0 \neq 0$$

$$a_{m+1} = -\frac{(\sigma+m)(\sigma+m-3)}{\sigma+m-1} a_m = -\frac{(m-1)(m+2)}{m+1} a_m$$

Hence

$$a_1 = 2a_0, \qquad a_2 = 0, \qquad a_n = 0 \quad \text{for } n \geq 2$$

Therefore, one solution of (124) is

$$y_1 = x^2(1+2x)$$

When $\lambda = 1$, y can be expressed as in (121a). Then, substituting (121a) into (124) and using the fact that $\lambda = 1$ yields

$$x^3 v'' - (x + 2x^2) v' + 2v = 0$$

Expressing v as in (127), we obtain

$$-\sum_{n=0}^{\infty} (\sigma + n - 2) a_n x^{\sigma+n} + \sum_{n=0}^{\infty} (\sigma + n)(\sigma + n - 3) a_n x^{\sigma+n+1} = 0$$

or

$$-(\sigma - 2) a_0 x^{\sigma} - \sum_{n=1}^{\infty} (\sigma + n - 2) a_n x^{\sigma+n} + \sum_{n=0}^{\infty} (\sigma + n)(\sigma + n - 3) a_n x^{\sigma+n+1} = 0$$

or

$$-(\sigma - 2) a_0 x^{\sigma} - \sum_{m=0}^{\infty} [(\sigma + m - 1) a_{m+1} - (\sigma + m)(\sigma + m - 3) a_m] x^{\sigma+m+1} = 0$$

Equating the coefficient of each power of x to zero yields

$$(\sigma - 2) a_0 = 0 \quad \text{or} \quad \sigma = 2 \quad \text{if } a_0 \neq 0$$

$$a_{m+1} = \frac{(\sigma + m)(\sigma + m - 3)}{\sigma + m - 1} a_m = \frac{(m-1)(m+2)}{m+1} a_m$$

Hence

$$a_1 = -2a_0, \qquad a_n = 0 \quad \text{for } n \geq 2$$

Therefore

$$v = x^2(1 - 2x) \quad \text{and} \quad y_2 = x^2(1 - 2x) e^{x^{-1}}$$

By knowing the solution of y_1, one can determine y_2 by letting

$$y_2(x) = y_1(x) u(x)$$

and determining a first-order linear equation for u'.

(b) Substituting (120b) and (120c) into (125) yields

$$\frac{\lambda^2 \nu^2}{x^{2\nu-2}} - \frac{2\lambda\nu}{x^{\nu-2}} - 1 + \cdots = 0$$

Balancing the dominant terms (the first and third), we have $\nu = 1$ and $\lambda = \pm 1$. Then y has the form in (121a). Substituting (121) into (125) yields

$$x^2 v'' + 2(x - \lambda) v' = 0$$

which is a first-order linear equation in v'. One of its solutions is $v = a$ constant. It follows from (121a) that

$$y = e^{\lambda x^{-1}} \quad \text{for } \lambda = \pm 1$$

Then two linearly independent solutions of (125) are

$$y_1 = \exp\left(\frac{1}{x}\right) \quad \text{and} \quad y_2 = \exp\left(-\frac{1}{x}\right)$$

(c) Substituting (120b) and (120c) into (126) yields

$$\frac{\lambda^2 \nu^2}{x^{2\nu}} - \frac{2\lambda\nu}{x^{\nu+1}} - 1 + \cdots = 0$$

Balancing the dominant terms (the first and second), we have $\nu = 1$ and $\lambda = 0$ or 2.
When $\lambda = 0$, y has the form (127). Then (126) yields

$$2\sum_{n=0}^{\infty} (\sigma + n) a_n x^{\sigma+n-1} + \sum_{n=0}^{\infty} \left[(\sigma + n)^2 - 3(\sigma + n) - 1\right] a_n x^{\sigma+n} = 0$$

or

$$2\sigma a_0 x^{\sigma-1} + 2\sum_{n=1}^{\infty} (\sigma + n) a_n x^{\sigma+n-1} + \sum_{n=0}^{\infty} \left[(\sigma + n)^2 - 3(\sigma + n) - 1\right] a_n x^{\sigma+n} = 0$$

or

$$2\sigma a_0 x^{\sigma-1} + \sum_{m=0}^{\infty} \left\{2(\sigma + m + 1) a_{m+1} + \left[(\sigma + m)^2 - 3(\sigma + m) - 1\right] a_m\right\} x^{\sigma+m} = 0$$

Equating the coefficient of each power of x to zero yields

$$\sigma a_0 = 0 \quad \text{or} \quad \sigma = 0 \quad \text{if } a_0 \neq 0$$

$$a_{m+1} = -\frac{(\sigma + m)^2 - 3(\sigma + m) - 1}{2(\sigma + m + 1)} a_m = -\frac{m^2 - 3m - 1}{2(m+1)} a_m$$

Hence one solution of (126) is

$$y_1 \sim 1 + \frac{1}{2}x + \frac{3}{2^3}x^2 + \frac{3x^3}{2^4} + \cdots$$

where an asymptotic sign was used because one can easily show that the series diverges.

When $\lambda = 2$, y has the form (121a). Substituting (121) into (126) yields

$$x^2 v'' - 2(1+x)v' + \left(\frac{8}{x} - 1\right)v = 0$$

Letting

$$v \sim \sum_{n=0}^{\infty} a_n x^{\sigma+n}$$

we have

$$-\sum_{n=0}^{\infty} 2(\sigma+n-4)a_n x^{\sigma+n-1} + \sum_{n=0}^{\infty}\left[(\sigma+n)^2 - 3(\sigma+n) - 1\right]a_n x^{\sigma+n} = 0$$

or

$$-2(\sigma-4)a_0 x^{\sigma-1} - \sum_{n=1}^{\infty} 2(\sigma+n-4)a_n x^{\sigma+n-1}$$

$$+\sum_{n=0}^{\infty}\left[(\sigma+n)^2 - 3(\sigma+n) - 1\right]a_n x^{\sigma+n} = 0$$

or

$$-2(\sigma-4)a_0 x^{\sigma-1} - \sum_{m=0}^{\infty}\left\{2(\sigma+m-3)a_{m+1}\right.$$

$$\left.-\left[(\sigma+m)^2 - 3(\sigma+m) - 1\right]a_m\right\}x^{\sigma+m} = 0$$

Equating the coefficient of each power of x to zero, we have

$$(\sigma-4)a_0 = 0 \quad \text{or} \quad \sigma = 4 \quad \text{if } a_0 \neq 0$$

$$a_{m+1} = \frac{(\sigma+m)^2 - 3(\sigma+m) - 1}{2(\sigma+m-3)}a_m = \frac{m^2+5m+3}{2(m+1)}a_m$$

Hence

$$v \sim x^4\left(1 + \tfrac{3}{2}x + \tfrac{27}{8}x^2 + \tfrac{153}{16}x^3 + \tfrac{4131}{128}x^4 + \cdots\right)$$

where an asymptotic sign was used because one can easily show that the series diverges. Substituting v into (121a) and recalling that $\lambda = 2$ yields the following solution of (126) that is linearly independent of y_1:

$$y_2 \sim e^{2x^{-1}} x^4\left(1 + \tfrac{3}{2}x + \tfrac{27}{8}x^2 + \tfrac{153}{16}x^3 + \tfrac{4131}{128}x^4 + \cdots\right)$$

13.16. Determine two linearly independent solutions for large x for each of the following equations:

(a)
$$16x^2y'' + 32xy' - (4x+5)y = 0 \tag{128}$$

(b)
$$xy'' + 2(1-x)y' - y = 0 \tag{129}$$

(c)
$$4x^2y'' + 8xy' - (4x^2+3)y = 0 \tag{130}$$

For each equation, we need to determine its behavior at infinity so that we can choose the appropriate form of the expansion. This is conveniently accomplished by mapping infinity into the origin using the transformation $z = x^{-1}$.

(a) Under the transformation $z = x^{-1}$, (128) becomes

$$\frac{d^2y}{dz^2} - \frac{4+5z}{16z^3}y = 0$$

Since $z = 0$ is an irregular singular point of this equation, $x = \infty$ is an irregular singular point of (128). Hence its solutions for large x have the form

$$y(x) \sim e^{\Lambda(x)}x^{\sigma}u(x) \tag{131}$$

where $u(x)$ can be expressed as a series in powers of $x^{-m/n}$ and Λ is a polynomial in $x^{m/n}$, where m and n are prime integers.

To determine the leading term in (131), we put

$$y(x) \sim e^{\lambda x^{\nu}}, \qquad \nu > 0 \tag{132a}$$

so that

$$y' \sim \lambda\nu x^{\nu-1}e^{\lambda x^{\nu}} \tag{132b}$$

$$y'' \sim \lambda^2\nu^2x^{2\nu-2}e^{\lambda x^{\nu}} \tag{132c}$$

Substituting (132) into (128) yields

$$16\lambda^2\nu^2x^{2\nu} + 32\lambda\nu x^{\nu} - 4x - 5 + \cdots = 0$$

Balancing the leading terms (the first and third), we have $\nu = \frac{1}{2}$ and $\lambda = \pm 1$. Then we put

$$y(x) = e^{\lambda x^{1/2}}v(x) \tag{133}$$

in (128) and obtain

$$v'' + \left(\frac{\lambda}{x^{1/2}} + \frac{2}{x}\right)v' + \left(\frac{3\lambda}{4x^{3/2}} - \frac{5}{16x^2}\right)v = 0 \tag{134}$$

Letting

$$v \sim \sum a_n x^{-(\sigma+n)/2}$$

in (134), we have

$$\sum_{n=0}^{\infty} \left[\tfrac{3}{4}\lambda - \tfrac{1}{2}\lambda(\sigma+n)\right] a_n x^{-(\sigma+n+3)/2}$$
$$+ \sum_{n=0}^{\infty} \tfrac{1}{4}\left(\sigma+n-\tfrac{5}{2}\right)\left(\sigma+n+\tfrac{1}{2}\right) a_n x^{-(\sigma+n+4)/2} = 0$$

or

$$\tfrac{1}{2}\lambda\left(\tfrac{3}{2}-\sigma\right) a_0 x^{-(\sigma+3)/2} + \sum_{m=0}^{\infty} \left\{\left[\tfrac{3}{4}\lambda - \tfrac{1}{2}\lambda(\sigma+m+1)\right] a_{m+1}\right.$$
$$\left. + \tfrac{1}{4}\left(\sigma+m-\tfrac{5}{2}\right)\left(\sigma+m+\tfrac{1}{2}\right) a_m\right\} x^{-(\sigma+m+4)/2} = 0$$

Equating the coefficient of each power of x to zero yields

$$\left(\tfrac{3}{2}-\sigma\right) a_0 = 0 \quad \text{or} \quad \sigma = \tfrac{3}{2} \quad \text{if } a_0 \neq 0$$

$$a_{m+1} = \frac{(\sigma+m-5/2)(\sigma+m+1/2)}{2\lambda(m+1)} a_m = \frac{(m-1)(m+2)}{2\lambda(m+1)} a_m \tag{135}$$

When $\lambda = 1$,

$$v(x) = \frac{1}{x^{3/4}}\left(1 - \frac{1}{x^{1/2}}\right)$$

and one solution of (128) is

$$y(x) = \frac{1}{x^{3/4}}\left(1 - \frac{1}{x^{1/2}}\right) e^{x^{1/2}}$$

When $\lambda = -1$,

$$v(x) = \frac{1}{x^{3/4}}\left(1 + \frac{1}{x^{1/2}}\right)$$

and a second solution of (128) that is linearly independent of y_1 is

$$y_2(x) = \frac{1}{x^{3/4}}\left(1 + \frac{1}{x^{1/2}}\right) e^{-x^{1/2}}$$

(b) Under the transformation $z = x^{-1}$, (129) becomes

$$\frac{d^2y}{dz^2} + \frac{2}{z^2}\frac{dy}{dz} - \frac{1}{z^3}y = 0$$

Since $z=0$ is an irregular singular point of this equation, $x=\infty$ is an irregular singular point of (129). Hence its solutions have the form (131).

To determine the leading term, we substitute (132) into (129) and obtain

$$\lambda^2\nu^2x^{2\nu-1}-2\lambda\nu x^{\nu}-1+\cdots=0$$

Balancing the leading terms (the first and second), we have $\nu=1$ and $\lambda=0$ or 2.

When $\lambda=0$, we let

$$y\sim\sum_{n=0}^{\infty}a_nx^{-(\sigma+n)} \tag{136}$$

in (129) and obtain

$$\sum_{n=0}^{\infty}(2\sigma+2n-1)a_nx^{-(\sigma+n)}+\sum_{n=0}^{\infty}(\sigma+n)(\sigma+n-1)a_nx^{-(\sigma+n+1)}=0$$

or

$$(2\sigma-1)a_0x^{-\sigma}+\sum_{n=1}^{\infty}(2\sigma+2n-1)a_nx^{-(\sigma+n)}$$
$$+\sum_{n=0}^{\infty}(\sigma+n)(\sigma+n-1)a_nx^{-(\sigma+n+1)}=0$$

or

$$(2\sigma-1)a_0x^{-\sigma}+\sum_{m=0}^{\infty}[(2\sigma+2m+1)a_{m+1}+(\sigma+m)(\sigma+m-1)a_m]x^{-(\sigma+m+1)}=0$$

Equating the coefficient of each power of x to zero yields

$$(2\sigma-1)a_0=0 \quad\text{or}\quad \sigma=\tfrac{1}{2}\quad\text{if } a_0\neq 0$$

$$a_{m+1}=-\frac{(2m+1)(2m-1)}{2^3(m+1)}a_m \tag{137}$$

Hence one of the solutions of (129) is

$$y_1(x)\sim x^{-1/2}\left(1+\tfrac{1}{8}x^{-1}-\frac{1\cdot 3}{8^2\cdot 2!}x^{-2}+\frac{1\cdot 3^2\cdot 5}{8^3\cdot 3!}x^{-3}-\frac{1\cdot 3^2\cdot 5^2\cdot 7}{8^4\cdot 4!}x^{-4}+\cdots\right)$$

When $\lambda=2$, we let

$$y=e^{2x}v(x) \tag{138}$$

in (129) and obtain

$$xv''+2(x+1)v'+3v=0 \tag{139}$$

We seek an asymptotic expansion of (139) in the form

$$v = \sum_{n=0}^{\infty} a_n x^{-(\sigma+n)} \tag{140}$$

Substituting (140) into (139) yields

$$\sum_{n=0}^{\infty} (3-2\sigma-2n) a_n x^{-(\sigma+n)} + \sum_{n=0}^{\infty} (\sigma+n)(\sigma+n-1) a_n x^{-(\sigma+n+1)} = 0$$

or

$$(3-2\sigma) a_0 x^{-\sigma} + \sum_{n=1}^{\infty} (3-2\sigma-2n) a_n x^{-(\sigma+n)}$$

$$+ \sum_{n=0}^{\infty} (\sigma+n)(\sigma+n-1) a_n x^{-(\sigma+n+1)} = 0$$

or

$$(3-2\sigma) a_0 x^{-\sigma} + \sum_{m=0}^{\infty} [(1-2\sigma-2m) a_{m+1}$$

$$+(\sigma+m)(\sigma+m-1) a_m] x^{-(\sigma+m+1)} = 0$$

Equating the coefficient of each power of x to zero yields

$$(3-2\sigma) a_0 = 0 \quad \text{or} \quad \sigma = \tfrac{3}{2} \quad \text{if } a_0 \neq 0$$

$$a_{m+1} = \frac{(2m+3)(2m+1)}{8(m+1)} a_m \tag{141}$$

Hence it follows from (140) and (141) that

$$v \sim x^{-3/2}\left(1 + \frac{1\cdot 3}{8} x^{-1} + \frac{1\cdot 3^2\cdot 5}{8^2\cdot 2!} x^{-2} + \frac{1\cdot 3^2\cdot 5^2\cdot 7}{8^3\cdot 3!} x^{-3} + \frac{1\cdot 3^2\cdot 5^2\cdot 7^2\cdot 9}{8^4\cdot 4!} x^{-4} + \cdots\right)$$

It follows from (138) that a second solution of (129) that is linearly independent of y_1 is

$$y_2 \sim x^{-3/2} e^{2x}\left(1 + \frac{1\cdot 3}{8} x^{-1} + \frac{1\cdot 3^2\cdot 5}{8^2\cdot 2!} x^{-2} + \frac{1\cdot 3^2\cdot 5^2\cdot 7}{8^3\cdot 3!} x^{-3} + \frac{1\cdot 3^2\cdot 5^2\cdot 7^2\cdot 9}{8^4\cdot 4!} x^{-4} + \cdots\right)$$

(c) Under the transformation $z = x^{-1}$, (130) becomes

$$\frac{d^2 y}{dz^2} - \frac{1}{4z^4}(4+3z^2) y = 0$$

Since $z = 0$ is an irregular singular point of this equation, $x = \infty$ is an irregular singular point of (130) and its solutions can be expressed in the form (131). To determine the leading term, we substitute (132) into (130) and obtain

$$4\lambda^2\nu^2 x^{2\nu} + 8\lambda\nu x^{\nu} - 4x^2 + \cdots = 0$$

Balancing the dominant terms (the first and third), we obtain $\nu = 1$ and $\lambda = \pm 1$. Hence y has the form

$$y = e^{\lambda x} v(x), \qquad \lambda = \pm 1 \tag{142}$$

which, when substituted into (130), yields

$$4x^2 v'' + 8x(\lambda x + 1)v' + (8\lambda x - 3)v = 0 \tag{143}$$

Next we seek solutions for (143) in the form (140). Substituting (140) into (143) we have

$$-\sum_{n=0}^{\infty} 8\lambda(\sigma + n - 1)a_n x^{-(\sigma+n-1)} + \sum_{n=0}^{\infty} (2\sigma + 2n - 3)(2\sigma + 2n + 1)a_n x^{-(\sigma+n)} = 0$$

or

$$-8\lambda(\sigma - 1)a_0 x^{1-\sigma} - 8\lambda \sum_{n=1}^{\infty} (\sigma + n - 1)a_n x^{-(\sigma+n-1)}$$
$$+ \sum_{n=0}^{\infty} (2\sigma + 2n - 3)(2\sigma + 2n + 1)a_n x^{-(\sigma+n)} = 0$$

or

$$-8\lambda(\sigma - 1)a_0 x^{1-\sigma} - \sum_{m=0}^{\infty} [8\lambda(\sigma + m)a_{m+1}$$
$$-(2\sigma + 2m - 3)(2\sigma + 2m + 1)a_m] x^{-(\sigma+m)} = 0$$

Equating the coefficient of each power of x to zero yields

$$(\sigma - 1)a_0 = 0 \quad \text{or} \quad \sigma = 1 \quad \text{if } a_0 \neq 0$$

$$a_{m+1} = \frac{(2\sigma + 2m - 3)(2\sigma + 2m + 1)}{8\lambda(\sigma + m)} a_m = \frac{(2m - 1)(2m + 3)}{8\lambda(m + 1)} a_m \tag{144}$$

Hence it follows from (140) that when $\lambda = 1$

$$v \sim x^{-1}\left(1 - \tfrac{3}{8}x^{-1} - \frac{3\cdot 5}{8^2\cdot 2!}x^{-2} - \frac{3^2\cdot 5\cdot 7}{8^3\cdot 3!}x^{-3} - \frac{3^2\cdot 5^2\cdot 7\cdot 9}{8^4\cdot 4!}x^{-4} + \cdots\right)$$

and hence it follows from (142) that one solution of (130) is

$$y_1 \sim e^x x^{-1}\left(1-\tfrac{3}{8}x^{-1}-\frac{3\cdot 5}{8^2\cdot 2!}x^{-2}-\frac{3^2\cdot 5\cdot 7}{8^3\cdot 3!}x^{-3}-\frac{3^2\cdot 5^2\cdot 7\cdot 9}{8^4\cdot 4!}x^{-4}+\cdots\right)$$

where an asymptotic sign was used because the series diverges. When $\lambda = -1$,

$$v \sim x^{-1}\left(1+\tfrac{3}{8}x^{-1}-\frac{3\cdot 5}{8^2\cdot 2!}x^{-2}+\frac{3^2\cdot 5\cdot 7}{8^3\cdot 3!}x^{-3}-\frac{3^2\cdot 5^2\cdot 7\cdot 9}{8^4\cdot 4!}x^{-4}+\cdots\right)$$

and hence a second solution of (130) that is linearly independent of y_1 is

$$y_2 \sim e^{-x} x^{-1}\left(1+\tfrac{3}{8}x^{-1}-\frac{3\cdot 5}{8^2\cdot 2!}x^{-2}+\frac{3^2\cdot 5\cdot 7}{8^3\cdot 3!}x^{-3}-\frac{3^2\cdot 5^2\cdot 7\cdot 9}{8^4\cdot 4!}x^{-4}+\cdots\right)$$

13.17. Consider the modified Bessel equation of zeroth order

$$xy'' + y' - xy = 0 \tag{145}$$

(a) Show that it has the following bounded solution at the origin:

$$I_0(x) = 1 + \frac{x^2}{2^2} + \frac{x^4}{2^2\cdot 4^2} + \frac{x^6}{2^2\cdot 4^2\cdot 6^2} + \cdots \tag{146}$$

Determine the second solution.

(b) Determine an asymptotic solution for large x; it involves an arbitrary constant.

(c) Show that

$$I_0(x) = \frac{1}{2\pi}\int_0^{2\pi} e^{x\sin\theta}\,d\theta \tag{147}$$

(d) Determine the asymptotic expansion for large x of (147) and use it to determine the constant in (b).

(a) Since the origin is a regular singular point of (145), it possesses solutions of the Frobenius type. Substituting (52) into (145) yields

$$\sum_{n=0}^{\infty} (\sigma+n)^2 a_n x^{\sigma+n-1} - \sum_{n=0}^{\infty} a_n x^{\sigma+n+1} = 0$$

or

$$\sigma^2 a_0 x^{\sigma-1} + (\sigma+1)^2 a_1 x^{\sigma} + \sum_{n=2}^{\infty} (\sigma+n)^2 a_n x^{\sigma+n-1} - \sum_{n=0}^{\infty} a_n x^{\sigma+n+1} = 0$$

or

$$\sigma^2 a_0 x^{\sigma-1} + (\sigma+1)^2 a_1 x^{\sigma} + \sum_{m=0}^{\infty} \left[(\sigma+m+2)^2 a_{m+2} - a_m \right] x^{\sigma+m+1} = 0$$

Equating the coefficient of each power of x to zero yields

$$\sigma^2 a_0 = 0 \quad \text{or} \quad \sigma = 0 \quad \text{if } a_0 \neq 0$$

$$(\sigma+1)^2 a_1 = 0 \quad \text{or} \quad a_1 = 0 \quad \text{if } \sigma = 0$$

$$a_{m+2} = \frac{a_m}{(\sigma+m+2)^2} \tag{148}$$

In this case the indices are equal; hence one solution of (145) is of the Frobenius type and the other involves a logarithm. It follows from (52) and (148), and because $a_1 = 0$, that

$$y(x;\sigma) = x^{\sigma}\left[1 + \frac{x^2}{(\sigma+2)^2} + \frac{x^4}{(\sigma+2)^2(\sigma+4)^2} + \frac{x^6}{(\sigma+2)^2(\sigma+4)^2(\sigma+6)^2} + \cdots\right]$$

Hence one solution of (145) is

$$y_1 = y(x;\sigma)|_{\sigma=0} = I_0(x)$$

Since the indices are equal, a second solution that is linearly independent of y_1 is

$$y_2 = \left.\frac{\partial y(x;\sigma)}{\partial \sigma}\right|_{\sigma=0}$$

$$= I_0(x)\ln x - 2\left[\frac{x^2}{2^3} + \frac{x^4}{2^2 \cdot 4^2}\left(\tfrac{1}{2}+\tfrac{1}{4}\right) + \frac{x^6}{2^2 \cdot 4^2 \cdot 6^2}\left(\tfrac{1}{2}+\tfrac{1}{4}+\tfrac{1}{6}\right) + \cdots\right]$$

(b) To determine a solution of (145) for large x, we first need to determine the behavior of (145) at $x = \infty$. To accomplish this, we introduce the transformation $z = x^{-1}$ in (145) and obtain

$$\frac{d^2y}{dz^2} + \frac{1}{z}\frac{dy}{dz} - \frac{1}{z^4}y = 0$$

Since $z = 0$ is an irregular singular point of this equation, $x = \infty$ is an irregular singular point of (145). Hence its solutions for large x are of the form (131). To determine the leading term in the asymptotic expansion of y for large x, we substitute (132) into (145) and obtain

$$\lambda^2 \nu^2 x^{2\nu-1} + \lambda \nu x^{\nu-1} - x + \cdots = 0$$

Balancing the dominant terms (the first and third), we have $\nu = 1$ and $\lambda = \pm 1$. Hence for large x, y has the form (142). Substituting (142) into (145) yields

$$xv'' + (2\lambda x + 1)v' + \lambda v = 0 \tag{149}$$

Substituting (140) into (149), we have

$$-\lambda \sum_{n=0}^{\infty} (2\sigma + 2n - 1) a_n x^{-(\sigma+n)} + \sum_{n=0}^{\infty} (\sigma + n)^2 a_n x^{-(\sigma+n+1)} = 0$$

or

$$-\lambda(2\sigma - 1) a_0 x^{-\sigma} - \lambda \sum_{n=1}^{\infty} (2\sigma + 2n - 1) a_n x^{-(\sigma+n)} + \sum_{n=0}^{\infty} (\sigma + n)^2 a_n x^{-(\sigma+n+1)} = 0$$

or

$$-\lambda(2\sigma - 1) a_0 x^{-\sigma} - \sum_{m=0}^{\infty} \left[(2\sigma + 2m + 1)\lambda a_{m+1} - (\sigma + m)^2 a_m \right] x^{-(\sigma+m+1)} = 0$$

Equating the coefficient of each power of x to zero yields

$$(2\sigma - 1) a_0 = 0 \quad \text{or} \quad \sigma = \tfrac{1}{2} \quad \text{if } a_0 \neq 0$$

$$a_{m+1} = \frac{(\sigma + m)^2}{\lambda(2\sigma + 2m + 1)} a_m = \frac{(2m+1)^2}{8\lambda(m+1)} a_m \tag{150}$$

Hence it follows from (140) and (150) that

$$v \sim x^{-1/2}\left(1 + \tfrac{1}{8}x^{-1} + \frac{3^2}{8^2 \cdot 2} x^{-2} + \cdots\right) \quad \text{for } \lambda = 1$$

$$v \sim x^{-1/2}\left(1 - \tfrac{1}{8}x^{-1} + \frac{3^2}{8^2 \cdot 2} x^{-2} + \cdots\right) \quad \text{for } \lambda = -1$$

Therefore, it follows from (142) that for large x

$$I_0(x) \sim c x^{-1/2} e^x \left(1 + \tfrac{1}{8}x^{-1} + \frac{3^2}{8^2 \cdot 2} x^{-2} + \cdots\right) \tag{151}$$

(c) Since

$$e^{x \sin\theta} = \sum_{n=0}^{\infty} \frac{(x \sin\theta)^n}{n!}$$

then

$$\frac{1}{2\pi}\int_0^{2\pi} e^{x\sin\theta}\,d\theta = \sum_{n=0}^{\infty} \frac{x^n}{n!}\frac{1}{2\pi}\int_0^{2\pi}\sin^n\theta\,d\theta \tag{152a}$$

But

$$\frac{1}{2\pi}\int_0^{2\pi}\sin^n\theta\,d\theta = 0 \quad \text{if } n \text{ is odd} \quad = \frac{(n-1)(n-3)\cdots 3\cdot 1}{n(n-2)\cdots 4\cdot 2} \quad \text{if } n \text{ is even} \tag{152b}$$

Hence (152a) becomes

$$\frac{1}{2\pi}\int_0^{2\pi} e^{x\sin\theta}\,d\theta = 1 + \frac{x^2}{2^2} + \frac{x^4}{2^2\cdot 4^2} + \frac{x^6}{2^2\cdot 4^2\cdot 6^2} + \cdots = I_0(x)$$

according to (146).

(d) To determine the leading term in the asymptotic expansion of (147), we use Laplace's method and observe that the maximum of $\sin\theta$ occurs at $\theta = \frac{1}{2}\pi$. Then, expanding $\sin\theta$ in the neighborhood of $\theta = \frac{1}{2}\pi$, we have

$$\sin\theta = 1 - \tfrac{1}{2}\left(\theta - \tfrac{1}{2}\pi\right)^2 + \cdots$$

Therefore

$$I_0(x) \sim \frac{1}{2\pi}\int_{-\infty}^{\infty} e^{x - x(\theta - \pi/2)^2/2}\,d\theta$$

Putting $\sqrt{x}\,(\theta - \frac{1}{2}\pi) = \sqrt{2}\,\tau$ gives

$$I_0(x) \sim \frac{\sqrt{2}}{2\pi\sqrt{x}}\,e^x\int_{-\infty}^{\infty} e^{-\tau^2}\,d\tau = \frac{1}{\sqrt{2\pi x}}\,e^x \tag{153}$$

Comparing the leading term in (151) with (153), we conclude that $c = (2\pi)^{-1/2}$. Hence (151) becomes

$$I_0(x) \sim \frac{1}{\sqrt{2\pi x}}\,e^x\left(1 + \tfrac{1}{8}x^{-1} + \frac{3}{8^2\cdot 2}x^{-2} + \cdots\right) \tag{154}$$

13.18. Consider Bessel's equations of order one

$$x^2y'' + xy' + (x^2 - 1)\,y = 0 \tag{155}$$

(a) Show that one of the solutions of this equation can be expressed as

$$J_1 = \frac{x}{2}\left[1 - \frac{x^2}{2\cdot 4} + \frac{x^4}{2\cdot 4^2\cdot 6} + \cdots\right] \tag{156}$$

which converges for all x. Determine the second solution.

(b) Expand $\sin\theta \exp(ix\sin\theta)$ in powers of x, integrate the result with respect to θ from 0 to 2π, and obtain

$$J_1(x) = -\frac{i}{2\pi}\int_0^{2\pi} \sin\theta\, e^{ix\sin\theta}\, d\theta \tag{157}$$

(c) Determine the asymptotic expansion of the solution of the equation for large x; it involves two arbitrary constants.

(d) Determine the leading term in the asymptotic expansion of the integral in (b) for large x and use it to determine the constants in (c).

(a) Since the origin is a regular singular point of (155), it possesses solutions of the Frobenius type. Substituting (52) into (155) yields

$$\sum_{n=0}^{\infty} \left[(\sigma+n)^2 - 1\right] a_n x^{\sigma+n} + \sum_{n=0}^{\infty} a_n x^{\sigma+n+2} = 0$$

or

$$(\sigma^2 - 1) a_0 x^\sigma + (\sigma^2 + 2\sigma) a_1 x^{\sigma+1} + \sum_{n=2}^{\infty} \left[(\sigma+n)^2 - 1\right] a_n x^{\sigma+n} + \sum_{n=0}^{\infty} a_n x^{\sigma+n+2} = 0$$

or

$$(\sigma^2 - 1) a_0 x^\sigma + (\sigma^2 + 2\sigma) a_1 x^{\sigma+1} + \sum_{m=0}^{\infty} \left[(\sigma+m+1)(\sigma+m+3) a_{m+2} + a_m\right] x^{\sigma+m+2} = 0$$

Equating the coefficient of each power of x to zero, we have

$$(\sigma^2 - 1) a_0 = 0 \quad \text{or} \quad \sigma = -1 \quad \text{or} \quad 1 \quad \text{if } a_0 \neq 0$$

$$(\sigma^2 + 2\sigma) a_1 = 0 \quad \text{or} \quad a_1 = 0$$

$$a_{m+2} = -\frac{a_m}{(\sigma+m+1)(\sigma+m+3)} \tag{158}$$

When $\sigma = 1$, it follows from (158) that

$$a_2 = -\frac{a_0}{2\cdot 4}, \qquad a_4 = \frac{a_0}{2\cdot 4^2\cdot 6}, \qquad a_6 = -\frac{a_0}{2\cdot 4^2\cdot 6^2\cdot 8}$$

which when substituted into (52), yields (156) if $a_0 = \frac{1}{2}$.

To determine a second solution of (155) that is linearly independent from J_1, we need to use the second index $\sigma = -1$. But a_2 is infinite according to (158). Hence the second solution involves a logarithm. It follows from (52) and (158) that

$$y(x;\sigma) = a_0 x^\sigma \left[1 - \frac{x^2}{(\sigma+1)(\sigma+3)} + \frac{x^4}{(\sigma+1)(\sigma+3)^2(\sigma+5)} - \cdots\right] \tag{159}$$

To circumvent the difficulty when $\sigma = -1$, we put $b = a_0(\sigma+1)$ in (159) and obtain

$$y(x;\sigma) = bx^\sigma\left[\sigma + 1 - \frac{x^2}{\sigma+3} + \frac{x^4}{(\sigma+3)^2(\sigma+5)} + \cdots\right]$$

Then if $b = -1$,

$$y_2 = \left.\frac{\partial y(x;\sigma)}{\partial\sigma}\right|_{\sigma=-1} = J_1(x)\ln x - x^{-1}\left[1 + \frac{x^2}{2^2} - \tfrac{5}{64}x^4 + \cdots\right]$$

(b) We note that

$$\sin\theta\, e^{ix\sin\theta} = \sum_{n=0}^{\infty}\frac{(ix)^n}{n!}(\sin\theta)^{n+1}$$

Hence

$$\frac{-i}{2\pi}\int_0^{2\pi}\sin\theta\, e^{ix\sin\theta}\,d\theta = \sum_{n=0}^{\infty}\frac{-i(ix)^n}{n!}\frac{1}{2\pi}\int_0^{2\pi}(\sin\theta)^{n+1}\,d\theta$$

Using (152b) we obtain

$$-\frac{i}{2\pi}\int_0^{2\pi}\sin\theta\, e^{ix\sin\theta}\,d\theta = \frac{x}{2} - \frac{x^3}{2^2\cdot 4} + \frac{x^5}{2^2\cdot 4^2\cdot 6} - \cdots = J_1(x)$$

according to (156).

(c) To determine an asymptotic expansion of the solution of (155) for large x, we investigate the behavior of (155) at $x = \infty$ by using the transformation $z = x^{-1}$. The result is

$$\frac{d^2y}{dz^2} + \frac{1}{z}\frac{dy}{dz} + \left(\frac{1}{z^4} - \frac{1}{z^2}\right)y = 0$$

Since $z = 0$ is an irregular singular point of this equation, $x = \infty$ is an irregular singular point of (155). Hence the solutions of (155) for large x are of the form (131). To determine the form of Λ, we substitute (132) into (155) and obtain

$$\lambda^2\nu^2x^{2\nu} + \lambda\nu x^\nu + x^2 + \cdots = 0$$

Balancing the dominant terms (the first and third), we have $\nu = 1$ and $\lambda = \pm i$. Hence y

has the form (142) when $\lambda = \pm i$. Substituting (142) into (155) when $\lambda = i$ yields

$$x^2 v'' + (2ix^2 + x) v' + (ix - 1) v = 0 \tag{160}$$

whose solutions have the form (140).

Substituting (140) into (160) gives

$$-i \sum_{n=0}^{\infty} (2\sigma + 2n - 1) a_n x^{-(\sigma+n-1)} + \sum_{n=0}^{\infty} \left[(\sigma + n)^2 - 1\right] a_n x^{-(\sigma+n)} = 0$$

or

$$-i(2\sigma - 1) a_0 x^{1-\sigma} - i \sum_{n=1}^{\infty} (2\sigma + 2n - 1) a_n x^{-(\sigma+n-1)}$$

$$+ \sum_{n=0}^{\infty} \left[(\sigma + n)^2 - 1\right] a_n x^{-(\sigma+n)} = 0$$

or

$$-i(2\sigma - 1) a_0 x^{1-\sigma} - \sum_{m=0}^{\infty} \left\{ i(2\sigma + 2m + 1) a_{m+1} - \left[(\sigma + m)^2 - 1\right] a_m \right\} x^{-(\sigma+m)} = 0$$

Equating the coefficient of each power of x to zero yields

$$(2\sigma - 1) a_0 = 0 \quad \text{or } \sigma = \tfrac{1}{2} \quad \text{if } a_0 \neq 0$$

$$a_{m+1} = -\frac{(2m-1)(2m+3)}{8(m+1)} i a_m \tag{161}$$

It follows from (140) and (161) that

$$v \sim x^{-1/2}\left[1 + \frac{1\cdot 3i}{8x} + \frac{1\cdot 3\cdot 5}{8^2\cdot 2!x^2} - \frac{1\cdot 3^2\cdot 5\cdot 7i}{8^3\cdot 3!x^3} - \frac{1\cdot 3^2\cdot 5^2\cdot 7\cdot 9}{8^4\cdot 4!x^4} + \right]$$

which when combined with (142) for $\lambda = i$, yields

$$y \sim e^{ix} x^{-1/2}\left[1 + \frac{1\cdot 3i}{8x} + \frac{1\cdot 3\cdot 5}{8^2\cdot 2!x^2} - \frac{1\cdot 3^2\cdot 5\cdot 7i}{8^3\cdot 3!x^3} - \frac{1\cdot 3^2\cdot 5^2\cdot 7\cdot 9}{8^4\cdot 4!x^4} + \cdots\right] \tag{162}$$

A second linearly independent solution of (155) can be obtained by using $\lambda = -i$, which turns out to be the complex conjugate of (162) since (155) is real. Then we separate (162) into real and imaginary parts and obtain the following two linearly independent solutions of (155):

$$\begin{aligned} y_1 &\sim x^{-1/2}[U(x)\cos x - V(x)\sin x] \\ y_2 &\sim x^{-1/2}[U(x)\sin x + V(x)\cos x] \end{aligned} \tag{163}$$

where

$$\begin{aligned} U(x) &= 1 + \frac{1\cdot 3\cdot 5}{8^2\cdot 2!x^2} - \frac{1\cdot 3^2\cdot 5^2\cdot 7\cdot 9}{8^4\cdot 4!x^4} + \cdots \\ V(x) &= \frac{1\cdot 3}{8x} - \frac{1\cdot 3^2\cdot 5\cdot 7}{8^3\cdot 3!x^3} + \cdots \end{aligned} \tag{164}$$

Hence

$$J_1(x) \sim x^{-1/2}[(c_1\cos x + c_2\sin x)U(x) - (c_1\sin x - c_2\cos x)V(x)] \tag{165}$$

where c_1 and c_2 are arbitrary constants to be determined from the asymptotic expansion of (157).

(d) To determine the leading term in the expansion of (157), we use the method of stationary phase and replace the interval of integration with $[-\pi, \pi]$. Since

$$\frac{d}{d\theta}(\sin\theta) = \cos\theta = 0$$

when $\theta = \frac{1}{2}\pi$ and $-\frac{1}{2}\pi$, the leading term in the asymptotic expansion of (157) arises from the neighborhoods of $\theta = \frac{1}{2}\pi$ and $\theta = -\frac{1}{2}\pi$. Thus

$$J_1(x) \sim -\frac{i}{2\pi}\int_{-\pi/2-\delta}^{-\pi/2+\delta} \sin\theta\, e^{ix\sin\theta}\, d\theta - \frac{i}{2\pi}\int_{\pi/2-\delta}^{\pi/2+\delta} \sin\theta\, e^{ix\sin\theta}\, d\theta \tag{166}$$

where δ is a small positive number. We expand $\sin\theta$ near $\theta = \pm\frac{1}{2}\pi$ and the result is

$$\begin{aligned} \sin\theta &= 1 - \tfrac{1}{2}\left(\theta - \tfrac{1}{2}\pi\right)^2 + \cdots = 1 - \tfrac{1}{2}\tau^2 + \cdots \\ \sin\theta &= -1 + \tfrac{1}{2}\left(\theta + \tfrac{1}{2}\pi\right)^2 + \cdots = -1 + \tfrac{1}{2}t^2 + \cdots \end{aligned} \tag{167}$$

Substituting (167) into (166) and taking $\delta = \infty$ yields

$$J_1(x) \sim -\frac{ie^{ix}}{2\pi}\int_{-\infty}^{\infty} e^{-ix\tau^2/2}\, d\tau + \frac{ie^{-ix}}{2\pi}\int_{-\infty}^{\infty} e^{ixt^2/2}\, dt$$

We use Cauchy's theorem, rotate the contour in the first integral by $-\frac{1}{4}\pi$ and in the second integral by $\frac{1}{4}\pi$, and obtain

$$\begin{aligned} J_1(x) \sim &-\frac{ie^{ix}}{2\pi}\int_{-\infty\exp(-i\pi/4)}^{\infty\exp(-i\pi/4)} e^{-ix\tau^2/2}\, d\tau \\ &+ \frac{ie^{-ix}}{2\pi}\int_{-\infty\exp(i\pi/4)}^{\infty\exp(i\pi/4)} e^{ixt^2/2}\, dt \end{aligned}$$

Letting

$$\tau = \sqrt{\frac{2}{x}}\, e^{-i\pi/4} u, \qquad t = \sqrt{\frac{2}{x}}\, e^{i\pi/4} w$$

we have

$$J_1(x) \sim \frac{1}{2\pi}\sqrt{\frac{2}{x}}\left[e^{i(x-3/4\pi)} + e^{-i(x-3/4\pi)}\right]\int_{-\infty}^{\infty} e^{-u^2}\,du$$

$$= \sqrt{\frac{2}{\pi x}}\cos\left(x - \tfrac{3}{4}\pi\right) \quad \text{as } x \to \infty \tag{168}$$

As $x \to \infty$, it follows from (164) that $U \to 1$ and $V \to 0$; hence it follows from (165) that

$$J_1 \sim x^{-1/2}(c_1 \cos x + c_2 \sin x) \tag{169}$$

Equating (168) and (169) yields

$$c_1 = \sqrt{\frac{2}{\pi}}\cos\tfrac{3}{4}\pi \quad \text{and} \quad c_2 = \sqrt{\frac{2}{\pi}}\sin\tfrac{3}{4}\pi$$

Hence (165) becomes

$$y_1 \sim \sqrt{\frac{2}{\pi x}}\left[U(x)\cos\left(x - \tfrac{3}{4}\pi\right) - V(x)\sin\left(x - \tfrac{3}{4}\pi\right)\right] \quad \text{as } x \to \infty \tag{170}$$

13.19. Determine approximations to three linearly independent solutions for large x for each of the following equations:

(a)
$$x^6 y''' + 6x^5 y'' - y = 0 \tag{171}$$

(b)
$$xy''' - (2x+1)y'' - (1+x)y' + (2x+3)y = 0 \tag{172}$$

(a) To determine the behavior of (171) at $x = \infty$, we introduce the transformation $z = x^{-1}$ and obtain

$$\frac{dy}{dx} = -z^2\frac{dy}{dz} \tag{173a}$$

$$\frac{d^2y}{dx^2} = z^4\frac{d^2y}{dz^2} + 2z^3\frac{dy}{dz} \tag{173b}$$

$$\frac{d^3y}{dx^3} = -z^6\frac{d^3y}{dz^3} - 6z^5\frac{d^2y}{dz^2} - 6z^4\frac{dy}{dz} \tag{173c}$$

Hence (171) becomes

$$\frac{d^3y}{dz^3} - \frac{6}{z^2}\frac{dy}{dz} + y = 0 \tag{173d}$$

Since $z = 0$ is a regular singular point of (173d), $x = \infty$ is a regular singular point of (171) and hence possesses solutions of the Frobenius type.

Letting

$$y = \sum_{n=0}^{\infty} a_n x^{-(\sigma+n)} \tag{174}$$

in (171), we have

$$-\sum_{n=0}^{\infty} (\sigma+n)(\sigma+n+1)(\sigma+n-4)a_n x^{-(\sigma+n-3)} - \sum_{n=0}^{\infty} a_n x^{-(\sigma+n)} = 0$$

or

$$-\sigma(\sigma+1)(\sigma-4)a_0 x^{3-\sigma} - (\sigma+1)(\sigma+2)(\sigma-3)a_1 x^{2-\sigma}$$
$$-(\sigma+2)(\sigma+3)(\sigma-2)a_2 x^{1-\sigma} - \sum_{n=3}^{\infty} (\sigma+n)(\sigma+n+1)$$
$$\times(\sigma+n-4)a_n x^{-(\sigma+n-3)} - \sum_{n=0}^{\infty} a_n x^{-(\sigma+n)} = 0$$

or

$$-\sigma(\sigma+1)(\sigma-4)a_0 x^{3-\sigma} - (\sigma+1)(\sigma+2)(\sigma-3)a_1 x^{2-\sigma}$$
$$-(\sigma+2)(\sigma+3)(\sigma-2)a_2 x^{1-\sigma} - \sum_{m=0}^{\infty} [(\sigma+m+3)(\sigma+m+4)$$
$$\times(\sigma+m-1)a_{m+3} + a_m]x^{-(\sigma+m)} = 0$$

Equating the coefficient of each power of x to zero yields

$$-\sigma(\sigma+1)(\sigma-4)a_0 = 0 \tag{175}$$

$$-(\sigma+1)(\sigma+2)(\sigma-3)a_1 = 0 \tag{176}$$

$$-(\sigma+2)(\sigma+3)(\sigma-2)a_2 = 0 \tag{177}$$

$$(\sigma+m+3)(\sigma+m+4)(\sigma+m-1)a_{m+3} = -a_m \tag{178}$$

It follows from (175) that the indices are $\sigma = -1$, 0, and 4. Since the difference between any two of them is an integer, all solutions of (171) may or may not be of the Frobenius type, depending on whether some of the a_n are infinite. When $\sigma = -1$, it follows from (176) that a_1 is indeterminate, from (177) that $a_2 = 0$, and from (178) that a_5 is indeterminate. Hence three linearly independent solutions of (171) are of the Frobenius type, corresponding to the indeterminate coefficients a_0, a_1, and a_5. To determine these solutions, we put $\sigma = -1$ in (178) and obtain

$$a_{m+3} = -\frac{a_m}{(m-2)(m+2)(m+3)}, \qquad m \neq 2 \tag{179}$$

Using (174) and (179), we have

$$y = a_0 x\left[1 + \frac{1}{12x^3} - \frac{1}{360x^6} + \cdots\right] + a_1 x\left[\frac{1}{x} + \frac{1}{12x^4} - \frac{1}{1008x^7} + \cdots\right]$$
$$+ a_5 x\left[\frac{1}{x^5} - \frac{1}{168x^8} + \frac{1}{110880x^{11}} + \cdots\right]$$

(b) Substituting (173a)–(173c) into (172) yields

$$\frac{d^3y}{dz^3} + \left(\frac{7}{z} + \frac{2}{z^2}\right)\frac{d^2y}{dz^2} + \left(\frac{8}{z^2} + \frac{3}{z^3} - \frac{1}{z^4}\right)\frac{dy}{dz} - \left(\frac{2}{z^6} + \frac{3}{z^5}\right)y = 0 \tag{180}$$

Since $z = 0$ is an irregular singular point of (180), $x = \infty$ is an irregular singular point of (172). Hence it possesses solutions of the form (131). To determine the form of Λ, we substitute (132) into (172) and obtain

$$\lambda^3\nu^3 x^{3\nu-2} - 2\lambda^2\nu^2 x^{2\nu-1} - \lambda\nu x^\nu + 2x + \cdots = 0$$

Balancing the dominant terms (all four terms), we have $\nu = 1$ and

$$\lambda^3 - 2\lambda^2 - \lambda + 2 = 0$$

whose solutions are $\lambda = 1, -1, 2$. Hence y has the form

$$y = e^{\lambda x}v(x), \qquad \lambda = 1, -1, 2 \tag{181}$$

Substituting (181) into (172) yields

$$xv''' + (3\lambda x - 2x - 1)v''$$
$$+ (3\lambda^2 x - 4\lambda x - 2\lambda - 1 - x)v' - (\lambda^2 + \lambda - 3)v = 0 \tag{182}$$

Substituting (140) into (182) yields

$$-\sum_{n=0}^{\infty}\left[(3\lambda^2 - 4\lambda - 1)(\sigma+n) + \lambda^2 + \lambda - 3\right]a_n x^{-(\sigma+n)}$$
$$+ \sum_{n=0}^{\infty}(\sigma+n)[(3\lambda-2)(\sigma+n+1) + 2\lambda + 1]a_n x^{-(\sigma+n+1)}$$
$$- \sum_{n=0}^{\infty}(\sigma+n)(\sigma+n+1)(\sigma+n+3)a_n x^{-(\sigma+n+2)} = 0$$

or

$$-\left[(3\lambda^2-4\lambda-1)\sigma+\lambda^2+\lambda-3\right]a_0x^{-\sigma}$$

$$-\left[(3\lambda^2-4\lambda-1)(\sigma+1)+\lambda^2+\lambda-3\right]a_1x^{-(\sigma+1)}$$

$$+\sigma[(3\lambda-2)(\sigma+1)+2\lambda+1]a_0x^{-(\sigma+1)}$$

$$-\sum_{n=2}^{\infty}\left[(3\lambda^2-4\lambda-1)(\sigma+n)+\lambda^2+\lambda-3\right]a_nx^{-(\sigma+n)}$$

$$+\sum_{n=1}^{\infty}(\sigma+n)[(3\lambda-2)(\sigma+n+1)+2\lambda+1]a_nx^{-(\sigma+n+1)}$$

$$-\sum_{n=0}^{\infty}(\sigma+n)(\sigma+n+1)(\sigma+n+3)a_nx^{-(\sigma+n+2)}=0$$

or

$$-\left[(3\lambda^2-4\lambda-1)\sigma+\lambda^2+\lambda-3\right]a_0x^{-\sigma}$$

$$-\left[(3\lambda^2-4\lambda-1)(\sigma+1)+\lambda^2+\lambda-3\right]a_1x^{-(\sigma+1)}$$

$$+\sigma[(3\lambda-2)(\sigma+1)+2\lambda+1]a_0x^{-(\sigma+1)}$$

$$-\sum_{m=0}^{\infty}\Big\{\left[(3\lambda^2-4\lambda-1)(\sigma+m+2)+\lambda^2+\lambda-3\right]a_{m+2}$$

$$-(\sigma+m+1)[(3\lambda-2)(\sigma+m+2)+2\lambda+1]a_{m+1}$$

$$+(\sigma+m)(\sigma+m+1)(\sigma+m+3)a_m\Big\}x^{-(\sigma+m+2)}=0$$

Equating the coefficient of each power of x to zero yields

$$\left[(3\lambda^2-4\lambda-1)\sigma+\lambda^2+\lambda-3\right]a_0=0 \quad \text{or} \quad \sigma=-\frac{\lambda^2+\lambda-3}{3\lambda^2-4\lambda-1} \quad \text{if } a_0\neq 0 \tag{183a}$$

$$\left[(3\lambda^2-4\lambda-1)(\sigma+1)+\lambda^2+\lambda-3\right]a_1=\sigma[(3\lambda-2)(\sigma+1)+2\lambda+1]a_0 \tag{183b}$$

$$a_{m+2}=\frac{(\sigma+m+1)[(3\lambda-2)(\sigma+m+2)+2\lambda+1]a_{m+1}-(\sigma+m)(\sigma+m+1)(\sigma+m+3)a_m}{(3\lambda^2-4\lambda-1)(\sigma+m+2)+\lambda^2+\lambda-3} \tag{183c}$$

When $\lambda = -1$, it follows from (183) that $\sigma = \frac{1}{2}$,

$$a_1 = -\tfrac{17}{24}a_0$$

and

$$a_{m+2} = -\frac{(2m+3)(10m+27)}{24(m+2)}a_{m+1} - \frac{(2m+1)(2m+3)(2m+7)}{48(m+2)}a_m$$

Hence

$$v \sim x^{-1/2}\left(1 - \frac{17}{24x} + \frac{125}{128x^2} + \cdots\right)$$

and one solution of (172) is

$$y_1(x) \sim e^{-x}x^{-1/2}\left(1 - \frac{17}{24x} + \frac{125}{128x^2} + \cdots\right)$$

When $\lambda = 1$, it follows from (183) that $\sigma = -\frac{1}{2}$,

$$a_1 = \tfrac{7}{8}a_0$$

and

$$a_{m+2} = -\frac{(2m+1)(2m+9)}{8(m+2)}a_{m+1} + \frac{(2m-1)(2m+1)(2m+5)}{16(m+2)}a_m$$

Hence

$$v \sim x^{1/2}\left(1 + \frac{7}{8x} - \frac{83}{128x^2} + \cdots\right)$$

and a second solution of (172) that is linearly independent of y_1 is

$$y_2 \sim e^{x}x^{1/2}\left(1 + \frac{7}{8x} - \frac{83}{128x^2} + \cdots\right)$$

When $\lambda = 2$, it follows from (183) that $\sigma = -1$,

$$a_1 = -\tfrac{5}{3}a_0$$

and

$$a_{m+2} = \frac{m(4m+9)}{3(m+2)}a_{m+1} - \frac{m(m-1)}{3}a_m$$

Hence

$$v = x\left(1 - \frac{5}{3x}\right)$$

and a third solution of (172) that is linearly independent of y_1 and y_2 is

$$y_3 = x\left(1 - \frac{5}{3x}\right)e^{2x}$$

13.20. Consider Bessel's equation of order ν

$$x^2y'' + xy' + (x^2 - \nu^2)y = 0 \tag{184}$$

(a) Show that

$$J_\nu(x) = \sum_{n=0}^{\infty} \frac{(-1)^n(x/2)^{2n+\nu}}{n!\Gamma(\nu+n+1)} \tag{185}$$

When ν is different from an integer, show that

$$J_{-\nu}(x) = \sum_{n=0}^{\infty} \frac{(-1)^n(x/2)^{2n-\nu}}{n!\Gamma(-\nu+n+1)} \tag{186}$$

(b) Show that for large x

$$y \sim Ay_1 + By_2 \tag{187}$$

where

$$y_1 \sim \frac{1}{\sqrt{x}}e^{ix}\left[1 + \frac{c_1}{x} + \frac{c_2}{x^2} + \cdots\right], \qquad y_2 = \bar{y}_1$$

(c) Use the integral representation

$$J_\nu(x) = \frac{2(\frac{1}{2}x)^\nu}{\sqrt{\pi}\,\Gamma(\nu+\frac{1}{2})}\int_0^1 (1-t^2)^{\nu-1/2}\cos xt\,dt, \qquad \nu > -\tfrac{1}{2} \tag{188}$$

and show that

$$J_\nu(x) \sim \sqrt{\frac{2}{\pi x}}\cos\left(x - \tfrac{1}{2}\nu\pi - \tfrac{1}{4}\pi\right) \quad \text{as } x \to \infty \tag{189}$$

(d) Use the result in (c) to determine A and B in (b).

Since the origin is a regular singular point of (184), it possesses solutions of the Frobenius type. Substituting (52) into (184) yields

$$\sum_{n=0}^{\infty}\left[(\sigma+n)^2-\nu^2\right]a_n x^{\sigma+n}+\sum_{n=0}^{\infty}a_n x^{\sigma+n+2}=0$$

or

$$(\sigma^2-\nu^2)a_0x^\sigma+\left[(\sigma+1)^2-\nu^2\right]a_1x^{\sigma+1}$$

$$+\sum_{n=2}^{\infty}\left[(\sigma+n)^2-\nu^2\right]a_n x^{\sigma+n}+\sum_{n=0}^{\infty}a_n x^{\sigma+n+2}=0$$

or

$$(\sigma^2-\nu^2)a_0x^\sigma+\left[(\sigma+1)^2-\nu^2\right]a_1x^{\sigma+1}$$

$$+\sum_{m=0}^{\infty}\left\{\left[(\sigma+m+2)^2-\nu^2\right]a_{m+2}+a_m\right\}x^{\sigma+m+2}=0$$

Equating the coefficient of each power of x to zero, we have

$$(\sigma^2-\nu^2)a_0=0 \quad \text{or} \quad \sigma=\pm\nu \quad \text{if } a_0=0$$

$$\left[(\sigma+1)^2-\nu^2\right]a_1=0 \quad \text{or} \quad a_1=0$$

$$\left[(\sigma+m+2)^2-\nu^2\right]a_{m+2}=-a_m \tag{190}$$

In this case, the indices are $-\nu$ and ν. If ν is an integer, only one solution of the Frobenius type can be found. When $\sigma=\nu$, it follows from (190) that

$$a_{m+2}=-\frac{a_m}{(m+2)(m+2+2\nu)} \tag{191}$$

Since $a_1=0$, it follows from (191) that

$$a_{2n+1}=0 \quad \text{for } n=0,1,2\cdots$$

Then it follows from (191) that

$$a_2=-\frac{a_0}{2^2(1+\nu)}, \qquad a_4=\frac{a_0}{2^4\cdot 2!(1+\nu)(2+\nu)}$$

$$a_6=-\frac{a_0}{2^6\cdot 3!(1+\nu)(2+\nu)(3+\nu)}$$

$$a_8=\frac{a_0}{2^8\cdot 4!(1+\nu)(2+\nu)(3+\nu)(4+\nu)}$$

Hence

$$y = a_0 x^{\nu}\left[1 - \frac{(x/2)^2}{1+\nu} + \frac{(x/2)^4}{2!(\nu+1)(\nu+2)} - \frac{(x/2)^6}{3!(\nu+1)(\nu+2)(\nu+3)}\right.$$

$$\left. + \frac{(x/2)^8}{4!(\nu+1)(\nu+2)(\nu+3)(\nu+4)} - \cdots\right]$$

Letting

$$a_0 = \frac{1}{2^{\nu}\Gamma(\nu+1)}$$

we have

$$y = \sum_{n=0}^{\infty} \frac{(-1)^n (x/2)^{2n+\nu}}{n!\Gamma(\nu+1)(\nu+1)(\nu+2)\cdots(\nu+n)}$$

which yields (185) because $\Gamma(z+1) = z\Gamma(z)$.

When ν is different from an integer, it follows from (190) that

$$a_{m+2} = -\frac{a_m}{(m+2-2\nu)(m+2)} \tag{192}$$

when $\sigma = -\nu$. Since $a_1 = 0$, it follows from (192) that

$$y = a_0 x^{-\nu}\left[1 - \frac{(x/2)^2}{1-\nu} + \frac{(x/2)^4}{2!(1-\nu)(2-\nu)} - \frac{(x/2)^6}{3!(1-\nu)(2-\nu)(3-\nu)}\right.$$

$$\left. + \frac{(x/2)^8}{4!(1-\nu)(2-\nu)(3-\nu)(4-\nu)} + \cdots\right]$$

Letting

$$a_0 = \frac{2^{\nu}}{\Gamma(1-\nu)}$$

we have

$$y = \sum_{n=0}^{\infty} \frac{(-1)^n (x/2)^{2n-\nu}}{n!\Gamma(1-\nu)(1-\nu)(2-\nu)\cdots(n-\nu)}$$

which yields (186).

(b) Under the transformation $z = x^{-1}$, (184) becomes

$$\frac{d^2y}{dz^2} + \frac{1}{z}\frac{dy}{dz} + \frac{1-\nu^2 z^2}{z^4} y = 0 \tag{193}$$

Since $z = 0$ is an irregular singular point of (193), $x = \infty$ is an irregular singular point of (184). For large x, the solutions of (184) have the form (131). To determine the form of Λ, we substitute (132) into (184) and obtain

$$\lambda^2 \nu^2 x^{2\nu} + \lambda \nu x^{\nu} + x^2 + \cdots = 0$$

Balancing the dominant terms (the first and third), we have $\nu = 1$ and $\lambda = \pm i$. Hence y has the form (142) for $\lambda = \pm i$. Substituting (142) when $\lambda = i$ in (184) yields

$$x^2 v'' + (2ix^2 + x) v' + (ix - \nu^2) v = 0 \tag{194}$$

Substituting (140) into (194) yields

$$-i \sum_{n=0}^{\infty} (2\sigma + 2n - 1) a_n x^{-(\sigma+n-1)} + \sum_{n=0}^{\infty} \left[(\sigma+n)^2 - \nu^2\right] a_n x^{-(\sigma+n)} = 0$$

or

$$-i(2\sigma - 1) a_0 x^{-\sigma+1} - i \sum_{n=1}^{\infty} (2\sigma + 2n - 1) a_n x^{-(\sigma+n-1)}$$

$$+ \sum_{n=0}^{\infty} \left[(\sigma+n)^2 - \nu^2\right] a_n x^{-(\sigma+n)} = 0$$

or

$$-i(2\sigma - 1) a_0 x^{-\sigma+1}$$

$$-i \sum_{m=0}^{\infty} \left\{(2\sigma + 2m + 1) a_{m+1} + i\left[(\sigma+m)^2 - \nu^2\right] a_m\right\} x^{-(\sigma+m)} = 0$$

Equating the coefficient of each power of x to zero yields

$$(2\sigma - 1) a_0 = 0 \quad \text{or } \sigma = \tfrac{1}{2} \quad \text{if } a_0 \neq 0$$

$$a_{m+1} = -\frac{i\left[(\sigma+m)^2 - \nu^2\right]}{2\sigma + 2m + 1} a_m = -\frac{i\left[(2m+1)^2 - 4\nu^2\right]}{8(m+1)} a_m$$

Hence

$$v \sim x^{-1/2}\left[1+\frac{i(4\nu^2-1^2)}{8x}-\frac{(4\nu^2-1^2)(4\nu^2-3^2)}{2!(8x)^2}\right.$$

$$-\frac{i(4\nu^2-1^2)(4\nu^2-3^2)(4\nu^2-5^2)}{3!(8x)^3}$$

$$\left.+\frac{(4\nu^2-1^2)(4\nu^2-3^2)(4\nu^2-5^2)(4\nu^2-7^2)}{4!(8x)^4}+\cdots\right]$$

and

$$y \sim \frac{1}{\sqrt{x}}e^{ix}[U(x)+iV(x)] \tag{195}$$

where

$$U(x) \sim 1-\frac{(4\nu^2-1^2)(4\nu^2-3^2)}{2!(8x)^2}$$

$$+\frac{(4\nu^2-1^2)(4\nu^2-3^2)(4\nu^2-5^2)(4\nu^2-7^2)}{4!(8x)^4}-\cdots \tag{196}$$

$$V(x) \sim \frac{4\nu^2-1^2}{8x}-\frac{(4\nu^2-1^2)(4\nu^2-3^2)(4\nu^2-5^2)}{3!(8x)^3}+\cdots \tag{197}$$

Since (184) is real, the complex conjugate of (195) is also a solution of (184). Separating (195) into real and imaginary parts yields the following two linearly independent solutions of (184):

$$y_1 \sim \frac{1}{\sqrt{x}}[U(x)\cos x - V(x)\sin x]$$

$$y_2 \sim \frac{1}{\sqrt{x}}[U(x)\sin x + V(x)\cos x] \tag{198}$$

Then, the general solution of (184) has the form (187).

(c) To determine the asymptotic expansion of the integral in (188), we note that

$$J_\nu(x) = \frac{2(\frac{1}{2}x)^\nu}{\sqrt{\pi}\,\Gamma(\nu+\frac{1}{2})}\text{Real}(I) \tag{199}$$

where

$$I(x)=\int_0^1 (1-t^2)^{\nu-1/2} e^{ixt}\, dt \tag{200}$$

To determine the asymptotic expansion of $I(x)$ for large x, we use the method of stationary phase and deform the contour of integration into one that consists of the three segments: C_1, which runs up from 0 to iY along the η axis $(t=\xi+i\eta)$; C_2, which runs parallel to the ξ axis from iY to $1+iY$; and C_3, which runs down from $1+iY$ to 1 along a straight line parallel to the η axis. By Cauchy's theorem,

$$\begin{aligned}I(x)&=\int_{C_1+C_2+C_3}(1-t^2)^{\nu-1/2} e^{ixt}\, dt\\&=\int_0^{iY}(1-t^2)^{\nu-1/2} e^{ixt}\, dt+\int_{iY}^{1+iY}(1-t^2)^{\nu-1/2} e^{ixt}\, dt\\&\quad+\int_{1+iY}^{1}(1-t^2)^{\nu-1/2} e^{ixt}\, dt\end{aligned}$$

Letting $t=i\eta$ in the first integral, $t=\xi+iY$ in the second integral, and $t=1+i\eta$ in the third integral, we obtain

$$\begin{aligned}I(x)&=i\int_0^{Y}(1+\eta^2)^{\nu-1/2} e^{-x\eta}\, d\eta+e^{-xY}\int_0^1(1-\xi^2-2i\xi Y+Y^2)^{\nu-1/2}\\&\quad\times e^{ix\xi}\, d\xi+ie^{ix}\int_Y^0(\eta^2-2i\eta)^{\nu-1/2} e^{-x\eta}\, d\eta\end{aligned}$$

When $x>0$, as $Y\to\infty$, the second integral vanishes and

$$I(x)=i\int_0^{\infty}(1+\eta^2)^{\nu-1/2} e^{-x\eta}\, d\eta-ie^{ix}\int_0^{\infty}(\eta^2-2i\eta)^{\nu-1/2} e^{-x\eta}\, d\eta \tag{201}$$

For large positive x, the major contributions of the integrals in (201) arise from the neighborhood of the origin. Hence, using Watson's lemma, we have

$$I(x)\sim i\int_0^{\infty} e^{-x\eta}\, d\eta-ie^{ix}(-2i)^{\nu-1/2}\int_0^{\infty}\eta^{\nu-1/2}e^{-x\eta}\, d\eta \tag{202}$$

Putting $x\eta=\tau$ in (202) yields

$$I(x)\sim\frac{i}{x}\int_0^{\infty}e^{-\tau}\, d\tau+\frac{2^{\nu-1/2}}{x^{\nu+1/2}}e^{i(x-\frac{1}{2}\nu\pi-\frac{1}{4}\pi)}\int_0^{\infty}\tau^{\nu-1/2}e^{-\tau}\, d\tau$$

Therefore

$$I(x)\sim\frac{2^{\nu-1/2}\Gamma(\nu+\frac{1}{2})}{x^{\nu+1/2}}e^{i(x-\frac{1}{2}\nu\pi-\frac{1}{4}\pi)}\quad\text{as } x\to\infty \tag{203}$$

because $\nu>-\frac{1}{2}$.

Substituting (203) into (199), we have

$$J_\nu(x) \sim \sqrt{\frac{2}{\pi x}} \cos\left(x - \tfrac{1}{2}\nu\pi - \tfrac{1}{4}\pi\right) \quad \text{as } x \to \infty \tag{204}$$

As $x \to \infty$, $U(x) \to 1$ and $V(x) \to 0$, and (198) becomes

$$y_1 \sim \frac{1}{\sqrt{x}} \cos x, \qquad y_2 \sim \frac{1}{\sqrt{x}} \sin x$$

It follows from (187) that

$$J_\nu(x) \sim \frac{1}{\sqrt{x}} (A \cos x + B \sin x) \quad \text{as } x \to \infty \tag{205}$$

Comparing (204) and (205), we conclude that

$$A = \sqrt{\frac{2}{\pi}} \cos\left(\tfrac{1}{2}\nu\pi + \tfrac{1}{4}\pi\right), \qquad B = \sqrt{\frac{2}{\pi}} \sin\left(\tfrac{1}{2}\nu\pi + \tfrac{1}{4}\pi\right)$$

Therefore

$$y \sim \frac{1}{\sqrt{x}} [(A \cos x + B \sin x) U(x) - (A \sin x - B \cos x) V(x)]$$

or

$$y \sim \sqrt{\frac{2}{\pi x}} \left[U(x) \cos\left(x - \tfrac{1}{2}\nu\pi - \tfrac{1}{4}\pi\right) - V(x) \sin\left(x - \tfrac{1}{2}\nu\pi - \tfrac{1}{4}\pi\right)\right] \quad \text{as } x \to \infty$$

Supplementary Exercises

13.21. Determine two linearly independent solutions of each of the following equations:

(a) $xy'' - y' = 0$
(b) $x^2y'' + 3xy' - 3y = 0$
(c) $x^2y'' + 5xy' + 3y = 0$
(d) $8x^2y'' + 10xy' - y = 0$
(e) $x^2y'' + 3xy' + y = 0$
(f) $2x^2y'' + 7xy' - 2y = 0$
(g) $x^2y'' - 3xy' + 4y = 0$
(h) $3x^2y'' - xy' + y = 0$
(i) $16x^2y'' + 8xy' + y = 0$
(j) $x^2y'' - 2xy' + 2y = 0$
(k) $x^2y'' - xy' - 3y = 0$
(l) $x^2y'' + 5xy' + 4y = 0$
(m) $x^2y'' + xy' + y = 0$
(n) $x^2y'' + 2xy' + 4y = 0$

(o) $x^2y'' - 4xy' + 4y = 0$
(p) $x^2y'' - 2xy' + 3y = 0$
(q) $x^2y'' - 5xy' + 9y = 0$

13.22. Determine three linearly independent solutions of each of the following equations:
(a) $x^2y''' + 4xy'' + 2y' = 0$
(b) $2x^2y''' + 7xy'' + 3y' = 0$
(c) $xy''' + y'' = 0$
(d) $x^3y''' + 4x^2y'' - 2xy' - 4y = 0$
(e) $x^3y''' - x^2y'' - 2xy' + 6y = 0$
(f) $x^3y''' + 3x^2y'' - 2xy' - 2 = 0$
(g) $x^3y''' + x^2y'' - 6xy' + 6y = 0$
(h) $x^3y''' + xy' - y = 0$

13.23. Classify each of the points $x = 0, 1, -1$, and ∞ as an ordinary point, a regular singular point, or an irregular singular point for each of the following equations:
(a) $y'' + xy = 0$
(b) $xy'' + 2y' + 3y = 0$
(c) $x^2y'' + 3xy' + (2+4x)y = 0$
(d) $(1-x^2)y'' + 2xy' + 3y = 0$
(e) $x^2y'' + 2xy' + (x^2 + n^2)y = 0$
(f) $y'' - 2xy' + \lambda y = 0$
(g) $x^3y'' + xy' + y = 0$
(h) $x^3y'' - 2y = 0$
(i) $x^2y'' + (1+2x)y' + y = 0$
(j) $x^2y'' + 2xy' + (x^2 - 9)y = 0$
(k) $x^3y'' - xy' + y = 0$
(l) $x^2y'' + (2x-1)y' + 2y = 0$
(m) $x^3(x-1)^2y'' + x^4(x-1)^3y' + y = 0$

13.24. Determine two linearly independent solutions of each of the following equations near the origin and determine the radius of convergence of each series:
(a) $y'' + xy' - 4y = 0$
(b) $y'' - xy' - 2y = 0$
(c) $(1-x)y'' + y' + y = 0$
(d) $y'' + x^2y' + 2xy = 0$
(e) $(1-x^2)y'' + 2xy' + y = 0$
(f) $(1+x^2)y'' - xy' + 2y = 0$
(g) $(1-x)y'' + y = 0$
(h) $y'' - x^3y = 0$
(i) $y'' - (1-x^2)y = 0$
(j) $(x-1)y'' - xy' + y = 0$

13.25. Determine two linearly independent solutions of each of the following equations for large x:
(a) $x^4y'' + x(2x^2 - 1)y' - 3y = 0$
(b) $x^4y'' + 2x^3y' + y = 0$
(c) $x^4y'' + (2x^3 - 1)y' + 2y = 0$
(d) $x^4y'' + x^2(2x-1)y' + y = 0$

13.26. Consider the equation

$$y'' + xy' + \lambda y = 0$$

Determine two linearly independent solutions of this equation in power series near the origin and show that one of them terminates if $\lambda = -n$, where $n = 0,1,2,\ldots$.

13.27. Consider the equation

$$(1 - x^2)y'' - 3xy' + \lambda y = 0$$

Determine two linearly independent solutions of this equation in power series near the origin and show that one of them terminates if $\lambda = n(n+2)$, where $n = 0,1,2,\ldots$.

13.28. Consider the equation

$$y'' - \alpha xy' + \lambda y = 0$$

where α and λ are constants. Determine two linearly independent solutions of this equation in power series near the origin and show that one of them terminates if $\lambda = \alpha n$, where $n = 0,1,2,\ldots$.

13.29. Consider the equation

$$(1 - \alpha x^2)y'' + \beta xy' + \lambda y = 0$$

where α, β, and λ are constants. Determine two linearly independent solutions of this equation in power series near the origin and show that one of them terminates if $\lambda = n(\alpha n - \alpha - \beta)$, where $n = 0,1,2\ldots$.

13.30. Determine two linearly independent solutions in power series near the origin for each of the following equations:

(a) $2x^2y'' + xy' - (1+x)y = 0$
(b) $3x(1-x)y'' + y' + y = 0$
(c) $6x^2y'' + x(1-x)y' + y = 0$
(d) $(2x + x^2)y'' + y' - y = 0$
(e) $x^2y'' + xy' + (x^2 - \frac{1}{4})y = 0$
(f) $2x^2y'' + 3x(1-x)y' - y = 0$
(g) $2x^2y'' + x(1-x)y' - 3y = 0$
(h) $8x^2y'' + 10x(1+x)y' - y = 0$
(i) $(2x^2 + x^3)y'' + 7xy' - 3y = 0$
(j) $3x^2y'' - xy' + (1+x)y = 0$
(k) $2x^2y'' - (x + x^2)y' + y = 0$
(l) $3x^2y'' - (2x + x^2)y' - 2y = 0$
(m) $4xy'' + 2y' + y = 0$
(n) $x^2y'' + 2xy' - 2xy = 0$
(o) $2xy'' + y' + xy = 0$
(p) $xy'' + y' - 2y = 0$
(q) $x^2y'' + xy' + (3x - 2)y = 0$

13.31. Determine two linearly independent solutions in power series for large x for each of the following equations:

(a) $2x^3y'' + x(3x-1)y' - y = 0$
(b) $3x^3y'' + 5x^2y' - y = 0$
(c) $4x^3y'' + 6x^2y' - y = 0$
(d) $6x^2y'' + (11x + 1)y' + y = 0$

13.32. Consider the hypergeometric equation

$$x(1-x)y''+[\gamma-(1+\alpha+\beta)x]y'-\alpha\beta y=0$$

where α, β, and γ are constants.

(a) When $1-\gamma$ is not a positive integer, show that

$$y_1(x)=1+\frac{\alpha\beta}{\gamma\cdot 1!}x+\frac{\alpha(\alpha+1)\beta(\beta+1)}{\gamma(\gamma+1)2!}x^2+\cdots$$

$$y_2(x)=x^{1-\gamma}\left[1+\frac{(\alpha-\gamma+1)(\beta-\gamma+1)}{(2-\gamma)1!}x\right.$$

$$\left.+\frac{(\alpha-\gamma+1)(\alpha-\gamma+2)(\beta-\gamma+1)(\beta-\gamma+2)}{(2-\gamma)(3-\gamma)2!}x^2+\cdots\right]$$

are two linearly independent solutions of this equation.

(b) Show that infinity is a regular singular point of this equation. Then show that the roots of the indicial equation are α and β. When $\beta-\alpha$ is not an integer, determine two linearly independent solutions of this equation for large x.

(c) Show that near $x=1$, the roots of the indicial equation are 0 and $\gamma-\alpha-\beta$. When $\gamma-\alpha-\beta$ is not an integer, determine two linearly independent solutions of this equation near $x=1$.

13.33. Determine two linearly independent solutions in power series near the origin for each of the following equations:

(a) $x^2y''-(3x+x^2)y'+(3+x)y=0$
(b) $x^2y''-(3x+2x^2)y'+(3+4x)y=0$
(c) $x^2y''-(4x-x^2)y'+(4-x)y=0$
(d) $x^2y''-(4x+3x^2)y'+(4+6x)y=0$
(e) $x^2y''-(4x-2x^2)y'+(4-6x)y=0$
(f) $x^2y''-(3x+2x^2)y'+(3+x)y=0$
(g) $x^2y''-(3x+x^2)y'+(3+4x)y=0$
(h) $x^2y''-(4x+x^2)y'+(4-x)y=0$
(i) $x^2y''-(4x+2x^2)y'+(4-6x)y=0$
(j) $x^2y''-(4x+2x^2)y'+(4-3x)y=0$
(k) $x^2y''-(3x+x^2)y'+(3+x)y=0$
(l) $x^2y''-(3x+x^2)y'+(3+2x)y=0$

13.34. Determine two linearly independent solutions in power series for large x for each of the following equations:

(a) $x^3y''+(6x^2+2x)y'+(4x+2)y=0$
(b) $x^3y''+(6x^2+x)y'+(4x+2)y=0$
(c) $x^3y''+(6x^2+x)y'+(4x+3)y=0$
(d) $x^3y''+(6x^2+x)y'+(4x+2)y=0$
(e) $x^3y''+(6x^2-x)y'+(4x+3)y=0$
(f) $x^3y''+(6x^2-5x)y'+(4x+1)y=0$

13.35. Determine two linearly independent solutions in power series near the origin for each of the following equations:

(a) $(x^2 + x^3)y'' + 3xy' + y = 0$
(b) $x^2y'' - (x + x^2)y' + y = 0$
(c) $x^2y'' + 5xy' + (4 + x)y = 0$
(d) $(x^2 - x^3)y'' - 3xy' + 4y = 0$
(e) $x^2y'' + (7x + 2x^2)y' + 9y = 0$
(f) $x^2y'' - 5xy' + 9(1 - x)y = 0$
(g) $x^2y'' + [(2\alpha + 1)x - x^2]y' + \alpha^2 y = 0$
(h) $x^2y'' + (1 - 2\alpha)xy' + (\alpha^2 + x)y = 0$

13.36. Show that the origin is an apparent singularity of the equation

$$x^2y'' - (3x + \lambda_1 x^2)y' + (3 - \lambda_2 x)y = 0$$

when (a) $\lambda_2 = -\lambda_1$, and (b) $\lambda_2 = -2\lambda_1$.

13.37. Show that the origin is an apparent singularity of the equation

$$x^2y'' + \alpha x^2 y'' - (2 + \beta x)y = 0$$

when (a) $\beta = -\alpha$, (b) $\beta = 0$, and (c) $\beta = \alpha$.

13.38. Show that the origin is an apparent singularity of the equation

$$xy'' - (1 + \alpha x)y' + \beta y = 0$$

when (a) $\beta = 0$, and (b) $\alpha = \beta$.

13.39. Determine three linearly independent solutions in power series near the origin of each of the following equations:

(a) $4x^2y''' + 12xy'' + 3y' - y = 0$
(b) $9x^2y''' + (27x + 2x^2)y'' + 8y' - y = 0$
(c) $x^2y''' + 4xy'' + 2y' + y = 0$
(d) $2x^2y''' + 7xy'' + 3y' - y = 0$
(e) $xy''' + y'' - y = 0$
(f) $x^3y''' + 4x^2y'' - 2xy' - (4 + x)y = 0$
(g) $x^3y''' - x^2y'' - (2x + x^2)y' + 6y = 0$
(h) $x^3y''' + 3x^2y'' - 2xy' - (2 - x)y = 0$
(i) $x^3y''' + (x^2 - x^3)y'' - 6xy' + 6y = 0$
(j) $x^3y''' + (x + x^2)y' - (1 - x)y = 0$

13.40. Determine asymptotic expansions for small x of two linearly independent solutions for each of the following equations:

(a) $x^3y'' - y = 0$
(b) $x^4y'' - y = 0$
(c) $x^3y'' + y = 0$
(d) $x^4y'' + y = 0$
(e) $x^2y'' + (1 + 2x)y' + y = 0$
(f) $x^3y'' - 2xy' + y = 0$

13.41. Determine asymptotic expansions for large x of two linearly independent solutions for each of the following equations:

(a) $y'' - x^2 y = 0$
(b) $y'' + x^2 y = 0$
(c) $y'' - x^3 y = 0$
(d) $y'' + x^3 y = 0$
(e) $y'' - x^4 y = 0$
(f) $y'' + x^4 y = 0$

13.42. Determine asymptotic expansions for large x of three linearly independent solutions for each of the following equations:

(a) $y''' - xy = 0$
(b) $y''' + xy = 0$
(c) $y''' - x^2 y = 0$
(d) $y''' + x^2 y = 0$
(e) $y''' - x^3 y = 0$
(f) $y''' + x^3 y = 0$

13.43. Consider Airy's equation

$$y'' - xy = 0$$

(a) Show that two linearly independent solutions of this equation can be represented in power series near the origin as

$$y_1 = \sum_{n=0}^{\infty} \frac{x^{3n}}{9^n n! \Gamma(n+2/3)}, \qquad y_2 = \sum_{n=0}^{\infty} \frac{x^{3n+1}}{9^n n! \Gamma(n+4/3)}$$

Airy's functions $Ai(x)$ and $Bi(x)$ are expressed in terms of y_1 and y_2 as

$$Ai(x) = 3^{-2/3} y_1(x) - 3^{-4/3} y_2(x)$$

$$Bi(x) = 3^{-1/6} y_1(x) + 3^{-5/6} y_2(x)$$

(b) For large x, show that

$$y^{(1)}(x) \sim x^{-1/4} \exp\left(-\tfrac{2}{3} x^{3/2}\right) \sum_{n=0}^{\infty} (-1)^n c_n x^{-3n/2}$$

$$y^{(2)}(x) \sim x^{-1/4} \exp\left(\tfrac{2}{3} x^{3/2}\right) \sum_{n=0}^{\infty} c_n x^{-3n/2}$$

where

$$c_n = \frac{(2n+1)(2n+3) \cdots (6n-1)}{144^n n!}$$

are asymptotic expansions of two linearly independent solutions of Airy's equation.

(c) Use the integral representation

$$Ai(x) = \frac{1}{\pi} \int_0^\infty \cos\left(\tfrac{1}{3}t^3 + xt\right) dt$$

and show that

$$Ai(x) \sim \tfrac{1}{2}\pi^{-1/2} x^{-1/4} \exp\left(-\tfrac{2}{3}x^{3/2}\right) \quad \text{as } x \to \infty$$

(d) Use the integral representation

$$Bi(x) = \frac{1}{\pi} \int_0^\infty \left[\exp\left(-\tfrac{1}{3}t^3 + xt\right) + \sin\left(\tfrac{1}{3}t^3 + xt\right)\right] dt$$

and show that

$$Bi(x) \sim \pi^{-1/2} x^{-1/4} \exp\left(\tfrac{2}{3}x^{3/2}\right) \quad \text{as } x \to \infty$$

Hence relate $Bi(x)$ to $y^{(2)}(x)$.

CHAPTER 14

Differential Equations with a Large Parameter

In this chapter we consider the approximations to the solutions of second-order linear homogeneous ordinary-differential equations having the standard form

$$y'' + q(x;\lambda)y = 0, \qquad \lambda \gg 1 \tag{1}$$

Any second-order linear homogeneous ordinary-differential equation

$$u'' + P(x;\lambda)u' + Q(x;\lambda)u = 0 \tag{2}$$

can be put into the standard form using the transformation

$$u(x) = y(x)\exp\left[-\tfrac{1}{2}\int P(x;\lambda)\,dx\right] \tag{3}$$

In particular, we consider the Liouville problem

$$y'' + \left[\lambda^2 q_1(x) + q_2(x)\right]y = 0, \qquad \lambda \gg 1 \tag{4}$$

A first approximation to the solution of the Liouville problem is given by the WKB approximation

$$y \approx \frac{c_1\cos\left[\lambda\int\sqrt{q_1}\,dx\right] + c_2\sin\left[\lambda\int\sqrt{q_1}\,dx\right]}{\sqrt[4]{q_1}} \tag{5}$$

when $q_1 > 0$ and

$$y \approx \frac{c_3\exp\left[\lambda\int\sqrt{-q_1}\,dx\right] + c_4\exp\left[-\lambda\int\sqrt{-q_1}\,dx\right]}{\sqrt[4]{-q_1}} \tag{6}$$

when $q_1 < 0$, where the c_n are constants. The WKB approximation breaks down at or near the zeros of $q_1(x)$. These zeros are called turning or transition points.

If $q_1(x)$ has a simple zero at $x = \mu$, one form of the WKB approximation is valid on one side of the turning point, whereas the other form is valid on the other side. Since the Liouville problem is a second-order differential equation, c_3 and c_4 are related to c_1 and c_2. The problem of relating them is called the connection problem. One way of connecting them is to introduce the stretching transformation

$$\xi = (x - \mu)\lambda^{2/3}$$

and determine an inner expansion valid at and near the turning point. Matching the inner expansion with the two forms of the WKB approximation (5) and (6) yields the desired connection formulas. As in Chapter 12, one can form a composite expansion to provide a single uniformly valid expansion.

Alternatively, we can use the Langer transformation to provide a single uniformly valid expansion without matching. To accomplish this, we introduce the transformation

$$z = \phi(x), \qquad v(z) = \psi y(x), \qquad \psi = \sqrt{\phi'} \tag{7}$$

into the Liouville problem and obtain

$$\frac{d^2v}{dz^2} + \left[\frac{\lambda^2 q_1}{\phi'^2} + \delta\right]v = 0 \tag{8a}$$

where

$$\delta = \frac{q_2}{\phi'^2} + \frac{3\phi''^2}{4\phi'^4} - \frac{\phi'''}{2\phi'^3} \tag{8b}$$

We set $\lambda^2 q_1 = \phi'^2 \zeta(z)$ so that the transformed equation becomes

$$\frac{d^2v}{dz^2} + \zeta(z)v = -\delta v \tag{9}$$

Next, we choose the simplest possible function $\zeta(z)$ that yields a nonsingular transformation (7) for $y(x)$ in terms of $v(z)$; that is, ϕ' must be regular and have no zeros or infinities in the interval of interest. Thus $\zeta(z)$ must have the same number, type, and order of singularities and zeros as q_1. For example, if $q_1(x)$ is regular and has a simple zero at $x = \mu$, $\zeta(z)$ must be regular and have only a simple zero; so that is, $\zeta(z) = z$. Thus

$$\lambda^2 q_1 = \phi\phi'^2$$

or

$$\tfrac{2}{3}\phi^{3/2} = \lambda \int_{\mu}^{x} \sqrt{q_1(\tau)}\, d\tau \tag{10}$$

If $q_1(x)$ is regular and has a double zero at $x = \mu$, $\zeta(z)$ must be regular and have a double zero; that is, $\zeta(z) = z^2$. Thus

$$\lambda^2 q_1 = \phi^2 \phi'^2$$

or

$$\tfrac{1}{2}\phi^2 = \lambda \int_{\mu}^{x} \sqrt{q_1(\tau)}\, d\tau \tag{11}$$

Having chosen $\zeta(z)$ so that the transformation is regular, we find that δ is small. Hence it can be neglected to the first approximation and v is given by

$$\frac{d^2 v}{dz^2} + \zeta(z) v = 0 \tag{12a}$$

Solving for $v(z)$, we find that to the first approximation

$$y(x) \approx \frac{v[\phi(x)]}{\sqrt{\phi'(x)}} \tag{12b}$$

which is a single uniformly valid expansion in an interval that includes the turning point.

Solved Exercises

14.1. Consider Bessel's function of order $\frac{1}{2}$

$$x^2 y'' + xy' + \left(x^2 - \tfrac{1}{4}\right) y = 0 \tag{13}$$

Introduce a transformation to eliminate the first derivative and obtain

$$u'' + u = 0 \tag{14}$$

Then show that

$$J_{1/2}(x) = x^{-1/2}(c_1 \sin x + c_2 \cos x) \tag{15}$$

We let

$$y(x) = P(x) u(x) \tag{16a}$$

in (13) and obtain

$$x^2(P''u+2P'u'+Pu'')+x(P'u+Pu')+\left(x^2-\tfrac{1}{4}\right)Pu=0 \tag{16b}$$

Setting the coefficient of u' equal to zero yields

$$2x^2P'+xP=0$$

which, upon separating variables, becomes

$$\frac{dP}{P}=-\frac{dx}{2x}$$

Then

$$\ln P=-\tfrac{1}{2}\ln x \quad \text{or} \quad P=x^{-1/2} \tag{17a}$$

and

$$P'=-\tfrac{1}{2}x^{-3/2}, \qquad P''=\tfrac{3}{4}x^{-5/2} \tag{17b}$$

Substituting (17) into (16b) yields (14) whose solution is

$$u=c_1\sin x+c_2\cos x \tag{18}$$

Substituting (17a) and (18) into (16a) yields (15).

14.2. Consider the differential equation

$$x^2\zeta''+x\zeta'+(x^2-\nu^2)\zeta=0 \tag{19}$$

governing the cylindrical function $\zeta_\nu(x)$. Let

$$x=\gamma z^\beta, \qquad \zeta_\nu=z^{\alpha-\beta\nu}u(z) \tag{20}$$

so that

$$u=z^{\beta\nu-\alpha}\zeta_\nu(\gamma z^\beta) \tag{21}$$

Show that u satisfies the differential equation

$$z^2\frac{d^2u}{dz^2}+(2\alpha-2\beta\nu+1)z\frac{du}{dz}+\left[\beta^2\gamma^2z^{2\beta}+\alpha(\alpha-2\nu\beta)\right]u=0 \tag{22}$$

hence u can be expressed in terms of cylindrical functions as above.

It follows from (20) that

$$\frac{d\zeta}{dx}=\frac{d\zeta}{dz}\frac{dz}{dx}=\left[z^{\alpha-\beta\nu}\frac{du}{dz}+(\alpha-\beta\nu)z^{\alpha-\beta\nu-1}u\right]\frac{dz}{dx} \tag{23a}$$

Also, it follows from (20) that

$$\frac{dx}{dz} = \gamma\beta z^{\beta-1}$$

so that

$$\frac{dz}{dx} = \frac{z^{1-\beta}}{\gamma\beta}$$

and (23a) becomes

$$\frac{d\zeta}{dx} = \frac{z^{1-\beta}}{\gamma\beta}\left[z^{\alpha-\beta\nu}\frac{du}{dz} + (\alpha-\beta\nu)z^{\alpha-\beta\nu-1}u\right] \tag{23b}$$

Differentiating (23a) with respect to x gives

$$\frac{d^2\zeta}{dx^2} = \frac{d}{dz}\left[z^{\alpha-\beta\nu}\frac{du}{dz} + (\alpha-\beta\nu)z^{\alpha-\beta\nu-1}u\right]\left(\frac{dz}{dx}\right)^2$$
$$+\left[z^{\alpha-\beta\nu}\frac{du}{dz} + (\alpha-\beta\nu)z^{\alpha-\beta\nu-1}u\right]\frac{d}{dz}\left(\frac{dz}{dx}\right)\frac{dz}{dx}$$

or

$$\frac{d^2\zeta}{dx^2} = \left[z^{\alpha-\beta\nu}\frac{d^2u}{dz^2} + 2(\alpha-\beta\nu)z^{\alpha-\beta\nu-1}\frac{du}{dz}\right.$$
$$\left.+(\alpha-\beta\nu)(\alpha-\beta\nu-1)z^{\alpha-\beta\nu-2}u\right]\frac{z^{2-2\beta}}{\gamma^2\beta^2}$$
$$+\left[z^{\alpha-\beta\nu}\frac{du}{dz} + (\alpha-\beta\nu)z^{\alpha-\beta\nu-1}u\right]\frac{(1-\beta)z^{1-2\beta}}{\gamma^2\beta^2} \tag{24}$$

Substituting (20), (23b), and (24) into (19), we have

$$\frac{z^2}{\beta^2}\left[z^{\alpha-\beta\nu}\frac{d^2u}{dz^2} + 2(\alpha-\beta\nu)z^{\alpha-\beta\nu-1}\frac{du}{dz} + (\alpha-\beta\nu)(\alpha-\beta\nu-1)z^{\alpha-\beta\nu-2}u\right]$$
$$+\frac{(1-\beta)z}{\beta^2}\left[z^{\alpha-\beta\nu}\frac{du}{dz} + (\alpha-\beta\nu)z^{\alpha-\beta\nu-1}u\right]$$
$$+\frac{z}{\beta}\left[z^{\alpha-\beta\nu}\frac{du}{dz} + (\alpha-\beta\nu)z^{\alpha-\beta\nu-1}u\right] + (\gamma^2z^{2\beta}-\nu^2)z^{\alpha-\beta\nu}u = 0$$

which can be rearranged to produce (22).

14.3. Consider the Airy equation

$$\frac{d^2u}{dz^2} + zu = 0 \tag{25}$$

Use the results of the preceding exercise to express the general solution of this equation as

$$u = \sqrt{z}\left[c_1 J_{-1/3}\left(\tfrac{2}{3}z^{3/2}\right) + c_2 J_{1/3}\left(\tfrac{2}{3}z^{3/2}\right)\right] \tag{26}$$

To use the results of the preceding exercise, we multiply (25) by z^2 and obtain

$$z^2\frac{d^2u}{dz^2} + z^3u = 0 \tag{27}$$

Thus (27) has the same form as (22) with

$$2\alpha - 2\beta\nu + 1 = 0 \tag{28a}$$

$$\beta^2\gamma^2 = 1 \tag{28b}$$

$$2\beta = 3 \quad \text{or} \quad \beta = \tfrac{3}{2} \tag{28c}$$

$$\alpha(\alpha - 2\nu\beta) = 0 \tag{28d}$$

The solutions of (28) are either

$$\alpha = 0, \qquad \beta = \tfrac{3}{2}, \qquad \gamma = \tfrac{2}{3}, \qquad \nu = \tfrac{1}{3} \tag{29a}$$

or

$$\alpha = -1, \qquad \beta = \tfrac{3}{2}, \qquad \gamma = \tfrac{2}{3}, \qquad \nu = -\tfrac{1}{3} \tag{29b}$$

Substituting (29a) into (21) yields

$$u = \sqrt{z}\, J_{1/3}\left(\tfrac{2}{3}z^{3/2}\right) \tag{30a}$$

whereas substituting (29b) into (21) yields

$$u = \sqrt{z}\, J_{-1/3}\left(\tfrac{2}{3}z^{3/2}\right) \tag{30b}$$

Equations (30a) and (30b) represent two linearly independent solutions of (25). Hence (26) is its general solution.

14.4. Use the results of Exercise 14.2 to show that the general solution of

$$\frac{d^2u}{dz^2} + z^n u = 0 \tag{31a}$$

is

$$u = \sqrt{z}\left[c_1 J_\nu\left(\frac{2}{n+2} z^{(n+2)/2}\right) + c_2 J_{-\nu}\left(\frac{2}{n+2} z^{(n+2)/2}\right)\right] \tag{31b}$$

where $\nu = (n+2)^{-1}$.

To apply the results of Exercise 14.2, we multiply (31a) by z^2 and obtain

$$z^2 \frac{d^2 u}{dz^2} + z^{n+2} u = 0 \tag{32}$$

Comparing (32) and (22), we have

$$2\alpha - 2\beta\nu + 1 = 0 \tag{33}$$

$$\beta^2 \gamma^2 = 1 \tag{34}$$

$$2\beta = n + 2 \tag{35}$$

$$\alpha(\alpha - 2\nu\beta) = 0 \tag{36}$$

The solutions of (33)–(36) are either

$$\alpha = 0, \qquad \beta = \frac{n+2}{2}, \qquad \gamma = \frac{2}{n+2}, \qquad \nu = \frac{1}{n+2} \tag{37}$$

or

$$\alpha = -1, \qquad \beta = \frac{n+2}{2}, \qquad \gamma = \frac{2}{n+2}, \qquad \nu = -\frac{1}{n+2} \tag{38}$$

Substituting (37) into (21) yields

$$u = \sqrt{z}\, J_\nu\left[\frac{2}{n+2} z^{(n+2)/2}\right] \tag{39}$$

whereas substituting (38) into (21) yields

$$u = \sqrt{z}\, J_{-\nu}\left[\frac{2}{n+2} z^{(n+2)/2}\right] \tag{40}$$

where $\nu = (n+2)^{-1}$. Equations (39) and (40) represent two linearly independent solutions of (31a). Hence (31b) is its general solution.

14.5. Consider the problem

$$\varepsilon^2 y'' + (x^2 + 2x + 2) y = 0, \qquad \varepsilon \ll 1 \tag{41}$$

$$y(0) = 0, \qquad y(1) = 1 \tag{42}$$

Show that

$$\varepsilon_n = \frac{1}{n\pi}\int_0^1 \sqrt{x^2+2x+2}\, dx \tag{43}$$

Putting $q_1 = x^2+2x+2$ and $\lambda = \varepsilon^{-1}$ in (5), we have

$$y \approx \frac{c_1 \cos\left(\dfrac{1}{\varepsilon}\displaystyle\int_0^x \sqrt{\tau^2+2\tau+2}\, d\tau\right) + c_2 \sin\left(\dfrac{1}{\varepsilon}\displaystyle\int_0^x \sqrt{\tau^2+2\tau+2}\, d\tau\right)}{\sqrt[4]{x^2+2x+2}} \tag{44}$$

where the lower limit of integration is taken to be zero to facilitate satisfaction of the boundary conditions. Imposing the boundary condition $y(0)=0$ yields $c_1 = 0$, whereas imposing the boundary condition $y(1)=0$ yields

$$\sin\left(\frac{1}{\varepsilon}\int_0^1 \sqrt{x^2+2x+2}\, dx\right) = 0 \tag{45}$$

Hence

$$\frac{1}{\varepsilon}\int_0^1 \sqrt{x^2+2x+2}\, dx = n\pi, \qquad n = 1,2,\ldots$$

which yields (43).

14.6. Show that the eigenvalues of

$$u'' + \lambda^2 f(x) u = 0, \qquad f(x) > 0 \tag{46}$$

$$u(0) = 0, \qquad u(1) = 0 \tag{47}$$

are given by

$$\lambda_n = n\pi\left[\int_0^1 \sqrt{f(x)}\, dx\right]^{-1} \tag{48}$$

Putting $q_1 = f(x)$ in (5), we have

$$u \approx \frac{c_1 \cos\left(\lambda\displaystyle\int_0^x \sqrt{f(\tau)}\, d\tau\right) + c_2 \sin\left(\lambda\displaystyle\int_0^x \sqrt{f(\tau)}\, d\tau\right)}{\sqrt[4]{f(x)}} \tag{49}$$

Imposing the boundary condition $u(0)=0$ yields $c_1 = 0$, whereas imposing the boundary condition $u(1)=0$ yields

$$\sin\left(\lambda\int_0^1 \sqrt{f(x)}\, dx\right) = 0 \tag{50}$$

Hence

$$\lambda \int_0^1 \sqrt{f(x)}\, dx = n\pi, \qquad n = 1,2,3,\ldots$$

which yields (48).

14.7. Show that the large eigenvalues of (46) subject to the boundary conditions

$$u(0) = 0, \qquad u'(1) = 0 \tag{51}$$

are given by

$$\lambda_n = \left(n + \tfrac{1}{2}\right)\pi \left[\int_0^1 \sqrt{f(x)}\, dx\right]^{-1} \tag{52}$$

As in the preceding exercise, an approximate expression to the general solution of (46) is given by (49). Imposing the boundary condition $u(0) = 0$ yields $c_1 = 0$ so that (49) becomes

$$u \approx \frac{c_2 \sin\left(\lambda \int_0^x \sqrt{f(\tau)}\, d\tau\right)}{\sqrt[4]{f(x)}} \tag{53}$$

Differentiating (53) with respect to x, we have

$$u' \approx c_2 \lambda \sqrt[4]{f(x)} \cos\left(\lambda \int_0^x \sqrt{f(\tau)}\, d\tau\right) \tag{54}$$

Imposing the boundary condition $u'(1) = 0$ yields

$$\cos\left(\lambda \int_0^1 \sqrt{f(x)}\, dx\right) = 0 \tag{55}$$

Hence

$$\lambda \int_0^1 \sqrt{f(x)}\, dx = \left(n + \tfrac{1}{2}\right)\pi, \qquad n = 0,1,2,\ldots$$

which yields (52).

14.8. Consider the problem

$$y'' + \lambda^2 x^2 y = 0 \tag{56}$$

$$y(1) = 0, \qquad y(2) = 0 \tag{57}$$

Show that

$$\lambda_n = \tfrac{2}{3} n\pi \tag{58}$$

In this case, $q_1 = x^2$ and it does not vanish in the interval of interest [1,2]. Hence an approximate expression of the general solution of (56) is given by the WKB approximation (5), which in this case becomes

$$y \approx \frac{c_1 \cos\left(\lambda \int_1^x \tau\, d\tau\right) + c_2 \sin\left(\lambda \int_1^x \tau\, d\tau\right)}{\sqrt{x}} \tag{59}$$

The lower limit of integration was taken to be the left end of the interval of interest to facilitate satisfaction of the boundary conditions. Imposing the boundary condition $y(1) = 0$ yields $c_1 = 0$, whereas imposing the boundary condition $u(2) = 0$ yields

$$c_2 \sin\left(\lambda \int_1^2 \tau\, d\tau\right) = 0 \quad \text{or} \quad \sin \tfrac{3}{2}\lambda = 0$$

Hence

$$\tfrac{3}{2}\lambda = n\pi, \qquad n = 1,2,3,\ldots$$

which yields (58).

14.9. Show that the large eigenvalues of

$$u'' + \lambda^2 f(x) u = 0, \qquad f(x) > 0 \tag{60}$$

$$u'(a) = 0, \qquad u(b) = 0, \qquad b > a \tag{61}$$

are given by

$$\lambda_n = \left(n + \tfrac{1}{2}\right)\pi \left[\int_a^b \sqrt{f(x)}\, dx\right]^{-1} \tag{62}$$

Putting $q_1 = f(x)$ and taking the lower limit of integration to be a (i.e., the left end of the interval), we rewrite (5) as

$$u \approx \frac{c_1 \cos\left(\lambda \int_a^x \sqrt{f(\tau)}\, d\tau\right) + c_2 \sin\left(\lambda \int_a^x \sqrt{f(\tau)}\, d\tau\right)}{\sqrt[4]{f(x)}} \tag{63}$$

Differentiating (63) with respect to x, we have

$$u' \approx \lambda \sqrt[4]{f(x)} \left[-c_1 \sin\left(\lambda \int_a^x \sqrt{f(\tau)}\, d\tau\right) + c_2 \cos\left(\lambda \int_a^x \sqrt{f(\tau)}\, d\tau\right)\right] \tag{64}$$

Imposing the boundary condition $u'(a) = 0$ yields $c_2 = 0$. Then, imposing the boundary condition $u(b) = 0$ yields

$$c_1 \cos\left(\lambda \int_a^b \sqrt{f(x)}\, dx\right) = 0 \tag{65}$$

Hence

$$\lambda \int_a^b \sqrt{f(x)}\,dx = \left(n + \tfrac{1}{2}\right)\pi, \qquad n = 0,1,2,\ldots$$

which yields (62).

14.10. Consider the problem

$$y'' + \lambda^2(1 - x^2)\,y = 0 \tag{66}$$

$$y'(0) = 0, \qquad y(1) = 0 \tag{67}$$

Show that the eigenvalues are given by

$$\lambda_n = 4\left(n + \tfrac{5}{12}\right) \tag{68}$$

In this case, $q_1 = 1 - x^2$ has a simple zero at $x = 1$ in the interval of interest. Thus (66) has a turning point of order one. Using the transformation (7) we transform (66) into (8a). We put

$$\lambda^2(1 - x^2) = \phi'^2\phi \tag{69}$$

so that (8a) becomes

$$\frac{d^2v}{dz^2} + zv = -\,\delta v \tag{70}$$

It follows from (69) that

$$\tfrac{2}{3}z^{3/2} = \tfrac{2}{3}\phi^{3/2} = \lambda \int_x^1 \sqrt{1 - \tau^2}\,d\tau \tag{71}$$

where the upper limit of integration is taken to correspond to the location of the turning point so that z will be positive when $0 \le x \le 1$. Since the interval of interest is bounded, we represent the solution of the dominant part of (70), which can be obtained by dropping δv, in terms of Bessel's functions as

$$v = \sqrt{z}\left[\tilde{c}_1 J_{-1/3}\left(\tfrac{2}{3}z^{3/2}\right) + \tilde{c}_2 J_{1/3}\left(\tfrac{2}{3}z^{3/2}\right)\right] \tag{72}$$

according to Exercise 14.4. Since $\psi = \sqrt{\phi'}$, it follows from (71) that

$$\psi = \sqrt{\lambda}\,\sqrt[4]{\frac{1 - x^2}{z}} \tag{73}$$

Substituting (71)–(73) into (7), we have

$$y \approx \left[\int_x^1 \sqrt{1 - \tau^2}\,d\tau\right]^{1/2} (1 - x^2)^{-1/4}$$

$$\cdot\left\{c_1 J_{-1/3}\left[\lambda \int_x^1 \sqrt{1 - \tau^2}\,d\tau\right] + c_2 J_{1/3}\left[\lambda \int_x^1 \sqrt{1 - \tau^2}\,d\tau\right]\right\} \tag{74}$$

To impose the boundary condition at $x=1$, we need the limit of y as $x \to 1$ in (74). It follows from (71) that as $x \to 1$

$$\tfrac{2}{3}z^{3/2} \to \lambda\sqrt{2}\int_x^1 (1-\tau)^{1/2}\,d\tau = \tfrac{2}{3}\lambda\sqrt{2}\,(1-x)^{3/2} \tag{75}$$

so that

$$z = O\left[\lambda^{2/3}(1-x)\right] \quad \text{as } x \to 1 \tag{76}$$

Therefore, as $x \to 1$

$$y = O\left[(1-x)^{3/4}(1-x)^{-1/4}\right]\left\{c_1 O\left[\lambda^{-1/3}(1-x)^{-1/2}\right] + c_2 O\left[\lambda^{1/3}(1-x)^{1/2}\right]\right\} \tag{77}$$

because $J_\nu(\alpha) = O(\alpha^\nu)$ as $\alpha \to 0$. Hence as $x \to 1$

$$y = c_1 O(1) + c_2 O(1-x) \tag{78}$$

and the boundary condition $y(1)=0$ demands that $c_1 = 0$. Imposing the boundary condition $y'(0)=0$ in (74) yields

$$J'_{1/3}\left[\lambda\int_0^1 \sqrt{1-\tau^2}\,d\tau\right] = 0 \tag{79}$$

But

$$J_{1/3}(\alpha) \sim \sqrt{\frac{2}{\pi\alpha}}\cos\left(\alpha - \tfrac{5}{12}\pi\right) \quad \text{as } \alpha \to \infty$$

hence it follows from (79) that

$$\sin\left[\lambda\int_0^1 \sqrt{1-\tau^2}\,d\tau - \tfrac{5}{12}\pi\right] = 0$$

or

$$\lambda\int_0^1 \sqrt{1-\tau^2}\,d\tau - \tfrac{5}{12}\pi = n\pi, \qquad n = 0,1,2,\ldots \tag{80}$$

Since

$$\int_0^1 \sqrt{1-\tau^2}\,d\tau = \int_0^{\pi/2} \cos\theta\cdot\cos\theta\,d\theta = \tfrac{1}{4}\pi$$

it follows from (80) that

$$\tfrac{1}{4}\pi\lambda = \left(n + \tfrac{5}{12}\right)\pi$$

which yields (68).

14.11. Apply the WKB method directly to

$$xy'' + y' + \lambda^2 x(1 - x^2)\, y = 0, \qquad \lambda \gg 1 \tag{81}$$

Indicate the validity of the resulting expansion.

We seek an expansion of the solution of (81) in the form

$$y = e^{\lambda G(x;\,\lambda)} \tag{82}$$

where G has a straightforward expansion in inverse powers of λ. It follows from (82) that

$$y' = \lambda G' e^{\lambda G}, \qquad y'' = (\lambda^2 G'^2 + \lambda G'')\, e^{\lambda G}$$

Substituting for y, y', and y'' in (81) yields

$$\lambda^2 x G'^2 + \lambda x G'' + \lambda G' + \lambda^2 x(1 - x^2) = 0$$

or

$$G'^2 + (1 - x^2) + \frac{1}{\lambda} G'' + \frac{1}{x\lambda} G' = 0 \tag{83}$$

We seek an expansion of G in the form

$$G(x;\lambda) = G_0(x) + \frac{1}{\lambda} G_1(x) + \cdots \tag{84}$$

Substituting (84) into (83) and equating the coefficients of λ^0 and λ^{-1} to zero, we obtain

$$G_0'^2 + 1 - x^2 = 0 \tag{85}$$

$$2G_0'G_1' + G_0'' + \frac{1}{x} G_0' = 0 \tag{86}$$

It follows from (85) that

$$G_0' = \pm i \int_x^1 \sqrt{1 - \tau^2}\, d\tau \tag{87}$$

Separating variables in (86) yields

$$G_1' + \frac{G_0''}{2G_0'} + \frac{1}{2x} = 0$$

which, upon integration, gives

$$G_1 + \tfrac{1}{2} \ln(xG_0') = \ln c$$

where c is a constant. Hence

$$y \approx \exp\left[\pm i\lambda \int_x^1 \sqrt{1-\tau^2}\, d\tau + \ln c - \tfrac{1}{2}\ln\left(xG_0'\right)\right]$$

$$\approx \frac{c}{\sqrt{xG_0'}} \exp\left[\pm i\lambda \int_x^1 \sqrt{1-\tau^2}\, d\tau\right]$$

Therefore, the general solution of (81) can be approximated by

$$y \approx \frac{1}{\sqrt[4]{x^2(1-x^2)}}\left[c_1 \cos\left(\lambda \int_x^1 \sqrt{1-\tau^2}\, d\tau\right) + c_2 \sin\left(\lambda \int_x^1 \sqrt{1-\tau^2}\, d\tau\right)\right] \tag{88}$$

where c_1 and c_2 are constants. Inspection of (88) shows that it is invalid when x is close to 0, 1, or -1.

14.12. Consider the problem

$$y'' + \lambda^2(1-x^2)f(x)y = 0 \tag{89}$$

$$y(0) = 0, \qquad y(1) = 0 \tag{90}$$

where $f(x) = f(-x) > 0$ in $[0,1]$. This problem describes heat transfer in a two-dimensional duct carrying fully developed turbulent flow. Show that

$$\lambda_n = \left(n + \tfrac{11}{12}\right)\pi \left[\int_0^1 \sqrt{(1-\tau^2)f(\tau)}\, d\tau\right]^{-1} \tag{91}$$

In this case, $q_1 = (1-x^2)f(x)$ has a simple zero at $x = 1$ in the interval of interest. Thus (89) has a turning point of order one. Using the transformation (7) we transform (89) into (8a). We put

$$\lambda^2(1-x^2)f(x) = \phi'^2\phi \tag{92}$$

so that (8a) becomes

$$\frac{d^2v}{dz^2} + zv = -\delta v \tag{93}$$

It follows from (92) that

$$\tfrac{2}{3}z^{3/2} = \tfrac{2}{3}\phi^{3/2} = \lambda \int_x^1 \sqrt{(1-\tau^2)f(\tau)}\, d\tau \tag{94}$$

where the upper limit of integration is taken to correspond to the location of the turning point so that z will be positive when $0 \le x \le 1$. Since the interval of interest is bounded, we represent the solution of the dominant part of (93), which can be obtained by dropping δv, in terms of Bessel's functions, as in (72). Since $\psi = \sqrt{\phi'}$, it follows from (94) that

$$\psi = \sqrt{\lambda}\left[(1-x^2)f(x)\right]^{1/4} z^{-1/4} \tag{95}$$

Substituting (72), (94), and (95) into (7), we obtain

$$y \approx \left[\int_x^1 \sqrt{(1-\tau^2)f(\tau)}\, d\tau\right]^{1/2} \left[(1-x^2)f(x)\right]^{-1/4}$$

$$\cdot\left\{ c_1 J_{-1/3}\left[\lambda \int_x^1 \sqrt{(1-\tau^2)f(\tau)}\, d\tau\right] + c_2 J_{1/3}\left[\lambda \int_x^1 \sqrt{(1-\tau^2)f(\tau)}\, d\tau\right]\right\} \quad (96)$$

To impose the boundary condition $x = 1$, we need the limit of y as $x \to 1$ in (96). It follows from (94) that as $x \to 1$

$$\tfrac{2}{3} z^{3/2} \to \lambda\sqrt{2f(1)} \int_x^1 (1-\tau)^{1/2}\, d\tau = \tfrac{2}{3}\lambda\sqrt{2f(1)}\,(1-x)^{3/2} \quad (97)$$

so that

$$z = O\left[\lambda^{2/3}(1-x)\right] \quad \text{as } x \to 1 \quad (98)$$

Therefore, as $x \to 1$

$$y = O\left[(1-x)^{3/4}(1-x)^{-1/4}\right]\left\{ c_1 O\left[\lambda^{-1/3}(1-x)^{-1/2}\right] + c_2 O\left[\lambda^{1/3}(1-x)^{1/2}\right]\right\} \quad (99)$$

where use has been made of $J_\nu(\alpha) = O(\alpha^\nu)$ as $\alpha \to 0$. As $x \to 1$

$$y = c_1 O(1) + c_2 O(1-x) \quad (100)$$

and the boundary condition $y(1) = 0$ demands that $c_1 = 0$. Imposing the boundary condition $y(0) = 0$ in (96) yields

$$J_{1/3}\left[\lambda \int_0^1 \sqrt{(1-\tau^2)f(\tau)}\, d\tau\right] = 0 \quad (101)$$

But

$$J_{1/3}(\alpha) \sim \sqrt{\frac{2}{\pi\alpha}} \cos\left(\alpha - \tfrac{5}{12}\pi\right) \quad \text{as } \alpha \to \infty$$

hence it follows from (101) that

$$\cos\left(\lambda \int_0^1 \sqrt{(1-\tau^2)f(\tau)}\, d\tau - \tfrac{5}{12}\pi\right) = 0$$

or

$$\lambda \int_0^1 \sqrt{(1-\tau^2)f(\tau)}\, d\tau - \tfrac{5}{12}\pi = \left(n + \tfrac{1}{2}\right)\pi, \qquad n = 0,1,2,\ldots \quad (102)$$

which yields (91).

14.13. Consider the problem in the preceding exercise, but with the boundary conditions

$$y'(0) = 0, \qquad y(1) = 0 \tag{103}$$

Show that

$$\lambda_n = \left(n + \tfrac{5}{12}\right)\pi\left(\int_0^1 \sqrt{(1-\tau^2)f(\tau)}\, d\tau\right)^{-1} \tag{104}$$

The steps up to (100) in the preceding exercise are valid for this problem, including the condition $c_1 = 0$ demanded by the boundary condition $y(1) = 0$. Then (96) becomes

$$y \approx c_2\left(\int_x^1 \sqrt{(1-\tau^2)f(\tau)}\, d\tau\right)^{1/2}\left[(1-x^2)f(x)\right]^{-1/4} J_{1/3}\left(\lambda\int_x^1 \sqrt{(1-\tau^2)f(\tau)}\, d\tau\right) \tag{105}$$

Differentiating (105) with respect to x, we have

$$y' \approx -c_2\lambda\left[\int_x^1 \sqrt{(1-\tau^2)f(\tau)}\, d\tau\right]^{1/2}\left[(1-x^2)f(x)\right]^{1/4} J'_{1/3}\left[\lambda\int_x^1 \sqrt{(1-\tau^2)f(\tau)}\, d\tau\right]$$

Imposing the condition $y'(0) = 0$ demands that

$$J'_{1/3}\left(\lambda\int_0^1 \sqrt{(1-\tau^2)f(\tau)}\, d\tau\right) = 0 \tag{106}$$

But

$$J'_{1/3}(\alpha) \sim -\sqrt{\frac{2}{\pi\alpha}}\,\sin\left(\alpha - \tfrac{5}{12}\pi\right) \quad \text{as } \alpha \to \infty$$

Hence it follows from (106) that

$$\sin\left(\lambda\int_0^1 \sqrt{(1-\tau^2)f(\tau)}\, d\tau - \tfrac{5}{12}\pi\right) = 0$$

or

$$\lambda\int_0^1 \sqrt{(1-\tau^2)f(\tau)}\, d\tau - \tfrac{5}{12}\pi = n\pi, \qquad n = 1,2,3,\ldots$$

which yields (104).

14.14. Consider the equation

$$y'' + \lambda^2(1-x^2)^2 y = 0 \tag{107}$$

Show that as $\lambda \to \infty$,

$$y \sim H^{1/2}(1-x^2)^{-1/2}\left[c_1 J_{1/4}(\lambda H) + c_2 J_{-1/4}(\lambda H)\right] \tag{108}$$

for $x > -1$ where

$$H = \int_1^x (1-\tau^2)\, d\tau = (x-1) - \tfrac{1}{3}(x^3 - 1) \tag{109}$$

In this case, $q_1 = (1-x^2)^2$ has a zero of order two at $x = 1$ in the interval $x > -1$. Thus (107) has a turning point of order two at $x = 1$. Using the transformation (7) we transform (107) into (8a). We put

$$\lambda^2(1-x^2)^2 = \phi'^2\phi^2 \tag{110}$$

so that (8a) becomes

$$\frac{d^2v}{dz^2} + z^2 v = -\delta v \tag{111}$$

It follows from (110) that

$$\tfrac{1}{2}z^2 = \tfrac{1}{2}\phi^2 = \lambda \int_1^x (1-\tau^2)\, d\tau = \lambda H \tag{112}$$

where H is defined in (109). As $x \to 1$, it follows from (112) that

$$\phi = O\left[\sqrt{\lambda}\,(x-1)\right] \quad \text{as } \lambda \to \infty$$

and from (7) that

$$\psi = \sqrt{\phi'} = O(\lambda^{1/4}) \quad \text{as } \lambda \to \infty$$

Substituting for ϕ and ψ in (8b) yields

$$\delta = O(\lambda^{-1}) \quad \text{as } \lambda \to \infty$$

Therefore, the dominant part of (111) is

$$\frac{d^2v}{dz^2} + z^2 v = 0 \tag{113}$$

whose solution is

$$v = \sqrt{z}\left[\tilde{c}_1 J_{1/4}\left(\tfrac{1}{2}z^2\right) + \tilde{c}_2 J_{-1/4}\left(\tfrac{1}{2}z^2\right)\right] \tag{114}$$

according to Exercise 14.4. Since $\psi = \sqrt{\phi'}$, it follows from (110) that

$$\psi = \sqrt{\frac{\lambda(1-x^2)}{\phi}} \tag{115}$$

Substituting (112), (114), and (115) into (7) yields (108).

14.15. Consider the equation

$$y'' + \lambda^2 q(x) y = 0 \tag{116}$$

where $q(\mu) = q'(\mu) = 0$ but $q''(\mu) \neq 0$. Show that as $\lambda \to \infty$,

$$y \sim H^{1/2} q^{-1/4} \left[c_1 J_{1/4}(\lambda H) + c_2 J_{-1/4}(\lambda H) \right] \tag{117}$$

where

$$H = \int_\mu^x \sqrt{q}\, dx \tag{118}$$

In this case (116) has a turning point of order two at $x = \mu$ because $q(x)$ has a zero of order two there. Using the transformation (7) we transform (116) into (8a). We put

$$\lambda^2 q(x) = \phi'^2 \phi^2 \tag{119}$$

so that (116) becomes (111). The solution of (119) can be expressed as

$$\tfrac{1}{2} z^2 = \tfrac{1}{2} \phi^2 = \lambda \int_\mu^x \sqrt{q(\tau)}\, d\tau = \lambda H \tag{120}$$

It follows from (7) and (119) that

$$\psi = \sqrt{\phi'} = \sqrt[4]{\frac{\lambda^2 q(x)}{\phi^2(x)}} \tag{121}$$

As $x \to \mu$, it follows from (120) that

$$\phi = O\left[\lambda^{1/2} (x - \mu) \right] \quad \text{as } \lambda \to \infty$$

Substituting for ϕ and ψ in (8b) yields

$$\delta = O(\lambda^{-1}) \quad \text{as } \lambda \to \infty$$

Therefore, the dominant part of (111) is given by (113), whose solution can be expressed as in (114). Substituting (114), (120), and (121) into (7) yields (117).

14.16. Consider

$$\varepsilon y'' + (2x+1) y' + 2y = 0, \qquad 0 < \varepsilon \ll 1 \tag{122}$$

Introduce a transformation to eliminate the middle term and then apply the WKB approximation to the resulting equation.

We let

$$y(x) = P(x) u(x) \tag{123}$$

in (122) and obtain

$$\varepsilon P u'' + \left[2\varepsilon P' + (2x+1) P \right] u' + \left[\varepsilon P'' + (2x+1) P' + 2P \right] u = 0 \tag{124}$$

Setting the coefficient of u' equal to zero, we have

$$2\varepsilon P' + (2x+1)P = 0 \tag{125}$$

Then (124) becomes

$$\varepsilon P u'' + \left[\varepsilon P'' + (2x+1)P' + 2P\right] u = 0 \tag{126}$$

Using separation of variables, we find that (125) has the solution

$$P = e^{-(x^2+x)/2\varepsilon} \tag{127}$$

and (126) becomes

$$u'' - \left[\frac{(2x+1)^2}{4\varepsilon^2} - \frac{1}{\varepsilon}\right] u = 0 \tag{128}$$

We seek an approximate solution of (128) in the form

$$u = e^{G(x,\varepsilon)/\varepsilon} \tag{129}$$

Substituting (129) into (128) yields

$$G'^2 - \tfrac{1}{4}(2x+1)^2 + \varepsilon(G''+1) = 0 \tag{130}$$

Next we see a straightforward expansion of G in the form

$$G(x;\varepsilon) = G_0(x) + \varepsilon G_1(x) + \cdots \tag{131}$$

Substituting (131) into (130) and equating each of the coefficients of ε^0 and ε to zero, we have

$$G_0'^2 = \tfrac{1}{4}(2x+1)^2 \tag{132}$$

$$2G_0'G_1' + G_0'' + 1 = 0 \tag{133}$$

It follows from (132) that

$$G_0 = \pm\tfrac{1}{2}(x^2+x) \tag{134}$$

Then (133) yields

$$G_1 = -\ln(2x+1) \quad \text{when } G_0 = \tfrac{1}{2}(x^2+x) \tag{135}$$

and

$$G_1 = 1 \quad \text{when } G_0 = -\tfrac{1}{2}(x^2+x) \tag{136}$$

Therefore, the general solution of (128) can be approximated by

$$u = \frac{c_1}{2x+1} e^{(x^2+x)/2\varepsilon} + c_2 e^{-(x^2+x)/2\varepsilon} + \cdots \tag{137}$$

where c_1 and c_2 are constants. Substituting (127) and (137) into (123) yields

$$y = \frac{c_1}{2x+1} + c_2 e^{-(x^2+x)/\varepsilon} + \cdots \tag{138}$$

where c_1 and c_2 are constants.

14.17. Consider

$$\varepsilon y'' + (2x+1) y' + 2y = 0 \quad \text{for } 0 < \varepsilon \ll 1 \tag{139}$$

Seek a solution in the form $y = \exp[\varepsilon^{-1} G(x; \varepsilon)]$ and determine two terms in the expansion of G.

In terms of G, (139) becomes

$$G'^2 + (2x+1) G' + \varepsilon (G'' + 2) = 0 \tag{140}$$

Substituting (131) into (140) and equating coefficients of like powers of ε, we obtain

$$G_0'^2 + (2x+1) G_0' = 0 \tag{141}$$

$$2G_0'G_1' + (2x+1) G_1' + G_0'' + 2 = 0 \tag{142}$$

Equation (141) has the following solutions:

$$G_0 = 1 \quad \text{and} \quad G_0 = -(x^2 + x) \tag{143}$$

The solutions of (142) can be expressed as

$$G_1 = -\ln(2x+1) \quad \text{when } G_0 = 1 \tag{144}$$

and

$$G_1 = 1 \quad \text{when } G_0 = -(x^2 + x) \tag{145}$$

Therefore

$$G = 1 - \varepsilon \ln(2x+1) + \cdots$$

or

$$G = -(x^2 + x) + \varepsilon + \cdots$$

Since $y = \exp(G/\varepsilon)$, the general solution of (139) can be expressed as in (138).

Supplementary Exercises

14.18. Show that $x^2 J_2(x)$ is a solution of

$$xy'' - 3y' + xy = 0$$

14.19. Show that $xJ_n(x)$ is a solution of

$$x^2 y'' - xy' + (x^2 + 1 - n^2)\, y = 0$$

14.20. Show that $x^{3/2} J_{3/2}(x)$ is a solution of

$$xy'' - 2y' + xy = 0$$

14.21. Show that a solution of

$$x^2 y'' + (\beta^2 \gamma^2 x^{2\beta} + \tfrac{1}{4} - \nu^2 \beta^2)\, y = 0$$

is given by $y = \sqrt{x}\, J_\nu(\gamma x^\beta)$.

14.22. Show that the solutions of

$$y'' + \left(\lambda^2 x + \frac{1}{4x^2} \right) y = 0$$

are related to the cylindrical function ζ_0 by

$$y = \sqrt{x}\, \zeta_0(\tfrac{2}{3} \lambda x^{3/2})$$

14.23. Show that the solutions of

$$y'' + \left(\lambda^2 x - \frac{2}{x^2} \right) y = 0$$

are related to the cylindrical function ζ_1 by

$$y = \sqrt{x}\, \zeta_1(\tfrac{2}{3} \lambda x^{3/2})$$

14.24. Show that the solutions of

$$y'' + \left(\lambda^2 x^\alpha + \frac{r_0}{x^2} \right) y = 0$$

are related to the cylindrical function ζ_ν by

$$y = \sqrt{x}\, \zeta_\nu(\gamma x^\beta)$$

where

$$\beta = \frac{\alpha + 2}{2}, \qquad \beta\gamma = \lambda, \qquad \nu = \frac{\sqrt{1 - 4r_0}}{2 + \alpha}$$

14.25. Use the WKB method to show that the leading behaviors of the solutions of

$$y'' = xy$$

for large x are

$$y_1 \sim x^{-1/4}\exp\left(-\tfrac{2}{3}x^{3/2}\right), \qquad y_2 \sim x^{-1/4}\exp\left(\tfrac{2}{3}x^{3/2}\right)$$

14.26. Use the WKB method to show that the leading behaviors of the solutions of the parabolic cylinder equation

$$y'' + \left(\nu + \tfrac{1}{2} - \tfrac{1}{4}x^2\right)y = 0$$

for large x are

$$y_1 \sim x^{-\nu-1}e^{\frac{1}{4}x^2}, \qquad y_2 \sim x^{\nu}e^{-\frac{1}{4}x^2}$$

14.27. Use the WKB method to show that the leading behaviors of the solutions of

$$y''' - x^3y = 0$$

for large x are

$$y \sim x^{-1}e^{\omega x^2/2}$$

where $\omega = \frac{2}{3}n\pi$, $n = 0, 1$, and 2.

14.28. Consider the equation

$$x^2y'' + axy' + (c + dx^{2s})\,y = 0$$

Use Exercise 14.2 to show that its solutions are given by

$$y = x^{(1-a)/2}\zeta_\nu\left(\frac{\sqrt{d}}{s}x^s\right)$$

where

$$\nu = \frac{\sqrt{(1-a)^2 - 4c}}{2s}$$

14.29. Consider the differential equation

$$x^2\zeta'' + x\zeta' + (x^2 - \nu^2)\zeta = 0$$

governing the cylindrical function $\zeta_\nu(x)$. Let

$$x = \gamma z^{\beta}, \qquad \zeta_\nu = u(z)\,z^{\alpha - \beta\nu}$$

Then $u(z)$ is governed by

$$z^2 u'' + (2\alpha - 2\beta\nu + 1) zu' + \left[\beta^2\gamma^2 z^{2\beta} + \alpha(\alpha - 2\beta\nu) \right] u = 0$$

according to Exercise 14.2. Let $u(z) = v(z)\exp(\delta z^{\lambda})$ and then show that $v(z)$ is governed by

$$z^2 v'' + (2\alpha - 2\beta\nu + 1 + 2\delta\lambda z^{\lambda}) zv'$$

$$+ \left[\alpha(\alpha - 2\beta\nu) + \beta^2\gamma^2 z^{2\beta} + \delta\lambda(2\alpha - 2\beta\nu + \lambda) z^{\lambda} + \delta^2\lambda^2 z^{2\lambda} \right] v = 0$$

Hence

$$v(z) = \zeta_\nu(\gamma z^{\beta}) z^{\beta\nu - \alpha} \exp(-\delta z^{\lambda})$$

14.30. Use the preceding exercise to show that $\zeta_0(x)e^{-x}$ is a solution of

$$xy'' + (2x + 1)(y' + y) = 0$$

14.31. Use Exercise 14.29 to express the solutions of

$$x^2 y'' + (a + 2bx^r) xy' + \left[c + dx^{2s} - b(1 - a - r) x^r + b^2 x^{2r} \right] y = 0$$

in terms of ζ_ν, where

$$\nu = \frac{\sqrt{(1-a)^2 - 4c}}{2s}$$

14.32. Use the WKB approximation to show that the general solution of

$$y'' + \lambda^2(1+x)^4 y = 0, \qquad \lambda \gg 1 \quad \text{and} \quad x > -1$$

is

$$y \approx \frac{c_1 \cos\left[\frac{1}{3}\lambda(1+x)^3\right] + c_2 \sin\left[\frac{1}{3}\lambda(1+x)^3\right]}{1+x}$$

14.33. Use the WKB approximation to show that the general solution of

$$y'' - \lambda^2(1+x)^4 y = 0, \qquad \lambda \gg 1 \quad \text{and} \quad x > -1$$

is

$$y \sim \frac{c_1 \exp\left[\frac{1}{3}\lambda(1+x)^3\right] + c_2 \exp\left[-\frac{1}{3}\lambda(1+x)^3\right]}{1+x}$$

14.34. Show that the eigenvalues of

$$y'' + \lambda^2(1+x)^4 y = 0, \qquad \lambda \gg 1$$

$$y(0) = 0, \qquad y(1) = 0$$

are given by

$$\lambda_n = \tfrac{3}{7} n\pi$$

14.35. Show that the eigenvalues of

$$y'' + \lambda^2(1+x)^{2\alpha} y = 0, \qquad \lambda \gg 1$$

$$y(0) = 0, \qquad y(1) = 0$$

are given by

$$\lambda_n = \frac{(\alpha+1) n\pi}{2^{\alpha+1} - 1}$$

14.36. Show that the eigenvalues of

$$y'' + \lambda^2(1+x)^4 y = 0, \qquad \lambda \gg 1$$

$$y'(0) = 0, \qquad y(1) = 0$$

are given by

$$\lambda_n = \tfrac{3}{7}\left(n + \tfrac{1}{2}\right)\pi$$

14.37. Show that the eigenvalues of

$$y'' + \lambda^2(1+x)^{2\alpha} y = 0, \qquad \lambda \gg 1$$

$$y'(0) = 0, \qquad y(1) = 0$$

are given by

$$\lambda_n = \frac{(\alpha+1)\left(n + \frac{1}{2}\right)\pi}{2^{\alpha+1} - 1}$$

14.38. Show that the large eigenvalues of

$$y'' + \lambda^2 x^{2\alpha} y = 0$$

$$y(1) = 0, \qquad y(2) = 0$$

are given by

$$\lambda_n = \frac{(\alpha+1)n\pi}{2^{\alpha+1}-1}$$

14.39. Show that the large eigenvalues of

$$y'' + \lambda^2 x^{2\alpha} y = 0$$

$$y'(1) = 0, \qquad y(2) = 0$$

are given by

$$\lambda_n = \frac{(\alpha+1)(n+\frac{1}{2})\pi}{2^{\alpha+1}-1}$$

14.40. Use the WKB method to show that, to the first approximation, the solution of

$$\varepsilon y'' - (2x+1)y' + 2y = 0, \qquad 0 < \varepsilon \ll 1$$

$$y(0) = \alpha, \qquad y(1) = \beta$$

is

$$y = \alpha(2x+1) + \frac{9(\beta - 3\alpha)}{(2x+1)^2} e^{-(2-x-x^2)/\varepsilon} + \cdots$$

14.41. Use the WKB method to show that, to the first approximation, the solution of

$$\varepsilon y'' - (3x^2+2x+1)y' + 2(3x+1)y = 0, \qquad 0 < \varepsilon \ll 1$$

$$y(0) = \alpha, \qquad y(1) = \beta$$

is

$$y = \alpha(3x^2+2x+1) + \frac{36(\beta - 6\alpha)}{(3x^2+2x+1)^2} e^{-(3-x-x^2-x^3)/\varepsilon} + \cdots$$

14.42. Use the WKB method to show that, to the first approximation, the solution of

$$\varepsilon y'' + (3x^2+2x+1)y' + 2(3x+1)y = 0, \qquad 0 < \varepsilon \ll 1$$

$$y(0) = \alpha, \qquad y(1) = \beta$$

is

$$y = \frac{6\beta}{3x^2+2x+1} + (\alpha - 6\beta)e^{-(x^3+x^2+x)/\varepsilon} + \cdots$$

14.43. Use the WKB method to show that, to the first approximation, the solution of

$$\varepsilon y'' + a(x)\, y' + b(x)\, y = 0, \qquad 0 < \varepsilon \ll 1$$

$$y(0) = \alpha, \qquad y(1) = \beta$$

where $a(x) > 0$ in $[0,1]$, is

$$y = \beta \exp\left[-\int_1^x \frac{b(t)}{a(t)}\, dt \right] + \frac{a(0)\left\{ \alpha - \beta \exp\left[-\int_1^0 \frac{b(t)}{a(t)}\, dt \right] \right\}}{a(x)}$$

$$\cdot \exp\left[-\frac{1}{\varepsilon} \int_0^x a(t)\, dt + \int_0^x \frac{b(t)}{a(t)}\, dt \right] + \cdots$$

14.44. Use the WKB method to show that, to the first approximation, the solution of

$$\varepsilon y'' - a(x)\, y' + b(x)\, y = 0, \qquad 0 < \varepsilon \ll 1$$

$$y(0) = \alpha, \qquad y(1) = \beta$$

where $a(x) > 0$ in $[0,1]$, is

$$y = \alpha \exp\left[\int_0^x \frac{b(t)}{a(t)}\, dt \right] + \frac{a(x)\left\{ \beta - \alpha \exp\left[\int_0^1 b(t)/a(t)\, dt \right] \right\}}{a(1)}$$

$$\cdot \exp\left[-\frac{1}{\varepsilon} \int_x^1 a(t)\, dt + \int_x^1 \frac{b(t)}{a(t)}\, dt \right] + \cdots$$

14.45. Use the WKB method to show that, to the first approximation, the general solution of

$$\varepsilon^2 y'' + \varepsilon(3x+2)\, y' + (2x^2+3x+1)\, y = 0, \qquad 0 < \varepsilon \ll 1$$

is

$$y = c_1 x \exp\left[-\frac{1}{\varepsilon}\left(x + \tfrac{1}{2}x^2\right) \right] + \frac{c_2}{x^2} \exp\left[-\frac{1}{\varepsilon}\left(x^2 + x\right) \right] + \cdots$$

provided that x is away from zero.

14.46. Use the WKB method to show that, to the first approximation, the general solution of

$$\varepsilon^2 y'' + \varepsilon[\omega_1(x) + \omega_2(x)]\, y' + \omega_1(x)\, \omega_2(x)\, y = 0, \qquad 0 < \varepsilon \ll 1$$

is

$$y = c_1 \exp\left[-\frac{1}{\varepsilon}\int^x \omega_1(t)\,dt + \int^x \frac{\omega_1'(t)\,dt}{\omega_2(t)-\omega_1(t)}\right]$$

$$+ c_2 \exp\left[-\frac{1}{\varepsilon}\int^x \omega_2(t)\,dt + \int^x \frac{\omega_2'(t)\,dt}{\omega_1(t)-\omega_2(t)}\right] + \cdots$$

14.47. Use the WKB method to show that, to the first approximation, the general solution of

$$y''' - \lambda^3 f(x)\,y = 0, \qquad f(x) > 0, \qquad \lambda \gg 1$$

is

$$y = \frac{1}{[f(x)]^{1/3}}\left\{c_1 e^{\lambda\int[f(x)]^{1/3}\,dx} + c_2 e^{\lambda\omega\int[f(x)]^{1/3}\,dx} + c_3 e^{\lambda\omega^2\int[f(x)]^{1/3}\,dx}\right\} + \cdots$$

where

$$\omega = e^{2i\pi/3}$$

14.48. Use the WKB method to show that, to the first approximation, the general solution of

$$y^{iv} - \lambda^4 f(x)\,y = 0, \qquad f(x) > 0, \qquad \lambda \gg 1$$

is

$$y = \frac{1}{[f(x)]^{3/8}}\left\{c_1 e^{\lambda\int[f(x)]^{1/4}\,dx} + c_2 e^{\lambda\omega\int[f(x)]^{1/4}\,dx}\right.$$

$$\left. + c_3 e^{\lambda\omega^2\int[f(x)]^{1/4}\,dx} + c_4 e^{\lambda\omega^3\int[f(x)]^{1/4}\,dx}\right\} + \cdots$$

where

$$\omega = e^{i\pi/2}$$

14.49. Consider the problem

$$y'' + \lambda^2(1-x^2)^n y = 0$$

Show that as $\lambda \to \infty$,

$$y \sim H^{1/2}(1-x^3)^{-n/4}[c_1 J_\nu(\lambda H) + c_2 J_{-\nu}(\lambda H)]$$

for $x > -1$, where

$$H = \int_1^x (1-\tau^2)^{n/2}\,d\tau \quad \text{and} \quad \nu = \frac{1}{n+2}$$

14.50. Consider the problem

$$y'' + \lambda^2 (x-\mu)^n f(x) y = 0, \qquad f(x) > 0$$

Show that as $\lambda \to \infty$,

$$y \sim H^{1/2} [(x-\mu)^n f(x)]^{-1/4} [c_1 J_\nu(\lambda H) + c_2 J_{-\nu}(\lambda H)]$$

where

$$H = \int_\mu^x \sqrt{(\tau-\mu)^n f(\tau)}\, d\tau \quad \text{and} \quad \nu = \frac{1}{n+2}$$

14.51. Consider the problem

$$y'' + \lambda^2 (\mu - x) f(x) y = 0, \qquad \mu > 0 \text{ and } f(x) > 0$$

$$y(0) = 0, \qquad y(\mu) = 0$$

where $f(x) > 0$ in $[0, \mu]$. Show that

$$\lambda \sim (n + \tfrac{11}{12})\pi \left[\int_0^\mu \sqrt{(\mu-\tau) f(\tau)}\, d\tau \right]^{-1} \quad \text{as } n \to \infty$$

14.52. Consider the equation in the preceding exercise subject to the boundary conditions

$$y'(0) = 0, \qquad y(\mu) = 0$$

Show that

$$\lambda \sim (n + \tfrac{5}{12})\pi \left[\int_0^\mu \sqrt{(\mu-\tau) f(\tau)}\, d\tau \right]^{-1} \quad \text{as } n \to \infty$$

14.53. Consider the problem

$$y'' + \lambda^2 (x-a)(b-x) f(x) y = 0$$

$$y \to 0 \quad \text{as } x \to \pm\infty$$

where $b > a$ and $f(x) > 0$ in $[a, b]$. Show that

$$y_1 \sim \frac{c_1}{\sqrt{H'}} Ai[-\lambda^{2/3} H] \quad \text{for } x < b$$

$$y_2 \sim \frac{c_2}{\sqrt{G'}} Ai(-\lambda^{2/3} G) \quad \text{for } x > a$$

where

$$\tfrac{2}{3} H^{3/2} = \int_a^x \sqrt{(\tau-a)(b-\tau) f(\tau)}\, d\tau$$

$$\tfrac{2}{3} G^{3/2} = \int_x^b \sqrt{(\tau-a)(b-\tau) f(\tau)}\, d\tau$$

Expand y_1 for large H and y_2 for large G, equate the resulting expansions, and obtain

$$\lambda \sim \left(n-\tfrac{1}{2}\right)\pi\left[\int_a^b \sqrt{(\tau-a)(b-\tau)f(\tau)}\,d\tau\right]^{-1} \quad \text{as } n\to\infty$$

14.54. Consider the problem

$$y''+\left[\lambda^2-V(x)\right]y=0$$

$$y\to 0 \quad \text{as } x\to\pm\infty$$

where $V(x)$ rises monotonically as $x\to\pm\infty$. Follow steps similar to those in the preceding exercise and show that

$$\int_a^b \sqrt{\lambda^2-V(x)}\,dx=\left(n-\tfrac{1}{2}\right)\pi$$

where a and b are the two roots of $\lambda^2-V(x)=0$.

14.55. Consider the problem

$$y''+(\lambda^2-x^2)y=0$$

$$y\to 0 \quad \text{as } x\to\pm\infty$$

Apply the result of the preceding exercise to show that

$$\int_{-\lambda}^{\lambda} \sqrt{\lambda^2-x^2}\,dx=\left(n-\tfrac{1}{2}\right)\pi$$

Let $x=\lambda\tau$ and then show that $\lambda=\sqrt{2n-1}$

14.56. Consider the problem

$$y''+(\lambda^2-x^4)y=0$$

$$y\to 0 \quad \text{as } x\to\pm\infty$$

Apply the results in Exercise (14.54) and obtain

$$\int_{-\sqrt{\lambda}}^{+\sqrt{\lambda}} \sqrt{\lambda^2-x^4}\,dx=\left(n-\tfrac{1}{2}\right)\pi$$

Let $x=\sqrt{\lambda}\,\tau$ and obtain

$$\lambda^{3/2}\int_{-1}^{+1}\sqrt{1-\tau^4}\,d\tau=\left(n-\tfrac{1}{2}\right)\pi$$

Then show that

$$\lambda\sim\left[\frac{3\Gamma(\frac{3}{4})(n-\frac{1}{2})\sqrt{\pi}}{\Gamma(\frac{1}{4})}\right]^{2/3} \quad \text{as } n\to\infty$$

CHAPTER 15

Solvability Conditions

In applying perturbation methods to physical problems, one encounters problems that need to be solved in succession. Often the zeroth-order problem is homogeneous, whereas the higher-order problems are linear and inhomogeneous. In the case of vibration problems, the inhomogeneous problems usually lead to secular terms. In boundary-value problems, the inhomogeneous problems may not possess solutions, unless certain conditions (usually called solvability, consistency, or compatability conditions) are satisfied.

In the case of single-degree-of-freedom vibratory systems, secular terms can be eliminated by setting the coefficients of the terms that lead to secular terms equal to zero. As an example, application of the method of multiple scales to

$$\ddot{u} + \omega_0^2 u = -2\varepsilon\mu\dot{u} - \varepsilon u^3$$

leads to the first-order problem

$$D_0^2 u_1 + \omega_0^2 u_1 = -2i\omega_0\left(A' + \mu A\right)e^{i\omega_0 T_0} - 3A^2\bar{A}e^{i\omega_0 T_0} - A^3 e^{3i\omega_0 T_0} + \text{cc}$$

Setting the coefficient of $\exp(i\omega_0 T_0)$ on the right-hand side of this equation equal to zero eliminates the secular terms in u_1 and provides the necessary equation for the determination of A. In multidegree-of-freedom systems, eliminating secular terms may not be as straightforward, especially if the equations are coupled. For example, application of the method of multiple scales to

$$\ddot{u}_1 + 2\dot{u}_2 + 3u_1 + 2\varepsilon f u_1 \cos 2t = 0$$

$$\ddot{u}_2 - 2\dot{u}_1 + 3u_2 + 2\varepsilon g u_2 \cos 2t = 0$$

leads to the first-order problem

$$D_0^2 u_{11} + 2D_0 u_{21} + 3u_{11} = -\left(4iA_1' + f\bar{A}_1 + fA_2\right)e^{iT_0} - \left(4iA_2' + fA_1\right)e^{3iT_0} - fA_2 e^{5iT_0} + \text{cc}$$

$$D_0^2 u_{21} - 2D_0 u_{11} + 3u_{21} = \left(4A_1' + ig\bar{A}_1 + igA_2\right)e^{iT_0} - \left(4A_2' + igA_1\right)e^{3iT_0} + igA_2 e^{5iT_0} + \text{cc}$$

In this case, one cannot eliminate the secular terms from u_{11} and u_{21} by simply setting the coefficients of $\exp(iT_0)$ and $\exp(3iT_0)$ equal to zero. Instead, one can seek a particular solution free of secular terms in the form

$$u_{11} = P_{11}e^{iT_0} + P_{12}e^{3iT_0} + P_{13}e^{5iT_0}$$

$$u_{21} = Q_{21}e^{iT_0} + Q_{21}e^{3iT_0} + Q_{23}e^{5iT_0}$$

This leads to linear inhomogeneous algebraic equations for the P_{mn} and Q_{mn} of the form

$$Fv = b$$

where F is a 2×2 matrix and v and b are 2×1 matrices. The matrix F is singular and hence the inhomogeneous system of equations has a solution if and only if the components of b are related; that is, unless a solvability condition is satisfied.

As an example of inhomogeneous problems arising from boundary-value problems, we consider

$$u'' + \pi^2 u = 2\pi \sin \pi x$$

$$u(0) = 0, \qquad u(1) = \alpha$$

The general solution of the differential equation is

$$u = c_1 \cos \pi x + c_2 \sin \pi x - x \cos \pi x$$

The boundary condition $u(0) = 0$ yields $c_1 = 0$, whereas the boundary condition $u(1) = \alpha$ yields $\alpha = 1$. Therefore, the inhomogeneous problem has a solution if and only if $\alpha = 1$.

Solved Exercises

15.1. Determine the solvability conditions for

$$\begin{aligned} \ddot{u}_1 + \dot{u}_2 + 2u_1 &= \sum_{n=1}^{4} P_n e^{int} \\ \ddot{u}_2 - \dot{u}_1 + 2u_2 &= \sum_{n=1}^{4} Q_n e^{int} \end{aligned} \tag{1}$$

where the P_n and Q_n are constants.

We seek a particular solution in the form

$$u_1 = ae^{int}, \qquad u_2 = be^{int} \tag{2}$$

Substituting (2) into (1) and equating the coefficients of $\exp(int)$ on both sides, we obtain

$$(2-n^2)a + inb = P_n$$

$$-ina + (2-n^2)b = Q_n$$

whose solution is

$$a = \frac{\begin{vmatrix} P_n & in \\ Q_n & 2-n^2 \end{vmatrix}}{\begin{vmatrix} 2-n^2 & in \\ -in & 2-n^2 \end{vmatrix}} = \frac{(2-n^2)P_n - inQ_n}{(n^2-1)(n^2-4)} \tag{3}$$

$$b = \frac{(2-n^2)Q_n + inP_n}{(n^2-1)(n^2-4)} \tag{4}$$

We note that the denominators vanish when $n=1$ or 2. Hence, when $n=3$ or 4, there is a unique particular solution for (1) and (2). However, when $n=1$ or 2, a and b are infinite and (1) and (2) are not solvable unless the numerators vanish. This condition demands that

$$P_1 = iQ_1 \quad \text{when } n=1$$

$$P_2 = -iQ_2 \quad \text{when } n=2$$

15.2. Determine the solvability conditions for

$$\begin{aligned} \ddot{x} + \dot{y} + x &= P_1 e^{2it} + P_2 e^{it/\sqrt{2}} \\ \ddot{y} - \tfrac{3}{2}\dot{x} + 2y &= Q_1 e^{2it} + Q_2 e^{it/\sqrt{2}} \end{aligned} \tag{5}$$

where the P_n and Q_n are constants.

We seek a particular solution for (5) in the form

$$\begin{aligned} x &= a_1 e^{2it} + a_2 e^{it/\sqrt{2}} \\ y &= b_1 e^{2it} + b_2 e^{it/\sqrt{2}} \end{aligned} \tag{6}$$

Substituting (6) into (5) and equating each of the coefficients of $\exp(2it)$ and $\exp(it/\sqrt{2})$

on both sides, we obtain

$$\begin{aligned} -3a_1 + 2ib_1 &= P_1 \\ -3ia_1 - 2b_1 &= Q_1 \end{aligned} \tag{7}$$

$$\begin{aligned} \tfrac{1}{2}a_2 + \frac{i}{\sqrt{2}}b_2 &= P_2 \\ -\frac{3i}{2\sqrt{2}}a_2 + \tfrac{3}{2}b_2 &= Q_2 \end{aligned} \tag{8}$$

Since the determinant of the coefficient matrix in (7) vanishes, the solvability condition of (7) is

$$\begin{vmatrix} P_1 & 2i \\ Q_1 & -2 \end{vmatrix} = 0 \quad \text{or} \quad P_1 = -iQ_1$$

Since the determinant of the coefficient matrix in (8) vanishes, the solvability condition of (8) is

$$\begin{vmatrix} P_2 & \dfrac{i}{\sqrt{2}} \\ Q_2 & \tfrac{3}{2} \end{vmatrix} = 0 \quad \text{or} \quad 3P_2 = i\sqrt{2}\,Q_2$$

15.3. Determine the solvability conditions for

(a)
$$\begin{gathered} u'' + \tfrac{1}{4}u = f(x) \\ u'(0) = a, \qquad u(\pi) = b \end{gathered} \tag{9}$$

(b)
$$\begin{gathered} u'' + \frac{1}{x}u' + k^2 u = f(x) \\ u(a) = c_1, \qquad u(b) = c_2 \end{gathered} \tag{10}$$

where $b > a > 0$ and the homogeneous problem has a nontrivial solution.

(a) We multiply the equation in (9) by the adjoint $v(x)$, integrate the result by parts from $x = 0$ to $x = \pi$, and obtain

$$\int_0^\pi u\left(v'' + \tfrac{1}{4}v\right) dx + \left[u'v - uv'\right]_0^\pi = \int_0^\pi f(x)v(x)\, dx \tag{11}$$

Then the adjoint equation is

$$v'' + \tfrac{1}{4}v = 0 \tag{12}$$

To determine the adjoint boundary conditions, we consider the homogeneous case and obtain from (11) that

$$[u'v - uv']_0^\pi = 0$$

which, upon using the homogeneous boundary conditions in (9), yields

$$u'(\pi)v(\pi) + u(0)v'(0) = 0 \tag{13}$$

We choose the adjoint boundary conditions such that the coefficients of $u'(\pi)$ and $u(0)$ in (13) vanish independently. The result is

$$v'(0) = 0, \qquad v(\pi) = 0 \tag{14}$$

Thus the solution of the adjoint problem, (12) and (14), is $v = \cos\frac{1}{2}x$.

Having defined the adjoint, we return to the inhomogeneous problem and obtain from (9) and (11) that

$$\int_0^\pi f(x)\cos\tfrac{1}{2}x\,dx = \tfrac{1}{2}b - a \tag{15}$$

(b) We note that the equation in (10) can be made self-adjoint by multiplying it with x. Hence we multiply the equation in (10) with $xv(x)$, integrate the result by parts from $x = a$ to $x = b$, and obtain

$$\int_a^b v\left[(xu')' + k^2xu\right]dx = \int_a^b xf(x)v(x)\,dx$$

or

$$\int_a^b u\left[(xv')' + k^2xv\right]dx + [xu'v - xuv']_a^b = \int_a^b xf(x)v(x)\,dx \tag{16}$$

The adjoint equation is

$$(xv')' + k^2xv = 0 \tag{17}$$

Considering the homogeneous problem, we obtain from (16) that

$$bu'(b)v(b) - bu(b)v'(b) - au'(a)v(a) + au(a)v'(a) = 0$$

But the homogeneous boundary conditions are $u(a) = u(b) = 0$, hence

$$bu'(b)v(b) - au'(a)v(a) = 0 \tag{18}$$

We define the adjoint boundary conditions by requiring that each of the coefficients of $u'(a)$ and $u'(b)$ in (17) vanish independently. Hence

$$v(a) = v(b) = 0 \tag{19}$$

Having defined the adjoint problem, we return to the inhomogeneous problem and obtain from (16), (10), (17), and (19) that

$$\int_a^b xf(x)\,v(x)\,dx = ac_1v'(a) - bc_2v'(b) \tag{20}$$

15.4. Determine the solvability condition for the problem

$$u'' + \frac{1}{x}u' + \left(\lambda^2 - \frac{n^2}{x^2}\right)u = f(x) \tag{21}$$

$$u(0) < \infty, \qquad u'(a) = g \tag{22}$$

where a and g are constants and λ is a root of $J_n'(\lambda a)$.

Equation (21) can be made self-adjoint by multiplying it with x. Thus we multiply (21) by xv, integrate the result by parts from $x = 0$ to $x = a$ to transfer the derivatives from u to v, and obtain

$$[xu'v - xuv']_0^a + \int_0^a u\left[(xv')' + \left(\lambda^2 x - \frac{n^2}{x}\right)v\right]dx = \int_0^a xf(x)\,v(x)\,dx \tag{23}$$

Then the adjoint equation is

$$(xv')' + \left(\lambda^2 x - \frac{n^2}{x}\right)v = 0 \tag{24}$$

To define the adjoint boundary conditions, we consider the homogeneous problem (i.e., $g = 0$ and $f = 0$). Then (23) becomes

$$-au(a)\,v'(a) - [xu'v - xuv']_{x=0} = 0 \tag{25}$$

The last term in (25) vanishes if $v(0) < \infty$, whereas the first term vanishes if $v'(a) = 0$. Thus the adjoint boundary conditions are

$$v(0) < \infty, \qquad v'(a) = 0 \tag{26}$$

Hence

$$v(x) = J_n(\lambda x) \tag{27}$$

To determine the solvability condition, we return to the inhomogeneous problem. Using (22), (24), and (26) in (23), we have

$$agJ_n(\lambda a) = \int_0^a xJ_n(\lambda x)\,f(x)\,dx \tag{28}$$

15.5. Determine the solvability conditions for

(a)
$$\phi'' + \frac{n^2\pi^2}{d^2}\phi = f(x) \qquad \phi'(0) = 0, \quad \phi'(d) = \beta \tag{29}$$

(b)
$$\phi'' + \gamma_n^2\phi = f(x) \qquad \phi'(0) = 0, \quad \phi'(1) - \alpha\phi(1) = \beta \tag{30}$$

where $\gamma_n \tan \gamma_n = -\alpha$.

(a) We multiply the equation in (29) by the adjoint $v(x)$, integrate the result by parts from $x = 0$ to $x = d$, and obtain

$$\int_0^d \phi\left[v'' + \frac{n^2\pi^2}{d^2}v\right] dx + [\phi' v - \phi v']_0^d = \int_0^d v(x) f(x)\, dx \tag{31}$$

Thus the adjoint equation is

$$v'' + \frac{n^2\pi^2}{d^2}v = 0 \tag{32}$$

To determine the adjoint boundary conditions, we consider the homogeneous problem and obtain from (31) that

$$[\phi v' - \phi' v]_0^d = 0$$

which, upon using the homogeneous boundary conditions in (29), yields

$$\phi(d)\, v'(d) - \phi(0)\, v'(0) = 0$$

We define the adjoint boundary conditions by demanding that the coefficients of $\phi(d)$ and $\phi(0)$ vanish independently. The result is

$$v'(0) = v'(d) = 0 \tag{33}$$

It follows from (32) and (33) that $v = \cos[(n\pi/d)x]$, where n is an integer.

Having defined the adjoint, we return to the inhomogeneous problem and obtain from (29) and (31) that

$$\int_0^d f(x) \cos\left(\frac{n\pi}{d}x\right) dx = \beta \cos n\pi \tag{34}$$

(b) We multiply the equation in (30) by the adjoint $v(x)$, integrate the result by parts from $x = 0$ to $x = 1$, and obtain

$$\int_0^1 \phi[v'' + \gamma_n^2 v]\, dx + [\phi' v - \phi v']_0^1 = \int_0^1 f(x) v(x)\, dx \tag{35}$$

Then the adjoint equation is

$$v'' + \gamma_n^2 v = 0 \tag{36}$$

To determine the adjoint boundary conditions, we consider the homogeneous problem and obtain from (35) and (30) that

$$\phi(1)\,v'(1) - \phi'(1)\,v(1) - \phi(0)\,v'(0) + \phi'(0)\,v(0) = 0 \tag{37}$$

$$\phi'(0) = 0, \qquad \phi'(1) - \alpha\phi(1) = 0 \tag{38}$$

Substituting for $\phi'(0)$ and $\phi'(1)$ in (37) and collecting the coefficients of $\phi(0)$ and $\phi(1)$, we obtain

$$-\phi(0)\,v'(0) + \phi(1)\left[v'(1) - \alpha v(1)\right] = 0 \tag{39}$$

We choose the adjoint boundary conditions such that the coefficients of $\phi(0)$ and $\phi(1)$ vanish independently; that is,

$$v'(0) = 0 \quad \text{and} \quad v'(1) - \alpha v(1) = 0 \tag{40}$$

It follows from (36) that

$$v = a_1 \cos \gamma_n x + a_2 \sin \gamma_n x$$

which, when substituted into (40), yields $a_2 = 0$ and

$$-\gamma_n \sin \gamma_n - \alpha \cos \gamma_n = 0 \quad \text{or} \quad \gamma_n \tan \gamma_n = -\alpha$$

Therefore

$$v = \cos \gamma_n x$$

Having defined the adjoint, we return to the inhomogeneous problem. It follows from (30) that $\phi'(0) = 0$ and $\phi'(1) = \alpha\phi(1) + \beta$ which, when substituted together with $v = \cos \gamma_n x$ into (35), yields the following solvability condition:

$$\int_0^1 f(x) \cos \gamma_n x \, dx = \beta \cos \gamma_n \tag{41}$$

15.6. Determine the solvability condition for

$$\frac{\partial^2 \phi}{\partial y^2} + \frac{\partial^2 \phi}{\partial z^2} + \left(\frac{n^2\pi^2}{d^2} + \frac{m^2\pi^2}{b^2} \right)\phi = f(y, z) \tag{42}$$

$$\phi_y(0, z) = 0, \qquad \phi_z(y, 0) = 0 \tag{43}$$

$$\phi_y(d, z) = \beta_1, \qquad \phi_z(y, b) = \beta_2 \tag{44}$$

We multiply (42) by the adjoint $u(y,z)$ and integrate the result over the domain of interest; that is, $0 \le y \le d$ and $0 \le z \le b$. The result is

$$\int_0^b \int_0^d \left[u\frac{\partial^2\phi}{\partial y^2} + u\frac{\partial^2\phi}{\partial z^2} + \left(\frac{n^2\pi^2}{d^2} + \frac{m^2\pi^2}{b^2}\right) u\phi \right] dy\,dz = \int_0^b \int_0^d u(y,z) f(y,z)\,dy\,dz$$

which, upon integration by parts to transfer the derivatives from ϕ to u, yields

$$\int_0^b \int_0^d \phi \left[\frac{\partial^2 u}{\partial y^2} + \frac{\partial^2 u}{\partial z^2} + \left(\frac{n^2\pi^2}{d^2} + \frac{m^2\pi^2}{b^2}\right) u \right] dy\,dz + \int_0^b \left[u\frac{\partial\phi}{\partial y} - \frac{\partial u}{\partial y}\phi \right]_0^d dz$$

$$+ \int_0^d \left[u\frac{\partial\phi}{\partial z} - \frac{\partial u}{\partial z}\phi \right]_0^b dy = \int_0^b \int_0^d uf\,dy\,dz \tag{45}$$

We define the adjoint equation by setting the coefficient of ϕ in the integrand of the double integral in (45) equal to zero. The result is

$$\frac{\partial^2 u}{\partial y^2} + \frac{\partial^2 u}{\partial z^2} + \left(\frac{n^2\pi^2}{d^2} + \frac{m^2\pi^2}{b^2}\right) u = 0 \tag{46}$$

To define the adjoint boundary conditions, we consider the homogeneous problem. Thus we let $f = 0$ and $\beta_1 = \beta_2 = 0$ so that (44) and (45) become

$$\phi_y(d,z) = 0, \qquad \phi_z(y,b) = 0 \tag{47}$$

$$\int_0^b \left[u\frac{\partial\phi}{\partial y} - \frac{\partial u}{\partial y}\phi \right]_0^d dz + \int_0^d \left[u\frac{\partial\phi}{\partial z} - \frac{\partial u}{\partial z}\phi \right]_0^b dy = 0 \tag{48}$$

Using (43) and (47) in (48), we obtain

$$-\int_0^b \frac{\partial u}{\partial y}(d,z)\phi(d,z)\,dz + \int_0^b \frac{\partial u}{\partial y}(0,z)\phi(0,z)\,dz - \int_0^d \frac{\partial u}{\partial z}(y,b)\phi(y,b)\,dy$$

$$+ \int_0^d \frac{\partial u}{\partial z}(y,0)\phi(y,0)\,dz = 0$$

We define the adjoint boundary conditions by setting each of the coefficients of $\phi(d,z)$, $\phi(0,z)$, $\phi(y,b)$, and $\phi(y,0)$ equal to zero. The result is

$$\begin{aligned} u_y(0,z) &= 0, \qquad u_z(y,0) = 0 \\ u_y(d,z) &= 0, \qquad u_z(y,b) = 0 \end{aligned} \tag{49}$$

It follows from (46) and (49) that

$$u = \cos\left(\frac{n\pi}{d}y\right)\cos\left(\frac{m\pi}{b}z\right) \tag{50}$$

Having defined the adjoint, we return to the inhomogeneous problem. It follows from (45), (43), (44), and (50) that

$$\int_0^b \int_0^d f(y,z)\cos\left(\frac{n\pi}{d}y\right)\cos\left(\frac{m\pi}{b}z\right)\,dy\,dz = \cos n\pi \int_0^b \beta_1 \cos\left(\frac{m\pi}{b}z\right)\,dz + \cos m\pi \int_0^d \beta_2 \cos\left(\frac{n\pi}{d}y\right)\,dy \tag{51}$$

is the solvability condition.

15.7. Determine the solvability condition for

$$\frac{d^2\phi}{dr^2} + \frac{1}{r}\frac{d\phi}{dr} + \left(\gamma_{nm}^2 - \frac{n^2}{r^2}\right)\phi = f(r) \tag{52}$$

$$\phi(0) < \infty, \qquad \phi'(1) - \beta\phi(1) = \alpha \tag{53}$$

where γ_{nm} is a root of

$$\gamma J_n'(\gamma) - \beta J_n(\gamma) = 0 \tag{54}$$

The bounded solution at the origin of the homogeneous equation is

$$\phi = J_n(\gamma_{nm} r)$$

which satisfies the homogeneous boundary conditions in (53) on account of (54). Therefore, the homogeneous problem has a nontrivial solution, and hence the inhomogeneous problem has a solution only if a solvability condition is satisfied.

We note that the homogeneous equation (52) can be made self-adjoint by multiplying it with r. Therefore, to determine the solvability condition of the inhomogeneous problem, we multiply (52) with $ru(r)$, integrate the result by parts from $r=0$ to $r=1$ to transfer the derivatives from ϕ to u, and obtain

$$\int_0^1 \phi\left[(ru')' + \left(r\gamma_{nm}^2 - \frac{n^2}{r}\right)u\right]dr + \left[r\phi' u - r\phi u'\right]_0^1 = \int_0^1 rf(r)\,u(r)\,dr \tag{55}$$

The adjoint equation is defined by setting the coefficient of ϕ in the integrand in (55) equal to zero. The result is

$$(ru')' + \left(r\gamma_{nm}^2 - \frac{n^2}{r}\right)u = 0 \tag{56}$$

To define the adjoint boundary conditions, we consider the homogeneous problem. Then (55) and (53) become

$$\begin{gathered} \left[r\phi' u - r\phi u'\right]_0^1 = 0 \\ \phi(0) < \infty, \qquad \phi'(1) = \beta\phi(1) \end{gathered} \tag{57}$$

Substituting for $\phi'(1)$ in (57) yields

$$-\phi(1)[u'(1)-\beta u(1)]-[r\phi' u - r\phi u']_{r=0}=0 \tag{58}$$

We note that the first term vanishes if

$$u'(1)-\beta u(1)=0 \tag{59a}$$

whereas the second term vanishes if

$$u(0)<\infty \tag{59b}$$

because $\phi(0)<\infty$. Hence $u=J_n(\gamma_{nm}r)$.

Having defined the adjoint, we return to the inhomogeneous problem. We solve for $\phi'(1)$ from (53), substitute the result into (55), use (56) and (59), recall that $u=J_n(\gamma_{nm}r)$, and obtain the following solvability condition:

$$\int_0^1 rf(r)\,J_n(\gamma_{nm}r)\,dr=\alpha J_n(\gamma_{nm}) \tag{60}$$

15.8. Determine the solvability condition for

$$\frac{\partial^2\phi}{\partial r^2}+\frac{1}{r}\frac{\partial\phi}{\partial r}+\frac{1}{r^2}\frac{\partial^2\phi}{\partial\theta^2}+\gamma_{nm}^2\phi=f(r)\,e^{in\theta} \tag{61}$$

$$\phi(0,\theta)<\infty, \qquad \phi_r(1,\theta)=\alpha e^{in\theta} \tag{62}$$

where γ_{nm} is a root of

$$J_n'(\gamma)=0 \tag{63}$$

We first separate the θ variation by letting

$$\phi=u(r)\,e^{in\theta}$$

in (61) and (62) and obtain

$$\frac{d^2u}{dr^2}+\frac{1}{r}\frac{du}{dr}+\left(\gamma_{nm}^2-\frac{n^2}{r^2}\right)u=f(r) \tag{64}$$

$$u(0)<\infty, \qquad u'(1)=\alpha \tag{65}$$

Thus the problem of determining the solvability condition of (61) and (62) is transformed into the problem of determining the solvability condition of (64) and (65). The bounded solution at the origin of the homogeneous equation (64) is

$$u=J_n(\gamma_{nm}r)$$

It satisfies the homogeneous boundary conditions (65) on account of (63). Hence the homogeneous problem has a nontrivial solution and, therefore, the inhomogeneous problem has a solution only if a solvability condition is satisfied.

As in the preceding exercise, we multiply (64) by $rv(r)$, integrate the result by parts from $r=0$ to $r=1$ to transfer the derivatives from u to v, and obtain

$$\int_0^1 u\left[(rv')' + \left(r\gamma_{nm}^2 - \frac{n^2}{r}\right)v\right] dr + [ru'v - ruv']_0^1 = \int_0^1 rf(r)v(r)\, dr \tag{66}$$

The adjoint equation is defined by setting the coefficient of u in the integrand in (66) equal to zero. The result is

$$(rv')' + \left(r\gamma_{nm}^2 - \frac{n^2}{r}\right)v = 0 \tag{67}$$

To define the adjoint boundary conditions, we consider the homogeneous problem and obtain from (65) and (66) that

$$-u(1)v'(1) - [ru'v - ruv']_{r=0} = 0$$

The first term vanishes if $v'(1) = 0$, whereas the second term vanishes if $v(0) < \infty$ because $u(0) < \infty$. Therefore, $v = J_n(\gamma_{nm} r)$.

Having defined the adjoint, we return to the inhomogeneous problem. Substituting $u'(1) = \alpha$ into (66) and using the definition of the adjoint, we obtain the following solvability condition:

$$\int_0^1 rf(r) J_n(\gamma_{nm} r)\, dr = \alpha J_n(\gamma_{nm}) \tag{68}$$

15.9. Determine the solvability condition for

$$\frac{d^2u}{dr^2} + \frac{1}{r}\frac{du}{dr} + \lambda u = F(r) \tag{69}$$

$$u(a) = u_a, \qquad u(b) = u_b \tag{70}$$

where λ is an eigenvalue of the homogeneous problem.

Since λ is an eigenvalue of the homogeneous problem (69) and (70), it has a nontrivial solution. Consequently, the inhomogeneous problem has a solution only if a solvability condition is satisfied.

As in the preceding two exercises, the homogeneous equation (69) can be made self-adjoint by its multiplication by r. Hence we multiply (69) by $rv(r)$, integrate the result by parts from $r=a$ to $r=b$ to transfer the derivatives from u to v, and obtain

$$\int_a^b u[(rv')' + \lambda rv]\, dr + [ru'v - ruv']_a^b = \int_a^b rF(r)v(r)\, dr \tag{71}$$

The adjoint equation is defined by setting the coefficient of u in the integrand in (71) equal to zero. The result is

$$(rv')' + \lambda rv = 0 \tag{72}$$

To define the adjoint boundary conditions, we consider the homogeneous problem and obtain from (70) and (71) that

$$bu'(b)v(b) - au'(a)v(a) = 0 \tag{73}$$

The adjoint boundary conditions are defined by setting each of the coefficients of $u'(b)$ and $u'(a)$ equal to zero in (73). The result is

$$v(a) = 0, \qquad v(b) = 0 \tag{74}$$

Having defined the adjoint, we return to the inhomogeneous problem. Substituting (70) into (71) and using the definition of the adjoint, we obtain the following solvability condition:

$$\int_a^b rF(r)v(r)\,dr = au_a v'(a) - bu_b v'(b) \tag{75}$$

where $v(r)$ is a solution of (72) and (74).

15.10. Determine the solvability condition for

$$\nabla^4 w - \lambda w = F(r) \tag{76}$$

$$w(0) < \infty, \qquad w(1) = \beta_1, \qquad w'(1) = \beta_2 \tag{77}$$

where λ is an eigenvalue of the homogeneous problem and

$$\nabla^2 = \frac{d^2}{dr^2} + \frac{1}{r}\frac{d}{dr}$$

Since λ is an eigenvalue of the homogeneous problem, it has a nontrivial solution. Consequently, the inhomogeneous problem has a solution only if a solvability condition is satisfied.

The homogeneous equation (76) can be made self-adjoint by its multiplication by r. Hence, to determine the solvability condition, we multiply (76) by $ru(r)$, integrate the result by parts from $r = 0$ to $r = 1$ to transfer the derivatives from w to u, and obtain

$$\int_0^1 rw[\nabla^4 u - \lambda u]\,dr + \left[ru\frac{d}{dr}\left(\frac{d^2w}{dr^2} + \frac{1}{r}\frac{dw}{dr}\right) - r\frac{du}{dr}\left(\frac{d^2w}{dr^2} + \frac{1}{r}\frac{dw}{dr}\right)\right.$$

$$\left. + r\frac{dw}{dr}\left(\frac{d^2u}{dr^2} + \frac{1}{r}\frac{du}{dr}\right) - rw\frac{d}{dr}\left(\frac{d^2u}{dr^2} + \frac{1}{r}\frac{du}{dr}\right)\right]_0^1 = \int_0^1 rF(r)u(r)\,dr \tag{78}$$

The adjoint equation is defined by setting the coefficient of w in the integrand in (78) equal to zero. The result is

$$\nabla^4 u - \lambda u = 0 \tag{79}$$

To determine the adjoint boundary conditions, we consider the homogeneous problem and obtain from (77) and (78) that

$$w'''(1)\,u(1)+w''(1)\left[u(1)-u'(1)\right]$$

$$-\left[ru\frac{d}{dr}(\nabla^2 w)-ru'\nabla^2 w+rw'\nabla^2 u-rw\frac{d}{dr}(\nabla^2 u)\right]_{r=0}=0 \tag{80}$$

The first term in (80) vanishes if $u(1)=0$, then the second term vanishes if $u'(1)=0$, and the last term vanishes if $u(0)<\infty$ because $w(0)<\infty$. Therefore, the problem governing u is the same as the homogeneous problem (76) and (77).

Having defined the adjoint, we return to the inhomogeneous problem. Using (77) in (78) and using the definition of the adjoint, we obtain the following solvability condition:

$$\int_0^1 rF(r)\,u(r)\,dr=\beta_2 u''(1)-\beta_1\left[u''(1)+u'''(1)\right] \tag{81}$$

where u is a solution of (79) subject to the boundary conditions

$$u(0)<\infty,\qquad u(1)=0,\qquad u'(1)=0$$

15.11. Determine the solvability condition for

$$\frac{d^2\phi_1}{dr^2}+\frac{1}{r}\frac{d\phi_1}{dr}+\left(\alpha_n^2-\frac{1}{r^2}\right)\phi_1=f_1(r) \tag{82}$$

$$\frac{d^2\phi_2}{dr^2}+\frac{1}{r}\frac{d\phi_2}{dr}-\left(\gamma_n^2+\frac{1}{r^2}\right)\phi_2=f_2(r) \tag{83}$$

$$\phi_1'=\mu_1\phi_1+\mu_2\phi_2,\qquad \phi_2'=\mu_3\phi_1+\mu_4\phi_2\quad \text{at } r=1 \tag{84}$$

$$\phi_1(0)<\infty,\qquad a\phi_2'(a)-\phi_2(a)=0,\qquad a>1 \tag{85}$$

where the homogeneous problem has a nontrivial solution.

We note that each of the homogeneous equations (82) and (83) can be made self-adjoint by its multiplication by r. Therefore, to determine the solvability condition, we multiply (82) by ru_1 and integrate the result by parts from $r=0$ to $r=1$ to transfer the derivative from ϕ_1 to u_1. Moreover, we multiply (83) by ru_2 and integrate the result by parts from $r=1$ to $r=a$ to transfer the derivatives from ϕ_2 to u_2. Finally, we add both results and obtain

$$\int_0^1 \phi_1\left[(ru_1')'+\left(r\alpha_n^2-\frac{1}{r}\right)u_1\right]dr+\int_1^a \phi_2\left[(ru_2')'-\left(r\gamma_n^2+\frac{1}{r}\right)u_2\right]dr$$

$$+\left[ru_1\phi_1'-ru_1'\phi_1\right]_0^1+\left[r\phi_2'u_2-r\phi_2u_2'\right]_1^a=\int_0^1 rf_1(r)\,u_1(r)\,dr+\int_1^a rf_2(r)\,u_2(r)\,dr \tag{86}$$

The adjoint equations are defined by setting each of the coefficients of ϕ_1 and ϕ_2 in the integrands in (86) equal to zero. The result is

$$\left(ru_1'\right)' + \left(r\alpha_n^2 - \frac{1}{r}\right)u_1 = 0 \tag{87}$$

$$\left(ru_2'\right)' - \left(r\gamma_n^2 + \frac{1}{r}\right)u_2 = 0 \tag{88}$$

To define the adjoint boundary conditions, we consider the homogeneous problem and obtain from (86) that

$$\left[u_1\phi_1' - u_1'\phi_1 - \phi_2'u_2 + \phi_2 u_2'\right]_{r=1} + a\left[\phi_2'u_2 - \phi_2 u_2'\right]_{r=a} - \left[ru_1\phi_1' - ru_1'\phi_1\right]_{r=0} = 0 \tag{89}$$

We note that the last term in (89) vanishes if

$$u_1(0) < \infty \tag{90}$$

because $\phi_1(0) < \infty$. Then, substituting for $\phi_1'(1)$, $\phi_2'(1)$, and $\phi_2'(a)$ from (84) and (85) into the remainder of (89) and collecting the coefficients of $\phi_1(1)$, $\phi_2(1)$, and $\phi_2(a)$, we obtain

$$\phi_1(1)\left[\mu_1 u_1 - \mu_3 u_2 - u_1'\right]_{r=1} + \phi_2(1)\left[\mu_2 u_1 - \mu_4 u_2 + u_2'\right]_{r=1}$$
$$+ \phi_2(a)\left[u_2 - au_2'\right]_{r=a} = 0 \tag{91}$$

The adjoint boundary conditions are defined by setting each of the coefficients of $\phi_1(1)$, $\phi_2(1)$, and $\phi_2(a)$ in (91) equal to zero. The result is

$$u_1' = \mu_1 u_1 - \mu_3 u_2, \qquad u_2' = -\mu_2 u_1 + \mu_4 u_2 \quad \text{at } r = 1 \tag{92}$$

$$au_2' - u_2 = 0 \quad \text{at } r = a \tag{93}$$

Having defined the adjoint, we return to the inhomogeneous problem. Substituting (84) and (85) into (86) and using the definition of the adjoint, we obtain the following solvability condition:

$$\int_0^1 rf_1(r)\,u_1(r)\,dr + \int_1^a rf_1(r)\,u_2(r)\,dr = 0 \tag{94}$$

where u_1 and u_2 are any solutions of (87), (88), (90), (92), and (93).

15.12. Determine the solvability condition for

$$\frac{d^2\phi}{dr^2} + \left[\frac{1}{r} + \frac{T_0'}{T_0} + \frac{2ku_0'}{\omega - ku_0}\right]\frac{d\phi}{dr} + \left[\frac{\left(\omega - ku_0\right)^2}{T_0} - k^2 - \frac{m^2}{r^2}\right]\phi = f(r) \tag{95}$$

$$\phi(0) < \infty, \qquad \phi' - \beta\phi = \alpha \quad \text{at } r = 1 \tag{96}$$

when T_0, u_0, and $f(r)$ are known functions of r; ω, k, β, and α are constants; m is an integer; and the homogeneous problem has a nontrivial solution.

The homogeneous equation (95) is not self-adjoint. To make it self-adjoint, we multiply it by $P(r)$ and obtain

$$P\phi'' + \left[\frac{1}{r} + \frac{T_0'}{T_0} + \frac{2ku_0'}{\omega - ku_0}\right]P\phi' + P\left[\frac{(\omega - ku_0)^2}{T_0} - k^2 - \frac{m^2}{r^2}\right]\phi = Pf \tag{97}$$

In order that the homogeneous equation (97) be self-adjoint,

$$P' = \left[\frac{1}{r} + \frac{T_0'}{T_0} + \frac{2ku_0'}{\omega - ku_0}\right]P$$

which, upon separation of variables, becomes

$$\frac{dP}{P} = \left[\frac{1}{r} + \frac{T_0'}{T_0} + \frac{2ku_0'}{\omega - ku_0}\right]dr$$

Integrating once gives

$$\ln P = \ln r + \ln T_0 - 2\ln(\omega - ku_0)$$

Hence

$$P = \frac{rT_0}{(\omega - ku_0)^2}$$

and (97) can be rewritten as

$$\left[\frac{rT_0}{(\omega - ku_0)^2}\phi'\right]' + \frac{rT_0}{(\omega - ku_0)^2}\left[\frac{(\omega - ku_0)^2}{T_0} - k^2 - \frac{m^2}{r^2}\right]\phi = \frac{rT_0 f}{(\omega - ku_0)^2} \tag{98}$$

To determine the solvability condition, we multiply (98) by $\psi(r)$, integrate the result by parts from $r = 0$ to $r = 1$ to transfer the derivatives from ϕ to ψ, and obtain

$$\int_0^1 \phi\left\{\left[\frac{rT_0}{(\omega - ku_0)^2}\psi'\right]' + \frac{rT_0}{(\omega - ku_0)^2}\left[\frac{(\omega - ku_0)^2}{T_0} - k^2 - \frac{m^2}{r^2}\right]\psi\right\}dr$$

$$+\left[\frac{rT_0}{(\omega - ku_0)^2}(\psi\phi' - \psi'\phi)\right]_0^1 = \int_0^1 \frac{rT_0\psi f}{(\omega - ku_0)^2}dr \tag{99}$$

The adjoint equation is defined by setting the coefficient of ϕ in the integrand in (99) equal to zero. The result is

$$\left[\frac{rT_0}{(\omega - ku_0)^2}\psi'\right]' + \frac{rT_0}{(\omega - ku_0)^2}\left[\frac{(\omega - ku_0)^2}{T_0} - k^2 - \frac{m^2}{r^2}\right]\psi = 0 \tag{100}$$

which is the same as the homogeneous equation (95). To define the adjoint boundary conditions, we consider the homogeneous problem and obtain from (96) and (99) that

$$\left[\frac{rT_0}{(\omega - ku_0)^2}(\psi\phi' - \psi'\phi)\right]_{r=1} - \left[\frac{rT_0}{(\omega - ku_0)^2}(\psi\phi' - \psi'\phi)\right]_{r=0} = 0 \tag{101}$$

$$\phi' - \beta\phi = 0 \quad \text{at } r = 1 \tag{102}$$

The second term in (101) vanishes if $\psi(0) < \infty$ because $\phi(0) < \infty$. Then (101) becomes

$$[\psi\phi' - \psi'\phi]_{r=1} = 0$$

which, upon substituting $\beta\phi$ for ϕ' from (102), becomes

$$\phi(1)[\beta\psi(1) - \psi'(1)] = 0$$

Hence

$$\psi'(1) - \beta\psi(1) = 0 \tag{103}$$

Having defined the adjoint, we return to the inhomogeneous problem. Using (96) in (99) and using the definition of the adjoint, we obtain the following solvability condition:

$$\int_0^1 \frac{rT_0\psi f}{(\omega - ku_0)^2}\,dr = \frac{\alpha T_0(1)\psi(1)}{[\omega - ku_0(1)]^2} \tag{104}$$

where $\psi(r)$ is any solution of (100) subject to (103) and $\psi(0) < \infty$.

15.13. In analyzing waves propagating in a duct, one might encounter the problem

$$\frac{\partial^2\phi}{\partial x^2} + \frac{\partial^2\phi}{\partial y^2} + 5\pi^2\phi = f(y)\sin 2\pi x \tag{105}$$

$$\frac{\partial\phi}{\partial y} = 0 \quad \text{at} \quad y = 0, \qquad \frac{\partial\phi}{\partial y} = \alpha\sin 2\pi x \quad \text{at} \quad y = 1 \tag{106}$$

Show that the solvability condition is

$$\int_0^1 \cos\pi y f(y)\,dy = -\alpha \tag{107}$$

We first separate the x-variation by letting

$$\phi(x, y) = u(y)\sin 2\pi x$$

in (105) and (106) and obtaining

$$u'' + \pi^2 u = f(y) \tag{108}$$

$$u'(0) = 0, \qquad u'(1) = \alpha \tag{109}$$

Thus the problem of determining the solvability condition of (105) and (106) is transformed into determining the solvability condition of (108) and (109). To determine the solvability condition of (108) and (109), we multiply (108) by $v(y)$, integrate the result by parts from $y = 0$ to $y = 1$, and obtain

$$\int_0^1 u[v'' + \pi^2 v]\, dy + [vu' - v'u]_0^1 = \int_0^1 f(y)\, u(y)\, dy \tag{110}$$

The adjoint equation is defined by setting the coefficient of u in the integrand in (110) equal to zero. The result is

$$v'' + \pi^2 v = 0 \tag{111}$$

To define the adjoint boundary conditions, we consider the homogeneous problem and obtain from (109) and (110) that

$$-v'(1)\, u(1) + v'(0)\, u(0) = 0$$

The adjoint boundary conditions are defined by setting each of the coefficients of $u(1)$ and $u(0)$ equal to zero. The result is

$$v'(0) = 0, \qquad v'(1) = 0$$

Hence $v = \cos \pi y$.

Having defined the adjoint, we return to the inhomogeneous problem. Using (109) in (110) and using the definition of the adjoint, we obtain (107) as the solvability condition.

15.14. Determine the solvability condition for

$$\frac{\partial}{\partial x}\left[p(x, y)\frac{\partial \phi}{\partial x}\right] + \frac{\partial}{\partial y}\left[q(x, y)\frac{\partial \phi}{\partial y}\right] + \lambda r(x, y)\phi = f(x, y) \tag{112}$$

$$\phi_x(0, y) = 0, \qquad \phi_y(x, 0) = 0 \tag{113}$$

$$\frac{\partial \phi}{\partial x}(a, y) - \alpha_1(y)\phi(a, y) = \beta_1(y) \tag{114}$$

$$\frac{\partial \phi}{\partial y}(x, b) - \alpha_2(x)\phi(x, b) = \beta_2(x) \tag{115}$$

We multiply (112) by $\chi(x, y)$ and integrate the result over the domain of interest; that is, $0 \leq x \leq a$ and $0 \leq y \leq b$. The result is

$$\int_0^b \int_0^a \left[\chi \frac{\partial}{\partial x}\left(p \frac{\partial \phi}{\partial x} \right) + \chi \frac{\partial}{\partial y}\left(q \frac{\partial \phi}{\partial y} \right) + \lambda r \chi \phi \right] dx\, dy = \int_0^b \int_0^a \chi f\, dx\, dy \tag{116}$$

Integrating (116) by parts to transfer the derivatives from ϕ to χ, we obtain

$$\int_0^b \left[p\chi \frac{\partial \phi}{\partial x}\Big|_0^a - p \frac{\partial \chi}{\partial x}\phi \Big|_0^a \right] dy + \int_0^a \left[q\chi \frac{\partial \phi}{\partial y} - q\phi \frac{\partial \chi}{\partial y} \right]_0^b dx$$

$$+ \int_0^b \int_0^a \phi \left[\frac{\partial}{\partial x}\left(p \frac{\partial \chi}{\partial x} \right) + \frac{\partial}{\partial y}\left(q \frac{\partial \chi}{\partial y} \right) + \lambda r \chi \right] dx\, dy = \int_0^b \int_0^a \chi f\, dx\, dy \tag{117}$$

We define the adjoint equation by setting the coefficient of ϕ in the integrand of the double integral in (117) equal to zero; that is,

$$\frac{\partial}{\partial x}\left(p \frac{\partial \chi}{\partial x} \right) + \frac{\partial}{\partial y}\left(q \frac{\partial \chi}{\partial y} \right) + \lambda r \chi = 0 \tag{118}$$

To determine the adjoint boundary conditions, we consider the homogeneous problem; that is, $f = 0$, $\beta_1 = 0$, and $\beta_2 = 0$. Thus it follows from (117) that

$$\int_0^b p\left(\chi \frac{\partial \phi}{\partial x} - \phi \frac{\partial \chi}{\partial x} \right)\Big|_0^a dy + \int_0^a q\left(\chi \frac{\partial \phi}{\partial y} - \phi \frac{\partial \chi}{\partial y} \right)\Big|_0^b dx = 0 \tag{119}$$

Substituting for the derivatives of ϕ from the homogeneous equations (113)–(115) into (119), we have

$$\int_0^b p\phi \left(\alpha_1 \chi - \frac{\partial \chi}{\partial x} \right)\Big|_{x=a} dy + \int_0^b p\phi \frac{\partial \chi}{\partial x}\Big|_{x=0} dy + \int_0^a q\phi \left(\alpha_2 \chi - \frac{\partial \chi}{\partial y} \right)\Big|_{y=b} dx$$

$$+ \int_0^a q\phi \frac{\partial \chi}{\partial y}\Big|_{y=0} dx = 0 \tag{120}$$

The adjoint boundary conditions are defined by setting each of the coefficients of $\phi(a, y)$, $\phi(0, y)$, $\phi(x, b)$, and $\phi(x,0)$ equal to zero, independently. The result is

$$\frac{\partial \chi}{\partial x} = 0 \quad \text{at } x = 0 \tag{121}$$

$$\frac{\partial \chi}{\partial y} = 0 \quad \text{at } y = 0 \tag{122}$$

$$\frac{\partial \chi}{\partial x} - \alpha_1 \chi = 0 \quad \text{at } x = a \tag{123}$$

$$\frac{\partial \chi}{\partial y} - \alpha_2 \chi = 0 \quad \text{at } y = b \tag{124}$$

Having defined the adjoint problem, we return to the inhomogeneous problem. Using (113)–(115), (118), and (121)–(124) in (117), we obtain the following solvability condition:

$$\int_0^b p(a,y)\beta_1(y)\chi(a,y)\,dy + \int_0^a q(x,b)\beta_2(x)\chi(x,b)\,dx$$

$$= \int_0^b \int_0^a \chi(x,y) f(x,y)\,dx\,dy \tag{125}$$

where χ is defined by (118) and (121)–(124).

15.15. Consider

$$\begin{aligned} \ddot{x} - \dot{y} + 2x + 3\varepsilon x^2 + 2\varepsilon y^2 &= 0 \\ \ddot{y} + \dot{x} + 2\delta y + 4\varepsilon xy &= 0 \end{aligned} \tag{126}$$

when $\delta = 1 + \varepsilon\sigma$, $\varepsilon \ll 1$. Use the method of multiple scales to show that

$$x = A_1(T_1)\,e^{iT_0} + A_2 e^{2iT_0} + \text{cc} + \cdots \tag{127}$$

where

$$\begin{aligned} A_2' &= \tfrac{1}{3} i\sigma A_2 - \tfrac{1}{2} i A_1^2 \\ A_1' &= \tfrac{1}{3} i\sigma A_1 - i\bar{A}_1 A_2 \end{aligned} \tag{128}$$

Let

$$\begin{aligned} x &= x_0(T_0, T_1) + \varepsilon x_1(T_0, T_1) + \cdots \\ y &= y_0(T_0, T_1) + \varepsilon y_1(T_0, T_1) + \cdots \end{aligned} \tag{129}$$

where

$$T_0 = t, \qquad T_1 = \varepsilon t$$

Substituting (129) into (126), recalling that $\delta = 1 + \varepsilon\sigma$, and equating coefficients of like powers of ε, we obtain

Order ε^0

$$\begin{aligned} D_0^2 x_0 - D_0 y_0 + 2x_0 &= 0 \\ D_0^2 y_0 + D x_0 + 2y_0 &= 0 \end{aligned} \tag{130}$$

Order ε

$$D_0^2 x_1 - D_0 y_1 + 2x_1 = -2D_0 D_1 x_0 + D_1 y_0 - 3x_0^2 - 2y_0^2$$
$$D_0^2 y_1 + D_0 x_1 + 2y_1 = -2D_0 D_1 y_0 - D_1 x_0 - 2\sigma y_0 - 4x_0 y_0 \tag{131}$$

The solution of (130) can be expressed as

$$x_0 = A_1(T_1)e^{iT_0} + A_2(T_1)e^{2iT_0} + \text{cc}$$
$$y_0 = -iA_1(T_1)e^{iT_0} + iA_2(T_1)e^{2iT_0} + \text{cc} \tag{132}$$

Then (131) become

$$D_0^2 x_1 - D_0 y_1 + 2x_1 = -2iA_1' e^{iT_0} - 4iA_2' e^{2iT_0} - iA_1' e^{iT_0}$$
$$+ iA_2' e^{2iT_0} + \text{cc} - 3\left(A_1 e^{iT_0} + A_2 e^{2iT_0} + \bar{A}_1 e^{-iT_0} + \bar{A}_2 e^{-2iT_0}\right)^2$$
$$+ 2\left(-A_1 e^{iT_0} + A_2 e^{2iT_0} + \bar{A}_1 e^{-iT_0} - \bar{A}_2 e^{-2iT_0}\right)^2$$

$$D_0^2 y_1 + D_0 x_1 + 2y_1 = -2A_1' e^{iT_0} + 4A_2' e^{2iT_0} - A_1' e^{iT_0} - A_2' e^{2iT_0}$$
$$+ 2i\sigma A_1 e^{iT_0} - 2i\sigma A_2 e^{2iT_0} + \text{cc}$$
$$-4i\left(A_1 e^{iT_0} + A_2 e^{2iT_0} + \bar{A}_1 e^{-iT_0} + \bar{A}_2 e^{-2iT_0}\right)$$
$$\times\left(-A_1 e^{iT_0} + A_2 e^{2iT_0} + \bar{A}_1 e^{-iT_0} - \bar{A}_2 e^{-2iT_0}\right)$$

or

$$D_0^2 x_1 - D_0 y_1 + 2x_1 = \left(-3iA_1' - 2A_2\bar{A}_1\right)e^{iT_0} + \left(-3iA_2' - A_1^2\right)e^{2iT_0}$$
$$+ \text{cc} + \text{NST} \tag{133}$$

$$D_0^2 y_1 + D_0 x_1 + 2y_1 = \left(-3A_1' + 2i\sigma A_1 - 8iA_2\bar{A}_1\right)e^{iT_0}$$
$$+ \left(3A_2' - 2i\sigma A_2 + 4iA_1^2\right)e^{2iT_0} + \text{cc} + \text{NST} \tag{134}$$

To determine the solvability conditions of (133) and (134), we seek a particular solution in the form

$$x_1 = P_1 e^{iT_0} + P_2 e^{2iT_0}$$
$$y_1 = Q_1 e^{iT_0} + Q_2 e^{2iT_0} \tag{135}$$

Substituting (135) into (133) and (134) and equating the coefficients of the exponents on both sides, we have

$$P_1 - iQ_1 = -3iA_1' - 2A_2\bar{A}_1 \tag{136}$$

$$-2P_2 - 2iQ_2 = -3iA_2' - A_1^2 \tag{137}$$

$$iP_1 + Q_1 = -3A_1' + 2i\sigma A_1 - 8iA_2\bar{A}_1 \tag{138}$$

$$2iP_2 - 2Q_2 = 3A_2' - 2i\sigma A_2 + 4iA_1^2 \tag{139}$$

The solvability condition of (136) and (138) is

$$\begin{vmatrix} 1 & -3iA_1' - 2A_2\bar{A}_1 \\ i & -3A_1' + 2i\sigma A_1 - 8iA_2\bar{A}_1 \end{vmatrix} = 0$$

or

$$A_1' = \tfrac{1}{3}i\sigma A_1 - iA_2\bar{A}_1 \tag{140}$$

The solvability condition of (137) and (139) is

$$\begin{vmatrix} -2 & -3iA_2' - A_1^2 \\ 2i & 3A_2' - 2i\sigma A_2 + 4iA_1^2 \end{vmatrix} = 0$$

or

$$A_2' = \tfrac{1}{3}i\sigma A_2 - \tfrac{1}{2}iA_1^2 \tag{141}$$

15.16. The free response of a two-degree-of-freedom system is governed by

$$\begin{aligned} \ddot{u}_1 + \tfrac{1}{2}\dot{u}_2 + \delta u_1 &= \varepsilon u_1 u_2 \\ \ddot{u}_2 - \tfrac{1}{2}\dot{u}_1 + \tfrac{1}{2}u_2 &= \varepsilon u_1^2 \end{aligned} \tag{142}$$

where $\varepsilon \ll 1$ and $\delta = \frac{1}{2} + \varepsilon\sigma$. Determine the equations describing the amplitudes and phases.

We let

$$\begin{aligned} u_1 &= u_{10}(T_0, T_1) + \varepsilon u_{11}(T_0, T_1) + \cdots \\ u_2 &= u_{20}(T_0, T_1) + \varepsilon u_{21}(T_0, T_1) + \cdots \end{aligned} \tag{143}$$

Substituting (143) into (142) and equating coefficients of like powers of ε, we obtain

Order ε^0

$$D_0^2 u_{10} + \tfrac{1}{2} D_0 u_{20} + \tfrac{1}{2} u_{10} = 0$$
$$D_0^2 u_{20} - \tfrac{1}{2} D_0 u_{10} + \tfrac{1}{2} u_{20} = 0 \tag{144}$$

Order ε

$$D_0^2 u_{11} + \tfrac{1}{2} D_0 u_{21} + \tfrac{1}{2} u_{21} = -\sigma u_{10} - 2 D_0 D_1 u_{10} - \tfrac{1}{2} D_1 u_{20} + u_{10} u_{20}$$
$$D_0^2 u_{21} - \tfrac{1}{2} D_0 u_{11} + \tfrac{1}{2} u_{21} = -2 D_0 D_1 u_{20} + \tfrac{1}{2} D_1 u_{10} + u_{10}^2 \tag{145}$$

We seek a solution to the zeroth-order problem in the form

$$u_{10} = c_1 e^{i\omega T_0}, \qquad u_{20} = c_2 e^{i\omega T_0}$$

Then

$$\left(\tfrac{1}{2} - \omega^2\right) c_1 + \tfrac{1}{2} i\omega c_2 = 0$$
$$-\tfrac{1}{2} i\omega c_1 + \left(\tfrac{1}{2} - \omega^2\right) c_2 = 0$$

The vanishing of the determinant of the coefficient matrix yields

$$\left(\tfrac{1}{2} - \omega^2\right)^2 - \tfrac{1}{4}\omega^2 = 0$$

or

$$\omega^4 - \tfrac{5}{4}\omega^2 + \tfrac{1}{4} = 0$$

or

$$\left(\omega^2 - \tfrac{1}{4}\right)\left(\omega^2 - 1\right) = 0$$

Hence

$$\omega = \tfrac{1}{2} \quad \text{or} \quad 1$$

and

$$c_2 = -\frac{\tfrac{1}{2} - \omega^2}{\tfrac{1}{2} i\omega} c_1$$

Then

$$c_2 = i c_1 \quad \text{when } \omega = \tfrac{1}{2}$$
$$c_2 = -i c_1 \quad \text{when } \omega = 1$$

Therefore, the solution of the zeroth-order problem can be written as

$$\begin{aligned} u_{10} &= A_1(T_1)\, e^{iT_0/2} + A_2(T_1)\, e^{iT_0} + \text{cc} \\ u_{20} &= iA_1 e^{iT_0/2} - iA_2 e^{iT_0} + \text{cc} \end{aligned} \tag{146}$$

Then Equations (145) become

$$\begin{aligned} D_0^2 u_{11} + \tfrac{1}{2} D_0 u_{21} + \tfrac{1}{2} u_{11} &= \left[-\sigma A_1 - iA_1' - \tfrac{1}{2} iA_1' - 2iA_2 \bar{A}_1 \right] e^{iT_0/2} \\ &\quad + \left[-\sigma A_2 - 2iA_2' + \tfrac{1}{2} iA_2' + iA_1^2 \right] e^{iT_0} + \text{cc} + \text{NST} \end{aligned} \tag{147}$$

$$\begin{aligned} D_0^2 u_{21} - \tfrac{1}{2} D_0 u_{11} + \tfrac{1}{2} u_{21} &= \left[A_1' + \tfrac{1}{2} A_1' + 2A_2 \bar{A}_1 \right] e^{iT_0/2} \\ &\quad + \left[-2A_2' + \tfrac{1}{2} A_2' + A_1^2 \right] e^{iT_0} + \text{cc} + \text{NST} \end{aligned} \tag{148}$$

Solvability Conditions. We let

$$u_{11} = \chi_1 e^{iT_0/2}, \qquad u_{21} = \chi_2 e^{iT_0/2}$$

Then

$$\tfrac{1}{4}\chi_1 + \tfrac{1}{4} i\chi_2 = -\sigma A_1 - \tfrac{3}{2} iA_1' - 2iA_2 \bar{A}_1 \tag{149}$$

$$-\tfrac{1}{4} i\chi_1 + \tfrac{1}{4}\chi_2 = \tfrac{3}{2} A_1' + 2A_2 \bar{A}_1 \tag{150}$$

Multiplying (149) by i and adding the result to (150) yields

$$3A_1' - i\sigma A_1 + 4A_2 \bar{A}_1 = 0$$

or

$$A_1' = \tfrac{1}{3} i\sigma A_1 - \tfrac{4}{3} A_2 \bar{A}_1 \tag{151}$$

We let

$$u_{11} = \chi_1 e^{iT_0}, \qquad u_{21} = \chi_2 e^{iT_0}$$

Then

$$-\tfrac{1}{2}\chi_1 + \tfrac{1}{2} i\chi_2 = -\tfrac{3}{2} iA_2' - \sigma A_2 + iA_1^2 \tag{152}$$

$$-\tfrac{1}{2} i\chi_1 - \tfrac{1}{2}\chi_2 = -\tfrac{3}{2} A_2' + A_1^2 \tag{153}$$

Multiplying (152) by $-i$ and adding the result to (153) yields

$$-3A_2' + i\sigma A_2 + 2A_1^2 = 0$$

or

$$A_2' = \tfrac{1}{3} i\sigma A_2 + \tfrac{2}{3} A_1^2 \tag{154}$$

15.17. Use the method of multiple scales to determine the equations describing the amplitudes and the phases of the system

$$\ddot{u}_1 + \omega_1^2 u_1 = \alpha_1 u_2 u_3 \tag{155}$$

$$\ddot{u}_2 + \omega_2^2 u_2 = \alpha_2 u_1 u_3 \tag{156}$$

$$\ddot{u}_3 + \omega_3^2 u_3 = \alpha_3 u_1 u_2 \tag{157}$$

when $\omega_3 \approx \omega_1 + \omega_2$.

We consider the case of small but finite amplitudes and introduce a small dimensionless parameter ε that is the order of the amplitudes as a bookkeeping device. Moreover, we seek a first-order uniform expansion in the form

$$u_n = \varepsilon u_{n_1}(T_0, T_1) + \varepsilon^2 u_{n_2}(T_0, T_1) + \cdots \tag{158}$$

Substituting (158) into (155)–(157) and equating coefficients of like powers of ε, we obtain

$$D_0^2 u_{n_1} + \omega_n^2 u_{n_1} = 0 \tag{159}$$

$$D_0^2 u_{12} + \omega_1^2 u_{12} = -2 D_0 D_1 u_{11} + \alpha_1 u_{21} u_{31} \tag{160}$$

$$D_0^2 u_{22} + \omega_2^2 u_{22} = -2 D_0 D_1 u_{21} + \alpha_2 u_{11} u_{31} \tag{161}$$

$$D_0^2 u_{32} + \omega_3^2 u_{32} = -2 D_0 D_1 u_{31} + \alpha_3 u_{11} u_{21} \tag{162}$$

The solution of (159) can be expressed as

$$u_{n_1} = A_n(T_1) e^{i\omega_n T_0} + \bar{A}_n e^{-i\omega_n T_0} \tag{163}$$

Then (160)–(162) become

$$D_0^2 u_{12} + \omega_1^2 u_{12} = -2i\omega_1 A_1' e^{i\omega_1 T_0} + \alpha_1 A_3 A_2 e^{i(\omega_3+\omega_2)T_0} + \alpha_1 A_3 \bar{A}_2 e^{i(\omega_3-\omega_2)T_0} + \text{cc} \tag{164}$$

$$D_0^2 u_{22} + \omega_2^2 u_{22} = -2i\omega_2 A_2' e^{i\omega_2 T_0} + \alpha_2 A_3 A_1 e^{i(\omega_3+\omega_1)T_0} + \alpha_2 A_3 \bar{A}_1 e^{i(\omega_3-\omega_1)T_0} + \text{cc} \tag{165}$$

$$D_0^2 u_{32} + \omega_3^2 u_{32} = -2i\omega_3 A_3' e^{i\omega_3 T_0} + \alpha_3 A_2 A_1 e^{i(\omega_2+\omega_1)T_0} + \alpha_3 A_2 \bar{A}_1 e^{i(\omega_2-\omega_1)T_0} + \text{cc} \tag{166}$$

To express the nearness of ω_3 to $\omega_2 + \omega_1$, we introduce the detuning parameter σ defined by

$$\omega_3 = \omega_2 + \omega_1 + \varepsilon\sigma$$

so that we can write

$$(\omega_3 - \omega_2)T_0 = \omega_1 T_0 + \sigma T_1$$

$$(\omega_3 - \omega_1)T_0 = \omega_2 T_0 + \sigma T_1$$

$$(\omega_2 + \omega_1)T_0 = \omega_3 T_0 - \sigma T_1$$

Then the solvability conditions (in this case, elimination of the secular terms) of (164)–(166) yield

$$2i\omega_1 A_1' = \alpha_1 A_3 \bar{A}_2 e^{i\sigma T_1}$$

$$2i\omega_2 A_2' = \alpha_2 A_3 \bar{A}_1 e^{i\sigma T_1}$$

$$2i\omega_3 A_3' = \alpha_3 A_2 A_1 e^{-i\sigma T_1}$$

Expressing the A_n in the polar form $\frac{1}{2}\, a_n \exp(i\beta_n)$ and separating real and imaginary parts, we obtain

$$a_1' = \alpha_1 \frac{a_3 a_2}{4\omega_1} \sin\gamma \tag{167a}$$

$$\beta_1' = -\alpha_1 \frac{a_3 a_2}{4\omega_1 a_1} \cos\gamma \tag{167b}$$

$$a_2' = \alpha_2 \frac{a_3 a_1}{4\omega_2} \sin\gamma \tag{168a}$$

$$\beta_2' = -\alpha_2 \frac{a_3 a_1}{4\omega_2 a_2} \cos\gamma \tag{168b}$$

$$a_3' = -\alpha_3 \frac{a_2 a_1}{4\omega_3} \sin\gamma \tag{169a}$$

$$\beta_3' = -\alpha_3 \frac{a_2 a_1}{4\omega_3 a_3} \cos\gamma \tag{169b}$$

where

$$\gamma = \beta_3 - \beta_2 - \beta_1 + \sigma T_1 \tag{170}$$

15.18. Consider

$$\ddot{u}_1 + \dot{u}_2 + 2u_1 + 2\varepsilon \cos\Omega t(f_{11}u_1 + f_{12}u_2) = 0 \tag{171}$$

$$\ddot{u}_2 - \dot{u}_1 + 2u_2 + 2\varepsilon \cos\Omega t(f_{21}u_1 + f_{22}u_2) = 0 \tag{172}$$

Determine the equations describing the complex amplitudes when (a) $\Omega \approx 1$, (b) $\Omega \approx 2$, and (c) $\Omega \approx 4$.

(a) When $\Omega \approx 1$, we let

$$\Omega = 1 + \varepsilon\sigma$$

and thus write $\Omega T_0 = T_0 + \sigma T_1$. Substituting (143) into (171) and (172) and equating coefficients of like powers of ε, we obtain equations for the determination of u_{mn}. The solution of the zeroth-order problem is

$$u_{10} = A_1(T_1)\, e^{iT_0} + A_2(T_1)\, e^{2iT_0} + \text{cc} \tag{173a}$$

$$u_{20} = iA_1(T_1)\, e^{iT_0} - iA_2(T)\, e^{2iT_0} + \text{cc} \tag{173b}$$

Then the first-order problem becomes

$$D_0^2 u_{11} + D_0 u_{21} + 2u_{11} = \left[-3iA_1' - (f_{11} - if_{12})A_2 e^{-i\sigma T_1}\right] e^{iT_0}$$
$$+ \left[-3iA_2' - (f_{11} + if_{12})A_1 e^{i\sigma T_1}\right] e^{2iT_0} + \text{cc} + \text{NST} \tag{174}$$

$$D_0^2 u_{21} - D_0 u_{11} + 2u_{21} = \left[3A_1' - (f_{21} - if_{22})A_2 e^{-i\sigma T_1}\right] e^{iT_0}$$
$$+ \left[-3A_2' - (f_{21} + if_{22})A_1 e^{i\sigma T_1}\right] e^{2iT_0} + \text{cc} + \text{NST} \tag{175}$$

To determine the solvability conditions for (174) and (175), we seek a particular solution in the form

$$\begin{aligned} u_{11} &= P_1 e^{iT_0} + P_2 e^{2iT_0} \\ u_{21} &= Q_1 e^{iT_0} + Q_2 e^{2iT_0} \end{aligned} \tag{176}$$

Substituting (176) into (174) and (175) and equating the coefficients of the exponentials on both sides, we obtain

$$\begin{aligned} P_1 + iQ_1 &= -3iA_1' - (f_{11} - if_{12})A_2 e^{-i\sigma T_1} \\ -iP_1 + Q_1 &= 3A_1' - (f_{21} - if_{22})A_2 e^{-i\sigma T_1} \end{aligned} \tag{177}$$

$$\begin{aligned} -2P_2 + 2iQ_2 &= -3iA_2' - (f_{11} + if_{12})A_1 e^{i\sigma T_1} \\ -2iP_2 - 2Q_2 &= -3A_2' - (f_{21} + if_{22})A_1 e^{i\sigma T_1} \end{aligned} \tag{178}$$

The solvability condition of (177) is

$$\begin{vmatrix} 1 & -3iA_1' - (f_{11} - if_{12})A_2 e^{-i\sigma T_1} \\ -i & 3A_1' - (f_{21} - if_{22})A_2 e^{-i\sigma T_1} \end{vmatrix} = 0$$

or

$$A_1' = \tfrac{1}{6}(f_{21} + f_{12} + if_{11} - if_{22})A_2 e^{-i\sigma T_1} \tag{179}$$

The solvability condition of (178) is

$$\begin{vmatrix} -2 & -3iA_2' - (f_{11} + if_{12}) A_1 e^{i\sigma T_1} \\ -2i & -3A_2' - (f_{21} + if_{22}) A_1 e^{i\sigma T_1} \end{vmatrix} = 0$$

or

$$A_2' = -\tfrac{1}{6}(f_{21} + f_{12} - if_{11} + if_{22}) e^{i\sigma T_1} \tag{180}$$

(b) When $\Omega \approx 2$, we let

$$\Omega = 2 + \varepsilon\sigma$$

and write

$$\Omega T_0 = 2T_0 + \sigma T_1$$

Then the first-order problem becomes

$$D_0^2 u_{11} + D_0 u_{21} + 2u_{11} = \left[-3iA_1' - (f_{11} - if_{12})\bar{A}_1 e^{i\sigma T_1}\right] e^{iT_0} - 3iA_2' e^{2iT_0} + \text{cc} + \text{NST} \tag{181}$$

$$D_0^2 u_{21} - D_0 u_{11} + 2u_{21} = \left[3A_1' - (f_{21} - if_{22})\bar{A}_1 e^{i\sigma T_1}\right] e^{iT_0} - 3A_2' e^{2iT_0} + \text{cc} + \text{NST} \tag{182}$$

To determine the solvability conditions for (181) and (182), we seek a particular solution in the form (176) and obtain

$$\begin{aligned} P_1 + iQ_1 &= -3iA_1' - (f_{11} - if_{12})\bar{A}_1 e^{i\sigma T_1} \\ -iP_1 + Q_1 &= 3A_1' - (f_{21} - if_{22})\bar{A}_1 e^{i\sigma T_1} \end{aligned} \tag{183}$$

$$\begin{aligned} -2P_2 + 2iQ_2 &= -3iA_2' \\ -2iP_2 - 2Q_2 &= -3A_2' \end{aligned} \tag{184}$$

The solvability condition of (183) is

$$\begin{vmatrix} 1 & -3iA_1' - (f_{11} - if_{12})\bar{A}_1 e^{i\sigma T_1} \\ -i & 3A_1' - (f_{21} - if_{22})\bar{A}_1 e^{i\sigma T_1} \end{vmatrix} = 0$$

or

$$A_1' = \tfrac{1}{6}(f_{21} + f_{12} + if_{11} - if_{22})\bar{A}_1 e^{i\sigma T_1} \tag{185}$$

The solvability condition of (184) is $A_2' = 0$.

(c) When $\Omega \approx 4$, we let

$$\Omega = 4 + \varepsilon\sigma$$

and thus write

$$\Omega T_0 = 4T_0 + \sigma T_1$$

Then the first-order problem becomes

$$D_0^2 u_{11} + D_0 u_{21} + 2u_{11} = -3iA_1' e^{iT_0} + \left[-3iA_2' - (f_{11} + if_{12})\bar{A}_2 e^{i\sigma T_1}\right] e^{2iT_0} + \text{cc} + \text{NST} \tag{186}$$

$$D_0^2 u_{21} - D_0 u_{11} + 2u_{21} = 3A_1' e^{iT_0} + \left[-3A_2' - (f_{21} + if_{22})\bar{A}_2 e^{i\sigma T_1}\right] e^{2iT_0} + \text{cc} + \text{NST} \tag{187}$$

We seek a particular solution for (186) and (187) in the form (176) and obtain

$$\begin{aligned} P_1 + iQ_1 &= -3iA_1' \\ -iP_1 + Q_1 &= 3A_1' \end{aligned} \tag{188}$$

$$\begin{aligned} -2P_2 + 2iQ_2 &= -3iA_2' - (f_{11} + if_{12})\bar{A}_2 e^{i\sigma T_1} \\ -2iP_2 - 2Q_2 &= -3A_2' - (f_{21} + if_{22})\bar{A}_2 e^{i\sigma T_1} \end{aligned} \tag{189}$$

The solvability condition of (188) is $A_1' = 0$, whereas the solvability condition of (189) is

$$\begin{vmatrix} -2 & -3iA_2' - (f_{11} + if_{12})\bar{A}_2 e^{i\sigma T_1} \\ -2i & -3A_2' - (f_{21} + if_{22})\bar{A}_2 e^{i\sigma T_1} \end{vmatrix} = 0$$

or

$$A_2' = -\tfrac{1}{6}(f_{12} + f_{21} + if_{22} - if_{11})\bar{A}_2 e^{i\sigma T_1} \tag{190}$$

15.19. Determine a first-order expansion for the eigenvalues of

$$\begin{aligned} u'' + \lambda u &= \varepsilon f(x) u \\ u(0) = 0, \qquad & u(1) = 0 \end{aligned} \tag{191}$$

where $\varepsilon \ll 1$.

We let

$$\begin{aligned} u &= u_0(x) + \varepsilon u_1(x) + \cdots \\ \lambda &= \lambda_0 + \varepsilon\lambda_1 + \cdots \end{aligned} \tag{192}$$

Substituting (192) into (191) and equating coefficients of like powers of ε, we obtain

$$u_0'' + \lambda_0 u_0 = 0$$
$$u_0(0) = 0, \qquad u_0(1) = 0 \tag{193}$$

$$u_1'' + \lambda_0 u_1 = f(x) u_0 - \lambda_1 u_0$$
$$u_1(0) = 0, \qquad u_1(1) = 0 \tag{194}$$

The general solution of the equation in (193) is

$$u_0 = c_1 \cos\left(\sqrt{\lambda_0}\, x\right) + c_2 \sin\left(\sqrt{\lambda_0}\, x\right)$$

Using the boundary conditions, we obtain $c_1 = 0$ and

$$\sin\sqrt{\lambda_0} = 0 \quad \text{or} \quad \lambda_0 = n^2\pi^2$$

where n is an integer. Hence

$$u_0 = \sin(n\pi x) \tag{195}$$

Since the homogeneous problem (194) is the same as (193) and since the latter problem has the nontrivial solution (195), the inhomogeneous problem (194) has a solution only if a solvability condition is satisfied. To determine this solvability condition, we multiply the equation in (194) by $v(x)$, integrate the result by parts from $x = 0$ to $x = 1$ to transfer the derivatives from u_1 to v, and obtain

$$\int_0^1 u_1[v'' + n^2\pi^2 v]\, dx + \left[u_1'v - u_1 v'\right]_0^1 = \int_0^1 v(x)[f(x) - \lambda_1]\sin n\pi x\, dx \tag{196}$$

The adjoint equation is

$$v'' + n^2\pi^2 v = 0 \tag{197}$$

To define the adjoint boundary conditions, we consider the homogeneous problem and obtain from (194) and (196) that

$$u_1'(1)\, v(1) - u_1'(0)\, v(0) = 0$$

Then the adjoint boundary conditions are chosen so that each of the coefficients of $u_1'(1)$ and $u_1'(0)$ vanishes independently. The result is

$$v(0) = 0, \qquad v(1) = 0 \tag{198}$$

Hence $v = \sin n\pi x$.

Having defined the adjoint, we return to the inhomogeneous problem. Using (196) and the definition of the adjoint, we obtain

$$\int_0^1 f(x)\sin^2 n\pi x\, dx = \lambda_1 \int_0^1 \sin^2 n\pi x\, dx$$

or

$$\lambda_1 = 2\int_0^1 f(x)\sin^2 n\pi x\,dx$$

Therefore

$$\lambda = n^2\pi^2 + 2\varepsilon\int_0^1 f(x)\sin^2 n\pi x\,dx + \cdots \tag{199}$$

15.20. Consider the eigenvalue problem

$$\begin{gathered} u'' + [\lambda + \varepsilon f(x)]u = 0, \qquad \varepsilon \ll 1 \\ u(0) = 0, \qquad u'(1) = 0 \end{gathered} \tag{200}$$

Determine a first-order expansion for λ.

We let

$$\begin{gathered} u = u_0(x) + \varepsilon u_1(x) + \cdots \\ \lambda = \lambda_0 + \varepsilon\lambda_1 + \cdots \end{gathered} \tag{201}$$

Substituting (201) into (200) and equating coefficients of like powers of ε, we obtain

Order ε^0

$$\begin{gathered} u_0'' + \lambda_0 u_0 = 0 \\ u_0(0) = 0, \qquad u_0'(1) = 0 \end{gathered} \tag{202}$$

Order ε

$$\begin{gathered} u_0'' + \lambda_0 u_1 = -\lambda_1 u_0 - u_0 f(x) \\ u_1(0) = 0, \qquad u_1'(1) = 0 \end{gathered} \tag{203}$$

The solution of (202) is

$$u_0 = c_n \sin(n + \tfrac{1}{2})\pi x, \qquad \lambda_0 = (n + \tfrac{1}{2})^2\pi^2 \tag{204}$$

where n is an integer and

$$c_n^2\int_0^1 \sin^2(n + \tfrac{1}{2})\pi x\,dx = 1 \Rightarrow \tfrac{1}{2}c_n^2 = 1, \quad \text{or} \quad c_n = \sqrt{2}$$

Substituting (204) into the equation in (203) yields

$$u_1'' + (n + \tfrac{1}{2})^2\pi^2 u_1 = -\lambda_1\sqrt{2}\sin(n + \tfrac{1}{2})\pi x - \sqrt{2}\sin(n + \tfrac{1}{2})\pi x f(x) \tag{205}$$

To determine the solvability condition for the problem governing u_1, we multiply (205) by $v(x)$, integrate the result by parts from $x=0$ to $x=1$ to transfer the derivatives from u_1 to v, and obtain

$$\left[u_1'v - u_1v'\right]_0^1 + \int_0^1 u_1\left[v'' + \left(n+\tfrac{1}{2}\right)^2\pi^2 v\right] dx$$

$$= -\lambda_1\int_0^1 \sqrt{2}\, v \sin\left(n+\tfrac{1}{2}\pi\right)x\, dx - \int_0^1 \sqrt{2}\, v(x)\sin\left(n+\tfrac{1}{2}\right)\pi x f(x)\, dx \tag{206}$$

Since the problem is self-adjoint, we choose

$$v = \sqrt{2}\sin\left(n+\tfrac{1}{2}\right)\pi x$$

Then

$$-\lambda_1 - 2\int_0^1 f(x)\sin^2\left(n+\tfrac{1}{2}\right)\pi x\, dx = 0$$

or

$$\lambda_1 = -2\int_0^1 f(x)\sin^2\left(n+\tfrac{1}{2}\right)\pi x\, dx$$

Therefore

$$\lambda = \left(n+\tfrac{1}{2}\right)^2\pi^2 - 2\varepsilon\int_0^1 f(x)\sin^2\left(n+\tfrac{1}{2}\right)\pi x\, dx \tag{207}$$

15.21. Consider the problem

$$u^{iv} + 5u'' + [\lambda + \varepsilon f(x)]u = 0, \qquad \varepsilon \ll 1$$
$$u(0) = u''(0) = u(\pi) = u''(\pi) = 0 \tag{208}$$

Show that when $\lambda \approx 4$,

$$\lambda = 4 + \varepsilon\lambda_1 + \cdots \tag{209}$$

where

$$\lambda_1^2 + (f_{11} + f_{22})\lambda_1 + f_{11}f_{22} - f_{12}^2 = 0 \tag{210}$$

$$f_{11} = \frac{2}{\pi}\int_0^\pi f(x)\sin^2 x\, dx, \qquad f_{22} = \frac{2}{\pi}\int_0^\pi f(x)\sin^2 2x\, dx,$$
$$f_{12} = \frac{2}{\pi}\int_0^\pi f(x)\sin x \sin 2x\, dx \tag{211}$$

We seek an expansion for u in the form

$$u = u_0(x) + \varepsilon u_1(x) + \cdots \tag{212}$$

and for λ as in (209). Substituting (209) and (212) into (208) and equating coefficients of like powers of ε, we obtain

Order ε^0

$$u_0^{iv} + 5u_0'' + 4u_0 = 0$$
$$u_0(0) = u_0''(0) = u_0(\pi) = u_0''(\pi) = 0 \tag{213}$$

Order ε

$$u_1^{iv} + 5u_1'' + 4u_1 = -[\lambda_1 + f(x)]u_0 \tag{214}$$

$$u_1(0) = u_1''(0) = u_1(\pi) = u_1''(\pi) = 0 \tag{215}$$

The solution of (213) is

$$u_0 = a_1 \sin x + a_2 \sin 2x \tag{216}$$

showing that there are two eigenfunctions corresponding to the same eigenvalue; that is, the problem is degenerate. The constants a_1 and a_2 are arbitrary at this order. The relationship between them may be determined and hence the degeneracy may be removed at the next order.

Since the homogeneous first-order problem, (214) and (215), is the same as (213) and since the latter has a nontrivial solution, the inhomogeneous problem has a solution only if a solvability condition is satisfied. To determine the solvability condition, we multiply (214) by $v(x)$, integrate the result by parts from $x=0$ to $x=\pi$ to transfer the derivatives from u_1 to v, and obtain

$$\int_0^\pi u_1[v^{iv} + 5v'' + 4v]\,dx + [u_1'''v - u_1''v' + u_1'v'' - u_1v''' + 5u_1'v - 5u_1v']_0^\pi$$
$$= -\int_0^\pi [\lambda_1 + f(x)]vu_0\,dx \tag{217}$$

The adjoint equation is defined by setting the coefficient of u_1 in the integrand in (217) equal to zero. The result is

$$v^{iv} + 5v'' + 4v = 0 \tag{218}$$

To define the adjoint boundary conditions, we consider the homogeneous problem and obtain from (217) and (215) that

$$u_1'''(\pi)v(\pi) + u_1'(\pi)[v''(\pi) + 5v(\pi)] - u_1'''(0)v(0) - u_1'(0)[v''(0) + 5v(0)] = 0 \tag{219}$$

We choose the adjoint boundary conditions by setting each of the coefficients of $u_1'''(\pi)$, $u_1'(\pi)$, $u_1'''(0)$, and $u_1'(0)$ equal to zero. The result is

$$v(0) = v''(0) = v(\pi) = v''(\pi) = 0 \tag{220}$$

Thus the problem is self-adjoint and

$$v = \sin x \quad \text{or} \quad \sin 2x \tag{221}$$

Having defined the adjoint, we return to the inhomogeneous problem. Using (215), (218), and (220) in (217), we obtain

$$\int_0^\pi [\lambda_1 + f(x)]\, v u_0(x)\, dx = 0 \tag{222}$$

Equation (222) should be satisfied for all possible solutions of the adjoint problem; that is, $v = \sin x$ and $v = \sin 2x$. Substituting (216) into (222) and letting $v = \sin x$, we have

$$\lambda_1 \int_0^\pi (a_1 \sin x + a_2 \sin 2x) \sin x\, dx + \int_0^\pi (a_1 \sin x + a_2 \sin 2x) f(x) \sin x\, dx = 0$$

Hence

$$(\lambda_1 + f_{11}) a_1 + f_{12} a_2 = 0 \tag{223}$$

where f_{11} and f_{12} are defined by (211). Substituting (216) into (222) and letting $v = \sin 2x$, we have

$$\lambda_1 \int_0^\pi (a_1 \sin x + a_2 \sin 2x) \sin 2x\, dx + \int_0^\pi (a_1 \sin x + a_2 \sin 2x) f(x) \sin 2x\, dx = 0$$

Hence

$$f_{12} a_1 + (\lambda_1 + f_{22}) a_2 = 0 \tag{224}$$

where f_{22} is defined by (211). Setting the determinant of the coefficient matrix in (223) and (224) equal to zero yields (210). Moreover, it follows from (223) that

$$a_2 = -\frac{\lambda_1 + f_{11}}{f_{12}} a_1$$

giving the relationship between a_1 and a_2.

15.22. Determine the solvability conditions for

$$p_2(x) y'' + p_1(x) y' + p_0(x) y = f(x) \tag{225}$$

subject to the boundary conditions

(a) $$y(0) = \beta_1, \qquad y(1) = \beta_2 \tag{226}$$

(b) $$y'(0) = \alpha_1 y(0) + \beta_1, \qquad y'(1) = \alpha_2 y(1) + \beta_2 \tag{227}$$

(c) $$y(0) = \alpha_{11} y'(0) + \alpha_{12} y'(1)$$

$$y(1) = \alpha_{21} y'(0) + \alpha_{22} y'(1) \tag{228}$$

We multiply (225) by $u(x)$, integrate the result by parts from $x=0$ to $x=1$ to transfer the derivatives from y to u, and obtain

$$\int_0^1 y\left[p_2u''+(2p_2'-p_1)u'+(p_0+p_2''-p_1')u\right]dx$$

$$+\left\{p_2uy'+\left[(p_1-p_2')u-p_2u'\right]y\right\}_0^1=\int_0^1 fu\,dx \tag{229}$$

The adjoint equation is

$$p_2u''+(2p_2'-p_1)u'+(p_0+p_2''-p_1')u=0 \tag{230}$$

To define the adjoint boundary conditions, we consider the homogeneous problem and obtain from (229) that

$$\left\{p_2uy'+\left[(p_1-p_2')u-p_2u'\right]y\right\}_0^1=0 \tag{231}$$

(a) In this case, it follows from (226) that the homogeneous boundary conditions are

$$y(0)=0 \quad \text{and} \quad y(1)=0$$

Then (231) becomes

$$p_2(1)u(1)y'(1)-p_2(0)u(0)y'(0)=0$$

The adjoint boundary conditions are chosen by setting each of the coefficients of $y'(1)$ and $y'(0)$ equal to zero. The result is

$$u(0)=0, \qquad u(1)=0 \tag{232}$$

Returning to the inhomogeneous problem, we use (226) in (229) and the definition of the adjoint and obtain

$$\int_0^1 f(x)u(x)\,dx=\beta_1p_2(0)u'(0)-\beta_2p_2(1)u'(1) \tag{233}$$

where $u(x)$ is any solution of (230) and (232).

(b) In this case, it follows from (227) that the homogeneous boundary conditions are

$$y'(0)=\alpha_1y(0) \quad \text{and} \quad y'(1)=\alpha_2y(1)$$

which, upon substitution into (231), yields

$$\left[(\alpha_2p_2+p_1-p_2')u-p_2u'\right]_{x=1}y(1)-\left[(\alpha_1p_2+p_1-p_2')u-p_2u'\right]_{x=0}y(0)=0$$

The adjoint boundary conditions are defined by setting each of the coefficients of $y(0)$

and $y(1)$ equal to zero. The result is

$$\begin{aligned} p_2 u' &= \left(\alpha_1 p_2 + p_1 - p_2'\right) u \quad \text{at } x=0 \\ p_2 u' &= \left(\alpha_2 p_2 + p_1 - p_2'\right) u \quad \text{at } x=1 \end{aligned} \tag{234}$$

Returning to the inhomogeneous problem, we substitute (227) into (229), use the definition of the adjoint, and obtain

$$\int_0^1 f(x)\,u(x)\,dx = \beta_2 p_2(1)\,u(1) - \beta_1 p_2(0)\,u(0) \tag{235}$$

where $u(x)$ is any solution of (230) and (234).

(c) In this case, the boundary conditions are homogeneous. Substituting for $y(0)$ and $y(1)$ from (228) into (231) and collecting the coefficients of $y'(0)$ and $y'(1)$, we obtain

$$\begin{aligned} &\left\{\left[p_2 + \alpha_{22}\left(p_1 - p_2'\right)\right]_{x=1} u(1) - \alpha_{22} p_2(1)\,u'(1) - \alpha_{12}\left(p_1 - p_2'\right)_{x=0} u(0) \right. \\ &\qquad \left. + \alpha_{12} p_2(0)\,u'(0)\right\} y'(1) \\ &\qquad - \left\{\left[p_2 + \alpha_{11}\left(p_1 - p_2'\right)\right]_{x=0} u(0) - \alpha_{11} p_2(0)\,u'(0) - \alpha_{21}\left(p_1 - p_2'\right)_{x=1} u(1) \right. \\ &\qquad \left. + \alpha_{21} p_2(1)\,u'(1)\right\} y'(0) = 0 \end{aligned} \tag{236}$$

We choose the adjoint boundary conditions by setting each of the coefficients of $y'(1)$ and $y'(0)$ equal to zero. The result is

$$\begin{aligned} &\left[p_2 + \alpha_{22}\left(p_1 - p_2'\right)\right]_{x=1} u(1) - \alpha_{22} p_2(1)\,u'(1) - \alpha_{12}\left(p_1 - p_2'\right)_{x=0} u(0) \\ &\qquad + \alpha_{12} p_2(0)\,u'(0) = 0 \end{aligned} \tag{237}$$

$$\begin{aligned} &- \alpha_{21}\left(p_1 - p_2'\right)_{x=1} u(1) + \alpha_{21} p_2(1)\,u'(1) + \left[p_2 + \alpha_{11}\left(p_1 - p_2'\right)\right]_{x=0} u(0) \\ &\qquad - \alpha_{11} p_2(0)\,u'(0) = 0 \end{aligned} \tag{238}$$

Returning to the inhomogeneous problem, we substitute (228) into (229), use (230) and (236), and obtain

$$\int_0^1 f(x)\,u(x)\,dx = 0 \tag{239}$$

where $u(x)$ is any solution of (230) subject to the boundary conditions (237) and (238).

15.23. Determine the conditions under which

$$p_3 y''' + p_2 y'' + p_1 y' + p_0 y = 0 \tag{240}$$

is self-adjoint.

Multiplying (240) by $u(x)$ and integrating the result by parts from $x = a$ to $x = b$ to transfer the derivatives from y to u, we obtain

$$\int_a^b u[p_3 y''' + p_2 y'' + p_1 y' + p_0 y]\, dx$$

$$= \int_a^b y[-(p_3 u)''' + (p_2 u)'' - (p_1 u)' + p_0 u]\, dx$$

$$+ \left\{ p_3 u y'' - [(p_3 u)' - p_2 u] y' + [(p_3 u)'' - (p_2 u)' + p_1 u] y \right\}_a^b \tag{241}$$

Then the adjoint equation is

$$(p_3 u)''' - (p_2 u)'' + (p_1 u)' - p_0 u = 0 \tag{242}$$

or

$$p_3 u''' + (3p_3' - p_2) u'' + (3p_3'' - 2p_2' + p_1) u' + (p_3''' - p_2'' + p_1' - p_0) u = 0 \tag{243}$$

Comparing (240) and (250), we conclude that (240) is self-adjoint if and only if

$$3p_3' - p_2 = p_2, \qquad 3p_3'' - 2p_2' + p_1 = p_1, \qquad p_3''' - p_2'' + p_1' - p_0 = p_0$$

or

$$p_2 = \tfrac{3}{2} p_3', \qquad p_0 = \tfrac{1}{2} p_1' - \tfrac{1}{4} p_3''' \tag{244}$$

15.24. Determine the solvability conditions for

$$p_3 y''' + p_2 y'' + p_1 y' + p_0 y = f(x) \tag{245}$$

subject to the boundary conditions

(a) $$y(0) = \beta_1, \qquad y'(0) = \beta_2, \qquad y(1) = \beta_3 \tag{246}$$

(b) $$y''(0) = \alpha_{11} y(0) + \alpha_{12} y'(0) + \alpha_{13} y(1) + \alpha_{14} y'(1) \tag{247}$$

$$y''(1) = \alpha_{21} y(0) + \alpha_{22} y'(0) + \alpha_{23} y(1) + \alpha_{24} y'(1) \tag{248}$$

$$0 = \alpha_{31} y(0) + \alpha_{32} y'(0) + \alpha_{33} y(1) + \alpha_{34} y'(1) \tag{249}$$

We multiply (245) by $u(x)$, integrate the result by parts from $x=0$ to $x=1$ to transfer the derivatives from y to u, and obtain

$$\int_0^1 y\left[-(p_3u)''' + (p_2u)'' - (p_1u)' + p_0u\right] dx + \Big\{ p_3uy'' - \left[(p_3u)' - p_2u\right] y'$$

$$+\left[(p_3u)'' - (p_2u)' + p_1u\right] y\Big\}_0^1 = \int_0^1 f(x)\,u(x)\,dx \tag{250}$$

Then the adjoint equation is given by (242). To define the adjoint boundary conditions, we consider the homogeneous problem and obtain from (250) that

$$\Big\{ p_3uy'' - \left[(p_3u)' - p_2u\right] y' + \left[(p_3u)'' - (p_2u)' + p_1u\right] y\Big\}_0^1 = 0 \tag{251}$$

(a) In this case, it follows from (246) that the homogeneous boundary conditions are

$$y(0)=0, \qquad y'(0)=0, \qquad y(1)=0$$

Then (251) becomes

$$p_3(1)\,u(1)\,y''(1) - \left[p_3u' + p_3'u - p_2u\right]_{x=1} y'(1) - p_3(0)\,u(0)\,y''(0) = 0$$

The adjoint boundary conditions are defined by setting each of the coefficients of $y''(1)$, $y'(1)$, and $y''(0)$ equal to zero. The result is

$$u(1)=0, \qquad u'(1)=0, \qquad u(0)=0 \tag{252}$$

Returning to the inhomogeneous problem, we substitute (246) into (250), use the definition of the adjoint, and obtain

$$\int_0^1 f(x)\,u(x)\,dx = \beta_3 p_3(1)\,u''(1) + \beta_2 p_3(0)\,u'(0) - \beta_1\left[p_3u'' + (2p_3' - p_2)\,u'\right]_{x=0} \tag{253}$$

where $u(x)$ is any solution of (242) and (252).

(b) In this case, the boundary conditions are homogeneous and general. We note that (251) can be expressed in matrix and vector notation as

$$u_b^T P y_b = 0 \tag{254}$$

where

$$u_b = \begin{bmatrix} u''(0) \\ u'(0) \\ u(0) \\ u''(1) \\ u'(1) \\ u(1) \end{bmatrix}, \quad y_b = \begin{bmatrix} y''(0) \\ y'(0) \\ y(0) \\ y''(1) \\ y'(1) \\ y(1) \end{bmatrix}, \quad P = \begin{bmatrix} \Lambda_0 & 0 \\ 0 & -\Lambda_1 \end{bmatrix}$$

$$\Lambda_{0,1} = \begin{bmatrix} 0 & 0 & -p_3 \\ 0 & p_3 & -2p_3' + p_2 \\ -p_3 & p_3' - p_2 & -p_3'' + p_2' - p_1 \end{bmatrix}_{x=0,1}$$

We note that $|P| = |p_3(0)p_3(1)|^3 \neq 0$, hence P is a nonsingular matrix. We introduce a linear nonsingular transformation from y_b to Y according to

$$Y = Ry_b \tag{255}$$

where

$$R = \begin{bmatrix} 1 & -\alpha_{12} & -\alpha_{11} & 0 & -\alpha_{14} & -\alpha_{13} \\ 0 & -\alpha_{22} & -\alpha_{21} & 1 & -\alpha_{24} & -\alpha_{23} \\ 0 & -\alpha_{32} & -\alpha_{31} & 0 & -\alpha_{34} & -\alpha_{33} \\ 1 & 0 & 0 & 0 & 0 & 0 \\ 0 & 1 & 0 & 0 & 0 & 0 \\ 0 & 0 & 1 & 0 & 0 & 0 \end{bmatrix}$$

The last three rows in R are chosen to be linearly independent from the first three rows, but otherwise are arbitrary. Inverting (255) yields

$$y_b = R^{-1}Y$$

so that (254) can be rewritten as

$$u_b^T P R^{-1} Y = 0 \quad \text{or} \quad U^T Y = 0 \tag{256}$$

where

$$U^T = u_b^T P R^{-1} \quad \text{or} \quad U = (R^{-1})^T P^T u_b \tag{257}$$

It follows from (247)–(249), (255), and the definition of R that

$$U_4 Y_4 + U_5 Y_5 + U_6 Y_6 = 0$$

We choose the adjoint boundary conditions by setting each of the coefficients of Y_4, Y_5, and Y_6 equal to zero. The result is

$$U_4 = 0, \qquad U_5 = 0, \qquad U_6 = 0 \tag{258}$$

Having defined the adjoint, we return to the inhomogeneous problem. Using (242) and (251) in (250), we obtain the following solvability condition:

$$\int_0^1 f(x)\, u(x)\, dx = 0 \tag{259}$$

where $u(x)$ is any solution of (242) subject to the boundary conditions (258).

15.25. Prove that any homogeneous self-adjoint sixth-order ordinary-differential equation can be written in the form

$$\frac{d^3}{dx^3}\left(A_3\frac{d^3y}{dx^3}\right)+\frac{d^2}{dx^2}\left(A_2\frac{d^2y}{dx^2}\right)+\frac{d}{dx}\left(A_1\frac{dy}{dx}\right)+A_0 y=0 \tag{260}$$

The most general linear homogeneous sixth-order ordinary-differential equation can be written as

$$p_6 y^{vi} + p_5 y^{v} + p_4 y^{iv} + p_3 y'' + p_2 y'' + p_1 y' + p_0 y = 0 \tag{261}$$

where the p_n are functions of x. Multiplying (261) with $u(x)$ and integrating the result by parts to transfer the derivatives from y to u, we conclude that the adjoint equation is

$$(p_6 u)^{vi} - (p_5 u)^{v} + (p_4 u)^{iv} - (p_3 u)''' + (p_2 u)'' - (p_1 u)' + p_0 u = 0$$

or

$$\begin{aligned}
&p_6 u^{vi} + (6p_6' - p_5)\, u^{v} + (15p_6'' - 5p_5' + p_4)\, u^{iv} + (20p_6''' - 10p_5'' + 4p_4' - p_3)\, u'''\\
&\quad + (15p_6^{iv} - 10p_5''' + 6p_4'' - 3p_3' + p_2)\, u''\\
&\quad + (6p_6^{v} - 5p_5^{iv} + 4p_4''' - 3p_3'' + 2p_2' - p_1)\, u'\\
&\quad + (p_6^{vi} - p_5^{v} + p_4^{iv} - p_3''' + p_2'' - p_1' + p_0)\, u = 0
\end{aligned} \tag{262}$$

Comparing (261) and (262), we conclude that they are the same if and only if

$$p_5 = 6p_6' - p_5$$

$$p_4 = 15p_6'' - 5p_5' + p_4$$

$$p_3 = 20p_6''' - 10p_5'' + 4p_4' - p_3$$

$$p_2 = 15p_6^{iv} - 10p_5''' + 6p_4'' - 3p_3' + p_2$$

$$p_1 = 6p_6^{v} - 5p_5^{iv} + 4p_4''' - 3p_3'' + 2p_2' - p_1$$

$$p_0 = p_6^{vi} - p_5^{v} + p_4^{iv} - p_3''' + p_2'' - p_1' + p_0$$

Hence

$$p_5 = 3p_6'$$

$$p_3 = 10p_6''' - 5p_5'' + 2p_4' = -5p_6''' + 2p_4'$$

$$p_1 = 3p_6^{v} - \tfrac{5}{2}p_5^{iv} + 2p_4''' - \tfrac{3}{2}p_3'' + p_2' = 3p_6^{v} - p_4''' + p_2'$$

Then (261) becomes

$$p_6 y^{vi} + 3p_6' y^{v} + p_4 y^{iv} + (-5p_6''' + 2p_4')y''' + p_2 y'' + (3p_6^{v} - p_4''' + p_2')y' + p_0 y = 0$$

which can be rewritten as

$$\frac{d^3}{dx^3}\left(p_6\frac{d^3y}{dx^3}\right) + \frac{d^2}{dx^2}\left[(p_4 - 3p_6'')\frac{d^2y}{dx^2}\right] + \frac{d}{dx}\left[(p_2 - p_4'' + 3p_6^{iv})\frac{dy}{dx}\right] + p_0 y = 0 \tag{263}$$

Equation (263) has the same form as (260).

15.26. Prove that any homogeneous self-adjoint differential equation of order $2m$ can be written in the form

$$\frac{d^m}{dx^m}\left(A_m\frac{d^my}{dx^m}\right) + \frac{d^{m-1}}{dx^{m-1}}\left(A_{m-1}\frac{d^{m-1}y}{dx^{m-1}}\right) + \cdots + \frac{d}{dx}\left(A_1\frac{dy}{dx}\right) + A_0 y = 0 \tag{264}$$

We prove this theorem by induction. First, we prove that if the theorem holds for an equation of order $(2m-2)$, it must hold for an equation of order $2m$. To accomplish this, we note that the most general linear homogeneous differential equation of order $2m$ can be written as

$$p_{2m}D^{2m}y + p_{2m-1}D^{2m-1}y + p_{2m-2}D^{2m-2}y + \cdots + p_1 Dy + p_0 y = 0 \tag{265}$$

or

$$p_{2m}D^{2m}y + p_{2m-1}D^{2m-1}y + L_{2m-2}(y) = 0 \tag{266}$$

where the p_n are functions of x, $D = d/dx$, and $L_{2n}(y)$ stands for the most general linear homogeneous differential operator of order $2n$. To determine the adjoint of the differential equation (266), we multiply it by $u(x)$, integrate the result by parts over the interval of interest, say $[a, b]$, to transfer the derivatives from y to u, and obtain

$$\int_a^b y\left[D^{2m}(p_{2m}u) - D^{2m-1}(p_{2m-1}u) + L_{2m-2}^*(u)\right] dx = P(u, y)\Big|_a^b \tag{267}$$

where L_{2m-2}^* is an operator of order $2m-2$ and $P(u, y)$ is the bilinear concomitant.

Hence the adjoint equation is

$$D^{2m}(p_{2m}u) - D^{2m-1}(p_{2m-1}u) + L^{*}_{2m-2}(u) = 0 \tag{268}$$

Expanding the first two terms in (268), we obtain

$$p_{2m}D^{2m}u + \left(2mp'_{2m} - p_{2m-1}\right)D^{2m-1}u + K_{2m-2}(u) = 0 \tag{269}$$

where $K_{2m-2}(u)$ is a linear homogeneous differential operator of order $(2m-2)$. Comparing (266) and (269), we conclude that a necessary condition for (266) to be self-adjoint is

$$p_{2m-1} = 2mp'_{2m} - p_{2m-1}$$

or

$$p_{2m-1} = mp'_{2m} \tag{270}$$

Thus we rewrite (266) as

$$p_{2m}D^{2m}y + mp'_{2m}D^{2m-1}y + L_{2m-2}(y) = 0 \tag{271}$$

The first two terms can be rewritten as

$$p_{2m}D^{2m}y + mp'_{2m}D^{2m-1}y = \sum_{n=0}^{m-1}(-1)^n C_n^m D^{m-n}\left[p_{2m}^{(2n)}D^{m-n}y\right] - M_{2m-2}(y) \tag{272}$$

so that (271) can be rewritten as

$$\sum_{n=0}^{m-1}(-1)^n C_n^m \frac{d^{m-n}}{dx^{m-n}}\left(p_{2m}^{(2n)}\frac{d^{m-n}y}{dx^{m-n}}\right) + L_{2m-2}(y) - M_{2m-2}(y) = 0 \tag{273a}$$

In order that (273a) be self-adjoint, the operator $(L_{2m-2} - M_{2m-2})(y)$ must be self-adjoint. But the theorem is assumed to hold for an equation of order $2m-2$, hence $(L_{2m-2} - M_{2m-2})(y)$ has the form in (264). Therefore, (273a) can be rewritten as

$$\sum_{n=0}^{m-1}(-1)^n C_n^m \frac{d^{m-n}}{dx^{m-n}}\left(p_{2m}^{(2n)}\frac{d^{m-n}y}{dx^{m-n}}\right) + \frac{d^{m-1}}{dx^{m-1}}\left(A_{m-1}\frac{d^{m-1}y}{dx^{m-1}}\right) + \cdots + \frac{d}{dx}\left(A_1\frac{dy}{dx}\right) + A_0 y = 0 \tag{273b}$$

which has the same form as (264). Therefore, if (264) holds for an equation of order $(2m-2)$, it holds also for an equation of order $2m$. One can easily show that (264) holds for $m=1$ and 2. Moreover, (260) shows that (264) holds for $m=3$. Hence it holds for $m = 4, 5, 6, 7, \ldots$.

15.27. Determine the solvability condition for

$$p_5(x)y^v + p_4(x)y^{iv} + p_3(x)y''' + p_2(x)y'' + p_1(x)y' + p_0(x)y = f(x) \quad (274)$$

$$y(0) = \beta_1, \qquad y'(0) = \beta_2, \qquad y''(0) = \beta_3, \qquad y(1) = \beta_4, \qquad y'(1) = \beta_5 \quad (275)$$

We multiply (274) by $u(x)$, integrate the result by parts from $x = 0$ to $x = 1$ to transfer the derivatives from y to u, and obtain

$$\int_0^1 y\left[-(p_5u)^v + (p_4u)^{iv} - (p_3u)''' + (p_2u)'' - (p_1u)' + p_0u\right] dx$$

$$+\left[p_5uy^{iv} - (p_5u)'y''' + (p_5u)''y'' - (p_5u)'''y' + (p_5u)^{iv}y\right.$$

$$+ p_4uy''' - (p_4u)'y'' + (p_4u)''y' - (p_4u)'''y + p_3uy'' - (p_3u)'y'$$

$$\left. + (p_3u)''y + p_2uy' - (p_2u)'y + p_1uy\right]_0^1 = \int_0^1 f(x)u(x)\,dx \quad (276)$$

Hence the adjoint equation is

$$(p_5u)^v - (p_4u)^{iv} + (p_3u)''' - (p_2u)'' + (p_1u)' - p_0 = 0 \quad (277)$$

To define the adjoint boundary conditions, we consider the homogeneous problem. Then it follows from (275)–(277) that

$$p_5(1)u(1)y^{iv}(1) - \left[(p_5u)' - p_4u\right]_{x=1}y'''(1) + \left[(p_5u)'' - (p_4u)' + p_3u\right]_{x=1}y''(1)$$

$$- p_5(0)u(0)y^{iv}(0) + \left[(p_5u)' - p_4u\right]_{x=0}y'''(0) = 0$$

$$(278)$$

We define the adjoint boundary conditions by setting each of the coefficients of $y^{iv}(1)$, $y'''(1)$, $y''(1)$, $y^{iv}(0)$, and $y'''(0)$ equal to zero. The result is

$$u(0) = 0, \qquad u'(0) = 0, \qquad u(1) = 0, \qquad u'(1) = 0, \qquad u''(1) = 0 \quad (279)$$

Returning to the inhomogeneous problem, we substitute (275) into (276), use the definition of the adjoint, and obtain the following solvability condition:

$$\int_0^1 f(x)u(x)\,dx = -\beta_5p_5(1)u'''(1) - \beta_4\left[p_4(1)u'''(1) - p_5(1)u^{iv}(1) - 4p_5'(1)u'''(1)\right]$$

$$- \beta_3p_5(0)u''(0) + \beta_2\left[p_5(0)u'''(0) + 3p_5'(0)u''(0) - p_4(0)u''(0)\right]$$

$$- \beta_1\left[p_5(0)u^{iv}(0) + 4p_5'(0)u'''(0) + 6p_5''(0)u''(0)\right.$$

$$\left. - p_4(0)u'''(0) - 3p_4'(0)u''(0) + p_3(0)u''(0)\right] \quad (280)$$

where u is any solution of (277) and (279).

15.28. Consider the eigenvalue problem

$$\phi(s)=\lambda\int_0^\pi[\cos(s+t)+\varepsilon K_1(s,t)]\phi(t)\,dt,\qquad \varepsilon\ll 1 \tag{281}$$

Show that

$$\phi^{(1)}=\cos s+\cdots,\qquad \phi^{(2)}=\sin s+\cdots \tag{282}$$

$$\lambda^{(1)}=\frac{2}{\pi}+\varepsilon\lambda_1^{(1)}+\cdots,\qquad \lambda^{(2)}=-\frac{2}{\pi}+\varepsilon\lambda_1^{(2)}+\cdots \tag{283}$$

Determine $\lambda_1^{(n)}$.

We seek a first-order expansion for ϕ and λ in the form

$$\begin{aligned}\phi&=\phi_0(s)+\varepsilon\phi_1(s)+\cdots\\ \lambda&=\lambda_0+\varepsilon\lambda_1+\cdots\end{aligned} \tag{284}$$

Substituting (284) into (281) and equating coefficients of like powers of ε, we obtain

$$\phi_0(s)=\lambda_0\int_0^\pi\cos(s+t)\phi_0(t)\,dt \tag{285}$$

$$\begin{aligned}\phi_1=\lambda_0\int_0^\pi\cos(s+t)\phi_1(t)\,dt+\lambda_1\int_0^\pi\cos(s+t)\phi_0(t)\,dt\\ +\lambda_0\int_0^\pi K_1(s,t)\phi_0(t)\,dt\end{aligned} \tag{286}$$

The kernel in (285) can be rewritten as

$$\cos(s+t)=\cos s\cos t-\sin s\sin t \tag{287}$$

and thus it is degenerate. To solve (285), we let

$$x_1=\int_0^\pi\phi_0(t)\cos t\,dt,\qquad x_2=\int_0^\pi\phi_0(t)\sin t\,dt \tag{288}$$

Substituting (287) into (285) and using (288), we obtain

$$\phi_0(s)=\lambda_0x_1\cos s-\lambda_0x_2\sin s \tag{289}$$

Multiplying (289) by $\cos s$, integrating the result over the interval $[0,\pi]$, and using (288), we have

$$x_1=\tfrac{1}{2}\pi\lambda_0x_1 \tag{290}$$

Similarly, multiplying (289) by $\sin s$ and integrating the result over the interval $[0,\pi]$, we have

$$x_2=-\tfrac{1}{2}\pi\lambda_0x_2 \tag{291}$$

It follows from (290) and (291) that either

$$\lambda_0 = \frac{2}{\pi} \quad \text{and} \quad x_2 = 0$$

or

$$\lambda_0 = -\frac{2}{\pi} \quad \text{and} \quad x_1 = 0$$

In the first case,

$$\phi_0 = \frac{2x_1}{\pi}\cos s = \cos s$$

where x_1 is chosen to be $\frac{1}{2}\pi$, without loss of generality; whereas in the second case,

$$\phi_0 = \frac{2x_2}{\pi}\sin s = \sin s$$

where x_2 is chosen to be $\frac{1}{2}\pi$, without loss of generality.

Substituting (287) into (286), using (288), and letting

$$y_1 = \int_0^\pi \phi_1(t)\cos t\,dt, \qquad y_2 = \int_0^\pi \phi_1(t)\sin t\,dt$$

we obtain

$$\phi_1(s) = \lambda_0 y_1 \cos s - \lambda_0 y_2 \sin s + \lambda_1(x_1 \cos s - x_2 \sin s) + \lambda_0 \int_0^\pi K_1(s,t)\phi_0(t)\,dt \tag{292}$$

Multiplying (292) by $\cos s$ and integrating the result over the interval $[0, \pi]$, we have

$$y_1 = \tfrac{1}{2}\pi\lambda_0 y_1 + \tfrac{1}{2}\pi\lambda_1 x_1 + \lambda_0 \int_0^\pi \int_0^\pi K_1(s,t)\phi_0(t)\cos s\,dt\,ds$$

which, upon letting $\lambda_0 = 2/\pi$ and $\phi_0(s) = \cos s$, yields

$$\lambda_1^{(1)} = -\frac{8}{\pi^3}\int_0^\pi \int_0^\pi K_1(s,t)\cos t\cos s\,dt\,ds$$

Multiplying (292) by $\sin s$ and integrating the result over the interval $[0, \pi]$, we have

$$y_2 = -\tfrac{1}{2}\pi\lambda_0 y_2 - \tfrac{1}{2}\pi\lambda_1 x_2 + \lambda_0 \int_0^\pi \int_0^\pi K_1(s,t)\phi_0(t)\sin s\,dt\,ds$$

which, upon putting $\lambda_0 = 2/\pi$ and $\phi_0 = \sin s$, yields

$$\lambda_1^{(2)} = \frac{8}{\pi^3}\int_0^\pi \int_0^\pi K_1(s,t)\sin t\sin s\,dt\,ds$$

Therefore

$$\phi^{(1)}(s) = \cos s + \cdots \quad \text{and} \quad \phi^{(2)}(s) = \sin s + \cdots \tag{293}$$

corresponding to

$$\lambda^{(1)} = \frac{2}{\pi} - \frac{8}{\pi^3}\varepsilon\int_0^\pi\int_0^\pi K_1(s,t)\cos t\cos s\,dt\,ds + \cdots \tag{294}$$

and

$$\lambda^{(2)} = -\frac{2}{\pi} + \frac{8}{\pi^3}\varepsilon\int_0^\pi\int_0^\pi K_1(s,t)\sin t\sin s\,dt\,ds + \cdots \tag{295}$$

15.29. Determine a first-order uniform expansion for

$$\phi(s) = \lambda\int_{-1}^{+1}\left[st + s^2t^2 + \varepsilon K_1(s,t)\right]\phi(t)\,dt, \qquad \varepsilon \ll 1 \tag{296}$$

We let

$$\phi(s) = \phi_0(s) + \varepsilon\phi_1(s) + \cdots$$

$$\lambda = \lambda_0 + \varepsilon\lambda_1 + \cdots$$

in (296), equate coefficients of like powers of ε, and obtain

$$\phi_0(s) = \lambda_0\int_{-1}^{+1}(st + s^2t^2)\phi_0(t)\,dt \tag{297}$$

$$\phi_1(s) = \lambda_0\int_{-1}^{+1}(st + s^2t^2)\phi_1(t)\,dt + \lambda_1\int_{-1}^{+1}(st + s^2t^2)\phi_0(t)\,dt$$
$$+ \lambda_0\int_{-1}^{+1}K_1(s,t)\phi_0(t)\,dt \tag{298}$$

We let

$$x_1 = \int_{-1}^{+1} t\phi_0(t)\,dt, \qquad x_2 = \int_{-1}^{+1} t^2\phi_0(t)\,dt$$

and rewrite (297) as

$$\phi_0(s) = \lambda_0 x_1 s + \lambda_0 x_2 s^2 \tag{299}$$

Multiplying (299) by s and integrating the result from $s = -1$ to $s = +1$, we obtain

$$x_1 = \tfrac{2}{3}\lambda_0 x_1 \tag{300}$$

Multiplying (299) by s^2 and integrating the result from $s=-1$ to $s=+1$, we obtain

$$x_2 = \tfrac{2}{5}\lambda_0 x_2 \tag{301}$$

It follows from (300) and (301) that either

$$\lambda_0 = \tfrac{3}{2} \quad \text{and} \quad x_2 = 0$$

or

$$\lambda_0 = \tfrac{5}{2} \quad \text{and} \quad x_1 = 0$$

In the first case,

$$\phi_0 = \tfrac{3}{2} x_1 s = s$$

whereas in the second case,

$$\phi_0 = \tfrac{5}{2} x_2 s^2 = s^2$$

where x_1 and x_2 are chosen, without loss of generality, to be $\frac{2}{3}$ and $\frac{2}{5}$, respectively.
Letting

$$y_1 = \int_{-1}^{+1} t\phi_1(t)\,dt, \qquad y_2 = \int_{-1}^{+1} t^2\phi_1(t)\,dt$$

and using (297), we rewrite (298) as

$$\phi_1(s) = \lambda_0 y_1 s + \lambda_0 y_2 s^2 + \frac{\lambda_1}{\lambda_0}\phi_0(s) + \lambda_0 \int_{-1}^{+1} K_1(s,t)\,\phi_0(t)\,dt \tag{302}$$

Multiplying (302) by s and integrating the result from $s=-1$ to $s=+1$, we have

$$y_1 = \tfrac{2}{3}\lambda_0 y_1 + \frac{\lambda_1 x_1}{\lambda_0} + \lambda_0 \int_{-1}^{+1}\int_{-1}^{+1} K_1(s,t)\,\phi_0(t)\,s\,dt\,ds$$

which, upon putting $\lambda_0 = \frac{3}{2}$ and $\phi_0(s) = s$, yields

$$\lambda_1^{(1)} = -\tfrac{27}{8}\int_{-1}^{+1}\int_{-1}^{+1} K_1(s,t)\,ts\,dt\,ds$$

Similarly, multiplying (302) by s^2 and integrating the result from $s=-1$ to $s=+1$, we have

$$y_2 = \tfrac{2}{5}\lambda_0 y_2 + \frac{\lambda_1}{\lambda_0} x_2 + \lambda_0 \int_{-1}^{+1}\int_{-1}^{+1} K_1(s,t)\,\phi_0(t)\,s^2\,dt\,ds$$

which, upon putting $\lambda_0 = \frac{5}{2}$ and $\phi_0(s) = s^2$, yields

$$\lambda_1^{(2)} = -\tfrac{125}{8}\int_{-1}^{+1}\int_{-1}^{+1} K_1(s,t)\, t^2 s^2\, dt\, ds$$

Therefore

$$\phi^{(1)}(s) = s + \cdots, \qquad \phi^{(2)}(s) = s^2 + \cdots \tag{303}$$

corresponding to

$$\lambda^{(1)} = \tfrac{3}{2} - \tfrac{27}{8}\varepsilon\int_{-1}^{+1}\int_{-1}^{+1} K_1(s,t)\, ts\, dt\, ds + \cdots \tag{304}$$

$$\lambda^{(2)} = \tfrac{5}{2} - \tfrac{125}{8}\varepsilon\int_{-1}^{+1}\int_{-1}^{+1} K_1(s,t)\, t^2 s^2\, dt\, ds + \cdots \tag{305}$$

15.30. Determine a first-order uniform expansion for

$$\phi(s) = \lambda\int_0^1 [s - t + \varepsilon K_1(s,t)]\,\phi(t)\, dt, \qquad \varepsilon \ll 1 \tag{306}$$

We let

$$\phi(s) = \phi_0(s) + \varepsilon\phi_1(s) + \cdots$$

$$\lambda = \lambda_0 + \varepsilon\lambda_1 + \cdots$$

in (306), equate coefficients of like powers of ε, and obtain

$$\phi_0(s) = \lambda_0\int_0^1 (s-t)\,\phi_0(t)\, dt \tag{307}$$

$$\phi_1(s) = \lambda_0\int_0^1 (s-t)\,\phi_1(t)\, dt + \lambda_1\int_0^1 (s-t)\,\phi_0(t)\, dt$$

$$+ \lambda_0\int_0^1 K_1(s,t)\,\phi_0(t)\, dt \tag{308}$$

We put

$$x_1 = \int_0^1 \phi_0(t)\, dt, \qquad x_2 = \int_0^1 t\phi_0(t)\, dt$$

in (307) and obtain

$$\phi_0(s) = \lambda_0 s x_1 - \lambda_0 x_2 \tag{309}$$

Integrating (309) from $s = 0$ to $s = 1$, we have

$$x_1 = \tfrac{1}{2}\lambda_0 x_1 - \lambda_0 x_2$$

or

$$\left(1 - \tfrac{1}{2}\lambda_0\right)x_1 + \lambda_0 x_2 = 0 \tag{310}$$

Multiplying (309) by s and integrating the result from $s=0$ to $s=1$, we have

$$x_2 = \tfrac{1}{3}\lambda_0 x_1 - \tfrac{1}{2}\lambda_0 x_2$$

or

$$\tfrac{1}{3}\lambda_0 x_1 - \left(1+\tfrac{1}{2}\lambda_0\right) x_2 = 0 \tag{311}$$

It follows from (310) and (311) that

$$-\left(1-\tfrac{1}{4}\lambda_0^2\right) - \tfrac{1}{3}\lambda_0^2 = 0$$

or

$$\lambda_0 = \pm 2i\sqrt{3}$$

Then, it follows from (310) that

$$x_2 = \left(\tfrac{1}{2} + \tfrac{1}{6}i\sqrt{3}\right) x_1 \quad \text{when } \lambda_0 = 2i\sqrt{3}$$

$$x_2 = \left(\tfrac{1}{2} - \tfrac{1}{6}i\sqrt{3}\right) x_1 \quad \text{when } \lambda_0 = -2i\sqrt{3}$$

Consequently,

$$\begin{aligned} \phi_0(s) &= 2i\sqrt{3}\, x_1\left(s - \tfrac{1}{2} - \tfrac{1}{6}i\sqrt{3}\right) \quad \text{when } \lambda_0 = 2i\sqrt{3} \\ \phi_0(s) &= -2i\sqrt{3}\, x_1\left(s - \tfrac{1}{2} + \tfrac{1}{6}i\sqrt{3}\right) \quad \text{when } \lambda_0 = -2i\sqrt{3} \end{aligned} \tag{312}$$

Putting

$$y_1 = \int_0^1 \phi_1(t)\, dt, \qquad y_2 = \int_0^1 t\phi_1(t)\, dt$$

and using (307), we rewrite (308) as

$$\phi_1(s) = \lambda_0 s y_1 - \lambda_0 y_2 + \frac{\lambda_1}{\lambda_0}\phi_0(s) + \lambda_0 \int_0^1 K_1(s,t)\phi_0(t)\, dt \tag{313}$$

Integrating (313) from $s=0$ to $s=1$, we have

$$y_1 = \tfrac{1}{2}\lambda_0 y_1 - \lambda_0 y_2 + \frac{\lambda_1}{\lambda_0} x_1 + \lambda_0 \alpha_1$$

or

$$\left(1 - \tfrac{1}{2}\lambda_0\right) y_1 + \lambda_0 y_2 = \frac{\lambda_1}{\lambda_0} x_1 + \lambda_0 \alpha_1 \tag{314}$$

where

$$\alpha_1 = \int_0^1 \int_0^1 K_1(s,t)\phi_0(t)\, dt\, ds$$

Multiplying (313) by s and integrating the result from $s=0$ to $s=1$, we have

$$y_2 = \tfrac{1}{3}\lambda_0 y_1 - \tfrac{1}{2}\lambda_0 y_2 + \frac{\lambda_1}{\lambda_0} x_2 + \lambda_0 \alpha_2$$

or

$$\tfrac{1}{3}\lambda_0 y_1 - \left(1 + \tfrac{1}{2}\lambda_0\right) y_2 = -\frac{\lambda_1}{\lambda_0} x_2 - \lambda_0 \alpha_2 \tag{315}$$

where

$$\alpha_2 = \int_0^1 \int_0^1 K_1(s,t)\, s\phi_0(t)\, dt\, ds$$

Since the homogeneous equations (314) and (315) are the same as (310) and (311) and since the latter have a nontrivial solution, the inhomogeneous equations have a solution if and only if a solvability condition is satisfied. This condition can be expressed as

$$\begin{vmatrix} 1 - \tfrac{1}{2}\lambda_0 & \dfrac{\lambda_1}{\lambda_0} x_1 + \lambda_0 \alpha_1 \\ \tfrac{1}{3}\lambda_0 & -\dfrac{\lambda_1}{\lambda_0} x_2 - \lambda_0 \alpha_2 \end{vmatrix} = 0$$

or

$$\lambda_1 = \frac{6\alpha_1\left(1 + \tfrac{1}{2}\lambda_0\right)}{x_1} - \frac{6\lambda_0 \alpha_2}{x_1}$$

Therefore

$$\phi^{(1)} = s - \tfrac{1}{2} - \tfrac{1}{6} i\sqrt{3} + \cdots \tag{316}$$

corresponding to

$$\lambda^{(1)} = 2i\sqrt{3} - 12\epsilon\Big[\left(3 - i\sqrt{3}\right)\int_0^1\int_0^1 K_1(s,t)\left(t - \tfrac{1}{2} - \tfrac{1}{6}i\sqrt{3}\right) ds\, dt$$
$$- 6\int_0^1\int_0^1 K_1(s,t)\, s\left(t - \tfrac{1}{2} - \tfrac{1}{6}i\sqrt{3}\right) ds\, dt\Big] + \cdots \tag{317}$$

and

$$\phi^{(2)} = s - \tfrac{1}{2} + \tfrac{1}{6} i\sqrt{3} + \cdots \tag{318}$$

corresponding to

$$\lambda^{(2)} = -2i\sqrt{3} - 12\varepsilon\Big[(3+i\sqrt{3})\int_0^1\int_0^1 K_1(s,t)\big(t-\tfrac{1}{2}+\tfrac{1}{6}i\sqrt{3}\big)\,ds\,dt$$

$$-6\int_0^1\int_0^1 K_1(s,t)s\big(t-\tfrac{1}{2}+\tfrac{1}{6}i\sqrt{3}\big)\,dt\,dt\Big] + \cdots \tag{319}$$

15.31. Show that

$$L(y) = \frac{d}{v_3\,dx}\cdot\frac{d}{v_2\,dx}\cdot\frac{y}{v_1} = 0 \tag{320}$$

is self-adjoint if and only if $v_3 = \pm v_1$.

Since

$$\int_a^b uL(y)\,dx = \left[\left(\frac{u}{v_3}\right)\left(\frac{d}{v_2\,dx}\cdot\frac{y}{v_1}\right)\right]_a^b - \int_a^b\left(\frac{d}{dx}\cdot\frac{u}{v_3}\right)\left(\frac{d}{v_2\,dx}\cdot\frac{y}{v_1}\right)dx,$$

$$\int_a^b\left(\frac{d}{dx}\cdot\frac{u}{v_3}\right)\left(\frac{d}{v_2\,dx}\cdot\frac{y}{v_1}\right)dx = \left[\left(\frac{d}{v_2\,dx}\cdot\frac{u}{v_3}\right)\left(\frac{y}{v_1}\right)\right]_a^b - \int_a^b y\left[\frac{d}{v_1\,dx}\cdot\frac{d}{v_2\,dx}\cdot\frac{u}{v_3}\right]dx$$

it follows that

$$\int_a^b [uL(y) - yL^*(u)]\,dx = \left[\left(\frac{u}{v_3}\right)\left(\frac{d}{v_2\,dx}\cdot\frac{y}{v_1}\right) - \left(\frac{d}{v_2\,dx}\cdot\frac{u}{v_3}\right)\left(\frac{y}{v_1}\right)\right]_a^b \tag{321}$$

where

$$L^*(u) = \frac{d}{v_1\,dx}\cdot\frac{d}{v_2\,dx}\cdot\frac{u}{v_3} \tag{322}$$

Comparing (320) and (322), we conclude that $L(y)=0$ is identical with its adjoint $L^*(u)=0$ if and only if $v_3 = \pm v_1$.

15.32. Show that

$$L(y) = \frac{d}{v_4\,dx}\cdot\frac{d}{v_3\,dx}\cdot\frac{d}{v_2\,dx}\cdot\frac{y}{v_1} = 0 \tag{323}$$

is self-adjoint if and only if

$$v_4 = \pm v_1, \qquad v_3 = \pm v_2$$

Since

$$\int_a^b uL(y)\,dx = \left[\left(\frac{u}{v_4}\right)\left(\frac{d}{v_3\,dx}\cdot\frac{d}{v_2\,dx}\cdot\frac{y}{v_1}\right)\right]_a^b - \int_a^b\left(\frac{d}{dx}\cdot\frac{u}{v_4}\right)\left(\frac{d}{v_3\,dx}\cdot\frac{d}{v_2\,dx}\cdot\frac{y}{v_1}\right)dx$$

$$\int_a^b\left(\frac{d}{dx}\cdot\frac{u}{v_4}\right)\left(\frac{d}{v_3\,dx}\cdot\frac{d}{v_2\,dx}\cdot\frac{y}{v_1}\right)dx = \left[\left(\frac{d}{v_3\,dx}\cdot\frac{u}{v_4}\right)\left(\frac{d}{v_2\,dx}\cdot\frac{y}{v_1}\right)\right]_a^b - \int_a^b\left(\frac{d}{dx}\cdot\frac{d}{v_3\,dx}\cdot\frac{u}{v_4}\right)\left(\frac{d}{v_2\,dx}\cdot\frac{y}{v_1}\right)dx$$

$$\int_a^b\left(\frac{d}{dx}\cdot\frac{d}{v_3\,dx}\cdot\frac{u}{v_4}\right)\left(\frac{d}{v_2\,dx}\cdot\frac{y}{v_1}\right)dx = \left[\left(\frac{d}{v_2\,dx}\cdot\frac{d}{v_3\,dx}\cdot\frac{u}{v_4}\right)\left(\frac{y}{v_1}\right)\right]_a^b - \int_a^b y\left[\frac{d}{v_1\,dx}\cdot\frac{d}{v_2\,dx}\cdot\frac{d}{v_3\,dx}\cdot\frac{u}{v_4}\right]dx$$

it follows that

$$\int_a^b[uL(y) - yL^*(u)]\,dx = \left[\left(\frac{u}{v_4}\right)\left(\frac{d}{v_3\,dx}\cdot\frac{d}{v_2\,dx}\cdot\frac{y}{v_1}\right) - \left(\frac{d}{v_3\,dx}\cdot\frac{u}{v_4}\right)\left(\frac{d}{v_2\,dx}\cdot\frac{y}{v_1}\right) + \left(\frac{d}{v_2\,dx}\cdot\frac{d}{v_3\,dx}\cdot\frac{u}{v_4}\right)\left(\frac{y}{v_1}\right)\right]_a^b$$

where

$$L^*(u) = -\frac{d}{v_1\,dx}\cdot\frac{d}{v_2\,dx}\cdot\frac{d}{v_3\,dx}\cdot\frac{u}{v_4} \tag{324}$$

Comparing (323) and (324), we conclude that $L(y)=0$ is identical with its adjoint $L^*(u)=0$ if and only if $v_4 = \pm v_1$ and $v_3 = \pm v_2$.

15.33. Show that

$$L(y) = \frac{d}{v_{n+1}\,dx}\cdot\frac{d}{v_n\,dx}\cdots\frac{d}{v_3\,dx}\cdot\frac{d}{v_2\,dx}\cdot\frac{y}{v_1} = 0 \tag{325}$$

is self-adjoint if and only if

$$v_{n+1} = \pm v_1, \qquad v_n = \pm v_2, \qquad v_{n-1} = \pm v_3, \ldots$$

Since

$$\int_a^b uL(y)\,dx = \left[\left(\frac{u}{v_{n+1}}\right)\left(\frac{d}{v_n\,dx}\cdots\frac{d}{v_3\,dx}\cdot\frac{d}{v_2\,dx}\cdot\frac{y}{v_1}\right)\right]_a^b$$

$$-\int_a^b\left(\frac{d}{dx}\cdot\frac{u}{v_{n+1}}\right)\left(\frac{d}{v_n\,dx}\cdots\frac{d}{v_3\,dx}\cdot\frac{d}{v_2\,dx}\cdot\frac{y}{v_1}\right)dx$$

$$\int_a^b\left(\frac{d}{dx}\cdot\frac{u}{v_{n+1}}\right)\left(\frac{d}{v_n\,dx}\cdot\frac{d}{v_{n-1}\,dx}\cdots\frac{d}{v_3\,dx}\cdot\frac{d}{v_2\,dx}\cdot\frac{y}{v_1}\right)dx$$

$$=\left[\left(\frac{d}{v_n\,dx}\cdot\frac{u}{v_{n+1}}\right)\left(\frac{d}{v_{n-1}\,dx}\cdots\frac{d}{v_3\,dx}\cdot\frac{d}{v_2\,dx}\cdot\frac{y}{v_1}\right)\right]_a^b$$

$$-\int_a^b\left(\frac{d}{dx}\cdot\frac{d}{v_n\,dx}\cdot\frac{u}{v_{n+1}}\right)\left(\frac{d}{v_{n-1}\,dx}\cdots\frac{d}{v_3\,dx}\cdot\frac{d}{v_2\,dx}\cdot\frac{y}{v_1}\right)dx$$

and so on, it follows that

$$\int_a^b[uL(y)-yL^*(u)]\,dx=[P(u,y)]_a^b$$

where $P(u,y)$ is a bilinear concomitant and

$$L^*(u)=(-1)^n\frac{d}{v_1\,dx}\cdot\frac{d}{v_2\,dx}\cdot\frac{d}{v_3\,dx}\cdots\frac{d}{v_{n-1}\,dx}\cdot\frac{d}{v_n\,dx}\cdot\frac{u}{v_{n+1}} \qquad (326)$$

Comparing (325) and (326), we conclude that $L(y)=0$ is identical with its adjoint $L^*(u)=0$ if and only if

$$v_{n+1}=\pm v_1, \qquad v_n=\pm v_2, \qquad v_{n-1}=\pm v_3,\ldots$$

15.34. Consider the problem

$$\nabla^2\phi+\omega^2\phi=0 \qquad (327)$$

$$\frac{\partial\phi}{\partial r}-\alpha\phi=0 \quad \text{at} \quad r=a+\varepsilon f(\theta) \qquad (328)$$

$$\frac{\partial\phi}{\partial r}=0 \quad \text{at} \quad r=b+\varepsilon g(\theta) \qquad (329)$$

where $\varepsilon\ll1$. Determine a first-order expansion for ω when $b>a$.

We first use Taylor-series expansions to transfer the boundary conditions in (328) and (329) to $r=a$ and $r=b$, respectively. It follows from (328) that

$$\frac{\partial\phi}{\partial r}(a+\varepsilon f,\theta)-\alpha\phi(a+\varepsilon f,\theta)=0$$

or

$$\frac{\partial\phi}{\partial r}(a,\theta)+\varepsilon f\frac{\partial^2\phi}{\partial r^2}(a,\theta)-\alpha\phi(a,\theta)-\alpha\varepsilon f\frac{\partial\phi}{\partial r}(a,\theta)+\cdots=0 \tag{330}$$

It follows from (329) that

$$\frac{\partial\phi}{\partial r}(b+\varepsilon g,\theta)=0$$

or

$$\frac{\partial\phi}{\partial r}(b,\theta)+\varepsilon g\frac{\partial^2\phi}{\partial r^2}(b,\theta)+\cdots=0 \tag{331}$$

Let

$$\begin{aligned}\phi&=\phi_0(r,\theta)+\varepsilon\phi_1(r,\theta)+\cdots\\ \omega&=\omega_0+\varepsilon\omega_1+\cdots\end{aligned} \tag{332}$$

Substituting (332) into (327), (330), and (331) and equating coefficients of like powers of ε, we obtain

Order ε^0

$$\nabla^2\phi_0+\omega_0^2\phi_0=0 \tag{333}$$

$$\begin{aligned}\frac{\partial\phi_0}{\partial r}(a,\theta)-\alpha\phi_0(a,\theta)&=0\\ \frac{\partial\phi_0}{\partial r}(b,\theta)&=0\end{aligned} \tag{334}$$

Order ε

$$\nabla^2\phi_1+\omega_0^2\phi_1=-2\omega_0\omega_1\phi_0 \tag{335}$$

$$\begin{aligned}\frac{\partial\phi_1}{\partial r}(a,\theta)-\alpha\phi_1(a,\theta)&=-f\frac{\partial^2\phi_0}{\partial r^2}(a,\theta)+\alpha f\frac{\partial\phi_0}{\partial r}(a,\theta)\\ \frac{\partial\phi_1}{\partial r}(b,\theta)&=-g\frac{\partial^2\phi_0}{\partial r^2}(b,\theta)\end{aligned} \tag{336}$$

The solution of (333) and (334) can be written as

$$\phi_0=\psi_{mn}(\Omega_{mn}r)\left(A_{mn}e^{im\theta}+\bar{A}_{mn}e^{-im\theta}\right),\qquad \omega_0=\Omega_{mn} \tag{337}$$

where m is an integer

$$\psi_{mn}(\Omega_{mn}r)=J_m(\Omega_{mn}r)Y_m'(\Omega_{mn}b)-Y_m(\Omega_{mn}r)J_m'(\Omega_{mn}b) \tag{338}$$

and Ω_{mn} is a root of

$$\Omega\left[J_m'(\Omega a)Y_m'(\Omega b)-Y_m'(\Omega a)J_m'(\Omega b)\right]-\alpha\left[J_m(\Omega a)Y_m'(\Omega b)-Y_m(\Omega a)J_m'(\Omega b)\right]=0$$

Substituting (337) into (335) and (336), we have

$$\frac{\partial^2\phi_1}{\partial r^2}+\frac{1}{r}\frac{\partial\phi_1}{\partial r}+\frac{1}{r^2}\frac{\partial\phi_1}{\partial\theta^2}+\Omega_{mn}^2\phi_1=-2\omega_1\Omega_{mn}\psi_{mn}(r)\left(A_{mn}e^{im\theta}+\bar{A}_{mn}e^{-im\theta}\right) \tag{339}$$

$$\frac{\partial\phi_1}{\partial r}-\alpha\phi_1=f(\theta)\left(-\psi_{mn}''+\alpha\psi_{mn}'\right)\left(A_{mn}e^{im\theta}+\bar{A}_{mn}e^{-im\theta}\right)\quad \text{at } r=a \tag{340}$$

$$\frac{\partial\phi_1}{\partial r}=-g\psi_{mn}''\left(A_{mn}e^{im\theta}+\bar{A}_{mn}e^{-im\theta}\right)\quad \text{at } r=b \tag{341}$$

where the prime indicates the derivative with respect to r.

To determine ω_1, we impose the solvability condition on (339)–(341). To this end, we express $f(\theta)$, $g(\theta)$, and ϕ_1 in Fourier series as

$$f(\theta)=\sum_{q=-\infty}^{\infty}f_qe^{iq\theta},\qquad g(\theta)=\sum_{q=-\infty}^{\infty}g_qe^{iq\theta}$$
$$\phi_1=\sum_{q=-\infty}^{\infty}\Phi_qe^{iq\theta} \tag{342}$$

where f_0 and g_0 are taken to be zero because they can be absorbed in a and b. Substituting (342) into (339)–(341) and equating the coefficients of $\exp(im\theta)$ on both sides, we obtain

$$\Phi_m''+\frac{1}{r}\Phi_m'+\left(\Omega_{mn}^2-\frac{m^2}{r^2}\right)\Phi_m=-2\omega_1\Omega_{mn}A_{mn}\psi_{mn} \tag{343}$$

$$\Phi_m'-\alpha\Phi_m=\left(-\psi_{mn}''+\alpha\psi_{mn}'\right)\bar{A}_{mn}f_{2m}\quad \text{at } r=a \tag{344}$$

$$\Phi_m'=-\psi_{mn}''\bar{A}_{mn}g_{2m}\quad \text{at } r=b \tag{345}$$

To determine the solvability condition of (343)–(345), we multiply (343) by $rv(r)$, integrate the result by parts from $r=a$ to $r=b$, and obtain

$$\left[r\Phi_m'v-r\Phi_mv'\right]_a^b+\int_a^b\Phi_m\left[(rv')'+\left(r\Omega_{mn}^2-\frac{m^2}{r}\right)v\right]dr$$
$$=-2\omega_1\Omega_{mn}A_{mn}\int_a^b rv(r)\psi_{mn}(\Omega_{mn}r)\,dr \tag{346}$$

The adjoint equation is

$$(rv')' + \left(r\Omega_{mn}^2 - \frac{m^2}{r} \right) v = 0 \tag{347}$$

which is identical with the homogeneous equation (343). To define the adjoint boundary conditions, we consider the homogeneous problem. Thus (346) becomes

$$-b\Phi_m(b)v'(b) - av(a)\alpha\Phi_m(a) + a\Phi_m(a)v'(a) = 0$$

or

$$-bv'(b)\Phi_m(b) + a\Phi_m(a)[v'(a) - \alpha v(a)] = 0$$

We define the adjoint boundary conditions by setting each of the coefficients of $\Phi_m(a)$ and $\Phi_m(b)$ equal to zero. The result is

$$\begin{aligned} v'(a) - \alpha v(a) &= 0 \\ v'(b) &= 0 \end{aligned} \tag{348}$$

Therefore, the homogeneous problem (343)–(345) is self-adjoint and hence $v(r) = \psi_{mn}(\Omega_{mn}r)$.

Having defined the adjoint problem, we return to the inhomogeneous problem. Using (344), (345), and the definition of the adjoint, we obtain from (346) that

$$\begin{aligned} &-b\psi_{mn}(\Omega_{mn}b)\psi_{mn}''(\Omega_{mn}b)\bar{A}_{mn}g_{2m} - a\psi_{mn}(\Omega_{mn}a) \\ &\quad\times[-\psi_{mn}''(\Omega_{mn}a) + \alpha\psi_{mn}'(\Omega_{mn}a)]\bar{A}_{mn}f_{2m} = -2\omega_1\Omega_{mn}A_{mn}\int_a^b r\psi_{mn}^2(\Omega_{mn}r)\,dr \end{aligned} \tag{349}$$

Letting

$$A_{mn} = \tfrac{1}{2}a_{mn}e^{i\beta_{mn}}$$

and

$$\begin{aligned} &b\psi_{mn}(\Omega_{mn}b)\psi_{mn}''(\Omega_{mn}b)g_{2m} + a\psi_{mn}(\Omega_{mn}a)[-\psi_{mn}''(\Omega_{mn}a) + \alpha\psi_{mn}'(\Omega_{mn}a)]f_{2m} \\ &\quad= \left[2\Omega_{mn}\int_a^b r\psi_{mn}^2(\Omega_{mn}r)\,dr\right]F_{mn}e^{i\tau_{mn}} \end{aligned}$$

we rewrite (349) as

$$\omega_1 = F_{mn}e^{i(\tau_{mn} - 2\beta_{mn})} \tag{350}$$

Hence

$$\omega_1 = \pm F_{mn} \quad \text{and} \quad \beta_{mn} = \tfrac{1}{2}\tau_{mn} \quad \text{or } \tfrac{1}{2}(\tau_{mn} - \pi)$$

so that

$$\omega = \Omega_{mn} \pm \varepsilon F_{mn} + \cdots \tag{351}$$

15.35. The incompressible flow past a wavy wall is governed by the mathematical problem

$$\nabla^2 \phi = 0 \tag{352}$$

$$\frac{\partial \phi}{\partial y} = -\varepsilon k \frac{\partial \phi}{\partial x} \sin kx \quad \text{at } y = \varepsilon \cos kx \tag{353}$$

$$\phi \to Ux \quad \text{as } y \to \infty \tag{354}$$

Show that

$$\phi = U\left[x + \varepsilon \sin kx e^{-ky} + \tfrac{1}{2}\varepsilon^2 k \sin 2kx e^{-2ky} + \cdots\right] \tag{355}$$

Discuss the uniformity of this expansion.

We first need to transfer the boundary condition in (353) from $y = \varepsilon \cos kx$ to $y = 0$. It can be rewritten as

$$\frac{\partial \phi}{\partial y}(x, \varepsilon \cos kx) = -\varepsilon k \frac{\partial \phi}{\partial x}(x, \varepsilon \cos kx) \sin kx$$

which, upon expanding, becomes

$$\frac{\partial \phi}{\partial y}(x,0) + \frac{\partial^2 \phi}{\partial y^2}(x,0)\, \varepsilon \cos kx + \tfrac{1}{2} \frac{\partial^3 \phi}{\partial y^3}(x,0)\, \varepsilon^2 \cos^2 kx + \cdots$$

$$= -\varepsilon k \left[\frac{\partial \phi}{\partial x}(x,0) + \varepsilon \frac{\partial^2 \phi}{\partial x \partial y}(x,0)\, \varepsilon \cos kx + \cdots\right] \sin kx$$

or

$$\frac{\partial \phi}{\partial y} + \varepsilon\left[\frac{\partial^2 \phi}{\partial y^2} \cos kx + k \frac{\partial \phi}{\partial x} \sin kx\right] + \varepsilon^2\left[\tfrac{1}{2} \frac{\partial^3 \phi}{\partial y^3} \cos^2 kx + k \frac{\partial^2 \phi}{\partial x \partial y} \sin kx \cos kx\right] + \cdots$$

$$= 0 \quad \text{at } y = 0 \tag{356}$$

We let

$$\phi = \phi_0 + \varepsilon \phi_1 + \varepsilon^2 \phi_2 + \cdots$$

in (352), (354), and (356), equate coefficients of like powers of ε, and obtain

Order ε^0

$$\nabla^2 \phi_0 = 0 \tag{357}$$

$$\phi_0 \to Ux \quad \text{as } y \to \infty$$

$$\frac{\partial \phi_0}{\partial y} = 0 \quad \text{at } y = 0 \tag{358}$$

Order ε

$$\nabla^2\phi_1 = 0 \tag{359}$$

$$\phi_1 \to 0 \quad \text{as } y \to \infty \tag{360}$$

$$\frac{\partial \phi_1}{\partial y} = -\frac{\partial^2 \phi_0}{\partial y^2}\cos kx - k\frac{\partial \phi_0}{\partial x}\sin kx \quad \text{at } y = 0 \tag{361}$$

Order ε^2

$$\nabla^2\phi_2 = 0 \tag{362}$$

$$\phi_2 \to 0 \quad \text{as } y \to \infty \tag{363}$$

$$\frac{\partial \phi_2}{\partial y} = -\tfrac{1}{2}\frac{\partial^3 \phi_0}{\partial y^3}\cos^2 kx - \tfrac{1}{2}k\frac{\partial^2 \phi_0}{\partial x \partial y}\sin 2kx - \frac{\partial^2 \phi_1}{\partial y^2}\cos kx$$

$$- k\frac{\partial \phi_1}{\partial x}\sin kx \quad \text{at } y = 0 \tag{364}$$

The solution of the zeroth-order problem, (357) and (358), is

$$\phi_0 = Ux \tag{365}$$

Then (361) becomes

$$\frac{\partial \phi_1}{\partial y} = -Uk\sin kx \quad \text{at } y = 0 \tag{366}$$

The boundary condition (366) suggests seeking the solution of the first-order problem in the separated form

$$\phi_1 = Uf(y)\sin kx$$

which, upon substitution into (359), (360), and (366), yields

$$f'' - k^2 f = 0$$

$$f \to 0 \quad \text{as } y \to \infty$$

$$f'(0) = -k$$

Hence

$$f = e^{-ky}$$

and

$$\phi_1 = U\sin kx e^{-ky} \tag{367}$$

Substituting for ϕ_0 and ϕ_1 in (364), we have

$$\frac{\partial \phi_2}{\partial y} = -Uk^2 \sin 2kx \quad \text{at } y=0 \tag{368}$$

which suggests seeking the solution of the second-order problem in the separated form

$$\phi_2 = Ug(y)\sin 2kx \tag{369}$$

Substituting (369) into (362), (363), and (368) yields

$$g'' - 4k^2 g = 0$$

$$g \to 0 \quad \text{as } y \to \infty$$

$$g'(0) = -k^2$$

whose solution is

$$g = \tfrac{1}{2}ke^{-2ky}$$

Hence

$$\phi_2 = \tfrac{1}{2}k \sin 2kx e^{-2ky}$$

Therefore, ϕ is given to second order by (355).

Since ϕ_2/ϕ_1 and ϕ_1/ϕ_0 are bounded for all x and y, we conclude that the expansion (355) is uniform.

15.36. The inviscid incompressible flow past a slightly distorted circular body is governed by

$$\frac{\partial^2 \psi}{\partial r^2} + \frac{1}{r}\frac{\partial \psi}{\partial r} + \frac{1}{r^2}\frac{\partial^2 \psi}{\partial \theta^2} = 0 \tag{370}$$

$$\psi \to Ur\sin\theta \quad \text{as } r \to \infty \tag{371}$$

$$\psi = 0 \quad \text{at } r = a(1-\varepsilon \sin^2\theta) \tag{372}$$

Show that

$$\psi = U\left(r - \frac{a^2}{r}\right)\sin\theta + \tfrac{1}{2}\varepsilon U\left(\frac{3a^2}{r}\sin\theta - \frac{a^4}{r^3}\sin 3\theta\right) + \cdots \tag{373}$$

Discuss the uniformity of this expansion.

We first transfer the boundary condition in (372) to $r=a$ using a Taylor-series expansion; that is,

$$\psi[a - \varepsilon a \sin^2\theta, \theta] = \psi(a,\theta) - \varepsilon \frac{\partial \psi}{\partial r}(a,\theta)\, a\sin^2\theta + \cdots = 0 \tag{374}$$

We let

$$\psi = \psi_0 + \varepsilon \psi_1 + \cdots \tag{375}$$

in (370), (371), and (374), equate coefficients of equal powers of ε, and obtain

Order ε^0

$$\frac{\partial^2 \psi_0}{\partial r^2} + \frac{1}{r}\frac{\partial \psi_0}{\partial r} + \frac{1}{r^2}\frac{\partial^2 \psi_0}{\partial \theta^2} = 0 \tag{376}$$

$$\begin{gathered} \psi_0 \to Ur\sin\theta \quad \text{as } r \to \infty \\ \psi_0(a,\theta) = 0 \end{gathered} \tag{377}$$

Order ε

$$\frac{\partial^2 \psi_1}{\partial r^2} + \frac{1}{r}\frac{\partial \psi_1}{\partial r} + \frac{1}{r^2}\frac{\partial^2 \psi_1}{\partial \theta^2} = 0 \tag{378}$$

$$\psi_1 = o(r) \quad \text{as } r \to \infty \tag{379}$$

$$\psi_1(a,\theta) = a\frac{\partial \psi_0}{\partial r}(a,\theta)\sin^2\theta \tag{380}$$

The boundary conditions in (377) suggest seeking the solution of the zeroth-order problem in the separated form

$$\psi_0 = Uf(r)\sin\theta$$

which, when substituted into (376) and (377), yields

$$f'' + \frac{1}{r}f' - \frac{1}{r^2}f = 0 \tag{381}$$

$$\begin{gathered} f \to r \quad \text{as } r \to \infty \\ f(a) = 0 \end{gathered} \tag{382}$$

Equation (381) is of the Euler type and hence its solutions can be sought in the form

$$f = r^s$$

which, when substituted into (381), gives

$$s(s-1) + s - 1 = 0$$

or

$$s^2 - 1 = 0$$

Hence $s = 1$ or -1, and the general solution of (381) is

$$f = c_1 r + \frac{c_2}{r}$$

Using the boundary conditions (382), we find that $c_1 = 1$ and $c_2 = -a^2$. Thus

$$f = r - \frac{a^2}{r}$$

and

$$\psi_0 = U\left(r - \frac{a^2}{r}\right)\sin\theta \tag{383}$$

Substituting (383) into (380) yields

$$\psi_1(a, \theta) = 2Ua\sin^3\theta = \tfrac{1}{2}Ua(3\sin\theta - \sin 3\theta) \tag{384}$$

which suggests seeking the solution of the first-order problem in the form

$$\psi_1 = Uag_1(r)\sin\theta + Uag_2(r)\sin 3\theta \tag{385}$$

Substituting (385) into (378), (379), and (384) and separating the θ variation, we obtain

$$g_1'' + \frac{1}{r}g_1' - \frac{g_1}{r^2} = 0 \tag{386}$$

$$g_1 = o(r) \quad \text{as } r \to \infty \tag{387}$$

$$g_1(a) = \tfrac{3}{2} \tag{388}$$

and

$$g_2'' + \frac{1}{r}g_2' - \frac{9g_2}{r^2} = 0 \tag{389}$$

$$g_2 = o(r) \quad \text{as } r \to \infty \tag{390}$$

$$g_2(a) = -\tfrac{1}{2} \tag{391}$$

The general solution of (386) is

$$g_1 = c_1 r + \frac{c_2}{r}$$

which, upon imposing the boundary conditions (387) and (388), yields $c_1 = 0$ and c_2

$=\frac{3}{2}a$. The general solution of (389) is

$$g_2 = c_3 r^3 + \frac{c_4}{r^3}$$

which, upon imposing the boundary conditions (390) and (391), yields $c_3 = 0$ and $c_4 = -\frac{1}{2}a^3$. Hence

$$\psi_1 = \tfrac{1}{2}Ua\left[\frac{3a}{r}\sin\theta - \frac{a^3}{r^3}\sin 3\theta\right] \tag{392}$$

Substituting ψ_0 and ψ_1 into (375) yields (373).

Comparing the second and first-order terms, we find that ψ_1/ψ_0 is bounded for all θ and $r \geq a$. Hence the expansion (373) is uniform.

15.37. Determine the solvability condition for

$$\nabla^4\phi - \lambda\phi = F(r,\theta) \tag{393}$$

$$\phi(a,\theta) = 0, \qquad \phi(b,\theta) = 0 \tag{394}$$

$$\frac{\partial\phi}{\partial r}(a,\theta) = f(\theta), \qquad \frac{\partial\phi}{\partial r}(b,\theta) = g(\theta) \tag{395}$$

We multiply (393) by ψ, integrate the result over the domain of interest, and obtain

$$\int_0^{2\pi}\int_a^b (\psi\nabla^4\phi - \lambda\psi\phi)\, r\,dr\,d\theta = \int_0^{2\pi}\int_a^b rF(r,\theta)\psi(r,\theta)\,dr\,d\theta \tag{396}$$

Next, we need to integrate the left-hand side by parts to transfer the derivatives from ϕ to ψ. We have two choices. We either express $\nabla^4\phi$ in terms of r and θ and integrate the result term by term or use the general Green identity

$$\iint_S (v\nabla^2 u - u\nabla^2 v)\,dS = \oint_\Gamma \left(v\frac{\partial u}{\partial n} - u\frac{\partial v}{\partial n}\right) ds$$

where Γ is the boundary of S. Here we use the latter. Let $\nabla^2\phi = u$, then according to Green's identity,

$$\iint_S \psi\nabla^4\phi\,dS = \iint_S \psi\nabla^2 u\,dS = \iint_S u\nabla^2\psi\,dS + \oint_\Gamma \left(\psi\frac{\partial u}{\partial n} - u\frac{\partial\psi}{\partial n}\right) ds$$

$$= \iint_S u\nabla^2\psi\,dS + \oint_\Gamma \left[\psi\frac{\partial}{\partial n}(\nabla^2\phi) - \nabla^2\phi\frac{\partial\psi}{\partial n}\right] ds \tag{397}$$

Next we let $\nabla^2\psi = v$, then according to Green's identity,

$$\iint_S u\nabla^2\psi\,dS = \iint_S v\nabla^2\phi\,dS = \iint_S \phi\nabla^2 v\,dS + \oint_\Gamma \left(v\frac{\partial\phi}{\partial n} - \phi\frac{\partial v}{\partial n}\right) ds$$

$$= \iint_S \phi\nabla^4\psi\,dS + \oint_\Gamma \left[\nabla^2\psi\frac{\partial\phi}{\partial n} - \phi\frac{\partial}{\partial n}(\nabla^2\psi)\right] ds \tag{398}$$

It follows from (397) and (398) that

$$\iint_S [\psi\nabla^4\phi - \phi\nabla^4\psi]\, dS = \oint_\Gamma \left[\psi\frac{\partial}{\partial n}(\nabla^2\phi) - \nabla^2\phi\frac{\partial\psi}{\partial n} + \nabla^2\psi\frac{\partial\phi}{\partial n} - \phi\frac{\partial}{\partial n}(\nabla^2\psi)\right] ds \tag{399}$$

In the present case, S is the region bounded by $r = b$ and $r = a$; Γ consists of the circumferences of the two circles $r = b$ and $r = a$; and $s = b\,d\theta$ on $r = b$ and $s = a\,d\theta$ on $r = a$. Since $\partial u/\partial n$ stands for the outward normal derivative,

$$\frac{\partial u}{\partial n} = \frac{\partial u}{\partial r} \quad \text{on } r = b$$

and

$$\frac{\partial u}{\partial n} = -\frac{\partial u}{\partial r} \quad \text{on } r = a$$

Therefore, (399) becomes

$$\int_0^{2\pi}\int_a^b [\psi\nabla^4\phi - \phi\nabla^4\psi]\, r\,dr\,d\theta$$

$$= \int_0^{2\pi}\left[r\psi\frac{\partial}{\partial r}(\nabla^2\phi) - r\nabla^2\phi\frac{\partial\psi}{\partial r} + r\nabla^2\psi\frac{\partial\phi}{\partial r} - r\phi\frac{\partial}{\partial r}(\nabla^2\psi)\right]_a^b d\theta \tag{400}$$

Substituting for $\psi\nabla^4\phi$ from (396) into (400), we have

$$\int_0^{2\pi}\int_a^b \phi[\lambda\psi - \nabla^4\psi]\, r\,dr\,d\theta + \int_0^{2\pi}\int_a^b rF(r,\theta)\psi(r,\theta)\, dr\,d\theta$$

$$= \int_0^{2\pi}\left[r\psi\frac{\partial}{\partial r}(\nabla^2\phi) - r\nabla^2\phi\frac{\partial\psi}{\partial r} + r\nabla^2\psi\frac{\partial\phi}{\partial r} - r\phi\frac{\partial}{\partial r}(\nabla^2\psi)\right]_a^b d\theta \tag{401}$$

Then the adjoint equation is

$$\nabla^4\psi - \lambda\psi = 0 \tag{402}$$

To define the adjoint boundary conditions, we consider the homogeneous problem; that is, $F \equiv 0$ and $f = g \equiv 0$. Then (401) becomes

$$b\int_0^{2\pi}\left[\psi\frac{\partial^3\phi}{\partial r^3} + \left(\frac{\psi}{b} - \frac{\partial\psi}{\partial r}\right)\frac{\partial^2\phi}{\partial r^2}\right]_{r=b} d\theta - a\int_0^{2\pi}\left[\psi\frac{\partial^3\phi}{\partial r^3} + \left(\frac{\psi}{a} - \frac{\partial\psi}{\partial r}\right)\frac{\partial^2\phi}{\partial r^2}\right]_{r=a} d\theta = 0$$

We choose the adjoint boundary conditions by setting each of the coefficients of

$$\frac{\partial^3\phi}{\partial r^3} \quad \text{and} \quad \frac{\partial^2\phi}{\partial r^2} \quad \text{at } r=a \text{ and } b$$

equal to zero, independently; that is,

$$\psi(a,\theta)=0, \qquad \psi(b,\theta)=0, \qquad \frac{\partial\psi}{\partial r}(a,\theta)=0, \qquad \frac{\partial\psi}{\partial r}(b,\theta)=0 \tag{403}$$

Having defined the adjoint problem, we return to the inhomogeneous problem. Substituting (394) and (395) into (401) and using the definition of the adjoint, we obtain the following solvability condition:

$$\int_0^{2\pi}\int_a^b rF(r,\theta)\psi(r,\theta)\,dr\,d\theta = \int_0^{2\pi}\left[bg(\theta)\frac{\partial^2\psi}{\partial r^2}(b,\theta) - af(\theta)\frac{\partial^2\psi}{\partial r^2}(a,\theta)\right]d\theta \tag{404}$$

where $\psi(r,\theta)$ is any solution of the adjoint problem governed by (402) and (403).

15.38. In analyzing the nonparallel two-dimensional stability of incompressible flows past a sinusoidal wall, one encounters the inhomogeneous problem

$$i\alpha u + v' = f_1(y) \tag{405}$$

$$-i(\omega-\alpha U)u + U'v + i\alpha p - \frac{1}{R}(u''-\alpha^2 u) = f_2(y) \tag{406}$$

$$-i(\omega-\alpha U)v + p' - \frac{1}{R}(v''-\alpha^2 v) = f_3(y) \tag{407}$$

$$u(0)=c_1, \qquad v(0)=c_2 \tag{408}$$

$$u, v \to 0 \quad \text{as } y\to\infty \tag{409}$$

where U and f_n are known functions of y; where ω, α, R, k, and c_n are constants; and where the prime indicates the derivative with respect to y. Determine the solvability condition when the homogeneous problem has a nontrivial solution.

We multiply (405), (406), and (407) by $\phi_1(y)$, $\phi_2(y)$, $\phi_3(y)$, respectively. Then we add and integrate the result from $y=0$ to $y=\infty$ and obtain

$$\int_0^\infty\left\{\phi_1[i\alpha u+v'] + \phi_2\left[-i(\omega-\alpha U)u + U'v + i\alpha p - \frac{1}{R}(u''-\alpha^2 u)\right]\right.$$

$$\left. + \phi_3\left[-i(\omega-\alpha U)v + p' - \frac{1}{R}(v''-\alpha^2 v)\right]\right\}dy$$

$$= \int_0^\infty (f_1\phi_1 + f_2\phi_2 + f_3\phi_3)\,dy \tag{410}$$

Integrating (410) by parts to transfer the derivatives from u, v, and p to the ϕ_n and collecting the coefficients of u, v, and p, we obtain

$$\int_0^\infty \left\{ u\left[i\alpha\phi_1 - i(\omega - \alpha U)\phi_2 - \frac{1}{R}(\phi_2'' - \alpha^2\phi_2) \right] \right.$$

$$\left. + v\left[-\phi_1' + U'\phi_2 - i(\omega - \alpha U)\phi_3 - \frac{1}{R}(\phi_3'' - \alpha^2\phi_3) \right] + p\left[i\alpha\phi_2 - \phi_3' \right] \right\} dy$$

$$+ \left[\phi_1 v - \frac{1}{R}(\phi_2 u' - \phi_2' u) + \phi_3 p - \frac{1}{R}(\phi_3 v' - \phi_3' v) \right]_0^\infty$$

$$= \int_0^\infty (f_1\phi_1 + f_2\phi_2 + f_3\phi_3)\, dy \tag{411}$$

The adjoint equations are defined by setting each of the coefficients of u, v, and p in the integrand on the left-hand side of (411) equal to zero. The result is

$$i\alpha\phi_2 - \phi_3' = 0 \tag{412}$$

$$i\alpha\phi_1 - i(\omega - \alpha U)\phi_2 - \frac{1}{R}(\phi_2'' - \alpha^2\phi_2) = 0 \tag{413}$$

$$-\phi_1' + U'\phi_2 - i(\omega - \alpha U)\phi_3 - \frac{1}{R}(\phi_3'' - \alpha^2\phi_3) = 0 \tag{414}$$

To define the adjoint boundary conditions, we consider the homogeneous problem; that is, the f_n and c_n are put equal to zero. It follows from (408) and (411) that

$$\lim_{y\to\infty} \left[\phi_1 v - \frac{1}{R}(\phi_2 u' - \phi_2' u) + \phi_3 p - \frac{1}{R}(\phi_3 v' - \phi_3' v) \right]$$

$$- \left[-\frac{1}{R}\phi_2 u' + \phi_3 p - \frac{1}{R}\phi_3 v' \right]_{y=0} = 0 \tag{415}$$

We choose the adjoint boundary conditions at infinity as

$$\phi_n \to 0 \quad \text{as } y \to \infty \tag{416}$$

and at $y = 0$ by setting each of the coefficients of u', p, and v' equal to zero there; that is,

$$\phi_2(0) = 0, \qquad \phi_3(0) = 0 \tag{417}$$

Having defined the adjoint, we return to the inhomogeneous problem. Substituting (408) and (409) into (411) and using the definition of the adjoint, we obtain

$$\int_0^\infty (f_1\phi_1 + f_2\phi_2 + f_3\phi_3)\, dy = -c_2\left[\phi_1(0) + \frac{1}{R}\phi_3'(0)\right] - \frac{c_1}{R}\phi_2'(0) \tag{418}$$

where the ϕ_n are any solution of the adjoint problem governed by (412)–(414) subject to (416) and (417).

15.39. In analyzing the stability of growing three-dimensional incompressible boundary layers over flat surfaces, one encounters the inhomogeneous problem

$$i\alpha u + i\beta w + v' = f_1(y) \tag{419}$$

$$-i\hat{\omega} u + U'v + i\alpha p - \frac{1}{R}(u'' - k^2 u) = f_2(y) \tag{420}$$

$$-i\hat{\omega} v + p' - \frac{1}{R}(v'' - k^2 v) = f_3(y) \tag{421}$$

$$-i\hat{\omega} w + i\beta p + W'v - \frac{1}{R}(w'' - k^2 w) = f_4(y) \tag{422}$$

$$u = v = w = 0 \quad \text{at } y = 0 \tag{423}$$

$$u, v, w \to 0 \quad \text{as } y \to \infty \tag{424}$$

where $\hat{\omega} = \omega - \alpha U - \beta W$ and $k^2 = \alpha^2 + \beta^2$. Here U, W, and $f_n(y)$ are known functions of y, and α, β, k, and ω are constants. Determine the solvability condition when the homogeneous problem has a nontrivial solution.

We multiply (419), (420), (421), and (422) by $\phi_1(y)$, $\phi_2(y)$, $\phi_3(y)$, and $\phi_4(y)$, respectively, integrate each from $y = 0$ to $y = \infty$; add the resulting equations; and obtain

$$\begin{aligned}\int_0^\infty \Big\{ & \phi_1[i\alpha u + i\beta w + v'] + \phi_2\left[-i\hat{\omega} u + U'v + i\alpha p - \frac{1}{R}(u'' - k^2 u)\right] \\ & + \phi_3\left[-i\hat{\omega} v + p' - \frac{1}{R}(v'' - k^2 v)\right] \\ & + \phi_4\left[-i\hat{\omega} w + i\beta p + W'v - \frac{1}{R}(w'' - k^2 w)\right]\Big\}\, dy \\ & = \int_0^\infty (f_1\phi_1 + f_2\phi_2 + f_3\phi_3 + f_4\phi_4)\, dy\end{aligned}$$

Next, we integrate each term by parts to transfer the derivatives from u, v, w, and p to the ϕ_n; collect the coefficients of u, v, w, and p; and obtain

$$\int_0^\infty \left\{ p\left[i\alpha\phi_2 + i\beta\phi_4 - \phi_3' \right] + u\left[i\alpha\phi_1 - i\hat{\omega}\phi_2 - \frac{1}{R}\left(\phi_2'' - k^2\phi_2\right)\right] \right.$$

$$+ v\left[-\phi_1' + U'\phi_2 - i\hat{\omega}\phi_3 - \frac{1}{R}\left(\phi_3'' - k^2\phi_3\right) + W'\phi_4 \right]$$

$$\left. + w\left[i\beta\phi_1 - i\hat{\omega}\phi_4 - \frac{1}{R}\left(\phi_4'' - k^2\phi_4\right)\right]\right\} dy$$

$$+ \left[\phi_1 v - \frac{1}{R}\left(\phi_2 u' - \phi_2' u\right) + \phi_3 p - \frac{1}{R}\left(\phi_3 v' - \phi_3' v\right) - \frac{1}{R}\left(\phi_4 w' - \phi_4' w\right)\right]_0^\infty$$

$$= \int_0^\infty \sum_{n=1}^{4} f_n \phi_n \, dy \tag{425}$$

The adjoint equations are defined by setting to zero each of the coefficients of p, u, v, and w. The result is

$$i\alpha\phi_2 + i\beta\phi_4 - \phi_3' = 0 \tag{426}$$

$$i\alpha\phi_1 - i\hat{\omega}\phi_2 - \frac{1}{R}\left(\phi_2'' - k^2\phi_2\right) = 0 \tag{427}$$

$$-\phi_1' + U'\phi_2 - i\hat{\omega}\phi_3 + W'\phi_4 - \frac{1}{R}\left(\phi_3'' - k^2\phi_3\right) = 0 \tag{428}$$

$$i\beta\phi_1 - i\hat{\omega}\phi_4 - \frac{1}{R}\left(\phi_4'' - k^2\phi_4\right) = 0 \tag{429}$$

To define the adjoint boundary conditions, we consider the homogeneous problem. Substituting (423) into (425), setting $f_n = 0$, and using the definition of adjoint equations, we obtain

$$\lim_{y\to\infty} \left[\phi_1 v - \frac{1}{R}\left(\phi_2 u' - \phi_2' u\right) + \phi_3 p - \frac{1}{R}\left(\phi_3 v' - \phi_3' v\right) - \frac{1}{R}\left(\phi_4 w' - \phi_4' w\right)\right]$$

$$- \left[-\frac{1}{R}\phi_2 u' + \phi_3 p - \frac{1}{R}\phi_3 v' - \frac{1}{R}\phi_4 w' \right]_{y=0} = 0$$

The first term vanishes if

$$\phi_n \to 0 \quad \text{as } y \to \infty \tag{430}$$

on account of (424), whereas the second term vanishes if we set each of the coefficients of u', p, v', and w' equal to zero; that is,

$$\phi_2(0) = 0, \qquad \phi_3(0) = 0, \qquad \phi_4(0) = 0 \tag{431}$$

Having defined the adjoint, we return to the inhomogeneous problem. Substituting (423) and (424) into (425) and using the definition of the adjoint, we obtain the following solvability condition:

$$\int_0^\infty \sum_{n=1}^\infty f_n(y)\phi_n(y)\,dy = 0 \tag{432}$$

15.40. Consider the eigenvalue problem

$$\phi_{xx} + \phi_{yy} + \lambda\phi = \varepsilon x^2\phi, \qquad \varepsilon \ll 1 \tag{433}$$

$$\phi(x,0) = \phi(x,\pi) = \phi(0,y) = \phi(\pi,y) = 0 \tag{434}$$

Determine first-order uniform expansions when λ is near 2 and 5.

We seek first-order expansions for ϕ and λ in the form

$$\begin{aligned} \phi &= \phi_0(x,y) + \varepsilon\phi_1(x,y) + \cdots \\ \lambda &= \lambda_0 + \varepsilon\lambda_1 + \cdots \end{aligned} \tag{435}$$

Substituting (435) into (433) and (434) and equating coefficients of like powers of ε, we obtain

Order ε^0

$$\phi_{0xx} + \phi_{0yy} + \lambda_0\phi_0 = 0 \tag{436}$$

$$\phi_0(x,0) = \phi_0(x,\pi) = \phi_0(0,y) = \phi_0(\pi,y) = 0 \tag{437}$$

Order ε

$$\phi_{1xx} + \phi_{1yy} + \lambda_0\phi_1 = -\lambda_1\phi_0 + x^2\phi_0 \tag{438}$$

$$\phi_1(x,0) = \phi_1(x,\pi) = \phi_1(0,y) = \phi_1(\pi,y) = 0 \tag{439}$$

The solution of the zeroth-order problem is

$$\phi_0 = \sin nx \sin my, \qquad \lambda_0 = n^2 + m^2 \tag{440}$$

where n and m are integers. Since the homogeneous problem (438) and (439) is the same as (436) and (437) and since the latter has a nontrivial solution, the inhomogeneous problem has a solution only if a solvability condition is satisfied. To determine the solvability condition, we multiply (438) by $\psi(x,y)$, integrate the result over the domain of interest, and obtain

$$\int_0^\pi \int_0^\pi \psi(\nabla^2\phi_1 + \lambda_0\phi_1)\, dx\, dy = \int_0^\pi \int_0^\pi \psi[x^2\phi_0 - \lambda_1\phi_0]\, dx\, dy \tag{441}$$

To perform the integration by parts, we use Green's identity

$$\iint_S (\psi\nabla^2\phi_1 - \phi_1\nabla^2\psi)\, dS = \oint_\Gamma \left(\psi\frac{\partial\phi_1}{\partial n} - \phi_1\frac{\partial\psi}{\partial n}\right) ds$$

In the problem considered, S consists of the square $0 \le x \le \pi$ and $0 \le y \le \pi$, Γ consists of its perimeter, and

$$\frac{\partial\phi}{\partial n} = \frac{\partial\phi}{\partial x} \quad \text{at } x = \pi$$

$$\frac{\partial\phi}{\partial n} = -\frac{\partial\phi}{\partial x} \quad \text{at } x = 0$$

$$\frac{\partial\phi}{\partial n} = \frac{\partial\phi}{\partial y} \quad \text{at } y = \pi$$

$$\frac{\partial\phi}{\partial n} = -\frac{\partial\phi}{\partial y} \quad \text{at } y = 0$$

Hence Green's identity becomes

$$\begin{aligned}\int_0^\pi \int_0^\pi (\psi\nabla^2\phi_1 - \phi_1\nabla^2\psi)\, dx\, dy = & \int_0^\pi \left[\psi\frac{\partial\phi_1}{\partial x} - \phi_1\frac{\partial\psi}{\partial x}\right]_0^\pi dy \\ & + \int_0^\pi \left[\psi\frac{\partial\phi_1}{\partial y} - \phi_1\frac{\partial\psi}{\partial y}\right]_0^\pi dx\end{aligned} \tag{442}$$

Substituting for $\nabla^2\phi_1$ from (438) into (442) yields

$$\begin{aligned}-\int_0^\pi \int_0^\pi \phi_1[\nabla^2\psi + \lambda_0\psi]\, dx\, dy + \int_0^\pi \int_0^\pi (x^2\phi_0 - \lambda_1\phi_0)\psi\, dx\, dy \\ = \int_0^\pi \left[\psi\frac{\partial\phi_1}{\partial x} - \phi_1\frac{\partial\psi}{\partial x}\right]_0^\pi dy + \int_0^\pi \left[\psi\frac{\partial\phi_1}{\partial y} - \phi_1\frac{\partial\psi}{\partial y}\right]_0^\pi dx\end{aligned} \tag{443}$$

The adjoint equation is defined by setting to zero the coefficient of ϕ_1 in the integrand on the left-hand side of (443); that is,

$$\nabla^2\psi + \lambda_0\psi = 0 \tag{444}$$

To define the boundary conditions, we consider the homogeneous problem. Then substituting (439) and (444) into (443), we have

$$\int_0^\pi \left[\psi \frac{\partial \phi_1}{\partial x}\right]_0^\pi dy + \int_0^\pi \left[\psi \frac{\partial \phi_1}{\partial y}\right]_0^\pi dx = 0$$

The adjoint boundary conditions are chosen by setting to zero each of the coefficients of

$$\frac{\partial \phi_1}{\partial x}(0, y), \qquad \frac{\partial \phi_1}{\partial x}(\pi, y), \qquad \frac{\partial \phi_1}{\partial y}(x, 0), \qquad \frac{\partial \phi_1}{\partial y}(x, \pi)$$

The result is

$$\psi(x,0) = \psi(x,\pi) = \psi(0,y) = \psi(\pi,y) = 0 \tag{445}$$

Having defined the adjoint, we return to the inhomogeneous problem. Substituting (439) into (443) and using the definition of the adjoint, we obtain the following solvability condition:

$$\int_0^\pi \int_0^\pi x^2 \phi_0 \psi \, dx\, dy = \lambda_1 \int_0^\pi \int_0^\pi \phi_0 \psi \, dx\, dy \tag{446}$$

where ψ is any solution of the adjoint problem.

When $\lambda_0 \approx 2$, it follows from (440) that there is only one eigenfunction, namely,

$$\phi_0 = \sin x \sin y$$

Moreover, the adjoint problem has only the eigenfunction

$$\psi = \sin x \sin y$$

Then, substituting ϕ_0 and ψ into (446), we obtain

$$\lambda_1 = \frac{2}{\pi} \int_0^\pi x^2 \sin^2 x \, dx = \frac{1}{\pi} \left[\tfrac{1}{3}x^3 - \tfrac{1}{2}x^2 \sin 2x - \tfrac{1}{2}x \cos 2x + \tfrac{1}{4} \sin 2x\right]_0^\pi$$

$$= \frac{1}{\pi}\left(\tfrac{1}{3}\pi^3 - \tfrac{1}{2}\pi\right) = \tfrac{1}{3}\pi^2 - \tfrac{1}{2}$$

Therefore

$$\begin{aligned} \phi &= \sin x \sin y + \cdots \\ \lambda &= 2 + \varepsilon\left(\tfrac{1}{3}\pi^2 - \tfrac{1}{2}\right) + \cdots \end{aligned} \tag{447}$$

When $\lambda \approx 5$, it follows from (440) that there are two eigenfunctions corresponding to $\lambda_0 = 5$; namely,

$$\sin x \sin 2y \quad \text{and} \quad \sin 2x \sin y \tag{448}$$

Hence

$$\phi_0 = a_1 \sin x \sin 2y + a_2 \sin 2x \sin y$$

where a_2/a_1 is arbitrary to the zeroth order; it may be determined by imposing the solvability condition of the first-order problem. Substituting ϕ_0 into (446) yields

$$\lambda_1 \int_0^\pi \int_0^\pi \psi(a_1 \sin x \sin 2y + a_2 \sin 2x \sin y)\, dx\, dy$$

$$= \int_0^\pi \int_0^\pi x^2 \psi(a_1 \sin x \sin 2y + a_2 \sin 2x \sin y)\, dx\, dy \tag{449}$$

where ψ is any solution of the adjoint problem. In this case, ψ has either one of the solutions in (448). Letting $\psi = \sin x \sin 2y$ in (449), we have

$$\tfrac{1}{4}\pi^2 \lambda_1 a_1 = \tfrac{1}{2} a_1 \pi \int_0^\pi x^2 \sin^2 x\, dx$$

or

$$\left[\lambda_1 - \left(\tfrac{1}{3}\pi^2 - \tfrac{1}{2}\right)\right] a_1 = 0 \tag{450}$$

Letting $\psi = \sin 2x \sin y$ in (449), we have

$$\tfrac{1}{4}\pi^2 \lambda_1 a_2 = \tfrac{1}{2} a_2 \pi \int_0^\pi x^2 \sin^2 2x\, dx$$

$$= \tfrac{1}{4} a_2 \pi \left[\tfrac{1}{3}x^3 - \tfrac{1}{4}x^2 \sin 4x - \tfrac{1}{8} x \cos 4x + \tfrac{1}{32} \sin 4x\right]_0^\pi$$

or

$$\left[\lambda_1 - \left(\tfrac{1}{3}\pi^2 - \tfrac{1}{8}\right)\right] a_2 = 0 \tag{451}$$

It follows from (450) and (451) that either

$$\lambda_1 = \tfrac{1}{3}\pi^2 - \tfrac{1}{2} \quad \text{and} \quad a_2 = 0$$

or

$$\lambda_1 = \tfrac{1}{3}\pi^2 - \tfrac{1}{8} \quad \text{and} \quad a_1 = 0$$

Therefore, either

$$\begin{aligned} \phi^{(1)} &= \sin x \sin 2y + \cdots \\ \lambda^{(1)} &= 5 + \left(\tfrac{1}{3}\pi^2 - \tfrac{1}{2}\right)\varepsilon + \cdots \end{aligned} \tag{452}$$

or

$$\phi^{(2)} = \sin 2x \sin y + \qquad (453)$$
$$\lambda^{(2)} = 5 + \left(\tfrac{1}{3}\pi^2 - \tfrac{1}{8}\right)\varepsilon + \cdots$$

15.41. Consider the problem

$$\nabla^2\phi + \lambda\Phi = \varepsilon f(x, y, z)\phi \qquad (454)$$

with ϕ vanishing on the surfaces of a cube of length π. Determine first-order expansions when $\lambda \approx 3$ and 6 if (a) $f = x^2$ and (b) $f = x^2 y$.

We seek first-order uniform expansions for ϕ and λ in the form

$$\phi = \phi_0(x, y, z) + \varepsilon\phi_1(x, y, z) + \cdots \qquad (455)$$
$$\lambda = \lambda_0 + \varepsilon\lambda_1 + \cdots$$

Substituting (455) into (454) and the boundary conditions and equating coefficients of like powers of ε, we obtain

Order ε^0

$$\nabla^2\phi_0 + \lambda_0\phi_0 = 0 \qquad (456)$$

$$\phi_0 = 0 \quad \text{on } \Gamma \qquad (457)$$

Order ε

$$\nabla^2\phi_1 + \lambda_0\phi_1 = -\lambda_1\phi_0 + f\phi_0 \qquad (458)$$

$$\phi_1 = 0 \quad \text{on } \Gamma \qquad (459)$$

where Γ stands for the surfaces of a cube of length π.

The solution of the zeroth-order problem is

$$\phi_0 = \sin nx \sin my \sin kz, \qquad \lambda_0 = n^2 + m^2 + k^2 \qquad (460)$$

where n, m, and k are integers. When $\lambda_0 = 3$, there is only one eigenfunction, namely

$$\phi_0 = \sin x \sin y \sin z \qquad (461)$$

When $\lambda_0 = 6$, it follows from (460) that there are three eigenfunctions, namely,

$$\eta_1 = \sin x \sin y \sin 2z, \qquad \eta_2 = \sin 2x \sin y \sin z, \qquad \eta_3 = \sin x \sin 2y \sin z \qquad (462)$$

Hence

$$\phi_0 = a_1 \sin x \sin y \sin 2z + a_2 \sin 2x \sin y \sin z + a_3 \sin x \sin 2y \sin z \qquad (463)$$

where the relations among the a_n are arbitrary at this order; they are determined by imposing the solvability condition at the next order.

Since the homogeneous problem (458) and (459) is the same as the zeroth-order problem and since the latter has a nontrivial solution, the inhomogeneous problem has a solution only if a solvability condition is satisfied. To determine the solvability condition, we multiply (458) by $\psi(x, y, z)$, integrate the result by parts over the domain $0 \le x \le \pi$, $0 \le y \le \pi$, $0 \le z \le \pi$ to transfer the derivatives from ϕ_1 to ψ, and obtain

$$\int_0^\pi \int_0^\pi \int_0^\pi \phi_1 [\nabla^2 \psi + \lambda_0 \psi]\, dx\, dy\, dz + \int_0^\pi \int_0^\pi \left[\psi \frac{\partial \phi_1}{\partial x} - \phi_1 \frac{\partial \psi}{\partial x} \right]_0^\pi dy\, dz$$

$$+ \int_0^\pi \int_0^\pi \left[\psi \frac{\partial \phi_1}{\partial y} - \phi_1 \frac{\partial \psi}{\partial y} \right]_0^\pi dx\, dz + \int_0^\pi \int_0^\pi \left[\psi \frac{\partial \phi_1}{\partial z} - \phi_1 \frac{\partial \psi}{\partial z} \right]_0^\pi dx\, dy$$

$$= \int_0^\pi \int_0^\pi \int_0^\pi (f - \lambda_1) \phi_0 \psi\, dx\, dy\, dz \tag{464}$$

The adjoint equation is defined by setting the coefficient of ϕ_1 in the integrand on the left-hand side of (464) equal to zero; that is,

$$\nabla^2 \psi + \lambda_0 \psi = 0 \tag{465}$$

To define the adjoint boundary conditions, we consider the homogeneous problem. Substituting (459) and (465) into (464) yields

$$\int_0^\pi \int_0^\pi \left[\psi \frac{\partial \phi_1}{\partial x} \right]_0^\pi dy\, dz + \int_0^\pi \int_0^\pi \left[\psi \frac{\partial \phi_1}{\partial y} \right]_0^\pi dx\, dz + \int_0^\pi \int_0^\pi \left[\psi \frac{\partial \phi_1}{\partial z} \right]_0^\pi dx\, dy = 0$$

Then the adjoint boundary conditions are chosen to be

$$\psi = 0 \quad \text{on } \Gamma \tag{466}$$

Therefore, the solution of the adjoint problem is given by (461) when $\lambda_0 = 3$ and by any of the eigenfunctions in (462) when $\lambda_0 = 6$.

Having defined the adjoint, we return to the inhomogeneous problem. Substituting (459) into (464) and using the definition of the adjoint, we obtain the following solvability condition:

$$\lambda_1 \int_0^\pi \int_0^\pi \int_0^\pi \phi_0 \psi\, dx\, dy\, dz = \int_0^\pi \int_0^\pi \int_0^\pi f \phi_0 \psi\, dx\, dy\, dz \tag{467}$$

When $\lambda \approx 3$, $\lambda_0 = 3$, and ϕ_0 and ψ are given by (461). Hence (467) becomes

$$\lambda_1 = \frac{8}{\pi^3} \int_0^\pi \int_0^\pi \int_0^\pi f \sin^2 x \sin^2 y \sin^2 z\, dx\, dy\, dz \tag{468}$$

For the case $f = x^2$, (468) yields

$$\lambda_1 = \frac{2}{\pi}\int_0^{\pi} x^2 \sin^2 x\, dx = \tfrac{1}{3}\pi^2 - \tfrac{1}{2}$$

Therefore

$$\phi = \sin x \sin y \sin z + \cdots$$
$$\lambda = 3 + \left(\tfrac{1}{3}\pi^2 - \tfrac{1}{2}\right)\varepsilon + \cdots \tag{469}$$

For the case $f = x^2 y$, (468) yields

$$\lambda_1 = \frac{4}{\pi^2}\left[\int_0^{\pi} x^2 \sin^2 x\, dx\right]\left[\int_0^{\pi} y \sin^2 y\, dy\right]$$
$$= \frac{1}{\pi^2}\left[\tfrac{1}{3}x^3 - \tfrac{1}{2}x^2 \sin 2x - \tfrac{1}{2}x\cos 2x + \tfrac{1}{4}\sin 2x\right]_0^{\pi}$$
$$\times\left[\tfrac{1}{2}y^2 - \tfrac{1}{2}y\sin 2y - \tfrac{1}{4}\cos 2y\right]_0^{\pi} = \frac{\pi}{2}\left(\tfrac{1}{3}\pi^2 - \tfrac{1}{2}\right)$$

Therefore

$$\phi = \sin x \sin y \sin z + \cdots$$
$$\lambda = 3 + \tfrac{1}{2}\pi\left(\tfrac{1}{3}\pi^2 - \tfrac{1}{2}\right)\varepsilon + \cdots \tag{470}$$

When $\lambda \approx 6$, $\lambda_0 = 6$, ϕ_0 is given by (463), and ψ is given by one of the functions η_n in (462). Substituting (463) into (467) and letting $\psi = \eta_1$, we obtain

$$\tfrac{1}{8}\pi^3 \lambda_1 a_1 = a_1 f_{11} + a_2 f_{21} + a_3 f_{31} \tag{471}$$

where

$$f_{nm} = \int_0^{\pi}\int_0^{\pi}\int_0^{\pi} f \eta_n \eta_m\, dx\, dy\, dz \tag{472}$$

Similarly, substituting (463) into (467) and letting $\psi = \eta_2$, we obtain

$$\tfrac{1}{8}\pi^3 \lambda_1 a_2 = a_1 f_{21} + a_2 f_{22} + a_3 f_{32} \tag{473}$$

Finally, substituting (463) into (467) and letting $\psi = \eta_3$, we obtain

$$\tfrac{1}{8}\pi^3 \lambda_1 = a_1 f_{31} + a_2 f_{32} + a_3 f_{33} \tag{474}$$

For a nontrivial solution, the determinant of the coefficient matrix in (471), (473), and

(474) must vanish. This condition leads to the eigenvalue (characteristic) equation

$$\begin{vmatrix} f_{11}-\frac{1}{8}\pi^3\lambda_1 & f_{21} & f_{31} \\ f_{21} & f_{22}-\frac{1}{8}\pi^3\lambda_1 & f_{32} \\ f_{31} & f_{32} & f_{33}-\frac{1}{8}\pi^3\lambda_1 \end{vmatrix} = 0 \tag{475}$$

which is a cubic polynomial in λ_1. It provides three roots for λ_1. If f and hence the f_{nm} are such that (475) has three distinct roots, the degeneracy is removed at first; otherwise, the degeneracy is not removed.

When $f = x^2$,

$$f_{nm} = 0 \quad \text{if } n \neq m$$

$$f_{11} = f_{33} = \tfrac{1}{4}\pi^2 \int_0^\pi x^2 \sin^2 x\, dx = \tfrac{1}{8}\pi^3\left(\tfrac{1}{3}\pi^2 - \tfrac{1}{2}\right)$$

$$f_{22} = \tfrac{1}{4}\pi^2 \int_0^\pi x^2 \sin^2 2x\, dx = \tfrac{1}{8}\pi^3\left(\tfrac{1}{3}\pi^2 - \tfrac{1}{8}\right)$$

Then (471), (473), and (474) become

$$\left[\lambda_1 - \left(\tfrac{1}{3}\pi^2 - \tfrac{1}{2}\right)\right] a_1 = 0$$

$$\left[\lambda_1 - \left(\tfrac{1}{3}\pi^2 - \tfrac{1}{8}\right)\right] a_2 = 0$$

$$\left[\lambda_1 - \left(\tfrac{1}{3}\pi^2 - \tfrac{1}{2}\right)\right] a_3 = 0$$

Hence either

$$\lambda_1 = \tfrac{1}{3}\pi^2 - \tfrac{1}{2} \quad \text{and} \quad a_2 = 0$$

or

$$\lambda_1 = \tfrac{1}{3}\pi^2 - \tfrac{1}{8} \quad \text{and} \quad a_1 = a_3 = 0$$

Consequently, either

$$\begin{aligned} \phi^{(1)} &= a_1 \sin x \sin y \sin 2z + a_3 \sin x \sin 2y \sin z + \cdots \\ \lambda^{(1)} &= 6 + \left(\tfrac{1}{3}\pi^2 - \tfrac{1}{2}\right)\varepsilon + \cdots \end{aligned} \tag{476}$$

or

$$\begin{aligned} \phi^{(2)} &= \sin 2x \sin y \sin z + \cdots \\ \lambda &= 6 + \left(\tfrac{1}{3}\pi^2 - \tfrac{1}{8}\right)\varepsilon + \cdots \end{aligned} \tag{477}$$

Equations (476) and (477) show that the degeneracy is only partially removed at first order.

When $f = x^2 y$,

$$f_{11} = \tfrac{1}{2}\pi\left[\int_0^\pi x^2 \sin^2 x\,dx\right]\left[\int_0^\pi y \sin^2 y\,dy\right]$$

$$f_{21} = f_{31} = 0$$

$$f_{32} = \tfrac{1}{2}\pi\left[\int_0^\pi x^2 \sin x \sin 2x\,dx\right]\left[\int_0^\pi y \sin y \sin 2y\,dy\right]$$

$$f_{22} = \tfrac{1}{2}\pi\left[\int_0^\pi x^2 \sin^2 2x\,dx\right]\left[\int_0^\pi y \sin^2 y\,dy\right]$$

$$f_{33} = \tfrac{1}{2}\pi\left[\int_0^\pi x^2 \sin^2 x\,dx\right]\left[\int_0^\pi y \sin^2 2y\,dy\right]$$

It follows from (475) that

$$\lambda_1 = \frac{8}{\pi^3} f_{11}$$

and

$$\left(\frac{\pi^3 \lambda_1}{8}\right)^2 - (f_{22} + f_{33})\frac{\pi^3 \lambda_1}{8} + f_{22} f_{33} - f_{32}^2 = 0$$

Hence (475) provides three distinct roots for λ_1. Let us denote the roots of the quadratic equation as $\lambda_1^{(2)}$ and $\lambda_1^{(3)}$. When $\lambda_1 = 8f_{11}/\pi^3$, (471), (473), and (474) show that $a_2 = a_3 = 0$. Hence

$$\phi^{(1)} = \sin x \sin y \sin 2z + \cdots$$
$$\lambda = 6 + \frac{8 f_{11}}{\pi^3}\varepsilon + \cdots \tag{478}$$

When $\lambda_1 = \lambda_1^{(2)}$, it follows from (471) that $a_1 = 0$ and from (473) that

$$a_3 = \frac{\tfrac{1}{8}\pi^3 \lambda_1^{(2)} - f_{22}}{f_{32}} a_2$$

Hence

$$\phi^{(2)} = \sin 2x \sin y \sin z + \frac{\tfrac{1}{8}\pi^3 \lambda_1^{(2)} - f_{22}}{f_{32}} \sin x \sin 2y \sin z + \cdots$$
$$\lambda^{(2)} = 6 + \varepsilon \lambda_1^{(2)} + \cdots \tag{479}$$

Finally,

$$\phi^{(3)} = \sin 2x \sin y \sin z + \frac{\frac{1}{8}\pi^3\lambda_1^{(3)} - f_{22}}{f_{32}} \sin x \sin 2y \sin z + \cdots$$

$$\lambda^{(3)} = 6 + \varepsilon\lambda_1^{(3)} + \cdots \tag{480}$$

Supplementary Exercises

15.42. Show that

$$x_1 - 2x_2 = b_1$$

$$2x_1 - 4x_2 = b_2$$

has a solution if and only if $b_2 = 2b_1$.

15.43. Show that

$$x_1 + 2x_2 - x_3 = b_1$$

$$2x_1 + x_2 + x_3 = b_2$$

$$x_1 - x_2 + 2x_3 = b_3$$

has a solution if and only if $b_3 = b_2 - b_1$.

15.44. Show that

$$x_1 + 2x_2 - x_3 + x_4 = b_1$$

$$2x_1 + x_2 - 2x_3 + 2x_4 = b_2$$

$$x_1 + x_2 - x_3 + x_4 = b_3$$

$$x_1 - x_2 - x_3 + x_4 = b_4$$

has a solution if and only if $b_3 = \frac{1}{3}(b_2 + b_1)$ and $b_4 = b_2 - b_1$.

15.45. Show that

$$\ddot{u}_1 + 2\dot{u}_2 + 3u_1 = \sum_{n=1}^{4} P_n e^{int}$$

$$\ddot{u}_2 + \dot{u}_1 + 12u_2 = \sum_{n=1}^{4} Q_n e^{int}$$

has a solution if and only if $Q_2 = -2iP_2$ and $Q_3 = -\frac{1}{2}iP_3$.

15.46. Show that

$$\ddot{u}_1 + 4\dot{u}_2 + 3u_1 = \sum_{n=1}^{4} P_n e^{int}$$

$$\ddot{u}_2 - \dot{u}_1 + 3u_2 = \sum_{n=1}^{4} Q_n e^{int}$$

has a solution if and only if $Q_1 = -\frac{1}{2}iP_1$ and $Q_3 = \frac{1}{2}iP_3$.

15.47. Show that

$$\ddot{u}_1 + 5\dot{u}_2 + 6u_1 = \sum_{n=1}^{4} P_n e^{int}$$

$$\ddot{u}_2 + \dot{u}_1 + 24u_2 = \sum_{n=1}^{4} Q_n e^{int}$$

has a solution if and only if $Q_3 = -iP_3$ and $Q_4 = -\frac{2}{5}iP_4$.

15.48. Consider the system

$$\ddot{u}_1 + 2\dot{u}_2 + 3u_1 + 2\varepsilon\cos\Omega t(f_{11}u_1 + f_{12}u_2) = 0$$

$$\ddot{u}_2 + \dot{u}_1 + 12u_2 + 2\varepsilon\cos\Omega t(f_{21}u_1 + f_{22}u_2) = 0$$

where $\varepsilon \ll 1$. Show that to the first approximation

$$u_1 = A_1(t)e^{2it} + A_2(t)e^{3it} + \text{cc}$$

$$u_2 = -\tfrac{1}{4}iA_1(t)e^{2it} - iA_2(t)e^{3it} + \text{cc}$$

and determine the equations describing A_1 and A_2 when $\Omega \approx 5$ and $\Omega \approx 1$.

15.49. Consider the system

$$\ddot{u}_1 + 4\dot{u}_2 + 3u_1 + 2\varepsilon\cos\Omega t(f_{11}u_1 + f_{12}u_2) = 0$$

$$\ddot{u}_2 - \dot{u}_1 + 3u_2 + 2\varepsilon\cos\Omega t(f_{21}u_1 + f_{22}u_2) = 0$$

where $\varepsilon \ll 1$. Show that to the first approximation

$$u_1 = A_1(t)e^{it} + A_2(t)e^{3it} + \text{cc}$$

$$u_2 = \tfrac{1}{2}iA_1(t)e^{it} - \tfrac{1}{2}iA_2(t)e^{3it} + \text{cc}$$

and determine the equations describing A_1 and A_2 when $\Omega \approx 4$ and $\Omega \approx 2$.

15.50. Consider the system

$$\ddot{u}_1 + 5\dot{u}_2 + 6u_1 + 2\varepsilon \cos \Omega t (f_{11}u_1 + f_{12}u_2) = 0$$

$$\ddot{u}_2 + \dot{u}_1 + 24u_2 + 2\varepsilon \cos \Omega t (f_{21}u_1 + f_{22}u_2) = 0$$

where $\varepsilon \ll 1$. Show that to the first approximation

$$u_1 = A_1(t) e^{3it} + A_2(t) e^{4it} + \text{cc}$$

$$u_2 = -\tfrac{1}{5} i A_1(t) e^{3it} - \tfrac{1}{2} i A_2(t) e^{4it} + \text{cc}$$

and determine the equations describing A_1 and A_2 when $\Omega \approx 7$ and $\Omega \approx 1$.

15.51. Determine the solvability condition for the problem

$$u'' + n^2\pi^2 u = f(x)$$

$$u(0) = \beta_1, \qquad u(1) = \beta_2$$

where n is an integer.

15.52. Determine the solvability condition for the problem

$$u'' + \left(n + \tfrac{1}{2}\right)^2 \pi^2 u = f(x)$$

$$u(0) = \beta_1, \qquad u'(1) = 0$$

where n is an integer.

15.53. Determine the solvability condition for the problem

$$u'' + \frac{1}{x} u' + k^2 u = f(x)$$

$$u(0) < \infty \quad \text{and} \quad u(1) = \beta$$

where k is a root of $J_0(k) = 0$.

15.54. Determine the solvability condition for the problem

$$u'' + \frac{1}{x} u' + k^2 u = f(x)$$

$$u(a) = \beta_1 \quad \text{and} \quad u(b) = \beta_2$$

where $b > a > 0$ and k is a root of

$$J_0(ka) Y_0(kb) - J_0(kb) Y_0(ka) = 0$$

15.55. Determine the solvability condition for the problem

$$u'' + \frac{1}{x} u' + k^2 u = f(x)$$

$$u(a) = \beta_1 \quad \text{and} \quad u'(b) = \beta_2$$

where $b > a > 0$ and k is a root of

$$J_0(ka)Y_0'(kb) - J_0'(kb)Y_0(ka) = 0$$

15.56. Determine the solvability condition for the problem

$$u'' + \frac{1}{x}u' + \left(k^2 - \frac{n^2}{x^2}\right)u = f(x)$$

$$u(0) < \infty \quad \text{and} \quad u(a) = \beta$$

where k is a root of $J_n(ka) = 0$.

15.57. Determine the solvability condition for the problem

$$u'' + \frac{1}{x}u' + \left(k^2 - \frac{n^2}{x^2}\right)u = f(x)$$

$$u'(a) = \beta_1, \qquad u(b) = \beta_2$$

where $b > a > 0$ and k is a root of

$$J_n'(ka)Y_n(kb) - J_n(kb)Y_n'(ka) = 0$$

15.58. Determine the solvability condition for the problem

$$\frac{\partial^2\phi}{\partial x^2} + \frac{\partial^2\phi}{\partial y^2} + (n^2 + m^2)\pi^2\phi = F(x, y)$$

$$\phi(0, y) = \phi(1, y) = \phi(x, 0) = \phi(x, 1) = 0$$

where m and n are integers.

15.59. Determine the solvability condition for the problem

$$\frac{\partial^2\phi}{\partial x^2} + \frac{\partial^2\phi}{\partial y^2} + (n^2 + m^2)\pi^2\phi = f(y)\sin n\pi x$$

$$\phi(0, y) = \phi(1, y) = 0$$

$$\phi(x, 0) = \beta_1 \sin n\pi x, \qquad \phi(x, 1) = \beta_2 \sin n\pi x$$

where m and n are integers.

15.60. Determine the solvability condition for the problem

$$\frac{\partial^2\phi}{\partial x^2} + \frac{\partial^2\phi}{\partial y^2} + (n^2 + m^2)\pi^2\phi = f(x, y)$$

$$\phi(0, y) = \beta_1(y), \qquad \phi(1, y) = \beta_2(y)$$

$$\phi(x, 0) = \beta_3(x), \qquad \phi(x, 1) = \beta_4(x)$$

where m and n are integers.

15.61. Determine the solvability condition for the problem

$$\frac{\partial^2\phi}{\partial x^2}+\frac{\partial^2\phi}{\partial y^2}+\frac{\partial^2\phi}{\partial z^2}+(m^2+n^2+k^2)\pi^2\phi=f(x,y,z)$$

$$\phi(0,y,z)=\phi(1,y,z)=\phi(x,0,z)=\phi(x,1,z)$$

$$=\phi(x,y,0)=\phi(x,y,1)=0$$

where m, n, and k are integers.

15.62. Determine the solvability condition for the problem

$$\frac{\partial^2\phi}{\partial r^2}+\frac{1}{r}\frac{\partial\phi}{\partial r}+\frac{1}{r^2}\frac{\partial^2\phi}{\partial\theta^2}+k^2\phi=f(r)\,e^{in\theta}$$

$$\phi(a,\theta)=\alpha_1 e^{in\theta},\qquad \phi(b,\theta)=\alpha_2 e^{in\theta}$$

where $b>a>0$ and k is a root of

$$J_n(ka)Y_n(kb)-J_n(kb)Y_n(ka)=0$$

15.63. Determine the solvability condition for the problem

$$u''+\frac{2}{r}u'+n^2\pi^2u=f(r)$$

$$u(0)<\infty \quad\text{and}\quad u(1)=\beta$$

where n is an integer.

15.64. Determine the solvability condition for the problem

$$u''+\frac{2}{r}u'+\frac{n^2\pi^2}{(b-a)^2}u=f(r)$$

$$u(a)=\beta_1,\qquad u(b)=\beta_2$$

where $b>a>0$.

15.65. Determine the solvability condition for the problem

$$u''+\frac{2}{r}u'+\frac{\left(n+\frac{1}{2}\right)^2\pi^2}{(b-a)^2}u=f(r)$$

$$u(a)=\beta_1,\qquad u'(b)=\beta_2$$

where $b>a>0$.

15.66. Determine the solvability condition for the problem

$$\nabla^4 w - \lambda w = f(r)$$

$$w(a) = \beta_1, \qquad w'(a) = \beta_2, \qquad w(b) = \beta_3, \qquad w'(b) = \beta_4$$

where $b > a > 0$, λ is an eigenvalue of the homogeneous problem and

$$\nabla^2 = \frac{d^2}{dr^2} + \frac{1}{r}\frac{d}{dr}$$

15.67. Determine the equations describing the amplitudes and phases of the system

$$\ddot{u}_1 + 4\dot{u}_2 + 3u_1 = \varepsilon\alpha_1 u_1^3$$

$$\ddot{u}_2 - \dot{u}_1 + 3u_2 = \varepsilon\alpha_2 u_2^3$$

where $\varepsilon \ll 1$.

15.68. Consider the system

$$\ddot{u}_1 + \omega_1^2 u_1 + \alpha_1 u_1 u_2 + 2\varepsilon\cos\Omega t(f_{11}u_1 + f_{12}u_2) = 0$$

$$\ddot{u}_2 + \omega_2^2 u_2 + \alpha_2 u_1^2 + 2\varepsilon\cos\Omega t(f_{21}u_1 + f_{22}u_2) = 0$$

where $\varepsilon \ll 1$ and $\omega_2 \approx 2\omega_1$. Determine the equations describing the amplitudes and phases when (a) $\Omega \approx 2\omega_1$, (b) $\Omega \approx 2\omega_2$, (c) $\Omega \approx \omega_1 + \omega_2$ and (d) $\Omega \approx \omega_2 - \omega_1$.

15.69. Consider the system

$$\ddot{u}_1 + \omega_1^2 u_1 = \varepsilon\left[\alpha_1 u_1^3 + \alpha_2 u_1^2 u_2 + \alpha_3 u_1 u_2^2 + \alpha_4 u_2^3 + f_1\cos\Omega t\right]$$

$$\ddot{u}_2 + \omega_2^2 u_2 = \varepsilon\left[\delta_1 u_1^3 + \delta_2 u_1^2 u_2 + \delta_3 u_1 u_1^2 + \delta_4 u_2^3\right]$$

where $\varepsilon \ll 1$ and $\omega_2 \approx \omega_1$. Determine the equations describing the amplitudes and phases when $\Omega \approx \omega_1$.

15.70. Determine a first-order expansion for the eigenvalues of

$$u'' + [\lambda + \varepsilon f(x)]u = 0, \qquad \varepsilon \ll 1$$

$$u(a) = 0, \qquad u(b) = 0$$

where $b > a$.

15.71. Determine a first-order expansion for the eigenvalues of

$$u'' + [\lambda + \varepsilon f(x)]u = 0, \qquad \varepsilon \ll 1$$

$$u'(a) = 0, \qquad u(b) = 0$$

where $b > a$.

15.72. Determine a first-order expansion for the eigenvalues of

$$u'' + \frac{1}{r}u' + [\lambda + \varepsilon f(r)]u = 0, \qquad \varepsilon \ll 1$$

$$u(0) < \infty, \qquad u'(1) = 0$$

15.73. Determine a first-order expansion for the eigenvalues of

$$u'' + \frac{1}{r}u' + [\lambda + \varepsilon f(r)]u = 0, \qquad \varepsilon \ll 1$$

$$u(a) = 0, \qquad u(b) = 0$$

where $b > a > 0$.

15.74. Determine a first-order expansion for the eigenvalues of

$$u'' + \frac{2}{r}u' + [\lambda + \varepsilon f(r)]u = 0, \qquad \varepsilon \ll 1$$

$$u(0) < \infty, \qquad u'(1) = 0$$

15.75. Determine a first-order expansion for the eigenvalues of

$$u'' + \frac{2}{r}u' + [\lambda + \varepsilon f(r)]u = 0, \qquad \varepsilon \ll 1$$

$$u(a) = 0, \qquad u'(b) = 0$$

where $b > a > 0$.

15.76. Determine a first-order expansion for the eigenvalues of

$$u'' + [\lambda + \varepsilon f(x)]u = 0, \qquad \varepsilon \ll 1$$

$$u(0) = 0, \qquad u'(1) - \alpha u(1) = 0$$

15.77. Show that the solvability condition for the problem

$$\frac{d^2\phi_1}{dr^2} + \frac{1}{r}\frac{d\phi_1}{dr} + \left(\alpha_n^2 - \frac{1}{r^2}\right)\phi_1 = f_1(r)$$

$$\frac{d^2\phi_2}{dr^2} + \frac{1}{r}\frac{d\phi_2}{dr} - \left(\gamma_n^2 + \frac{1}{r^2}\right)\phi_2 = f_2(r)$$

$$\phi_1' = \mu_1\phi_1 + \mu_2\phi_2 + \beta_1, \qquad \phi_2' = \mu_3\phi_1 + \mu_4\phi_2 + \beta_2 \quad \text{at } r = 1$$

$$\phi_1(0) < \infty, \qquad a\phi_2'(a) - \phi_2(a) = \beta_3$$

is

$$\int_0^1 r f_1(r) u_1(r)\, dr + \int_1^a r f_2(r) u_2(r)\, dr = \beta_1 u_1(1) + \beta_3 u_2(a) - \beta_2 u_2(1)$$

where $u_1(r)$ and $u_2(r)$ are solutions of the adjoint problem.

15.78. Determine a first-order expansion for the eigenvalues of

$$\nabla^4\phi - \lambda\phi = \varepsilon f(r)\phi, \qquad \varepsilon \ll 1$$

$$\phi(0) < \infty, \qquad \phi(1) = 0, \qquad \phi'(1) = 0$$

where

$$\nabla^2 = \frac{d^2}{dr^2} + \frac{1}{r}\frac{d}{dr}$$

15.79. Determine a first-order expansion for the eigenvalues of

$$u^{iv} + 10u'' + [\lambda + \varepsilon f(x)]u = 0, \qquad \varepsilon \ll 1$$

$$u(0) = u''(0) = u(\pi) = u''(\pi) = 0$$

Note that a degeneracy exists when $\lambda \approx 9$.

15.80. Determine a first-order expansion for the eigenvalues of

$$\phi(s) = \lambda \int_0^1 [st + s^2t^2 + \varepsilon K_1(s,t)]\phi(t)\, dt, \qquad \varepsilon \ll 1$$

15.81. Determine a first-order expansion for the eigenvalues of

$$\phi(s) = \lambda \int_{-1}^1 [1 - st + \varepsilon K_1(s,t)]\phi(t)\, dt, \qquad \varepsilon \ll 1$$

Index